Human Symmetry Perception and its Computational Analysis

Human Symmetry Perception and its Computational Analysis

Editor

C.W. Tyler

///VSP///

Utrecht, The Netherlands, 1996

VSP BV
P.O. Box 346
3700 AH Zeist
The Netherlands

First published in 1996

ISBN 90-6764-210-X

Typeset in Lithuania by TEV, Vilnius

CONTENTS

Theoretical Issues in Symmetry Analysis

INTRODUCTION

Human Symmetry Perception, pp. 3–22
C. W. Tyler (Ed.)

Human symmetry perception

CHRISTOPHER W. TYLER*
Smith-Kettlewell Eye Research Institute, 2232 Webster St., San Francisco, CA 94115, USA

> O Nature, and O soul of Man! how far beyond all utterance are your linked analogies! not the smallest atom stirs or lives in matter, but has its cunning duplicate in mind.
>
> Herman Melville, *Moby Dick* (1850, p. 295).

At a time of increasing *rapprochement* between the analyses of local spatial properties and the global organization of Gestalt pattern relations, symmetry relations can provide a guiding principle to help connect the two. It therefore seems a good juncture to take stock of research in symmetry perception, which began over a century ago with the trenchant analysis of Mach (1886). The pace of investigation in the field has been accelerating over the past two decades and this collection provides a representative overview of current work.

Symmetry is a general concept that refers to any manner in which part of a pattern may be mapped on to another part (or the whole pattern onto itself). Symmetries of various kinds play a fundamental role in both the structure of the physical universe and, in a different form, our interactions with it on a human scale. Early art and architecture offer some insight into the role of symmetry in human perception, suggesting that repetition (translational symmetry) was perhaps the earliest to be aesthetically appreciated. This is evident as much as four thousand years ago in the ranks of repeated figures of terracotta warriors in the tomb of Emperor T'sin, the founder of China, in the repeated profiles of the ancient Persian bas-reliefs of Persepolis in the era of King Xerxes, in the wall-paintings of the Egyptian pyramids and in the Cambodian temple decorations such as at Angkor Wat (see Fig. 1). Remarkably, the mirror symmetry of the human face was hardly ever depicted in painting of this era (notable particularly in the uniform depiction of faces as profiles in Egyptian wall paintings), although the monumental architecture of its palaces and tombs are replete with both mirror symmetry in the structure of the buildings and repetition symmetry in their ornamentation such as colonnades and battlements. In general, then, symmetries of design have long been widely incorporated into human artefacts, as is amply documented in *Symmetries of Culture* by Washburn and Crowe (1988), for example.

---

*Supported by NIMH grant MH 49044.

Figure 1. An example of repetition symmetry from the temple walls at Angkor Wat, Combodia (courtesy Dr. Christopher Tori).

Nevertheless, there is significant evidence, reviewed by Wagemans (1995), that mirror symmetry has a special status in human perception in comparison with other types (such as translation or rotation symmetry, see Fig. 2). Why this should be the case may be evaluated in terms of the environment in which our visual system evolved. Mirror symmetry of the objects that we encounter usually betrays the presence of living organisms, for one of several reasons. Inanimate objects, consisting predominantly of rocks and geological formations, generally appear to be randomly distributed in the natural environment and exhibit none of the types of regular symmetry. There has been much recent interest in the tendency for the visual environment to exhibit the self-scaling symmetry of fractal self-similarity across size scale (i.e. viewing distance). However, this fractal property has a random component in the natural world; it is a statistical self-similarity rather than an accurate regularity. As such, this self-similarity may better be regarded as a textural property than a true symmetry.

Crystals are often cited as a special case of natural symmetry, but even then it is rare to find an isolated object consisting of a single crystal with an identifiable symmetry axis. Most natural occurrences of crystals are haphazard agglomerations with a pronounced random component. It is only on a non-human scale of analysis that inanimate objects are limited by symmetry constraints. Microscopically, of course, atoms have pervasive symmetries, although these are already delimited by the molecular level. Crystals then constitute an intermediate ('miniscopic'?) level at which the molecular symmetry is expressed in magnified form. On an astronomical scale, symmetry is resuscitated by the gravitational constraint in the form of pronounced spherical symmetry, together with examples of radial and spiral symmetry in conglomerations of astronomical objects. But these symmetries evident at either end of the inanimate scale spectrum break down into random structure at the scale of human perception.

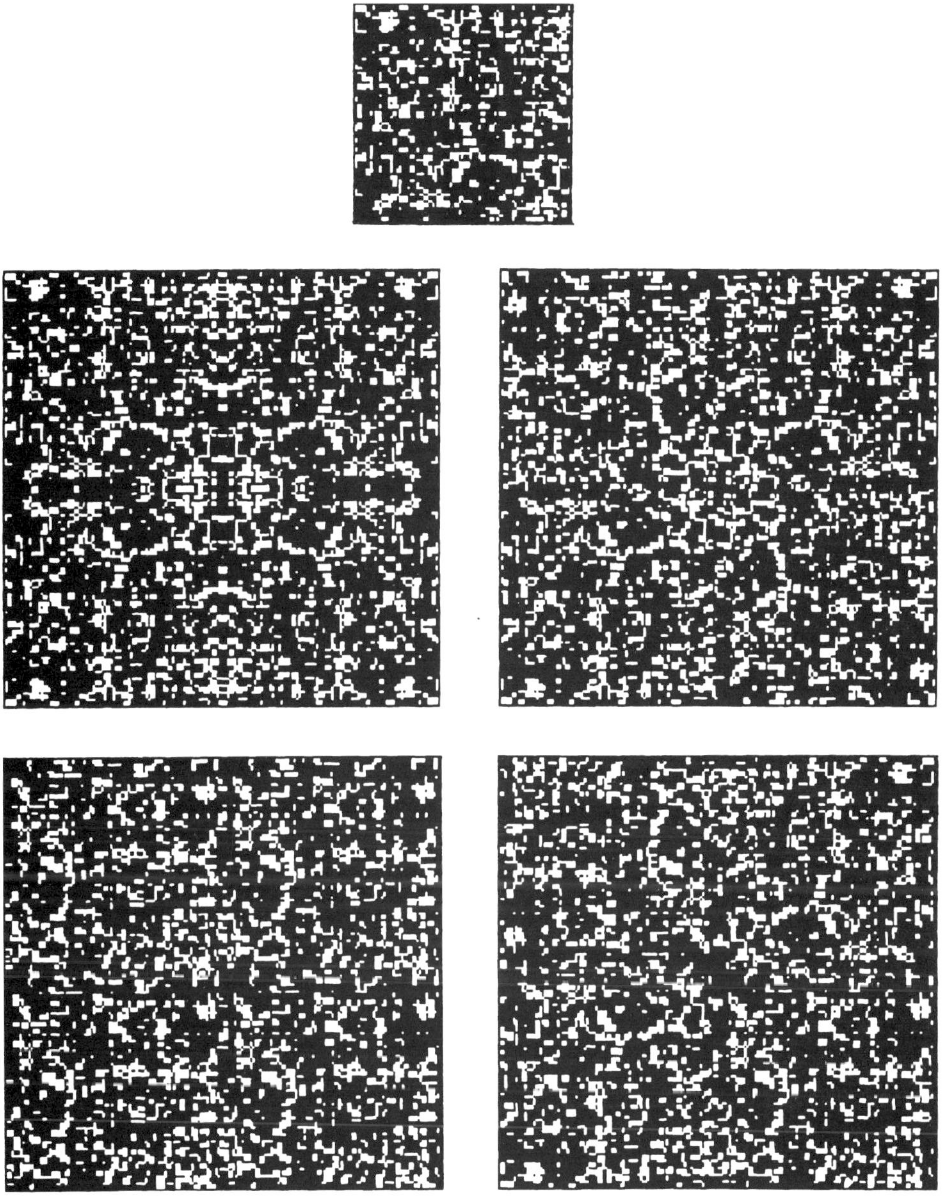

Figure 2. Representative examples of symmetry types in fractal noise, illustrating their perceptual differences. Each panel consists of the same four repeats of the same base pattern in different spatial arrangements. Upper left: bifold reflective symmetry; upper right: rotation symmetry; lower left: translation symmetry; lower right: centric symmetry. Note the perceptual immediacy of the reflection symmetry and, perhaps the rotation symmetry compared with the other two types, even though the predominant features have similar separations in all types.

Animals that move linearly through the environment, on the other hand, always are formed with an axis of mirror symmetry aligned with their typical direction of movement. This design is an obvious adaptation to avoid asymmetries that might cause a bias from the axis of movement. Any tendency to drift to the left or right would create difficulties in orienting and navigating during reduced cue situations (locomoting at night or in tall grass, for example). It seems that it is the motion constraint that is the limiting factor (rather than some inherent genetic property, for example) because only external parts of moving organisms are symmetrical. Internal organs, such as brain, heart, liver and intestines, may exhibit substantial asymmetry as long as the external envelope conforms to the constraint. It is interesting to note that the design of fast-moving vehicles such as cars and planes adheres to the same symmetry constraints, with exterior mirror symmetry but lateralized placement of internal components such as the steering wheel and driver. On the other hand, animals that do not locomote consistently through the environment, such as starfish and limpets, or that drift or pulse through it, as do plankton and jellyfish, do not conform to the bilateral symmetry constraint but typically have either cylindrical or multifold symmetry.

A striking feature of vertebrate animals is that, with rare exceptions such as fingers, their bodily structure exhibits no symmetries other than bilateral (absence of radial, rotational, repetitive, etc.). Almost every part of a vertebrate animal differs from every other except for its symmetrical mate on the other side. This rule is broken in the skeletal structure, which has strong repetition symmetries, such as the ribs, that are hidden by the musculature. Ironically, the rule is also broken by the kind of skin patterning that most clearly reveals the mirror symmetry, such as the repetitive stripes of the zebra and the tiger. They show a degree of local repetition symmetry, although the stripes tend to show idiosyncratic fine structure that deviates from local symmetry, one manifestation of which is the well-known anomaly that zebra stripes run vertically on the abdomen and horizontally on the rump (Fig. 3).

Figure 3. Bilateral animal symmetry on zebra hides. Note that phase symmetry is not always maintained on tails, legs, etc. (from Burton, J. (1978). '*Wild Animals*' *Crescent Books*. Crown Publishers, New York).

Gardner (1964) has explained the predominance of mirror symmetry in animals by reference to the forces that tend to disrupt the perfect spherical symmetry of the single cell; the vertical force of gravity and the fore/aft asymmetry of the motion direction. On this view, the prevalence of mirror symmetry is a default result of the absence of a perturbing lateralization force; since there are no lateral asymmetries in the locomoting environment, there is nothing to induce a lateral asymmetry. But the same argument would predict mirror symmetry for the internal organs, and probably also cylindrical symmetry for fish, who are essentially immune from gravitational pull. It also would make it difficult to explain the total lack of symmetry in amoebae and many other single-celled organisms. Thus, it seems that organismic symmetries may betray the operation of some active symmetrizing principle, such as the adaptive value of mirror symmetry for unbiased motion, rather than a default state to which organisms tend in the absence of asymmetric forces. The spherical symmetry often found in single-celled organisms would be attributable to the symmetrizing force of surface tension, as in bubbles in general.

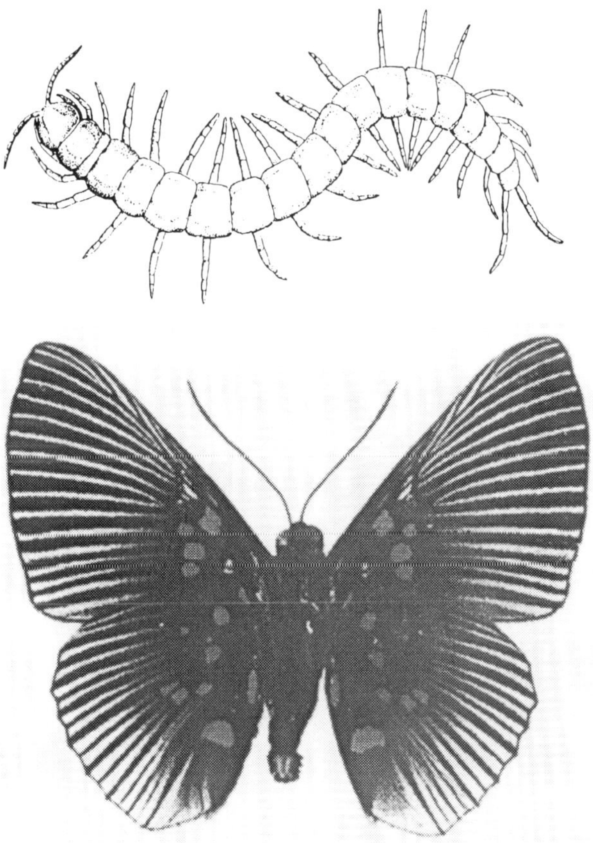

Figure 4. Examples of arthropod symmetry: a centiped (Chilopoda) and a butterfly (*Uraneis ucubis*) whose wing pattern reflects the repeated structure of the wing segments. (Centipede from Fig. 1.3 of Wootton, A. (1984). *Insects of the World*. Blanford Press, London; butterfly from Feltwell, J. (1993) *The Illustrated Encyclopedia of Butterflies*. Blanford Press, London, p. 252

Insects and other arthropods share the universality of a vertical symmetry axis but also exhibit manifold examples of repetition symmetry (Fig. 4), such as the repeated thoracic segments of the centipede, the iterated leg and antennae segments of most insects, spiders and crustaceans, and the multiple 'arms' of the octopus and squid. It is interesting that this repetition strikes us as inhuman and perhaps repulsive. It may be that some of the aversive quality of Hallowe'en skeletons derives from the way they reveal this insect-like quality of our repetitive bone structure.

Another prevalent feature of animals is that their long-distance sense organs are located close to the front as defined by their direction of movement. This arrangement implies that they become perceptually symmetric when facing in the direction of an observer. This orientation is one of high significance to the observer because it implies that the perceived organism has the observer at the centre of its sensory input, i.e. that it is maximally sensitive to the observer's presence. The significance may be either positive or negative according to whether the organism represents a threat or an advantage, but either role is maximized at the point of perceived symmetry. For example, a tiger passing by may be of concern, but when it turns to exhibit its proverbial symmetry by facing you, it is time to take action! Similarly, when the face of a human companion is turned to its position of full symmetry, the implication is strong that that person is maximally aware of your gestural signals. This association of perceived mirror symmetry with the maximal likelihood of another consciousness being focussed on oneself may help to account for the perceptual specialization for mirror symmetry and our fascination with it.

Figure 5. Plant symmetries. An maple leaf exhibits repetition symmetry, mirror symmetry and size-scaled symmetry (approximate with respect to both the outline and the vein structure in each case).

The plant kingdom also exhibits widespread mirror symmetry, but for a different reason. In plants, accurate symmetry is a local principle determining, for example, the shapes of leaves and petals rather than of the plant as a whole (see Fig. 5). In this case, mirror symmetry seems to be just one of many types of symmetry constraint that have evolved in the design of plants, perhaps based on the principle of economy of design. Just as it is ontologically efficient to replicate the design of one petal around the circumference of a flower, efficiency is gained by using the same template for both sides of a petal or a leaf. In addition, plants often have a rough cylindrical symmetry imposed by gravitational constraints, which appears as a bilateral symmetry when viewed from the side. This cylindrical symmetry appears in many aspects of plant structure, elaborated into helical and rotational symmetries. In addition, the replication of similar elements such as leaves all over the plant provides a prevalent example of repetition symmetry throughout the plant kingdom. The growth of such elements of plant structure from small to large also provides extensive examples of scale symmetry in plants. This diverse array of symmetries seen in plants may be a reflection of the varieties of symmetry evident in diatoms, the microscopic constituents of ocean vegetation, which exhibit bilateral, triangular and higher multiaxial symmetries in an astonishing display of crystalline regularity (Fig. 6).

It is worth noting that the repetition and scale symmetries often are combined in plants to generate a spiral symmetry in flower heads (Fig. 7). It is remarkable that this spiral symmetry is governed by the golden angle of 137.5 deg, an irrational angle that

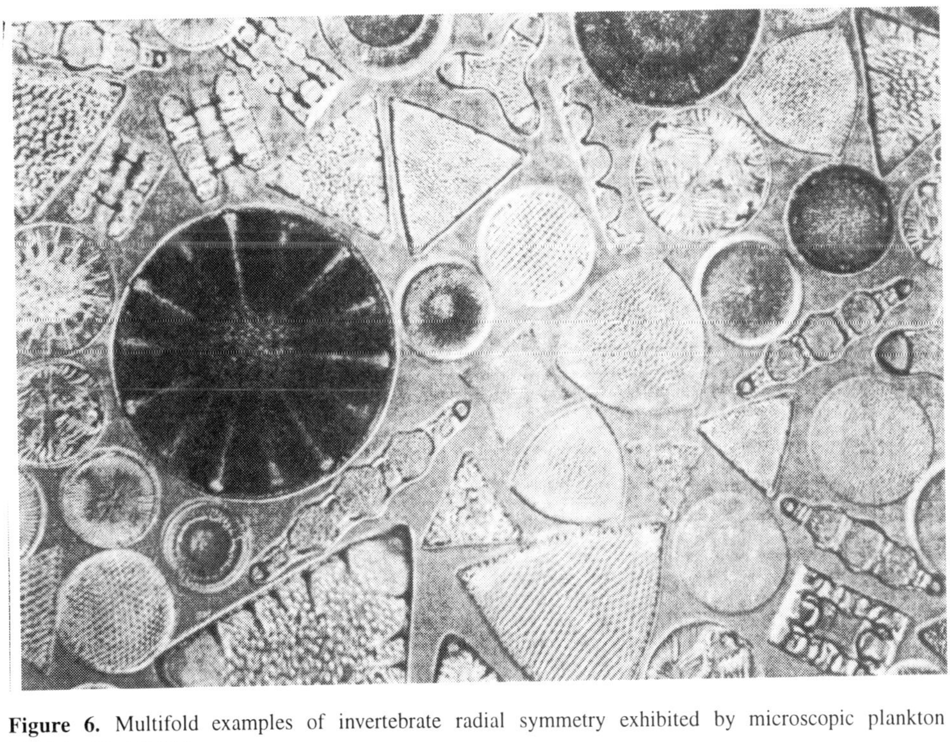

Figure 6. Multifold examples of invertebrate radial symmetry exhibited by microscopic plankton (Walker, D., NHPA).

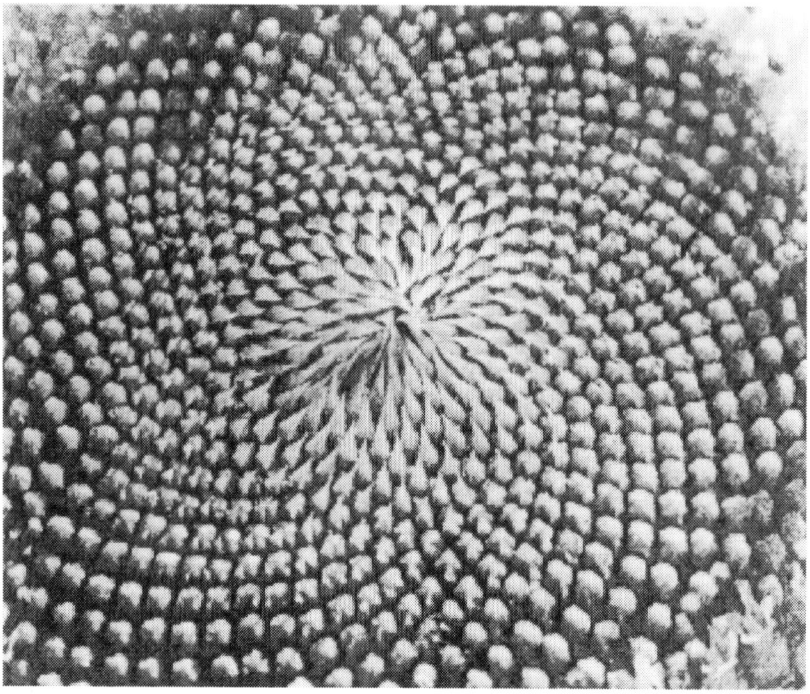

Figure 7. Spiral symmetry in sunflower head. Note that there are 34 clockwise and 55 counterclockwise spirals to make a ratio of 0.61818..., matching the golden ratio to three decimal places (Courtesy of Bruce Coleman Ltd. Photo by Leonard Lee Rue).

arises because the seeds pack wherever space become available in the growthing of the developing flower head and hence fill the space efficiently. The golden angle is related to the Greeks' golden ratio of $G = 0.618034$ by the expression $(1 - G) \times 360$ deg.

To delve into the numerology of this relationship, the golden ratio G is defined such that $1 + G = 1/G$. Spacing seeds around at this angle generates asymmetric spirals with different numbers of clockwise and counterclockwise whorls, where the numbers correspond to two adjacent members of the Fibonacci series (e.g. 34 and 55 in Fig. 7), whose ratio converges on G, which the Greeks discovered to be $(\sqrt{5}-1)/2$ (from one of the two roots of the equation $G^2 + G - 1 = 0$). Thus, $G^2 = 1 - G$, so the two values of G^2 are 2.618034 and 0.381966 $(= 1 - 0.618034)$. With the latter translation, both squares are seen to contain the same decimal portion as the values for G itself, which are 0.618034 and 1.618034!

A final class of objects in our developmental (rather than evolutionary) environment constitutes those of human construction. These artefacts often have extensive symmetries, both for practical and aesthetic reasons. Much of the constructed world is characterized by the two-fold symmetry of the elaborated rectangle, but all types of symmetry have found their place (see Fig. 8). The practical reasons for such symmetries include stability and rigidity requirements, the constraints of machine production (particularly promoting circular symmetry) and the need to match the symmetry of the human body in objects that constitute furniture. Aesthetic reasons for symmetry also

Figure 8. Symmetries of human construction: An Art Deco door (L), an ensemble from the Ziegfeld follies (R) (Hornak Picture Library; The Kobal Collection).

may derive from a variety of sources. One is the aforesaid symmetry of the natural environment, whose properties often are reflected in constructed objects to appeal to the perceptual specializations that have developed in this environment. A second is a resonance with the inherent symmetries of the optics, retina and cortical projection of the visual system, which has circular symmetry overlaid with a mirror symmetry. Designs such as those of Persian carpets may be particularly appealing because they harmonize with such symmetries. Lastly, symmetries may be represented as an aesthetic exploration of the combinatoric possibilities of design elements (Washburn and Crowe, 1988), visual structure for its own sake rather than for any relation to the inner or outer environment. Thus, symmetries of all kinds may be present in constructed objects, with an emphasis on two-fold symmetry.

In summary, varieties of symmetry tend to distinguish the components of our perceptual environment according to a simple scheme:

vertebrate animal:	mirror symmetry
invertebrate animal:	mirror and repetition symmetry
vegetable:	multiple symmetries (emphasizing repetition, scale, cylindrical, helical and multifold)
mineral:	none (at the macroscopic scale)
constructed:	multiple symmetries (emphasizing two-fold)

WHY STUDY SYMMETRY PERCEPTION?

The intent of the present collection of papers on symmetry is to provide insight into new analyses that are being developed for the perception of symmetries of various types. Taken together, the contributions, as diverse geographically as they are conceptually, reveal a multiplicity of perspectives on the processing of symmetry by the human brain.

The main kinds of symmetry of interest in spatial vision are reflection (or bilateral), rotation (or axial), translational (or repetition) and size-scaling (or fractal self-similarity), most of which are explored in the following papers. As attested by the preponderance of contributions, mirror symmetry is the type that seems to hold the most appeal for perceptual investigation, but does it offer a particular benefit for the analysis of human pattern recognition? Most studies in pattern recognition are based on a past memory of the recognized object and therefore deal with the nature of representation in memory. Symmetry perception is distinct, however, in that it is based on a comparison of representations in immediate perception rather than memory. A memory of one part of the image is not required to recognize the similarity of another part, although memory may be activated during the task. It is the representation of the matching patterns after taking the requisite symmetry transform into account that is the crucial aspect of the recognition process. The immediacy of this process in non-familiar (randomly generated) images, as reported in the literature from Julesz (1966) to the present collection, militates against a significant role for memory in the task. Thus, the symmetry paradigm allows analysis of an aspect of complex pattern recognition that is difficult to access by other techniques.

METHODS FOR STUDYING SYMMETRY PERCEPTION

In general, the empirical study of symmetry perception employs the same array of techniques as other perceptual tasks. Of course, most types of visual stimulus may be used for the study of symmetry, although reduced stimuli such as horizontal or vertical sinusoidal gratings do not permit a meaningful symmetry manipulation. The typical approach from the earliest studies has been to use non-familiar or randomized shapes on which the symmetry transforms are imposed. This approach was taken to its logical extreme by Julesz (1966, 1970), who introduced the computerized random-dot field as the base stimulus for symmetry manipulations and showed that symmetry could be perceived preattentively from very brief presentations. Many other variants have been utilized, including filtered random textures, dynamic random textures and random polygons, in addition to familiar images such as faces and outline figures.

Except for the use of eye movement patterns to investigate symmetry (Locher and Nodine, 1987), the experimental techniques employed all involve methods of pushing the visual system to its performance limit in order to quantify the characteristics of the perceptual processing. Limiting the time available for processing the presence of symmetry defines the **reaction time** task that has found common use. Determining the percent correct discrimination for a brief exposure constitutes the **accuracy** task.

Reducing the stimulus to the minimum duration at which the symmetry can be detected (with unlimited response time) defines the **duration threshold** task. Reducing the stimulus to the minimum contrast for the detection of symmetry defines the **contrast threshold** task. The accuracy in setting the orientation of the symmetry axis defines the **orientation discrimination** task. Each task has a role to play in exploring the properties of symmetry processing.

A revealing task that has not commonly been used in perceptual experiments is to introduce positional noise into the pattern elements until the symmetry is detectably degraded, or until its presence is just discriminable from a pure noise stimulus, defining types of **pattern discrimination** task employed by Bruce and Morgan (1975) and Barlow and Reeves (1979). Another task that was introduced for the study of stereopsis (Julesz and Tyler, 1976) but has been adapted to symmetry perception is the transition reciprocity paradigm, which addresses whether the change from stimulus A to stimulus B is perceptually equivalent to the change from stimulus B to stimulus A. The failure of this equivalence is evidence for a characteristic neural bias in processing the pattern information, requiring the definition of a neurally relevant definition of the available information, termed 'neurontropy'. A corresponding definition of the information available for symmetry perception has been termed 'symmetropy' by Yodogawa (1982), although he did not investigate transition reciprocity under this measure.

NEURAL REPRESENTATION OF SYMMETRY

It is hard to think about the perception of symmetry without considering the nature of its neural representation. How are physical object relations, such as their inherent symmetries, represented in the brain? This raises the classic chestnut, 'to what degree is the neural representation isomorphic with the physical properties of the object and to what degree is it abstracted?'.

The simplistic answer of the Gestalt school was that representation is isomorphic with the stimulus, so that a symmetric object would have a symmetric representation in the brain. Although now generally dismissed as begging the question of the representation code, this view has been resuscitated by Shepard (1981), who argues that an isomorphic representation is necessary to allow internal transformations of the representation to match manipulations of the object in the world. This stress on a veridical representation is an elaboration of Young's (1962) paradoxical insight that higher organisms maintain their difference from the environment only by *mirroring* the properties of the environment in the brain. This perspective emphasizes the structural symmetry between subject and object in the act of perception. The experiments of Shepard and colleagues on such transformations as mental rotation of random objects seem to provide good evidence for a 3D neural representation mirroring the 3D world. Indeed, the classic experiments by Penfield (1959) of the sequential readout of memory sequences, and our mental ability to replay songs and speeches in temporal order, would extend the isomorphic representation to the fourth dimension of time.

However, the problem of a representation code remains. The presence of an isomorphic copy of a relevant object in the brain may have value, but it is still subject to the criticism that the object has not been encoded into some form that captures the connotations of its features to the organism. Minsky (1975) and Pylyshyn (1986), for example, argue for an entirely propositional code for object features with no coherent spatial representation. Such a code would resemble lists of attributes associated with each object, where each item would be cross-referenced to other related properties that would constitute its meaning to the organism. This view of encoding arises from the position that much of our sense of meaning must be carried by such a propositional or cognitive code, so it is parsimonious to assume that perception is similarly coded. For symmetry, such a code would consist of a list of attributes for the base pattern motif and a specification of the transformations required to generate the complete pattern from operations on its base motif. Without such a list of transformations, the propositional code could not be said to have encoded the symmetry of the pattern.

Shepard (1981) has argued cogently that a propositional code alone is insufficient to account for many of the properties of object recognition under spatial transformations. In particular, the speed of matching objects that are rotated copies of each other is proportional to the angle of rotation but independent of the complexity of the objects. Neither result would be predicted from the inherent properties of a propositional code, but both are consistent with the idea of a neural representation of the objects in which elements that are close in physical space are represented as close in the neural connectivity space. The comparison of objects at different angles of rotation would then correspond to a matching process after the appropriate transformation has been applied to the neural representation. This transformation would correspond not to a physical rotation in the brain but to an adjustment of the local neural codes for each part of the object equivalent to such a rotation (as opposed to merely updating the propositional code for the transformation itself). One advantage of an isomorphic code is that it allows operations such as filtering, segmentation and spatial relations to be performed in a natural way by local neural operations. What is known of the neurophysiology supports the idea of the isomorphic mapping of 2D space to retinal space, depth via binocular disparity and time via velocity coding (DeYoe and van Essen, 1988).

However, the issue of how symmetry is encoded neatly underlines the limitations of an isomorphic code, which would require that the object be translated, rotated, dilated or otherwise transformed in all possible combinations for the symmetry properties to be discovered. Given the time established for mental rotation to take place (e.g. Cooper and Shepard, 1973, 1978), the isomorphic hypothesis would require that it would take many minutes to become aware of the possible structural symmetries of even 2D images, whereas these relations seem to be immediately evident for brief presentations of less than 100 ms (Corballis and Roldan, 1974; Hogben *et al.*, 1976). The implication is that there must be some means of encoding the similarity relations implicit in the similarity structure in a more direct manner than is available by an isomorphic representation. Is this the point of transition to a propositional code, or is there some more geometric code for these fundamentally geometric relationships?

PROPERTIES OF A PERCEPTUAL ENCODING HIERARCHY

The preceding considerations lead to the view that there are four basic aspects of perceptual encoding that form the structure within which any perceptual task, such as symmetry perception, should be viewed. To explain the properties of perceptual coding of three-dimensional objects in a three-dimensional world, a more elaborate scheme is required than is available in most models of the process (Fig. 9). Beginning with an elaboration of the full dimensionality of the stimulus properties, the information is passed to sets of neural analyzers that impose a variety of coding symmetries on the encoding signals. This stage seems to be followed by a parallel self-matching process that can segregate image features according to their similarities across a variety of encoding symmetries. The final process of perceptual coding is an active, manipulative recognition process that operates to identify object properties that are too complex to be characterized at previous stages.

Encoding the metric for the stimulus space

The first property of perceptual encoding is the metric within which the referent space is represented in the brain. In some cases, such as retinal space, the metric is a relatively faithful match to the stimulus space. In others, such as color, it may be drastically reduced in comparison to the available stimulus metric. Some examples of the metrics of perceptually relevant spaces are:

a 1D metric for time,
a 2D metric for color (hue),
a 3D metric for space,
a 6D metric for rigid objects in space (three dimensions of position and three dimensions of rotation around each position).

The encoding metric may fall far short of the stimulus metric in dimensionality; for example, a random dot field has an N-dimensional metric of the intensity of each dot, where N is the number of dots. We may encode these as mostly indiscriminable random textures, an example of failing to distinguish the metric dimensions themselves rather than just the positions along the dimension. Only idiosyncratic examples may evoke sufficient encoding response to form distinguishable dimensions. A less extreme example of encoding insufficiency is the dimensionality of perceived rotations. Intuitively, it is obvious that there are three dimensions of positional variation, but most people need to try it out to convince themselves that rotation around a point also has a dimensionality of three.

In recognition of their imperfect reflexivity, Shepard (1981) refers to the match between the encoding metric and the stimulus metric as one of complementarity, the relationship between a lock and key or a photographic negative and its print. Piaget (1969) originated a similar concept in referring to schemata, by which the organism accommodates itself to the nature of the object (in order to assimilate the object into the arrangement desired by the organism). The schema is a broader concept than Shepard's complementarity, including action components as well as the representative

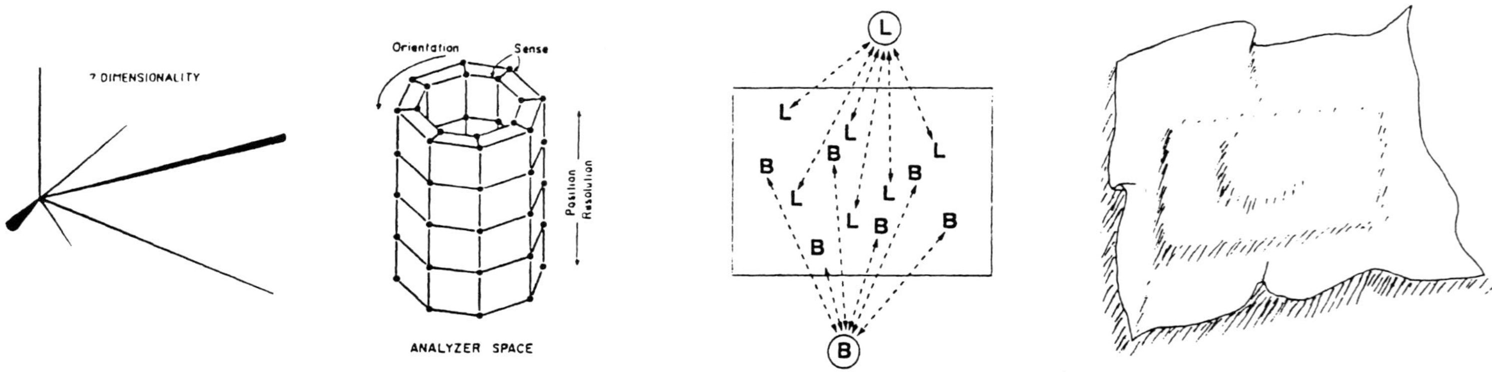

Figure 9. Overview of the encoding process. From left to right: 1. Encoding metric for the stimulus space. 2. Structure of the representation in the coding metric (analyzer space; from Palmer, 1982). 3. Coding of image symmetries by a self-matching or autocorrelation process. 4. Schematic representation of the object in the coding space.

aspects, but it carries the full sense of the reciprocity between the mental representation and the world that is to be brought under its hegemony.

The structure of the representation in the coding metric

After establishment of the internal space of the coding metric, the next question is the structure of the representation in each dimension of the metric. Is the metric a continuous one, like an intensity code along the metric dimension, or are there discrete channels with local preferences? This distinction has been emphasized by Foster (1982), although he included structured representations of higher-order features in the discrete category, whereas here such higher-order features are considered to form dimensions of the coding metric itself. This distinction relates to the question of whether each dimension is intensive (coded by the intensity of a single channel) or extensive (coded by an array of similar channels differing in one respect). But this is a subsidiary question that has little effect on the structure of the encoding metric.

Note that the presence of independent channels at one level of encoding may be converted into a continuous representation by a push-pull, or opponent, linkage between channels, as in the classic example of the opponent mechanisms of color vision. This conversion highlights the properties of the encoding mechanism in determining the encoding metric. The presence of more than one channel whose outputs make essentially independent contributions to the response results in a discrete channel code; linkage between channels to determine the output generates a continuous code for each pair of linked channels (assuming a graded response for each channel alone). Another way to obtain an essentially continuous code is to have a large number of independent channels across the dimension, forming a continuous extensive coding through the channel space as opposed to the continuous intensive coding of the push-pull arrangement.

If there is a discrete channel coding, a key question to answer is how many channels lie along the coding dimension (in each local retinal region). Many techniques have been developed to address this question but few of them have been worked out in sufficient detail to have full confidence in the answers (Tyler *et al.*, 1993, 1994). Historically, channel modeling in vision began with discrete channel models, as exemplified by the threshold elevation paradigm developed in color vision by Stiles (1939, 1959). Discrete channel analysis in spatial vision goes back to Wilson and Bergen (1979), followed by Wilson *et al.* (1983), Swanson and Wilson (1985) and Foley and Yang (1991), among many others. It has also been used for a variety of other stimulus domains, such as temporal frequency (Mandler and Makous, 1984; Anderson and Burr, 1985; Hess and Snowden, 1992) and stereomotion (Beverley and Regan, 1973).

A subsequent question, when the channel distribution on each dimension has been established, is the nature of the interactions between channels. Strictly speaking, a channel is an entity that is independent of its companion channels, so that interacting channels constitute an overall channel. However, the situation is analogous to that of neural receptive fields, where the excitatory region of the receptive field is considered to be the primary or 'classical' receptive field. Surrounding regions that modulate the response of the excitatory region without generating a response when stimulated by themselves form the integration field that presumably represents interactions (excitatory or inhibitory) with adjacent neurons (MacIlwain, 1964; Fischer and Kruger, 1974; DeAngelis *et al.*, 1992). In the same way, paradigms may be designed to distinguish between primary and interactive aspects of channel behavior. Color provides a canonical example of such a distinction, where the cone sensitivity functions for intensity increments may be determined by appropriate isolation techniques but these primary channels then interact in opponent fashion to provide chromatic information. As long as there is empirical access to both levels of organization, there is no categorical problem in distinguishing between the primary and interactive aspects of the channel structure.

The more concrete aspect of encoding structure, its physiological instantiation in the hardware of the brain, is tangential to the present discussion of its logical structure. For example, realizing an intensity code as a neural firing rate adds little to our appreciation of the properties of the code; it could equally well be a cellular voltage or a concentration of transmitter molecules, which indeed seem largely interchangeable with the firing rate code in the operation of neurons. Rather, it is the functional organization of the encoding process that determines its effectiveness to the organism.

The coding of image symmetries by a self-matching or autocorrelation process

Beyond the low-level structure of the representation, there arises the issue of the encoding of regularities or symmetries in the image by capitalizing on their redundancy to simplify the representation. These are issues that have been raised variously by Gibson (1950), Attneave (1954) and Garner (1962), among many venerable figures. The mechanisms for implementing this simplification remain unresolved, however.

A variety of schemes has been proposed to address the issue of how symmetries may be encoded. One of the most sophisticated is that of Palmer (1982), a transformational code based on the fact that the global structure of the coding space is determined by the symmetry relations in the local operators (cortical filters) that are generating the coding space. Thus, a symmetry relation that exists in the image will evoke multiple matching patterns in the response space. This operation reduces the complexity of the problem because all types of symmetry in the image space are converted simply to one type of symmetry in the coding space: translational symmetry. However, as Palmer points out, a second-order comparison mechanism is required to compare the outputs in the coding space and determine whether such matches exist.

The need for second- (and higher-) order comparison stages reveals that it is the connectivity relations in the comparison stage that determine what aspects of symmetry in the image are processed. In fact, Palmer postulates that the higher-order stages are local within the coding space, which means that global symmetries would be detected only by their local symmetry relations. If we had only local connectivity, for example, we could not detect that the ears are symmetrically located in a face unless we had detectors large enough to encompass both ears in a facial image.

Rather than postulating second-, third- and higher-order matching mechanisms for detecting symmetry relations (as does Palmer, 1992), one may postulate a general-purpose autocorrelation mechanism operating in the coding space to detect symmetries of any type at any range. Tyler and Miller (1994) have shown how the core element of repeated or translationally-symmetric patterns may emerge through an autocorrelation process of mutual reinforcement. One way that such a mechanism might be implemented in the nervous system is for the local pattern vector at the focus of attention (usually the fovea) to form a template that is correlated automatically with the local pattern vector at all other locations. By 'local pattern vector' is meant the profile of response strength through the array of local detectors of all sizes, orientations and types. A high correlation of this vector between two locations means that the same pattern of inputs was present at the two locations at some orientation and scale. Any mechanism that could make such comparisons therefore would act as a detector of all the types of symmetry present in the local pattern vector.

The schematic representation of the object in the coding space

Objects are represented by clusters of features with specifiable symmetry relations, but there is a kind of perceptual 'surface tension' that holds the features of a specific object together in the representation space. There must be a vocabulary of object forms or schemata of the type proposed by Piaget (1969) to which are attached the features specific to the object in the current focus of attention. The concept of a schema is understood as an action structure; e.g. a 'box' is a cuboid structure with an inside in which other objects may be placed. Part of this action structure is its three-dimensional symmetry relations, i.e. the aspects or regions of the object that may be made congruent by a symmetry transformation. Unlike the 2D symmetry structure (which, as I have argued in the previous section, is likely to be coded by inherent symmetries in the coding space), the complexity of generalized 3D symmetries probably require internal manipulation of the type proposed by Shepard (1981) to ascertain its symmetry structure.

I would suggest that the total process of object characterization may involve both structural and manipulative aspects in the symmetry encoding. For example, if a cube is viewed in perspective projection with one side facing the observer, the quartic symmetry of the front face might be perceived by inherent coding symmetries, implying that it does not need to be mentally rotated around the line of sight for the quartic symmetry to be appreciated. Conversely, the symmetrical identity of the six faces (or lack of identity if the cube is a rectangular cuboid), can be appreciated only by mental rotation of the front face into the other positions. In this way, the structural aspect of the symmetry coding process may be brought to bear in those regions of the image where manipulation is not required; the results of the structural analysis may then be carried into the manipulative phase so that features that are found to be congruent be a manipulative analysis are consequently perceived to have the same inherent symmetry structure.

There may also be a level of cognitive assumptions in the manipulative aspect of the encoding that, for symmetry operations, takes the form of avoiding the need for some mental rotations on the basis of an assumed symmetry relation. For example, when looking at a set of banisters in perspective projection, one may need only to perform the mental operation of matching one banister to the next to establish the similarity and then make the cognitive assumption that all the others are similarly matched. This assumption would economize on the number of mental manipulations that need to be made at the cost of missing deviations from the assumed symmetry in regions where manipulative testing is avoided.

Much of what we see may be only part of the total object schema, e.g. when sitting on a couch, we may have only an arm and part of the seat cushions projecting on to the retina, but perceive them as part of the complete couch. This process of amodal (or non-sensory) completion has been most fully explored by Kanisza (1976), although he emphasized regions hidden by other objects rather than regions outside the visual field. The concept of an abstract schema representing the whole object gives a direct instantiation to the process of amodal completion; if part of an object is sufficient to evoke unambiguously the full schema, the whole object is perceived in the implied position. This evocation is similar to the process known as 'lexical access' in language perception (Marslen-Wilson and Tyler, 1982), in which the full meaning of a spoken word is evoked as soon as it is unambiguously distinguishable from all other words. In the same way, the whole object may be perceived when the information from the visible parts is sufficient to identify it from the vocabulary of known objects or plausible object classes. There is the obvious proviso that the manner in which the non-visible parts are obscured must be consistent with their occlusion by an intervening object of some kind, so that attended region of the scene is itself a plausible arrangement. That there are counterexamples in which the attended region makes an 'impossible' object (Hochberg, 1981) does not reduce the general applicability of this proviso.

The four processing stages described constitute a framework in which most studies of perception in general and symmetry perception in particular may be viewed. One aspect that may be considered to be understressed is the role of attention in of this framework, but this is in accord with the view of perception as a largely parallel, subconscious process for which attention is the mechanism of access or gateway to the higher, serial processes of cognition. This gateway presumably occurs somewhere in the fourth of the stages described, the object representation stage, and channels the preprocessed symmetry information to consciousness to aid the organism in dealing with its environment. A fuller account of the role of attention as an organizing principle in 3D shape reconstruction is provided in Tyler (1995, 1996).

CONCLUSION

The present set of papers constitute more of a freeze-frame of current activities in the field of symmetry perception rather than any kind of unified approach. One interesting aspect of the collection is the connection between the experimental and the computational approaches, which promises to be an increasing feature of sensory research in general. Development of sophisticated analytic models should not be allowed to obscure the fact that very little is known as yet about the details of human symmetry perception. Instead, the models may provide a driving force for the formulation of new empirical studies to assess the mechanisms proposed, leading to an augmented exploration of symmetry perception. As in all fields, only by taking the models to their predictive limit can their validity be tested against the perceptual reality.

Most of the papers included here were previously published in one of two Special Issues of *Spatial Vision* (Vol. 8, No. 4, 1994 and Vol. 9, No. 1, 1995). Several original contributions, that were not able to be included for one reason or another, are now incorporated into the present set (one previously published elsewhere). Although all the papers pertain to human symmetry perception to some degree, the papers are grouped into those that emphasized theoretical issues of symmetry in general and those that had a stronger emphasis on the empirical properties of human symmetry perception (although many of the latter include cogent theoretical analysis of the issues raised by their data). It is hoped that publication in this fashion will help the material reach the wider audience of those interested in symmetry from diverse fields of science, particularly from the analytic disciplines of computer science.

REFERENCES

Anderson, S. J. and Burr, D. C. (1985). Spatial and temporal selectivity of the human motion detection system. *Vision Res.* **25**, 1147–1154.

Attneave, F. (1954). Some informational aspects of visual perception. *Psychol. Rev.* **61**, 183–193.

Barghout-Stein, L. and Tyler, C. W. (1994). A new computational analysis of spatial-frequency mechanisms. *Invest. Ophthalmol. Visual Sci.* **35** (Suppl.), 2007.

Barlow, H. B. and Reeves, B. C. (1979). The versatility and absolute efficiency of detecting mirror symmetry in random dot displays. *Vision Res.* **19**, 783–793.

Beverley, K. I. and Regan, D. (1973). Evidence for the existence of neural mechanisms selectively sensitive to the direction of movement in space. *J. Physiol.* **235**, 17–29.

Bruce, V. G. and Morgan, M. J. (1975). Violations of symmetry and repetition in visual patterns. *Perception* **4**, 239–249.

Cooper, L. A. and Shepard, R. N. (1973). The time required to prepare for a rotated stimulus. *Mem. Cognit.* **1**, 246–250.

Cooper, L. A. and Shepard, R. N. (1978). Transformations on representations of objects in space. In: *Handbook Perception.* (Vol. 8). E. C. Carterette and M. P. Friedman (Eds). Academic Press, New York.

Corballis, M. C. and Roldan, C. E. (1974). On the perception of symmetrical and repeated patterns. *Percept. Psychophys.* **16**, 136–142.

DeAngelis, G. C., Robson, J. G., Ohzawa, I. and Freeman, R. D. (1992). Organization of suppression in receptive fields of neurons in cat visual cortex. *J. Neurophysiol.* **68**, 144–163.

DeYoe, E. A. and van Essen, D. C. (1988). Concurrent processing streams in monkey visual cortex. *Trends Neurosci.* **11**, 219–226.

Fischer, B. and Kruger, J. (1974). The shift effect in the cat's lateral geniculate nucleus. *Exp. Brain Res.* **21**, 225–227.

Foley, J. M. and Yang, Y. (1991). Forward pattern masking: effects of spatial frequency and contrast. *J. Opt. Soc. Am.* **8**, 2026–2037.

Foster, D. H. (1982). Analysis of discrete internal representations of visual pattern stimuli. In: *Organization and Representation in Perception.* J. Beck (Ed.). Lawrence Erlbaum, Hillsdale, NJ, pp. 319–341.

Gibson, J. J. (1950). *The Perception of the Visual World.* Houghton Mifflin, Boston.

Gardner, M. (1964). *The Ambidextrous Universe*; *Left, Right and the Fall of Parity.* Mentor, New York.

Garner, W. (1962). *Uncertainty and Structure as Psychological Concepts.* Wiley, New York.

Hess, R. F. and Snowden, R. J. (1992). Temporal properties of human visual filters: Number, shapes and spatial covariation. *Vision Res.* **32**, 47–59.

Hochberg, J. (1978). *Perception.* Prentice-Hall, Englewood Cliffs, NJ.

Hogben, J. H., Julesz, B. and Ross, J. (1976). Short-term memory for symmetry. *Vision Res.* **16**, 861–866.

Julesz, B. (1966). Binocular disappearance of monocular symmetry. *Science* **153**, 657–658.

Julesz, B. (1970). *Foundations of Cyclopean Perception*. Chicago University Press, Chicago.

Julesz, B. and Tyler, C. W. (1976). Neurontropy, an entropy-like measure of neural correlation, in binocular fusion and rivalry. *Biol. Cybernet.* **23**, 25–32.

Kanisza, G. (1976). Subjective contours. *Scient. Amer.* **234**, 48–52.

Locher, P. J. and Nodine, C. F. (1987). Influence of stimulus symmetry on visual scanning patterns. *Percept. Psychophys.* **13**, 408–412.

Mach, E. (1887). *The Analysis of Sensations*. Dover, New York.

McIlwain, J. T. (1964). Receptive fields of optic tract axons and lateral geniculate cells: Peripheral extent and barbiturate sensitivity. *J. Neurophysiol.* **27**, 1154–1173.

Mandler, M. B. and Makous, W. (1984). A three channel model of temporal frequency perception. *Vision Res.* **24**, 1881–1887.

Marslen-Wilson, W. and Tyler, L. K. (1980). The temporal structure of spoken language understanding. *Cognition* **8**, 1–71.

Minsky, M. (1975). A framework for representing knowledge. In: *The Psychology of Computer Vision*. P. H. Winston (Ed.). McGraw-Hill, New York.

Palmer, S. E. (1982). Symmetry, transformation and the structure of perceptual systems. In: *Organization and Representation in Perception*. J. Beck (Ed.). Lawrence Erlbaum, Hillsdale, NJ, pp. 95–144.

Penfield, W. (1959). The interpretative cortex. *Science* **129**, 1719–1725.

Piaget, J. (1969). *The Mechanisms of Perception*. G.N. Geagrin (Trans.). Basic Books, New York.

Pylyshyn, Z. (1986). *Computation and Cognition*. MIT Press, Cambridge, MA.

Shepard, R. N. (1981). Psychophysical complementarity. In: *Perceptual Organization*. M. Kubovy and J. R. Pomerantz (Eds). Lawrence Erlbaum, Hillsdale, NJ, pp. 279–342.

Stiles, W. S. (1939). The directional sensitivity of the retina and the spectral sensitivities of the rods and cones. *Proc. R. Soc. London* **B127**, 64–105.

Stiles, W. S. (1959). Colour vision: The approach through increment threshold sensitivity. *Proc. Nat. Acad. Sci.* **45**, 100–114.

Swanson, W. H. and Wilson, H. R. (1985). Eccentricity dependence of contrast matching and oblique masking. *Vision Res.* **25**, 1285–1295.

Tyler, C. W., Barghout, L. and Kontsevich, L. L. (1993). Computational reconstruction of the mechanisms of human stereopsis. In: *Computational Vision Based on Neurobiology*. T. B. Lawton (Ed.). Proceedings of the SPIE **2054**, pp. 52–68.

Tyler, C. W. and Miller, R. H. (1994). Pattern identification by trajectory analysis in autocorrelation hyperspace. *Proc. of the World Congress on Neural Networks* **III**, 312–316.

Tyler, C. W. and Kontsevich, L. L. (1995). Mechanisms of stereoscopic processing: stereoattention and surface perception in depth reconstruction. *Perception* **24**, 127–153.

Tyler, C. W. (1996). Symmetries, structure and schemata in perceptual coding. In: *Human Perceptual Coding*. M. Jenkin and L. Harris (Eds). Cambridge University Press, Cambridge (in press).

Wagemans, J. (1995). Detection of visual symmetries. *Spatial Vision* **9**, 9–32. (Reprinted in the present collection on pages 25–48.)

Washburn, D. K. and Crowe, D. S. (1988). *Symmetries of Culture*. University of Washington Press, Seattle.

Wenderoth, P. (1995). The role of pattern outline in bilateral symmetry detection with briefly flashed patterns. *Spatial Vision* **9**, 57–77. (Reprinted in the present collection on pages 49–69.)

Wilson, H. R. and Bergen, J. (1979). A four-mechanism model for threshold spatial vision. *Vision Res.* **19**, 19–32.

Wilson, H. R., McFarlane, D. K. and Phillips, G. C. (1983). Spatial frequency tuning of orientation selective units estimated by oblique masking. *Vision Res.* **23**, 873–882.

Yodogawa, E. (1982). Symmetropy, an entropy-like measure of visual symmetry. *Percept. Psychophys.* **32**, 230–240.

Young, J. Z. (1962). Why do we have two brains? In: *Interhemispheric Relations and Cerebral Dominance*. V. B. Mountcastle (Ed.). Johns Hopkins University Press, Baltimore, MD.

Part One

EMPIRICAL EVALUATION OF SYMMETRY PERCEPTION

Human Symmetry Perception, pp. 25–48
C. W. Tyler (Ed.)

Detection of visual symmetries

JOHAN WAGEMANS
Laboratory of Experimental Psychology, University of Leuven, Tiensestraat 102, B-3000 Leuven, Belgium

Abstract—This paper reviews empirical evidence for the detection of visual symmetries and explanatory theories and models of symmetry detection. First, mirror symmetry is compared to other types of symmetry. The idea that symmetry detection is preattentive is then discussed and other roles that attention might play in symmetry detection are considered. The major part of the article consists of a critical examination of the extensive literature about the effects on symmetry detection of several major factors such as the orientation of the symmetry axis, the location of the stimulus in the visual field, grouping, and perturbations. Constraints on plausible models of symmetry detection are derived from this rich database and several proposals are evaluated against it. As a result of bringing this research together, open questions and remaining gaps to be filled by future research are identified.

1. INTRODUCTION

Symmetry is everywhere: in natural objects, from crystals to living organisms, in manufactured articles of many kinds, and in art works from all cultures throughout the world and at all times (Washburn and Crowe, 1988). As pointed out by the editor of this Special Issue, it is no surprise then that biological vision systems have evolved adaptive strategies for perceiving such symmetries and utilizing them in all kinds of tasks. Pigeons discriminate and classify shapes on the basis of symmetry (Delius and Nowak, 1982). Animals at many phylogenetic scales use symmetry in mate selection (Moeller, 1992; Swaddle and Cuthill, 1993). Experimental evidence from human infants and young children demonstrates that symmetry receives ontogenetic priority as well (Bornstein *et al.*, 1981; Fisher *et al.*, 1981). Finally, computer-vision techniques have also been developed to detect symmetry and exploit it in all sorts of ways (Stevens, 1980; Kanade and Kender, 1983; Marola, 1989; Nalwa, 1989; Van Gool *et al.*, 1990, 1995).

Because all this suggests that symmetry is a salient visual property that must be detected efficiently and rapidly by the visual system, considerable research effort has been devoted to the study of the detection of visual symmetries in artificial and biological vision, by scientists in various disciplines from art to zoology, from the early days of visual science (Mach, 1886/1959) until now, and inspired by all kinds of theoretical approaches. Although this research tradition has provided us with a rich database of empirical findings and theoretical proposals, it has left open many questions. This paper brings some of this research together, to review what we know about symmetry detection and where gaps remain to be filled by future research. This review will focus only on aspects of symmetry perception related to its detection as such, not on the usefulness of symmetry in other tasks.

2. DIFFERENT TYPES OF SYMMETRY

Usually, symmetry is implicitly equated with mirror symmetry and most of the empirical studies of symmetry detection have been restricted to mirror symmetry. However, from a mathematical point of view, the class of symmetries is much richer (Weyl, 1952; Armstrong, 1988). Informally, symmetry means self-similarity under a class of transformations, usually the group of Euclidean transformations in the plane, that is, translations, rotations, and reflections (also collectively denoted by 'isometries'). Surprisingly few studies have explicitly compared the three pure types of symmetries or investigated the ways they interact when present together. In this section, the evidence is reviewed that the researchers' preoccupation with mirror symmetry reflects the fact that it is a more salient property for the visual system than the other symmetries.

For example, Julesz (1971) observed that detection of symmetry created by repetition (i.e. translation) or by point-reflection (i.e. 180 deg rotation) requires scrutiny, whereas perception of mirror symmetry (i.e. created by reflection) is effortless, rapid, and spontaneous. He also pointed out that the opposite is true in the auditory modality, where repetitions of melodic sequences are very easy to notice but mirror symmetries are not. The advantage of mirror symmetry over repetition has been confirmed in other, more systematic studies.

Corballis and Roldan (1974) designed two different tasks with the same simple stimuli (i.e. sparse dot patterns, arrowheads, and C-shapes) presented for 100 ms. In one task, subjects had to judge the relationship between the two pattern halves (as 'same' for translation or 'mirror' for reflection), whereas subjects in the other task had to judge each pattern as a whole (as 'symmetrical' for reflection or 'asymmetrical' for translation). In most of the conditions, a mirror-symmetry advantage was obtained. Bruce and Morgan (1975) asked subjects to detect small violations in reflected or translated line patterns and found that response times were generally faster for the mirror symmetries, except when the violations were near the edge of the patterns. This suggests that the salience of mirror symmetry is probably based on the ease of comparing spatially contiguous elements near the axis. In a systematic investigation of the effect of point-pair separation on the detection of translational symmetry, Jenkins (1983a) confirmed this intuition. His results may be compared with those of Tyler and Chang (1977), who showed that detectability of repetitive patterns declined with the square root of the number of repeats, in accord with ideal-observer predictions.

However, even when distance is controlled, there is often an advantage of reflection over translation. For example, when subjects have to compare two pseudo-random contours, the task is much easier when they are reflected than when they are translated, regardless of whether the contours are part of the same object or of two different objects (Baylis and Driver, 1995). An even more striking result was obtained in another recent study by Baylis and Driver (1994). They varied the number of elements in each half-pattern from 4 to 16 and found that the response-time functions were relatively flat for the detection of reflectional symmetry (i.e. 3 to 6 ms per element pair) and quite steep for the detection of translational symmetry (i.e. 25 to 40 ms per element pair). Although I would not go as far as Baylis and Driver (1994), who claim that this implies parallel versus serial computation of reflectional versus translational symmetry, respectively, these results do suggest collectively that there is something special about mirror symmetry which makes it an intrinsically more salient stimulus attribute than translational symmetry.

The special status of mirror symmetry seems supported by a couple of studies which have compared it with rotational symmetry. For example, Royer (1981) showed that symmetries created by 90 or 180 deg rotations were always much harder to detect than those created by reflections, regardless of the specific display type (i.e. dots, blocks, diagonolinear, or rectilinear line segments) and degree of practice. However, it remains to be seen whether this mirror-symmetry advantage would still hold when compared to rotational symmetries with smaller angles. A mirror-symmetry advantage was also obtained by Palmer and Hemenway (1978) in a study with closed polygons instead of discrete patterns, although subjects had to respond negatively to rotational symmetries, which might have affected the response times for other reasons. In other words, it appears hard to design a fair comparison and more work along the same lines seems warranted.

As far as I know, Kahn and Foster (1986) were the first to report a systematic investigation of the different types of symmetries created by translation, rotation, and reflection in dot patterns. Because they were interested in pattern recognition more than in symmetry detection, Kahn and Foster designed their experiments as 'same-different' discriminations to be made between two dot patterns, instead of global symmetry judgments of the whole configuration (which may cause important differences; see Corballis and Roldan, 1974). Although some specific differences between the three types of symmetry were found, the overall performance levels for all of them were quite good (d' between 1 and 2), considering the short exposures of 100 ms.

These reasonable levels of detectability for symmetries created by translation, rotation, and reflection were also obtained by Wagemans *et al.* (1993) in a study designed to test a specific model of symmetry detection (see later). In three separate experiments, the detection of symmetry in dot patterns was tested as a function of some transformation parameters (e.g. orientation, distance, angle), as well as some factors specifically manipulated to introduce or destroy higher-order structures that were proposed as being important in supporting efficient symmetry detection. In all conditions, d's for regular-random discriminations in dot patterns presented for 100 ms were generally above 1. Although no direct statistical comparisons between the three types of symmetry were made, the trends indicated that reflections were easier to detect than translations and rotations, which did not differ much. The same pattern of results was also obtained by Zimmer (1984), who used pseudo-random line drawings and different degrees of asymmetry as distractors for each type of symmetry (an experimental design which requires careful stimulus construction).

In summary, the different types of symmetry created by reflection, translation, and rotation can all be detected, although they are not equally salient. However, more direct comparisons in systematic, parametric studies are clearly needed. Because mirror symmetry seems to have a special status for the visual system, the remainder of the paper will focus on this type of symmetry only.

3. SYMMETRY AND ATTENTION

The salience of symmetry created by reflection about a vertical axis suggests that symmetry might be one of the privileged properties that are detected preattentively. Indeed, most symmetry-detection researchers have implicitly or explicitly adopted this view (e.g. Barlow and Reeves, 1979; Wolfe and Friedman-Hill, 1992; Locher and Wagemans, 1993). Experimentalists have developed two operational definitions of 'preattentive' to test this idea about the processing of stimulus attributes such as symmetry. In addition, there is some pertinent neuropsychological evidence.

First, following Julesz' (1981, p. 28) operational definition of a preattentive process as one in which an observer is able 'to perceive certain structures in the stimulus array when the stimulus is briefly presented — say for less than 160 ms', symmetric displays have often been presented at short exposure durations. This research confirms that symmetry can be detected preattentively (in Julesz's sense of the word) over a wide range of stimulus and viewing conditions. One can perceive symmetry in brief presentations of simple random shapes (at 25 ms; see Carmody *et al.*, 1977), in dot patterns (at 100 ms; see Barlow and Reeves, 1979; Wagemans *et al.*, 1991, 1993), in dynamic dot textures (at 40–50 ms; see Julesz, 1971; Hogben *et al.*, 1976), in other discrete patterns consisting of line segments at different orientations (at 10–125 ms; see Locher and Wagemans, 1993), as well as in complex abstract art displays (at 50–100 ms; see Locher and Nodine, 1989).

A second, perhaps more powerful, technique to assess preattentive processing has become very popular after Treisman introduced her feature-integration theory and the visual search paradigm to test it (Treisman and Gelade, 1980). In this paradigm, subjects have to search for a target in a display with a variable number of distractors. Response times are generally plotted as a function of display size and the slope of these functions is taken as an index of search efficiency. The basic finding is that targets defined by primitive features such as color or line orientation can be detected in parallel (i.e. independent of the display size), whereas more complex combinations such as conjunctions of color and form, or specific spatial relations between line segments seem to require a serial search process (as inferred from a linear display-size effect on response times). Because symmetry is critically dependent on the spatial relations between more primitive elements (such as dots or line segments), it seems to follow that symmetry must require attention to be detected. However, preattentive grouping processes can produce emergent properties and this might give local symmetry the status of a special feature (Treisman and Patterson, 1984; Pomerantz and Pristach, 1989). Moreover, more recent research has made it clear that conjunction search is often more efficient than predicted by a serial model (e.g. Wolfe *et al.*, 1989; Treisman and Sato, 1990; Enns and Rensink, 1991).

Inspired by studies showing that more global spatial relations between elements in a display affect the search efficiency (e.g. Moraglia, 1989; Nothdurft, 1992), the possibility that symmetry detection is preattentive in the sense of producing flat search functions has been tested in two studies. First, Javadnia and Ruddock (1988) showed that targets could be discriminated from distractors by parallel processing if they differed in symmetry (e.g. an E against Es where the symmetry was perturbed). More recently, Wolfe and Friedman-Hill (1992) have examined the role of the symmetry relations among display elements, which were line segments at variable orientations. Search for a target, defined as an element of a third orientation, was more efficient when the orientations of the background elements were symmetrical about a vertical axis (e.g. a 50 deg clockwise target against a background of 20 deg clockwise, CW, and 20 deg counterclockwise, CCW, distractors) than when they were symmetrical about an oblique axis (e.g. 20 deg and 80 deg distractors for the same target). However, both of these studies have their own limitations and more research along the same lines seems warranted, especially in light of other recent research suggesting that attention might play a role after all.

Driver *et al.* (1992) reported some results from a neuropsychological case study which are interesting in this respect. They investigated symmetry detection by a patient who failed to attend to the left side of objects throughout the visual field, resulting from brain damage centered on the right parietal lobe (i.e. the 'visual neglect syndrome'). As one would expect, this patient could not detect vertical symmetry in shapes (presented until a response was made), while he was able to detect horizontal symmetry. Interestingly, when asked which of two sets of shapes were seen as figures against a background, this patient clearly preferred the symmetrical shapes in much the same way as normal observers do (Bahnsen, 1928). As stressed by Driver *et al.* (1992), this result indicates that both the right and the left of each shape must be represented at some stage in the patient's visual system in a form that supports symmetry detection, at least covertly. The more general implication seems to be that symmetry can be detected preattentively, that is, prior to the attentional stage at which the patient's impairment leads him to neglect the left side of each shape when he is asked to make overt judgments about them.

Despite the operational definitions of 'preattentive' used in the experiments mentioned earlier, attention is involved in the sense that subjects always know when the stimulus is coming and where it will be presented. Moreover, in the first set of studies, subjects were explicitly instructed to detect symmetry. Recently, Rock, Mack, and their colleagues (Mack *et al.*, 1992; Rock *et al.*, 1992) have shown that this direction of attention to the array and to the task at hand is important. They demonstrated that many of the phenomena of grouping and perceptual organization studied by Gestalt psychologists do not occur in conditions of inattention, although they have since then been thought of as largely preattentive and automatic. In the cited studies, subjects had to perform a foveal task — line-length discrimination — while the stimulus was embedded in a field of other elements, or while it had an array of dots or a simple shape in its immediate neighborhood. Surprisingly, subjects could not answer simple questions such as whether the field was homogeneous or heterogeneous, how dense the array of dots was, or what shape the additional element had. Of course, this question could be asked only once to each subject in conditions of complete inattention, that is, independent of the attentional mechanisms activated by the intention to look for a certain stimulus. Despite the intrinsic limitations of this paradigm, it seems important to investigate whether symmetry can be detected under these conditions of inattention.

Another demonstration of the role of attention in symmetry detection has to do with the effect of orientation (see later). In an attempt to investigate whether the salience of vertical symmetry is based on the fixed neural architecture that supports the oblique effect in many other perceptual tasks (e.g. Appelle, 1972), Wenderoth (1994) recently manipulated the relative frequencies of different orientations in different blocks of trials. He found that the detection of mirror symmetry is best at the orientation which is at the mean of the frequency distribution. For example, when the 16 different orientations which were tested were vertical, horizontal, the two diagonals (45 deg and 135 deg), and 5, 10, and 15 deg CW and CCW deviations from diagonal, performance was best at diagonal instead of vertical or horizontal. The obvious explanation of this result is that the range of stimuli in a block of trials affects subjects' scanning or attentional strategies so that they focus on the symmetry axis with the greatest likelihood. This interpretation is congruent with Pashler's (1990) finding that cueing the subjects in advance about the orientation of the axis of symmetry produced a considerable increase in speed and accuracy.

In summary, there is overwhelming evidence that the percept of symmetry emerges seemingly effortlessly and automatically in a wide variety of conditions. However, the implications in terms of the role of attention are far from clear. Further research employing visual-search and inattention paradigms or other methods of manipulating attention seems important in this respect.

4. CHARACTERISTICS OF SYMMETRY DETECTION

Most of the studies on symmetry detection have investigated the effects of some major factors on the efficiency and speed of symmetry detection. This kind of research has yielded important information about the general principles of possible underlying mechanisms and the constraints within which they operate. Although these studies do not corroborate any specific model of how exactly symmetry is detected, they do provide powerful indications against the plausibility of some proposed models of symmetry detection. Four such explorations will be discussed in turn: the effects on symmetry detection of the orientation of the axis, the location of the patterns in the visual field, the grouping of the elements in the patterns, and perturbations of all kinds.

4.1. Symmetry and orientation

Since Mach's (1886/1959) observation that symmetry about a vertical axis is more salient than symmetry about any other axis, a large number of studies have quantified this advantage experimentally. Two basically different paradigms have been used to this end although neither addresses the issue of whether the vertical preference is based on a structural bias in the neural array or attentional preference for the vertical axis. In the first, introduced by Goldmeier (1937/1972), subjects are presented with a pattern with two axes of mirror symmetry, vertical and horizontal, and they have to indicate which of two test patterns with single symmetry, vertical or horizontal, best resembles the vertical-horizontal-symmetric reference pattern. The results clearly indicate a vertical symmetry advantage (Rock and Leaman, 1963; Fisher and Fracasso, 1987).

The second paradigm investigates the detectability of symmetry created by reflection about axes in different orientations, usually only vertical and horizontal and the two diagonals in-between. Generally, the results indicate that vertical symmetry is easier to detect (i.e. faster or better) than horizontal symmetry, which is easier to detect than diagonal symmetry (e.g. Palmer and Hemenway, 1978; Royer, 1981). However, a significant number of results do not fit this simple summary. In some studies, horizontal symmetry was not harder than vertical symmetry (e.g. Fisher and Bornstein, 1982), or horizontal symmetry was even easier (e.g. Jenkins, 1983b, Experiment 4; Pashler, 1990, Experiment 4). In other studies, diagonal symmetry was not harder than horizontal symmetry (e.g. Jenkins, 1985), or diagonal symmetry was even easier (e.g. Corballis and Roldan, 1975). Other work shows that the effect of axis orientation interacts with the effect of other variables such as the orientation of the individual line segments or their spatial grouping (Locher and Wagemans, 1993). The diversity of these results makes it difficult to support the hypothesis of a structural bias in the neural filters processing the symmetrical pattern.

Only a few studies have tested other oblique orientations in addition to the main diagonals. Barlow and Reeves (1979) reported a response-bias free measure of detectability (d') for one subject tested with eight different orientations, which suggested that vertical symmetry was easiest, followed by horizontal symmetry, followed by diagonal symmetry, and followed by other obliques (30 and 60 deg from horizontal, both CW and CCW). The difference between diagonal and other oblique orientations is not congruent with the neurophysiological evidence on the oblique effect, which suggests that the visual system's sensitivity for orientations decreases with larger deviations from vertical and horizontal (Appelle, 1972). Nevertheless, a similar pattern of results was reported in three more recent studies (Zimmer, 1984; Wagemans *et al.*, 1992; Wenderoth, 1994). However, Wenderoth (1994), by manipulating the range of orientations presented within a block of trials, showed that orientation preferences in symmetry detection could be biased or reversed at will to any selected orientation. Wenderoth's results suggest that a wide range of orientation effects in symmetry detection can be explained by attentional selectivity for particular axis orientations rather than biases in the neural array of orientation detectors.

4.2. Symmetry and visual field

A second set of factors investigated in some detail have to do with the location of the symmetric stimulus pattern in the visual field. Two kinds of manipulations have been employed: first, the contribution of different zones in a pattern to the global impression of symmetry; and, second, the effect of noncentral presentation of the pattern. Together with the orientation effects, these manipulations are interesting in as far as they test the importance of a symmetric projection to the visual system, which must be critical if the salience of vertical mirror symmetry depends on the vertical mirror symmetry of the neural architecture. This assumption has pervaded many of the ideas about symmetry detection since Mach's (1886/1959) early proposals (e.g. Julesz, 1971).

Most studies show that a restricted area around the axis is the most important one (e.g. Julesz, 1971; Bruce and Morgan, 1975; Barlow and Reeves, 1979; Jenkins, 1982) and that symmetry is easier to detect when the axis of symmetry is located at the point of fixation (e.g. Barlow and Reeves, 1979; Saarinen, 1988; Locher and Nodine, 1989). As with the orientation effects, however, several findings urge a qualification of this general statement.

First, with respect to the contribution of different zones, Barlow and Reeves (1979) measured the detectability of symmetry in displays in which only pairs of vertical slices of the dot patterns were symmetrical. The results showed that the symmetry is also detectable when the symmetrical dots lie only near the edge of each half pattern. In other words, the contribution of different zones does not decrease linearly with increasing distance from the symmetry axis. Instead, a U-shaped function was obtained with increased performance near the axis and near the edge of the pattern (see Wenderoth, 1995, for more recent evidence along the same lines). This is, of course, what one would expect on the basis of the positive results obtained with symmetric line drawings (e.g. Palmer and Hemenway, 1978) and filled polygons (e.g. Carmody *et al.*, 1977), which contain no internal features in the immediate neighborhood of the axis.

Second, with respect to the issue of central presentation, Julesz (1971) noted that the detection of symmetry in simple patterns like random shapes does not require that the center of the symmetry coincides with the fixation point of the eyes, whereas the opposite seems true for more complex patterns like dot textures. Based on this observation, Julesz concluded that symmetry detection operates at two levels: for patterns with low spatial frequencies, the symmetric relations are extracted globally, whereas a point-by-point comparison seems required for patterns with high spatial frequencies. A similar distinction has been incorporated in most subsequent theoretical proposals about symmetry-detection mechanisms (e.g. Bruce and Morgan, 1975; Palmer and Hemenway, 1978; Zimmer, 1984). In one of the few studies focused on the role of different spatial frequencies, Julesz and Chang (1979) showed that the sum of two random-dot arrays, one with vertical symmetry and one with horizontal symmetry, could simultaneously be perceived as two separate symmetrical patterns instead of a single random array when they are spatially filtered so that their respective power spectra are far enough apart (e.g. low-pass vertical symmetry and high-pass horizontal symmetry). In addition, they demonstrated that the low-band frequency channels contribute more heavily to the symmetry percept than the high-band channels do.

4.3. Symmetry and grouping

These and other ideas about spatial filtering (e.g. Watt and Morgan, 1985; Watt, 1987) receive indirect support from studies with symmetric patterns consisting of oriented line segments. Locher and Wagemans (1993) obtained results suggesting that the spatial grouping (e.g. clustering) of line segments determines the detectability of symmetry more than their individual identities (e.g. orientation relative to the axis of symmetry). The perception of symmetry might be the conscious concomitant of the output of filtering operations executed in parallel on a symmetric display, which make the locations of large-scale tokens (blobs) available before the figural identity of the elements. Different versions of this idea are widely spread among symmetry-detection studies (e.g. Barlow and Reeves, 1979; Royer, 1981; Jenkins, 1983b; Pashler, 1990; Locher and Wagemans, 1993).

In addition to the evidence discussed earlier, two other results argue against the idea of a point-by-point comparison of all the elements in a display. First, Troscianko (1987) has demonstrated that isoluminance does not destroy perception of random-dot symmetry. In the light of his view that the main effect of isoluminance may be an introduction of a small positional uncertainty into the neural representation of the stimulus, Troscianko interpreted this finding as evidence that exact position information is not essential for symmetry detection. Second, by comparing human performance with an ideal observer mechanism looking for all pairwise matches, Tapiovaara (1990) has convincingly shown that only a modest number of all possible point-by-point comparisons are made. This finding is, of course, what one could expect on the basis of the wide range of patterns used in symmetry-detection research. The results obtained in these studies do not seem to agree with the simple prediction that detectability of symmetry takes more time with an increasing number of pattern elements (see Baylis and Driver, 1994, for more direct evidence with filled polygons). Apparently, elements become grouped together and only a restricted number of groups are compared.

Nevertheless, there is some evidence that at least one featural characteristic, the luminance of the elements in relation to the background, determines the matching of symmetrically positioned elements. Zhang and Gerbino (1992) studied the detection of vertical symmetry in different kinds of opposite-contrast dot patterns. The background was always grey and dot-background contrast was varied in four different conditions: (i) same contrast for all dots, either black or white; (ii) black dots on the left of the axis and white on the right, or vice versa; (iii) half of the dots white and half black, with positive correspondence (white-to-white or black-to-black); (iv) half of the dots white and half black, with negative correspondence (white-to-black or black-to-white). Discrimination from similar noise patterns was equal in the first and third conditions and better than the second and fourth, a result which argues against a contrast-insensitive mechanism based on abstract-token matching (see also Tyler *et al.*, 1993). In other words, what enters the symmetry-detection mechanism seems more than mere spatial positions of completely abstract place tokens.

4.4. Symmetry and perturbations

In many of the forementioned studies, perturbations of all kinds have been used to assess the operating characteristics of the symmetry-detection mechanism. We tend to think that symmetry is abundant in our perceptual world (see Introduction); yet few instances of symmetry in natural objects are really perfect. For example, human faces are never perfectly symmetrical. This can be demonstrated quite easily by creating two symmetric variants of a picture of a face by reflecting the right half and the left half. The differences from the original picture are striking. Moreover, bilateral symmetry can be detected from general viewpoints, which suggests that the symmetry-detection mechanism is robust to the skewing transformation associated with non-frontal viewing positions (see later). All this seems to imply that symmetry is a canonical property which tends to be exaggerated by the visual system (Freyd and Tversky, 1984), much like orientations slightly off vertical or horizontal seem to profit from the special status of the cardinal orientations (see earlier). On the other hand, if symmetry really is special for the visual system, it might be important to signal minor deviations from it (especially in animals where they might be correlated with gene deficiencies; see the evolutionary biology references cited earlier).

Somewhat surprisingly, psychophysical research supports both of these apparently conflicting intuitions. On the one hand, many studies have used perturbations of symmetry to show how robust the detection mechanism is, whereas other studies, on the other hand, have required subjects to discriminate perfect from imperfect symmetries, a task subjects could do just as well. Notice that the major difference between these two types of studies lies in the task given to the subjects, not the stimuli *per se*. In the first, subjects have to distinguish perfectly symmetric or imperfectly symmetric patterns from completely random ones; in the second, subjects have to respond 'symmetrical' only to the perfectly symmetric ones. It is also interesting to note that some of the specific perturbations that have been used were related to more-or-less detailed proposals about potential mechanisms.

Within the first category of studies (with random distractors), Barlow and Reeves (1979) have performed seminal work by systematically testing the discriminability of random dot patterns and dot patterns with variable proportions p of symmetric pairs. With 100 ms exposures and 100 dots, d's were still around 1.0 with $p = 0.3$ or 0.4. This means that in dense patterns 30 or 40% of pairwise correspondences are sufficient to trigger the preattentive percept of symmetry. In another experiment, Barlow and Reeves (1979) smeared symmetry by reducing the accuracy with which pairs are placed. This was done by positioning one of the two dots of a symmetric pair randomly within a square area centered on the symmetric position, the size of which was varied in different steps. As one would expect, the performance level decreased with increasing size of the tolerance area, but subjects still performed at 75% correct with a square of 0.4 deg visual angle, which was 20% of the total width of the patterns displayed. Both of these results suggest that the mechanism for symmetry detection is remarkably robust.

More recently, the author and his colleagues (Wagemans *et al.*, 1991, 1992, 1993) have introduced a different type of distortion away from perfect mirror symmetry,

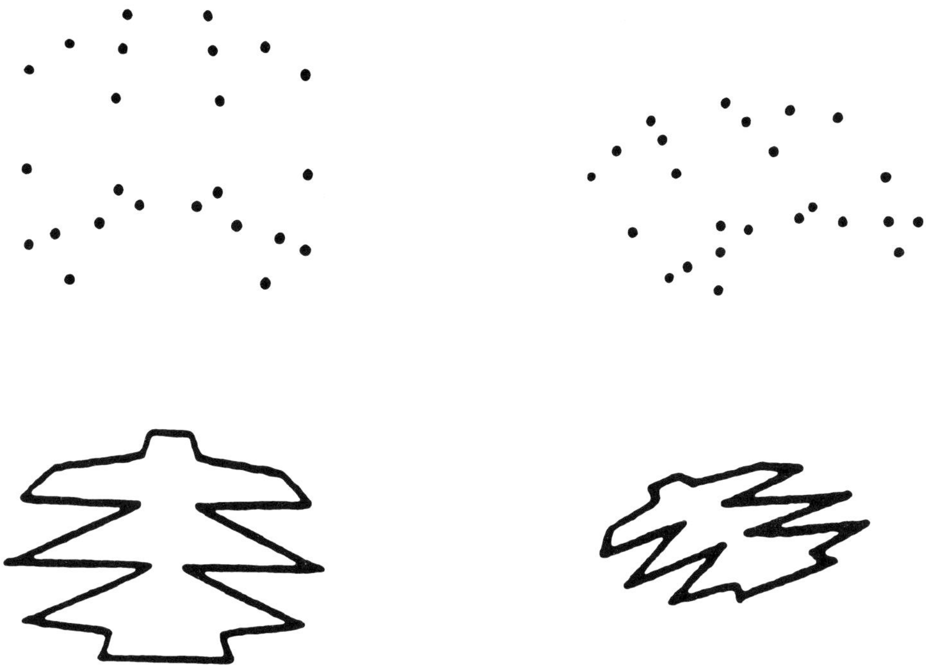

Figure 1. A dot pattern with bilateral symmetry as viewed head-on (left) and from a non-orthogonal viewpoint (right), giving rise to so-called 'skewed symmetry'. The bottom two panels show the same for a polygon. Skewing the symmetry is probably less disruptive here. From Wagemans (1993). Copyright 1993 by the American Psychological Association Inc. Adapted with permission.

which might have greater ecological significance. Whenever a perfectly bilateral symmetry is viewed from a nonorthogonal viewpoint, the actual projection on the retina is skewed symmetric (see Fig. 1 for some examples). This means that corresponding elements are not at orthogonal positions across the axis, but at an orientation which differs from 90 deg with an offset (i.e. the skewing angle) depending on the viewing position. In one set of experiments with unlimited presentation times, response times always increased dramatically with increasing skewing angle (Wagemans *et al.*, 1992) and in another set of experiments with 100 ms exposures, d' always dropped significantly, although seldomly to chance levels (Wagemans *et al.*, 1991). Similar effects of skewing were obtained in two other laboratories as well (Gerbino and Zhang, 1991; Locher and Smets, 1992). By specific additional manipulations in further research (see later), the impact of skewing could be reduced. Moreover, based on indications that skewed symmetry in polygons can be used for shape recovery much better than with dot patterns (Wagemans, 1992, 1993), it can be expected that skewing would be even less detrimental with polygons. Unfortunately, a symmetry-detection experiment with skewed symmetric versus random polygons still remains to be performed.

Within the second category of studies (with perturbed distractors), Barlow and Reeves (1979) reasoned that the remarkable degree of sensitivity to even small proportions of symmetry which they had obtained in the experiments mentioned earlier would be much less interesting if the visual system were unable to signal varying degrees of symmetry above that level (of say, 30 or 40%). This would indicate that the mechanism could only give ungraded, all-or-nothing symmetry responses to all stimuli with a sufficiently high degree of symmetry (i.e. above a certain threshold). Barlow and Reeves (1979) therefore repeated the experiment with variable degrees of symmetry but now asked subjects to discriminate them from one another instead of from completely random patterns. With a difference of 0.3 between the proportions p of symmetric pairs in the two stimulus categories to be discriminated (i.e. random from 0.3, 0.1 from 0.4, etc.), performance was reasonable ($d' = 1$) across the whole range. This result implies that symmetry is represented as a graded rather than a discrete all-or-nothing property. In a sense, this is the way it should be, if one considers the fact that completely random patterns are also quite rare. When symmetry is defined as self-identity under Euclidean transformations (see earlier), it is hard to design dot patterns or polygons in which the generalized autocorrelation (Uttal, 1975) would be zero. This approach of near symmetries has been fruitful (e.g. Farrell and Shepard, 1981; Zimmer, 1984).

Jenkins (1983b) introduced perturbations to test his proposal about the role of two types of regularities in bilaterally symmetric dot patterns. He observed that symmetric point pairs have the same orientation and collinear midpoints throughout the pattern. The purpose of his study was to determine the visual system's sensitivity to each of these factors. In one experiment, Jenkins required his observers to discriminate random-dot textures from equally dense patterns with uniformly oriented point pairs but a variable range within which the midpoints could be centered. As one would expect, performance dropped with increasing perturbation of midpoint collinearity, but even with a range of 4.4 deg of visual angle, it was still above 65%. This means that the visual system is able to detect orientational uniformity as such. In another experiment, subjects had to discriminate perfectly symmetric dot textures from patterns in which midpoint collinearity was perturbed. The results of this experiment showed discrimination performance of 85% correct with midpoint collinearity perturbations over a range as small as 0.07 deg of visual angle. This means that the visual system has a very high sensitivity to midpoint collinearity as well. As in Barlow and Reeves' (1979) study, these results indicate a remarkable robustness of the symmetry-detection mechanism against perturbations (as measured in an imperfect-random discrimination task), as well as a high sensitivity to minor deviations from perfect symmetry (as measured in a perfect-imperfect discrimination task). It is worth pointing out that a similar set of experiments with perturbations of orientational uniformity remains to be done.

In one study (Carmody *et al.*, 1977), subjects were given three response categories (symmetrical, random or mixed) and they were able to discriminate pseudo-random polygons of the three types surprisingly well, considering the very short exposure duration of 25 ms. However, the sensitivity for the mixed category was remarkably lower, which might have been due to biases to respond 'symmetrical' or 'random' more often. In a second experiment, the same patterns were presented but different two-alternative forced-choice tasks were used for two groups of subjects. One group had to respond 'target' only to the perfectly symmetric patterns and 'nontarget' to both other types of patterns. The second group had to respond 'nontarget' to the perfectly symmetric patterns and 'target' to both other types of patterns. The data showed higher detectability for symmetric patterns than for random patterns, which was in turn higher than for the mixed patterns. It would be interesting to repeat the same experiment with a slightly different task, that is, with perfectly symmetrical and mixed patterns as targets and only completely random patterns as nontargets. Additional manipulations of the exposure duration and the frequencies of each type of stimulus might reveal a pattern of detectabilities consistent with the idea that there is perhaps an initial bias to exaggerate symmetry (i.e. imperfect symmetry 'equals' symmetry), followed by more detailed processing which could signal deviations from symmetry if needed for the task at hand (see also Freyd and Tversky, 1984).

5. THEORIES AND MODELS OF SYMMETRY DETECTION

In the light of what we now know about symmetry detection from this review of the literature, how can we explain the way symmetry is detected by the visual system? Despite the long list of robust effects of several different variables known to affect symmetry detection, they have not led to a convergence upon one single, theoretically satisfying answer. However, several attempts to formulate possible mechanisms for symmetry detection in general or quite specific terms have been made in the past decades of symmetry research.

Palmer and Hemenway (1978) were among the first to propose a fairly general process model which seemed to be congruent with most of the empirical findings available at that time. In essence, they proposed a dual-process model consisting of a selection-evaluation cycle. First, a potential axis of symmetry would be selected by a rapid but crude analysis of symmetry in all orientations simultaneously. By establishing a perceptual frame of reference in the appropriate orientation, a detailed evaluation would then be performed by explicit comparison of the two pattern halves. This model explains why it takes longer to reject near and rotational symmetries than to identify perfect symmetries. Furthermore, by assuming that the initial axis selection is biased towards vertical and perhaps horizontal, this account is also able to explain orientational effects. Finally, by assuming a variable order of selection, the advantage of multiple symmetries (created by reflection about more than one axis) can be accounted for as well, since, on average, a symmetry axis would be selected sooner when there are more axes to choose from.

The distinction between two processes in symmetry detection, a fast, holistic process and a slow, point-by-point matching process, has been made by many other researchers too. For example, Julesz (1971) used it to explain the dissociations obtained between patterns dominated by low versus high spatial frequencies, respectively (see earlier). Bruce and Morgan (1975) used it to explain different types of symmetry-violation detection. Foster (1991, pp. 63–64) also proposed two kinds of operations for symmetry detection: one, a fast reversal of spatial-order information; the other, a potentially slower, progressive alteration of positional information. In addition, the possibility of two completely different processes, perhaps levels of processing, should always be borne in mind when considering possible conflicts between data obtained with different experimental procedures. For example, some paradigms seem to require subjects to perform a pointwise matching process (e.g. when response times are measured with sparse patterns and perturbed distractors), whereas others allow subjects to rely on their first impression (e.g. short exposure durations, dense displays, random distractors).

Despite its attractive characteristics, Palmer and Hemenway's model has two major problems. First, some effects that are explained by specific properties of the axis-selection process remain even when there is no need to select an axis. For example, the multiple-symmetry advantage was still present in an experiment in which subjects had to report on only one specific type of symmetry (e.g. vertical symmetry; see Palmer and Hemenway, 1978, Experiment 2). Palmer and Hemenway explained this by suggesting that the evaluation stage might be facilitated by the good Gestalt of the two pattern halves to be compared in the case of double or fourfold symmetry, but this seems to beg the question as to where the good Gestalt came from in the first place. Likewise, the vertical-symmetry advantage still remained when subjects were cued about the orientation of the symmetry axis (Pashler, 1990) and diagonal symmetry was still not easier to detect than vertical symmetry when the pattern was surrounded by a tilted frame which should bias the selection of the corresponding reference frame (Herbert *et al.*, 1994; Zimmer, 1984).

As noted by Royer (1981), a second, more serious, problem with the dual-process model is the difficulty in elaborating the nature of the crude but rapid analysis in the selection stage. In a sense, this kind of preattentive symmetry detection is the more basic process to be explained, because it seems most closely tied to the fixed functional architecture of the visual system and it is probably the one which affects other perceptual and cognitive processes. The pervasive role of symmetry suggests that we should look for an explanation of symmetry detection which has more general applicability than the selection-evaluation account, which is tailored only for symmetry-detection tasks. We know that symmetry affects many other processes, even if subjects are never asked to select and evaluate symmetry axes. An excellent example of this important distinction has been discussed earlier: Driver *et al.* (1992) have shown convincingly that symmetry can affect figure-ground organization in a patient with hemineglect, even though he was unable to perform above chance in an explicit symmetry-detection task.

The same problems appear to plague other proposed models as well. For example, based on a comparison between their subjects' performance levels against those of an ideal mechanism with absolute efficiency, Barlow and Reeves (1979) proposed that the symmetry-detection mechanism employed by the human visual system probably does not perform an exhaustive search through all possible pairs to find those that qualify as symmetric. Rather, the empirically obtained efficiency suggested that only 25% of the pairs were used in the discrimination tasks. In addition, the mechanism tolerated surprisingly large inaccuracy in the placing of the symmetric pairs, which led Barlow and Reeves to believe that the only thing the visual system does in detecting symmetry is to compare dot densities measured in relatively large areas placed symmetrically about the putative axis of symmetry, which reduces the number of comparisons to be made enormously (e.g. for 100 dots, from 4950 to 8). This seems to be an operation which can be performed quite easily by visual neurons with fixed receptive fields. However, this proposed mechanism cannot explain symmetry detection in patterns where density has been made homogeneous (e.g. Julesz, 1971, for high-density patterns and Wagemans *et al.*, 1991, for low-density patterns). In addition, the mechanism is not only tolerant and efficient, it is also quite versatile in the sense that it can also detect symmetry by reflection about axes that are not vertical and not central in the visual field. As admitted by Barlow and Reeves (1979), each different position and orientation of the axis seems to require a different set of comparisons and it is not at all clear how these are brought about. Again, to explain how all of these comparisons can be performed preattentively is quite difficult.

Another example is Jenkins' (1983b) distinction between three component processes involved in the detection of mirror symmetry in dense dot textures. Using perturbations of midpoint collinearity in three different discrimination tasks (see earlier), Jenkins obtained the following three major empirical results: (1) the orientational uniformity of dot textures can be detected, even when the pairs are distributed randomly over a large region; (2) the visual system can correlate points only when they are within a relatively narrow region around the axis; and (3) deviations from perfect midpoint collinearity can be detected quite accurately. In line with these findings, Jenkins proposed that the detection of mirror symmetry involves three different processes: (1) a process that detects orientational uniformity; (2) a process that fuses the most salient point-pairs around the axis into a larger feature; and (3) a process that determines whether that feature is symmetric. Assuming that each of these component processes responds differently to variations in axis orientation, Jenkins could explain what was known about symmetry detection without postulating the existence of a symmetrical neural organization centered about the fovea (as Mach, 1886/1959, and Julesz, 1971, had previously done). However, once more, these component processes could also be regarded as different strategies one can use depending on the discrimination task at hand, rather than a theoretically satisfying proposal of preattentive symmetry detection in general.

In order to understand the enormous efficiency, robustness, and versatility of the first, preattentive stage of global symmetry detection, one should take the pervasiveness of symmetry in other perceptual and cognitive processes seriously. This suggests that symmetry might be an integral part of the way the visual system encodes and represents visual patterns in general. Perhaps, perception of symmetry is nothing more than the conscious concomitant of the output of filtering and grouping operations which are executed in parallel on all visual displays, regardless of whether they are symmetric or not. This idea has been around for a while (e.g. Barlow and Reeves, 1979; Royer, 1981; Pashler, 1990; Foster, 1991; Locher and Wagemans, 1993), but it has never been spelled out in sufficient detail to be incorporated in a computational symmetry-detection model before Wagemans *et al.* (1993) did so in their so-called bootstrapping model.

Two robust or striking empirical findings have inspired this model. First, the fact that skewing affects symmetry detection so much (Gerbino and Zhang, 1991; Wagemans *et al.*, 1991, 1992; Locher and Smets, 1992) ruled out one class of symmetry-detection

(a) (b)

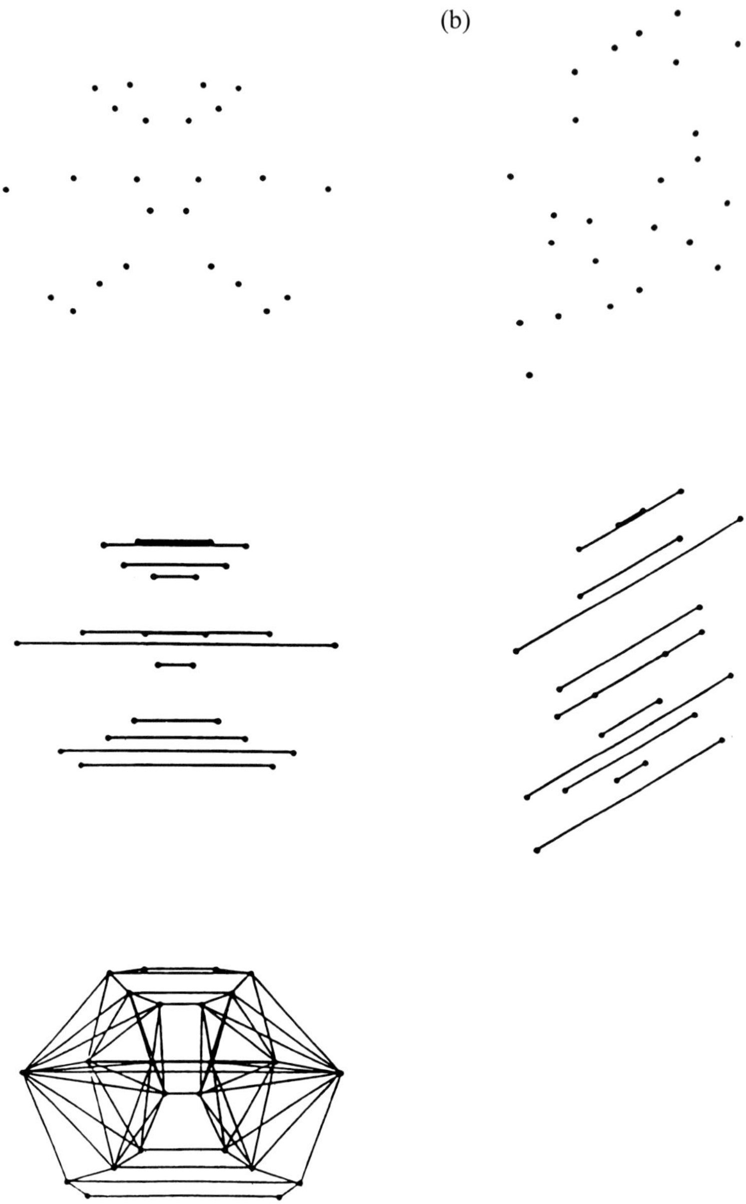

Figure 2. A dot pattern with (a) perfect bilateral symmetry and (b) skewed symmetry, together with their lower-order structure (i.e. virtual line parallelism) and, for the perfect bilateral case only, higher-order structure (i.e. correlation quadrangles). From Wagemans *et al.* (1993). Copyright 1993 by Pergamon Press Ltd.

models. Second, the fact that detection of single symmetry is easier when it is supported by a second symmetry along an orthogonal axis (Palmer and Hemenway, 1978, Experiment 2) suggested the plausibility of another class of models. The first class of models assumes that symmetry detection is based on first-order (or more generally, lower-order) regularities such as orientational uniformity and midpoint collinearity (Jenkins, 1983b), defined on point-pairs. Because skewed symmetry has exactly the same first-order regularities and is nevertheless much harder to detect preattentively, it is clear that symmetry detection relies on something else, perhaps in addition to it (see Fig. 2). Perfect bilateral symmetry differs from skewed symmetry in having regular second-order (or, more generally, higher-order) structures as well. The quadrilaterals formed between two symmetric point-pairs are symmetric trapezoids in perfect bilateral symmetry and irregular trapezoids in skewed symmetry. Likewise, double symmetry differs from single symmetry in having rectangles instead of trapezoids.

The basic assumption of the model by Wagemans *et al.* (1993) is that these pairwise correlations between the angles in these quadrilateral structures (which have been called 'correlation quadrangles' for that reason) facilitate the propagation of local pairwise groupings (called 'bootstrapping' for that reason). The idea is that quadrilaterals such as trapezoids specify a reference frame which suggests a unique direction within which other correspondences are much more likely to be found (see Fig. 3). In other words, the initial randomness in pairing elements in a pattern within some local neighborhood converges to systematicity much more easily, creating a coherent global structure more rapidly and more efficiently.

This proposal of a mechanism that allows local pairings to spread out throughout the whole pattern almost automatically seems to come close to capturing the essential nature of symmetry detection: In one sense, symmetry only exists as the total sum of pairwise correspondences in the complete pattern, whereas all available evidence suggests that the preattentive mechanism responsible for its detection is not performing such a point-by-point comparison. In addition, such a mechanism could explain the superiority of symmetry created by reflection compared to symmetry created by translation or rotation. Whereas the latter two have regular higher-order structures (in this case, parallelograms instead of trapezoids), they do not allow the same amount of bootstrapping, because there is no single direction of propagation.

The first study the author and his colleagues performed to investigate the plausibility of this account was to combine the two manipulations that inspired the proposal,

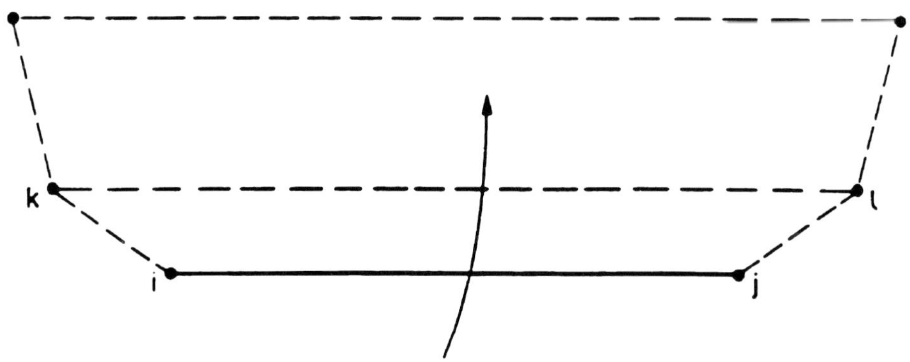

Figure 3. Bootstrapping. The correlation quadrangle formed between $(i, j)(k, l)$ suggests a direction in which to proceed. From Wagemans *et al.* (1993). Copyright 1993 by Pergamon Press Ltd.

introducing skewing and additional axes of symmetry at the same time. When dot patterns were flashed for 100 ms, increasing skewing angles caused decreasing detectability for single symmetries much more than it did for double symmetries, and skewing had almost no effect on quadruple symmetry (Wagemans *et al.*, 1991). In other words, what affects the preattentive symmetry-detection mechanism is not the degree of skewing as such, but the degree of remaining regular higher-order structures.

This idea was further tested in a study with three different types of symmetry in similar dot patterns (Wagemans *et al.*, 1993). In each of three experiments, parametric variations were introduced in the form of axis orientation and skewing angle for reflections, translation direction and distance for translations, and rotation angle for rotations. In addition, the presence of higher-order structures was always manipulated by introducing specific kinds of non-randomness in the pseudo-random dot patterns used to create the symmetries (e.g. equidistance in reflections, collinearity in rotations), or by introducing specific noise types (e.g. varying the translation distance within a pattern). In general, the results obtained in these three experiments provided strong support for the idea that detection of symmetry is easy when the pairwise correspondences are supported by regular higher-order structures and more difficult when lower-order regularity is all that is available.

In the same study, a possible implementation of the bootstrapping model was also proposed with two processes, a cost function to express the cost of different pairings and a probabilistic optimization procedure to converge on a globally optimal grouping solution. The general ideas of the model were incorporated in this specific implementation in the following way. First, the cost function used two components, one corresponding to first-order structure and one corresponding to second-order structure. Second, the optimization procedure allowed more rapid convergence based on bootstrapping. Although the implementation was not advocated as the only possible one, nice fits to the psychophysical results were obtained. Moreover, this implementation was spelled out in sufficient detail to qualify as a computational model, not just a conceptual approach.

Obviously, additional experiments are needed to test the model in its details and further elaboration is required to incorporate more of what we know about symmetry detection. For example, orientation effects are not part of the current model, but they could be incorporated quite easily, either as part of the cost function or as part of the optimization procedure. The same is true for effects of eccentricity, either as a simple increase in distance from fixation (different zones around the axis, see earlier) or as a displacement of the axis away from fixation (noncentral presentation, see earlier). Moreover, although the specific details of the model as they are now suggest that the model only works for high-spatial-frequency, low-density patterns consisting of discrete elements (preferably dots), something along the same lines could be developed for low-spatial-frequency patterns like dense dot textures or closed figures, by assuming that the model operates on the centroids of blobs or the vertices of polygons.

One other characteristic of this model which is worth mentioning is that it suggests a much wider applicability than has been assumed in previous symmetry-detection models. Wagemans *et al.* (1993) pointed out that bootstrapping based on higher-order structures appears a plausible mechanism in a wide variety of other tasks such as the detection of global structures in Glass patterns (e.g. Glass, 1969; Prazdny, 1986) and in vector graphs (Caelli and Dodwell, 1982). In addition, some specific results about perception of stereo (Akerstrom and Todd, 1988) and motion (Werkhoven *et al.*, 1990) suggest that a similar mechanism might underlie grouping or correspondence problems in many other areas as well. There are good reasons to suppose that the mechanism of symmetry detection is so general: simple transformations like translations, rotations, expansions and contractions occur each time when the observer or the object moves. It is not unlikely, therefore, that biological vision systems have evolved to detect the invariances in the resulting optic flow patterns (e.g. Lappin *et al.*, 1991). In any event, the potential scope of this mechanism stands in contrast to the specificity of previously available proposals which could count as 'models' of symmetry detection (Julesz, 1971; Palmer and Hemenway, 1978; Barlow and Reeves, 1979; Jenkins, 1983b; Foster, 1991).

Whether one single mechanism for all sorts of grouping tasks is possible, or whether, alternatively, the visual system might have developed multiple mechanisms, each devoted to one specific task, remains to be investigated in more detail. On the one hand, the special status which mirror symmetry seems to have for the human visual system (see above) suggests that we might have a special mechanism for its detection. Some researchers would go even further and propose several different mechanisms for the detection of reflection symmetry only. For example, Tyler *et al.* (1993) recently reported evidence that the perception of mirror symmetry imposed on a field of either static or dynamic noise, with either the same or opposite contrast in the two half-fields, seems to require three separate mechanisms, each with specific spatiotemporal properties. On the other hand, such a proliferation of multiple symmetry-detection mechanisms seems quite unlikely if one considers the many different types of symmetry to be detected and the large number of other grouping tasks which appear to employ similar correlational mechanisms (such as in motion or stereo). Clearly, more research is needed to find out whether the truth perhaps lies somewhere in the middle, with one general architecture for spatiotemporal grouping, the parameters of which can be fine-tuned or optimized for each specific task with which the visual system is confronted.

6. SUMMARY AND CONCLUSION

One way to summarize what is known about the detection of visual symmetries is by attempting to answer some specific questions. As one would expect with issues that have puzzled researchers for a long time, the answers are generally not simply yes or no. Probably the truth lies somewhere between the different extreme positions that have been taken in the past. Future studies should help us reveal the circumstances under which the visual system employs one strategy instead of another.

(1) Is mirror symmetry special? Yes, but this does not mean that visual symmetries created by other kinds of self-similarities such as translations or rotations cannot be

detected. The reason that symmetry created by reflection is more salient might have to do with the fact that it is the only type in which all local reference frames on a small number of elements are globally consistent. At least one model of symmetry detection suggests that it is this alignment that facilitates the fast and automatic propagation of local correspondences throughout the whole pattern, which appears to underlie the efficiency of symmetry detection in human vision.

(2) Can symmetry be detected preattentively? Yes, in the sense that symmetry can be detected easily in various displays presented very briefly (below 150 ms). Yes, in the sense that experiments with the visual search paradigm suggest that search efficiency is influenced by the symmetry relations between targets and distractors and among the distractors. However, these results do not mean that attention does not play a role in symmetry detection. For example, it is not clear what is the effect of directing the subject's attention to the array and the task at hand. Moreover, orientation effects can be attenuated or reversed by directing the subject's attention to particular orientations and scrutiny is often required for the detection of minor deviations from perfect symmetry.

(3) Is there an effect of the orientation of the axis of symmetry? Yes, but this does not mean that mirror symmetry can be detected efficiently only when the axis is oriented vertically. The general trends obtained in several studies suggest that, in the absence of biasing factors, symmetry is increasingly harder to detect when the axis is horizontal or near vertical, then diagonal or near horizontal, and, finally, in other oblique orientations. Note that this orientation function deviates from the classically reported oblique effect in two respects. First, near-vertical and near-horizontal axes are better, probably because they are close to the cardinal reference frame. Second, diagonals (45 deg and 135 deg) are better than other obliques, suggesting that they are references too. However, it is possible that many of the previously reported results are based on the mean orientation of the distribution of orientations used in a block of experimental trials.

(4) Is symmetric projection to the visual system necessary for the perception of symmetry? No. Although it becomes harder to detect mirror symmetry when the axis is displaced away from fixation, symmetry detection is still possible with non-central presentation. Together with the orientation effects, these results suggest that an explanation of the symmetry-detection mechanism need not be closely tied to the symmetric neural architecture that has sometimes been postulated.

(5) Does detection of mirror symmetry depend exclusively on pairwise correspondences between elements close to the axis? No. Although there is evidence to suggest that contiguous pointpairs contribute more heavily to the symmetry percept, other results indicate that they are not necessary. For example, symmetry still can be detected when there is no information about symmetry close to the axis, as in filled or wide polygons, or by replacing pairs close to the axis by random noise.

(6) Does symmetry detection rely on pairwise comparisons? Probably yes, but to a much lesser extent than assumed in some models of symmetry detection. For example, the symmetrically positioned elements do not have to match perfectly, either in their figural attributes, nor in their positions. Large perturbations of all kinds can

be tolerated. It is more likely that a small number of groups, possibly derived from low-pass filtering, are compared rather quickly and crudely. Alternatively, the massive parallelism of the system might allow a large number of 'comparisons' to be performed simultaneously, while still producing a stable global organization quickly and under serious degrees of local perturbation.

(7) Does this imply that symmetry detection is sloppy and gives a non-graded symmetry response to all stimuli with a certain level of symmetry? No. Despite its remarkable robustness against perturbations of all kinds, the symmetry-detection mechanism(s) also can be surprisingly precise, if the experimental task requires it. Small deviations from perfect symmetry can be detected. One way to compromise between these two apparently conflicting requirements of robustness and sensitivity is to have two mechanisms or operation modes: one, an initial, crude mechanism with a bias towards exaggerating symmetry, and the other, a secondary mechanism for checking the details. Whereas the first probably operates on the basis of a small number of large groups (clusters), the second requires intrinsically slower point-by-point comparisons. Of course, these two mechanisms (or modes of operation of the same mechanism) are at work in different viewing and task conditions. Such a two-mode account need not be in conflict with our bootstrapping model. It could well be the case that the first mode reflects an automatic, unconscious 'signal' based on the low cost of the grouping and the global characteristics of the minimal solution (acting like an emergent property), whereas the second mode would then reflect the conscious accessing of this representation to scrutinize its details.

(8) Do we have one single mechanism for the detection of all types of symmetry? Probably the answer is yes and no. Some studies suggest that we have several different mechanisms, even for the detection of mirror symmetry, and important dissociations have been reported. On the other hand, at least one mechanism of symmetry detection has been proposed which could also explain the detection of many other types of structure (e.g. in vector graphs, in Glass patterns), as well as more general grouping or correspondence problems (e.g. in stereo, in motion). It is worth looking at the possibility of one general architecture for spatiotemporal grouping with the potential of parameter-optimization for different tasks.

Acknowledgements

This article was written while the author was at the University of Virginia (Department of Psychology). This visit was made possible by grants from the National Fund for Scientific Research (Belgium) and the Academic Enhancement Program from the University of Virginia, as well as a NATO Fellowship and a Fulbright Hayes Award. I wish to thank Michael Kubovy, Dennis Proffitt, and Bennett Bertenthal for the stimulating environment and the continuous interest in my work. Helpful comments on a previous draft were made by Marco Bertamini, David Foster, Jay Friedenberg, Michael Kubovy, Paul Locher, Dennis Proffitt, Christopher Tyler, Luc Van Gool, Piet Vos, and two anonymous reviewers. The revisions were made while the author held a postdoctoral research position at the University of Leuven (PDM/94/32). The support from VIVA (EC-BRA-6448) is also gratefully acknowledged.

REFERENCES

Akerstrom, R. A. and Todd, J. T. (1988). The perception of stereoscopic transparency. *Percept. Psychophys.* **44**, 421–432.

Appelle, S. (1972). Perception and discrimination as a function of stimulus orientation: The 'oblique effect' in man and animals. *Psychol. Bull.* **78**, 266–278.

Armstrong, M. A. (1988). *Groups and Symmetry*. Springer-Verlag, New York.

Bahnsen, P. (1928). Eine Untersuchung über Symmetrie und Asymmetrie bei visuellen Wahrnehmungen. *Zeitschrift für Psychologie* **108**, 129–154.

Barlow, H. B. and Reeves, B. C. (1979). The versatility and absolute efficiency of detecting mirror symmetry in random dot displays. *Vision Res.* **19**, 783–793.

Baylis, G. C. and Driver, J. (1994). Parallel computation of symmetry but not repetition within single visual shapes. *Visual Cognit.* **1**, 377–400.

Baylis, G. C. and Driver, J. (1995). Obligatory edge-assignment in vision: The role of figure- and part-segmentation in symmetry detection. *J. Exp. Psychol: Human Percept. Perform.* **21**, (In press).

Bornstein, M. H., Ferdinandsen, K. and Gross, C. G. (1981). Perception of symmetry in infancy. *Develop. Psychol.* **17**, 82–86.

Bruce, V. G. and Morgan, M. J. (1975). Violations of symmetry and repetition in visual patterns. *Perception* **4**, 239–249.

Caelli, T. and Dodwell, P. (1982). The discrimination of structure in vectorgraphs: Local and global effects. *Percept. Psychophys.* **32**, 314–326.

Carmody, D. P., Nodine, C. F. and Locher, P. J. (1977). Global detection of symmetry. *Percept. Motor Skills* **45**, 1267–1273.

Corballis, M. C. and Roldan, C. E. (1974). On the perception of symmetrical and repeated patterns. *Percept. Psychophys.* **16**, 136–142.

Corballis, M. C. and Roldan, C. E. (1975). Detection of symmetry as a function of angular orientation. *J. Exp. Psychol: Human Percept. Perform.* **1**, 221–230.

Delius, J. D. and Nowak, B. (1982). Visual symmetry recognition by pigeons. *Psychol. Res.* **44**, 199–212.

Driver, J., Baylis, G. C. and Rafal, R. (1992). Preserved figure-ground segregation and symmetry detection in visual neglect. *Nature* **360**, 73–75.

Enns, J. T. and Rensink, R. A. (1991). Preattentive recovery of three-dimensional orientation from line drawings. *Psychol. Rev.* **98**, 335–351.

Farrell, J. E. and Shepard, R. N. (1981). Shape, orientation, and apparent rotational motion. *J. Exp. Psychol: Human Percept. Perform.* **7**, 477–486.

Fisher, C. B. and Bornstein, M. H. (1981). Identification of symmetry: Effects of stimulus orientation and head position. *Percept. Psychophys.* **32**, 443–448.

Fisher, C. B., Ferdinandsen, K. and Bornstein, M. H. (1981). The role of symmetry in infant form discrimination. *Child Develop.* **52**, 457–462.

Fisher, C. B. and Fracasso, M. P. (1987). The Goldmeier effect in adults and children: Environmental, retinal, and phenomenal influences on judgments of visual symmetry. *Perception* **16**, 29–39.

Foster, D. H. (1991). Operating on spatial operations. In: *Pattern Recognition by Man and Machine*. R. J. Watt (Ed.). Macmillan Press, Basingstoke, UK, pp. 50–68.

Freyd, J. J. and Tversky, B. (1984). Force of symmetry in form perception. *Am. J. Psychol.* **97**, 109–126.

Gerbino, W. and Zhang, L. (1991). Visual orientation and symmetry detection under affine transformations. *Bull. Psychonom. Soc.* **29**, 480.

Glass, L. (1969). Moiré effect from random dots. *Nature* **223**, 578–580.

Goldmeier, E. (1972). Similarity in visually perceived forms. *Psychol. Issues* **8** (1), Monograph 29. (Originally published in German, 1937.)

Herbert, A. M., Humphrey, G. K. and Jolicoeur, P. (1994). The detection of bilateral symmetry: Effects of surrounding frames. *Can. J. Exp. Psychol.* **48**, 140–148.

Hogben, J. H., Julesz, B. and Ross, J. (1976). Short-term memory for symmetry. *Vision Res.* **16**, 861–866.

Javadnia, A. and Ruddock, R. H. (1988). The limits of parallel processing in the visual discrimination of orientation and magnification. *Spatial Vision* **3**, 97–114.

Jenkins, B. (1982). Redundancy in the perception of bilateral symmetry in dot textures. *Percept. Psychophys.* **32**, 171–177.

Jenkins, B. (1983a). Spatial limits to the detection of transpositional symmetry in dynamic dot textures. *J. Exp. Psychol: Human Percept. Perform.* **9**, 258–269.

Jenkins, B. (1983b). Component processes in the perception of bilaterally symmetric dot textures. *Percept. Psychophys.* **34**, 433–440.

Jenkins, B. (1985). Orientational anisotropy in the human visual system. *Percept. Psychophys.* **37**, 125–134.

Julesz, B. (1971). *Foundations of Cyclopean Perception.* University of Chicago Press, Chicago, IL.

Julesz, B. (1981). Figure and ground perception in briefly presented isodipole textures. In: *Perceptual Organization.* M. Kubovy and J. Pomerantz (Eds). Erlbaum, Hillsdale, NJ, pp. 27–54.

Julesz, B. and Chang, J. J. (1979). Symmetry perception and spatial-frequency channels. *Perception* **8**, 711–718.

Kahn, J. I. and Foster, D. H. (1986). Horizontal-vertical structure in the visual comparison of rigidly transformed patterns. *J. Exp. Psychol: Human Percept. Perform.* **12**, 422–433.

Kanade, T. and Kender, J. R. (1983). Mapping image properties into shape constraints: Skewed symmetry, affine-transformable patterns, and the shape-from-texture paradigm. In: *Human and Machine Vision*, Vol. 1. J. Beck, B. Hope and A. Rosenfeld (Eds). Academic Press, New York, pp. 237–257.

Lappin, J. S., Norman, J. F. and Mowafy, L. (1991). The detectability of geometric structure in rapidly changing optical patterns. *Perception* **20**, 513–528.

Locher, P. J. and Nodine, C. F. (1989). The perceptual value of symmetry. *Comp. Math. Appl.* **17**, 475–484.

Locher, P. and Smets, G. (1992). The influence of stimulus dimensionality and viewing orientation on detection of symmetry in dot patterns. *Bull. Psychonom. Soc.* **30**, 43–46.

Locher, P. and Wagemans, J. (1993). The effects of element type and spatial grouping on symmetry detection. *Perception* **22**, 565–587.

Mach, E. (1959). *The Analysis of Sensations.* Dover, New York. (Originally published in German, 1906.)

Mack, A., Tang, B., Tuma, R., Kahn, S. and Rock, I. (1992). Perceptual organization and attention. *Cognit. Psychol.* **24**, 475–501.

Marola, G. (1989). On the detection of the axes of symmetry of symmetric and almost symmetric planar images. *IEEE Trans. Pattern Anal. Machine Intell.* **PAMI-11**, 104–108.

Moeller, A. P. (1992). Female swallow preference for symmetrical male sexual ornaments. *Nature* **357**, 238–240.

Moraglia, G. (1989). Display organization and the detection of horizontal line segments. *Percept. Psychophys.* **45**, 265–272.

Nalwa, V. S. (1989). Line-drawing interpretation: Bilateral symmetry. *IEEE Trans. Pattern Anal. Machine Intell.* **PAMI-11**, 1117–1120.

Nothdurft, H. C. (1992). Feature analysis and the role of similarity in preattentive vision. *Percept. Psychophys.* **52**, 355–375.

Palmer, S. E. and Hemenway, K. (1978). Orientation and symmetry: Effects of multiple, rotational, and near symmetries. *J. Exp. Psychol: Human Percept. Perform.* **4**, 691–702.

Pashler, H. (1990). Coordinate frame for symmetry detection and object recognition. *J. Exp. Psychol: Human Percept. Perform.* **16**, 150–163.

Pomerantz, J. R. and Pristach, E. A. (1989). Emergent features, attention, and perceptual glue in visual form perception. *J. Exp. Psychol: Human Percept. Perform.* **15**, 635–649.

Prazdny, K. (1986). Psychophysical and computational studies of random-dot Moiré patterns. *Spatial Vision* **1**, 153–158.

Rock, I. and Leaman, R. (1963). An experimental analysis of visual symmetry. *Acta Psychol.* **21**, 171–183.

Rock, I., Linnett, C. M., Grant, P. and Mack, A. (1992). Perception without attention: Results of a new method. *Cognit. Psychol.* **24**, 502–534.

Royer, F. L. (1981). Detection of symmetry. *J. Exp. Psychol: Human Percept. Perform.* **7**, 1186–1210.

Saarinen, J. (1988). Detection of mirror symmetry in random dot patterns at different eccentricities. *Vision Res.* **28**, 755–759.

Stevens, K. A. (1980). *Surface Perception by Local Analysis of Texture and Contour.* Tech. Rep. No. 512, MIT AI Lab, Cambridge, MA.

Swaddle, J. P. and Cuthill, I. C. (1993). Preference for symmetric males by female zebra finches. *Nature* **367**, 165–166.

Tapiovaara, M. (1990). Ideal observer and absolute efficiency of detecting mirror symmetry in random images. *J. Opt. Soc. Am.* **A7**, 2245–2253.

Treisman, A. and Gelade, G. (1980). A feature-integration theory of attention. *Cognit. Psychol.* **12**, 97–136.

Treisman, A. and Patterson, R. (1984). Emergent features, attention, and object perception. *J. Exp. Psychol: Human Percept. Perform.* **10**, 12–31.

Treisman, A. and Sato, S. (1990). Conjunction search revisited. *J. Exp. Psychol: Human Percept. Perform.* **16**, 459–478.

Troscianko, T. (1987). Perception of random-dot symmetry and apparent movement at and near isoluminance. *Vision Res.* **27**, 547–554.

Tyler, C. W. and Chang, J. J. (1977). Visual echoes: The perception of repetition in quasi-random patterns. *Vision Res.* **17**, 109–116.

Tyler, C. W., Hardage, L. and Miller, R. (1993). Multiple mechanisms for detection of reflection symmetry in brief exposures. *Perception* **22** (Suppl.), 47.

Uttal, W. R. (1975). *An Autocorrelation Theory of Form Detection.* Erlbaum, Hillsdale, NJ.

Van Gool, L., Wagemans, J., Vandeneede, J. and Oosterlinck, A. (1990). Similarity extraction and modelling. In: *Proc. Third Int. Conf. Computer Vision.* IEEE, Washington, DC, pp. 530–534.

Van Gool, L., Moens, T., Unganeanu, D. and Oosterlinck, A. (1995). The characterization and detection of steroed symmetry. *Computer Vision and Image Understanding* **61**, 138–150.

Wagemans, J. (1992). Perceptual use of nonaccidental properties. *Can. J. Psychol.* **46**, 236–279.

Wagemans, J. (1993). Skewed symmetry: A nonaccidental property used to perceive visual forms. *J. Exp. Psychol: Human Percept. Perform.* **19**, 364–380.

Wagemans, J., Van Gool, L. and d'Ydewalle, G. (1991). Detection of symmetry in tachistoscopically presented dot patterns: Effects of multiple axes and skewing. *Percept. Psychophys.* **50**, 413–427.

Wagemans, J., Van Gool, L. and d'Ydewalle, G. (1992). Orientational effects and component processes in symmetry detection. *Q. J. Exp. Psychol.* **44A**, 475–508.

Wagemans, J., Van Gool, L., Swinnen, V. and Van Horebeek, J. (1993). Higher-order structure in regularity detection. *Vision Res.* **33**, 1067–1088.

Washburn, D. K. and Crowe, D. W. (1988). *Symmetries of Culture.* University of Washington Press, Seattle, WA.

Watt, R. J. (1987). Scanning from coarse to fine spatial scales in the human visual system after the onset of a stimulus. *J. Opt. Soc. Am.* **A4**, 2006–2021.

Watt, R. J. and Morgan, M. (1985). A theory of the primitive spatial code in human vision. *Vision Res.* **23**, 1457–1477.

Wenderoth, P. (1994). The salience of vertical symmetry. *Perception* **23**, 221–236.

Wenderoth, P. (1995). The role of pattern outline in bilateral symmetry detection with briefly flashed dot patterns. *Spatial Vision* **9**, 57–77.

Werkhoven, P., Snippe, H. P. and Koenderink, J. J. (1990). Effects of element orientation on apparent motion perception. *Percept. Psychophys.* **47**, 509–525.

Weyl, H. (1952). *Symmetry.* Princeton University Press, Princeton, NJ.

Wolfe, J. M., Cave, K. R. and Franzel, S. J. (1989). Guided search: An alternative to the feature-integration model of visual search. *J. Exp. Psychol: Human Percept. Perform.* **15**, 419–433.

Wolfe, J. M. and Friedman-Hill, S. R. (1992). On the role of symmetry in visual search. *Psychol. Sci.* **3**, 194–198.

Zhang, L. and Gerbino, W. (1992). Symmetry in opposite-contrast dot patterns. *Perception* **21** (Suppl. 2), 95.

Zimmer, A. C. (1984). Foundations for the measurement of phenomenal symmetry. *Gestalt Theory* **6**, 118–157.

Human Symmetry Perception, pp. 49–69
C. W. Tyler (Ed.)

The role of pattern outline in bilateral symmetry detection with briefly flashed dot patterns

PETER WENDEROTH

Department of Psychology, School of Behavioural Sciences, Macquarie University, Sydney, Australia 2109

Abstract—Experiment 1 demonstrates that, while the outline shapes of bilaterally symmetrical dot patterns play a role in symmetry detection, the removal of the outline by a surrounding random-dot annulus merely reduces performance by a fixed amount. It does not affect the relative salience of different symmetry-axis orientations. The converse is also true: performance is almost as good when the symmetrical dot pattern is confined to the surrounding annulus but is disrupted similarly when the central area is filled with random dots (Experiment 2). In Experiment 3, any one or more of three areas — a central circular area and two surrounding annuli — could be filled with vertically symmetrical or random dots and symmetry was detected reliably only when the central circular area contained the symmetrical dots. A new paradigm was explored in Experiment 4: subjects judged the orientation — left or right of vertical — of 20-dot symmetrical patterns oriented 1–4 deg left or right of vertical and with or without surrounding random-outline masks. Surprisingly, the orientation judgments were extremely precise, with JNDs of the order of 1 deg or less, and outline masking dots made no difference. Experiment 5 showed that performance was even better with just two dots and it made no difference whether these were separated by just under 4 deg or just over 20 deg. It was concluded that while pattern outline plays some role in dot symmetry detection, the major mechanisms are located near the fovea. It is suggested that the relative salience of vertical and horizontal symmetry only in central vision is related to the preponderance of foveal cells tuned to those orientations (Mansfield, *Science* **186**, 1133–1135, 1974; Mansfield and Ronner, *Brain Res.* **149**, 229–234, 1978) but that attentional factors and scanning strategies also boost the salience of oblique symmetry and the salience of vertical over horizontal symmetry (Wenderoth, *Perception* **23**, 221–236, 1994).

1. INTRODUCTION

A pattern is said to be bilaterally symmetrical when one half of the pattern is a mirror image of the other half. Thus, the 38-dot pattern in Fig. 1(a) is bilaterally symmetrical about a vertical axis (represented by the dashed line which is not itself part of the pattern) because the 19 dots to the left of the vertical axis can be folded over, or reflected, on to the 19 dots to the right of it. Of course, patterns other than dot displays can also be symmetrical and a number of investigators have studied symmetry detection using closed, outline shapes of various kinds (e.g. Locher and Nodine, 1973; Carmody *et al.*, 1977; Palmer and Hemenway, 1978).

Several researchers who have used dot displays have concluded that the efficacy of symmetry detection depends upon the opportunity for the observer to compare elements of the pattern which both span the axis of symmetry and are also spatially contiguous to it. Thus, it has been reported that small violations of symmetry are easily detected when they are near the midline of a symmetrical pattern but not when they are near its periphery (Bruce and Morgan, 1975); that symmetry information in high-density dynamic dot displays is only usable when it falls within a 1 deg strip about the central axis of symmetry (Jenkins, 1982); and that symmetry detection falls off, as the axis moves from the fovea into the retinal periphery (Barlow and Reeves, 1979) even when the patterns are M-scaled, although the decrease in accuracy with eccentricity is reduced (Saarinen, 1988). There is evidence to suggest, however, that other factors affect symmetry detection.

For example, it has been frequently reported that the orientation of the symmetry axis is a critical variable with vertical symmetry most easily detected, followed by horizontal symmetry and then other orientations (Corballis and Roldan, 1975; Palmer and Hemenway, 1978; Barlow and Reeves, 1979; Royer, 1981; Fisher and Bornstein, 1982; Jenkins, 1985; Pashler, 1990; Wagemans *et al.*, 1992; Wenderoth, 1994). It also has been demonstrated recently that attentional factors influence symmetry detection: vertical symmetry in dot patterns which is detected with a probability near 0.9 when all axis orientations are equiprobable is detected with a probability closer to 0.1 when axis orientations are concentrated around horizontal, indicating that factors other than inherent properties of the stimulus need to be considered (Wenderoth, 1994).

One major investigation of bilateral symmetry detection using dot displays (Barlow and Reeves, 1979) drew attention to the possible role of the outline of the pattern in judging symmetry; and Fig. 1 is a typical example. Thus, Barlow and Reeves noted:

> Subjectively, one would also judge that the 'outline', if one can call it such, is important, for this can create the impression of a vase, or a butterfly, or some other symmetrical object.

To be explicit, the outline of a dot pattern refers to the shape defined by the outermost dots. It can be found roughly by an algorithm in which every dot is joined to its two nearest neighbours (see Fig. 1(b)) and this outline shape would exist whether the shape were filled with solid colour or with other dots.[1] To test for the role of the pattern outline, Barlow and Reeves logically divided their 2.1 deg display area into six vertical strips each containing 16 or 17 dots of the total 100-dot pattern and then attempted to discriminate patterns with no symmetrical dot pairs — completely random patterns — from partially symmetrical patterns which had symmetrical pairs *only* in the two abutting, central strips, or *only* in the flanking outer strips, or *only* in the two strips which were neither in the centre nor at the edge or outline of the pattern. They found that sensitivity (d') averaged 4.1 at the centre, 2.25 at the periphery and 0.86 when the symmetrical dots fell in-between the centre and the outline and concluded (p. 790) that '. . . the outer slices, containing the dots that form the outer contour of the figure, are important in generating the perception of symmetry' (Barlow and Reeves, 1979).

(a)

(b)

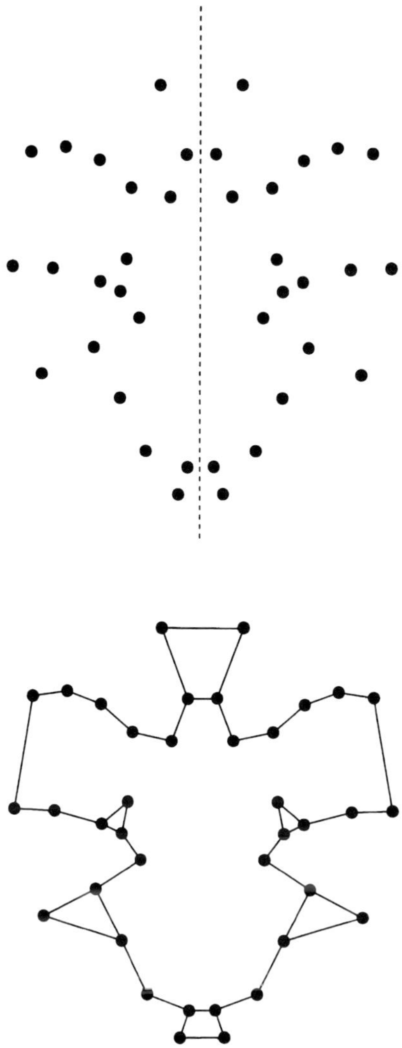

Figure 1. (a) Bilaterally symmetrical dot pattern with vertical axis of symmetry (dashed) indicated. (b) Application of the algorithm of joining each dot to its two nearest neighbours to specify pattern outline (see text).

One may, however, question this inference. First, although the authors do not explicitly describe the experiment, they do state that the estimates of d' were based on 200 trials each, giving the impression — as was the case with most of their other experiments — that trials were blocked so that the two subjects (who were the authors) always knew where the symmetrical dots would be when they were present as well as knowing that the symmetry would be about a vertical axis. Yet the role of the outline dots in symmetry detection when trials are not blocked and the subject has no prior knowledge of the location or the orientation of the symmetrical dots may be very different. Second, an experiment has been reported (Fisher and Bornstein, 1982) using

16-dot patterns but with 8 of these dots forming a constant symmetrical outline for *all* patterns whether they are otherwise internally symmetrical or not, thus removing any outline cue. The 8 fixed dots were located at 45 deg intervals around the circular pattern outline. Subjects had to judge whether the remaining 8 internal dots were symmetrical or random. Performance was found to be much as when the outline cue is available, with the usual pattern of vertical symmetry detection superior to horizontal symmetry detection which in turn is superior to oblique symmetry detection. This result suggests that the outline plays little role in symmetry detection in dot patterns.

The aim of the experiments reported here was to study the relative contributions to symmetry detection of symmetrical relations between dots internal to the pattern and cues to symmetry given by the shape of the pattern outline.[2]

2. EXPERIMENT 1

In this experiment, subjects were presented with four kinds of dot patterns, two of which are shown in Fig. 2. Figure 2(a) is a 28-dot pattern which is bilaterally symmetrical about vertical; Fig. 2(b) is a 56-dot pattern in which the Fig. 2(a) pattern is embedded but the outer 28 dots are randomly positioned. Thus, the outer 28 dots in Fig. 2(b) effectively mask the outline of the symmetrical dots. The other two kinds of pattern used in Experiment 1 were logically the same with the one difference that the central 28-dot pattern was random also. That is, any dot pattern was one of the following: (1) 28 symmetrical dots, (2) 28 random dots, (3) 56 dots with a symmetrical 28-dot centre and random 28-dot surround, or (4) 56 dots with all dots randomly positioned. The axis of symmetry of a pattern, when present, was at one of 8 orientations: 0 deg (horizontal), 22, 45, 67, 90, 112, 135 and 157 deg, with

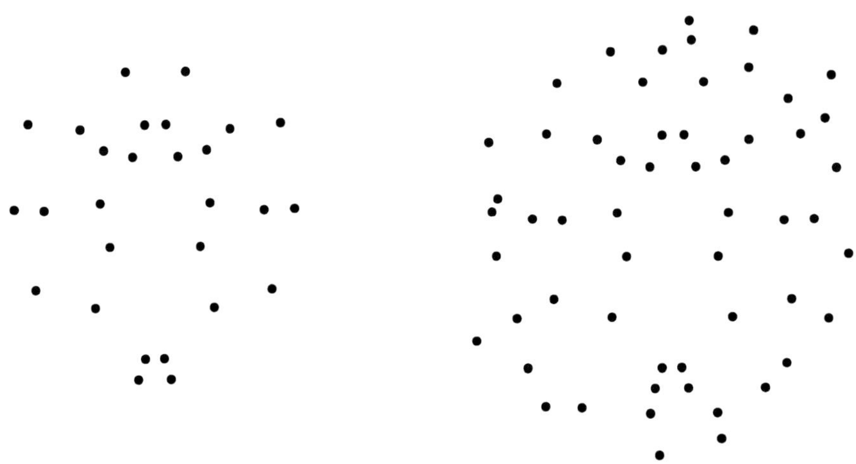

(a) (b)

Figure 2. Examples of the kinds of dot patterns used in Experiment 1. (a) 28-dot vertically symmetrical pattern. (b) Same pattern as in (a) but now surrounded by 28 random, outline masking dots.

orientations between 0 deg and 90 deg defining axes in the first (top right) and third (bottom left) quadrants.

2.1. Methods

2.1.1. Apparatus and stimuli. Stimulus displays were generated by a Deltacom 486 computer using a menu driven program written in Borland C++ and displayed on a NEC Multisync 1280 × 1024 5D monitor, via a Tseng Labs MegaEva 4000 VGA graphics card. Stimulus displays consisted of 28 or 56 black dots presented on a grey background and confined to a 20.5 cm circular area of the screen (20.5 deg visual angle at the viewing distance of 57 cm). Each dot had a diameter of 2 mm (6 pixels). Each symmetrical dot pair was generated as follows. First, a distance from the centre of the screen between 1 and 307 pixels (10.25 cm) was randomly selected and multiplied by the square root of a randomly chosen fraction between 0 and 1. This last transformation ensured that the distribution of dots was fairly uniform over the screen without concentrating dots at the centre. Without this transformation, interdot distances would increase markedly with distance from the origin because the arcs joining dots would be longer. Second, the dot was placed at the calculated distance at a randomly chosen orientation from the centre of the screen, provided that the dot was not within 12 pixels (4 mm) of the centre of any neighbouring dot. Otherwise, a new distance and a new direction were selected. Third, once the dot position had been selected, its partner's position across the axis of symmetry was automatically generated. Subjects viewed the screen from a distance of 57 cm in a windowless laboratory in which a standard lamp was positioned indirectly to place a veiling luminance on the screen. Dot luminance was 10.34 cd m^{-2} and with the grey background luminance 24.02 cd m^{-2}. Dot contrast, defined as $[L_{MAX} - L_{MIN}]/[L_{MAX} + L_{MIN}]$, was 0.43. There were no obvious afterimages at the offset of a flashed stimulus when the screen reverted to homogeneous grey. A padded chinrest with forehead rest and temple clamps was used to restrict head movements and subjects were also instructed not to try to tilt their heads. A 24-cm diameter circular hole was cut in a large oval milk perspex screen, which was painted matte black and could be mounted in front of the display using velcro strips to remove vertical and horizontal reference cues.

The pattern display area was logically divided into two areas — a central circular area with diameter 10.25 deg and a surrounding annulus with inside diameter 10.25 deg and outside diameter 20.5 deg. The experimental pattern was displayed in the former while the random mask, when present, appeared in the latter.

2.1.2. Design and procedure. Trials on which the outline was or was not masked were blocked so that alternate subjects completed that the no-mask or mask conditions first, with each consisting of 160 trials. On 80 of these trials the pattern of 28 or 56 dots was completely random and on the other 80 trials the 28 dots (or the central 28 or 56) were bilaterally symmetrical. Each of the 8 axis orientations was presented 10 times.

The 10 replications of any given axis orientation were randomly selected patterns so that no one stimulus was ever re-presented and all 160 trials in a block were presented in a different random order to each subject. The stimulus flashed for 150 ms and, if no response had occurred 2000 ms after pattern onset, the trial was automatically aborted and a randomly drawn new pattern was presented at the end of the 160 trials. Between trials, a small red fixation point was drawn on the otherwise grey screen, at the centre of the area in which the patterns would appear. The subject controlled the pace of the experiment, initiating each trial by fixating the red spot and then pressing the spacebar of the keyboard with the left hand to present the stimulus. Responses were made by pressing two of the adjacent numeric keypad keys with the index and middle fingers of the right hand, '*' to indicate 'symmetrical' and '-' to indicate 'random'. For each response, both correctness and reaction time were recorded. The subject read printed instructions which defined mirror-image symmetry and gave examples of symmetrical and random-dot patterns as well as masked and unmasked outlines, which were screen dumps of actual stimuli and had axes at various orientations. The instructions were those usually employed in such experiments, giving the subject the impossible task of being both fast and accurate. The written instructions also stressed that the axis of symmetry could be at any angle whatsoever around the clock and that the pattern would be random on exactly half of all trials and perfectly symmetrical on the other half. The exact instructions were:

> Your task in this experiment is to look at patterns of dots and to decide whether the dots do or do not have a particular axis of symmetry. A pattern is said to have an axis of mirror-image symmetry when you can draw a line through it and divide it into two halves which are exact mirror images of each other, like patterns shown here. Exactly half of the patterns you see will be 100% symmetrical, the other half will be completely random, like this one. However, in the experiment proper the axis of symmetry can be at *any angle whatsoever*, right around the clock. Here are some examples printed out from the actual experimental screen.
>
> Your task is to decide on any trial whether the dots are symmetrical or not. The pattern will flash very briefly but you will have 2 s to make a response. If you have not responded by then, the pattern will be presented again later. However, try to respond *as quickly as you can* but without making too many errors. Rest the index finger of your right hand on the '*' key and the middle finger on the '-' key. Put a finger of your left hand on the spacebar and let's do some examples. OK. There's a red dot on the screen which tells you where the centre of the pattern will be. When you are ready, pressing the spacebar will bring up the pattern and you should then press * (for 'Symmetrical') or - (for 'Random') as quickly as you can without making errors. You can take a break at any time but we'll give you one anyhow at various times to make sure you are not too fatigued. Any questions?
>
> OK. Now you will do the experiment twice with one slight difference between the two runs. On one run, the patterns you see will be like those I have shown you already [A, B]. On the other run, the task will be a bit harder, as you can see here [C, D]. Although there are symmetrical dot patterns in this one (vertical symmetry) and this one (oblique symmetry) they are surrounded by other, random dots. On this next sheet, we have put a line around the symmetrical pattern so that you can see it. Remember, in each run your task is the same. Try to respond 'symmetrical' (by pressing '*') to the 50% of patterns which contain symmetry and 'random' (by pressing '-') to the other 50% which have no symmetry. Before we start each run, you'll get a few practice trials. OK?
>
> Right: you will be judging the C, D (A, B) type stimuli first, then the A, B (C, D) type stimuli.

Subjects were given familiarization trials, usually about twenty, and were told they could take a break whenever they wished, although a rest of about a minute was always given after 100 trials. Testing in this and subsequent experiments lasted about 50 min to 1 hour. Subjects were 25 voluntar undergraduate students.

2.2. *Results*

The average number of correct responses out of 10 are shown as a function of symmetry axis orientation in Fig. 3(a) which also shows the number correct on random-pattern trials (converted to a number out of 10 for comparative purposes). Figure 3(b) shows reaction times obtained on correct-response trials. A two-way repeated measures analysis of variance on the number correct data, excluding the random trials, indicated that the effect of outline masking was significant with $F(1, 24) = 89.64$, $p < 0.0001$: the mean numbers correct under the outline present and outline masked conditions were 8.06 and 5.70, respectively. The effect of axis orientation was also significant with $F(8, 192) = 4.67$, $p < 0.0001$: planned contrasts to test trend showed that quadratic, 6th-order and 8th-order trends were significant, with respective F values of 8.28, 5.63 and 16.07 and p values of 0.005, 0.02 and < 0.0001. The interaction between outline or no outline and axis orientation was not significant with $F(8, 192) = 0.25$ and $p = 0.98$.

An identical analysis of the reaction time data, after one subject was excluded because there were several conditions in which no correct responses were made and the reaction times were recorded as zero, showed that the effect of outline masking was significant with $F(1, 23) = 54.26$, $p < 0.0001$: the mean reaction times under the outline present and outline masked conditions were 793.49 ms and 901.51 ms, respectively. The effect of axis orientation was not significant with $F(8, 184) = 1.70$ and $p = 0.10$. The interaction between outline or no outline and axis orientation was not significant with $F(8, 184) = 0.40$ and $p = 0.93$.

2.3. *Discussion*

Several results of Experiment 1 are generally consistent with those reported in previous work. First, in both the proportion correct and the reaction time data obtained mean performance was best with vertical symmetry, although the next best performances were at the obliques rather than at horizontal. The relative performances at the different axes were not explicitly tested here apart from the significant trend effects: the eighth-order trend in the proportion correct data indicates that every change in direction of the average of the functions in Fig. 3 is significant. The tendency for relatively good performance at the obliques has been reported and discussed elsewhere (Wenderoth, 1994).

The new results of interest are, first, that removing the outline shape of the dot pattern by surrounding it with random dots significantly reduces proportion correct and increases reaction time; and, second, that the random-dot mask merely reduces performance by a fixed amount, effectively preserving the ordering of performance for the various axis orientations, as indicated by the nonsignificant interaction between outline or no outline and axis orientation. This finding suggests that while the pattern outline is a useful cue to symmetry detection it does not play a role in the relative salience of the different axis orientations. This in turn suggests that the factor which does determine the relative salience of different axis orientations is the availability of dot pairs close to the axis itself, in this case near the fixation point.

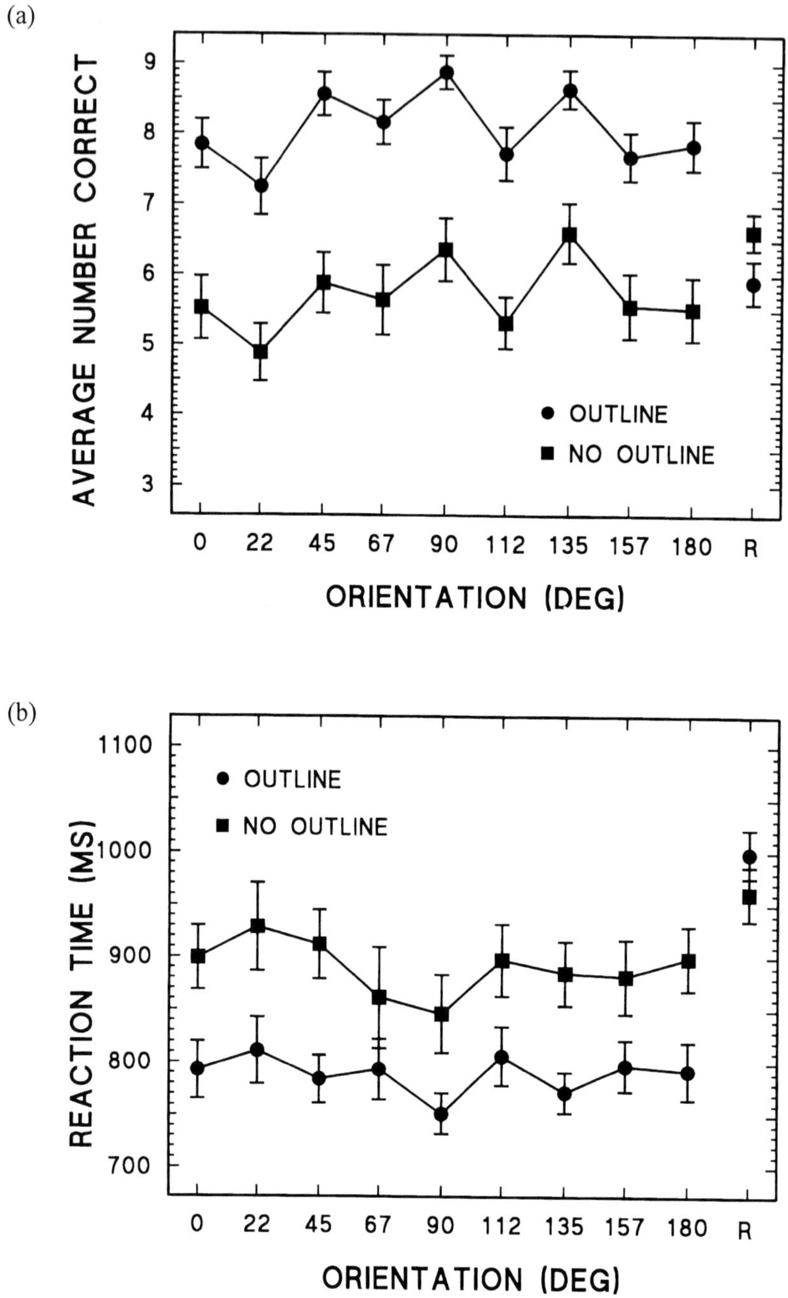

Figure 3. (a) Average number of correct responses (out of 10) as a function of axis orientation and the presence or absence of an outline shape, Experiment 1. Error bars show ±1 standard error. R = random dot pattern. (b) Reaction times, Experiment 1. Otherwise, as for (a).

To test this hypothesis, Experiment 2 was designed to repeat Experiment 1 with one major variation: the symmetrical or random pattern to be judged was presented in the *outer* annulus while the random masking pattern, when present, appeared in the *inner* circular area.

3. EXPERIMENT 2

The predictions for Experiment 2 were as follows. First, it was predicted that no statistically significant effect of symmetry axis orientation would be obtained because the symmetrical dot pairs, when present, would always be at least 5.25 deg either side of the fixation point and so would not provide the necessary cue of dot pairs near the axis. Second, if a major cue to symmetry detection is the presence of symmetrical pairs close to fixation, the random masking dots should have an effect similar to that in Experiment 1, if not greater, because the presence of random rather than symmetrical dots near fixation should disrupt the detection of symmetry in the periphery.

3.1. Methods

These were as in Experiment 1. Subjects were 23 observers from the same population as those in Experiment 1.

3.2. Results

The average number of correct responses out of 10 are shown as a function of symmetry axis orientation in Fig. 4(a) which also shows the number correct on random-pattern trials. Figure 4(b) shows reaction times obtained on correct-response trials, symmetrical and random. A two-way repeated measures analysis of variance on the number correct data, excluding the random trials, indicated that the effect of outline masking was significant with $F(1, 22) = 92.22$, $p < 0.0001$: the mean numbers correct under the outline present and outline masked conditions were 7.33 and 4.13, respectively. The effect of axis orientation was not significant with $F(8, 176) = 1.90$ and $p = 0.06$. The interaction between outline or no outline and axis orientation was also not significant with $F(8, 176) = 0.72$ and $p = 0.67$.

The identical analysis of the reaction time data, after five subjects were excluded because of zero correct trials and thus zero recorded reaction times, showed that the effect of outline masking was significant with $F(1, 17) = 24.49$, $p < 0.0001$: the mean reaction times under the outline present and outline masked conditions were 869.09 ms and 955.01 ms, respectively. The effect of axis orientation was not significant with $F(8, 136) = 1.88$ and $p = 0.07$. The interaction between outline or no outline and axis orientation was not significant with $F(8, 136) = 1.78$ and $p = 0.09$.

It is noticeable that, in Figs 3 to 6, performance on random trials was always better when the masking dots were present, suggesting a response bias in favour of the 'random' response on those trials, more significantly in Experiment 2 than in Experiment 1. Thus, in Experiment 1 the proportion correct was greater ($F(1, 24) = 8.75$; $p = 0.0068$) and reaction time was shorter ($F(1, 23) = 8.50$; $p = 0.0078$) on trials on which the extra random dots were present; and the same was true in Experiment 2 with $F(1, 22) = 47.70$, $p < 0.0001$ and $F(1, 17) = 53.87$, $p < 0.0001$, respectively.

(a)

(b)

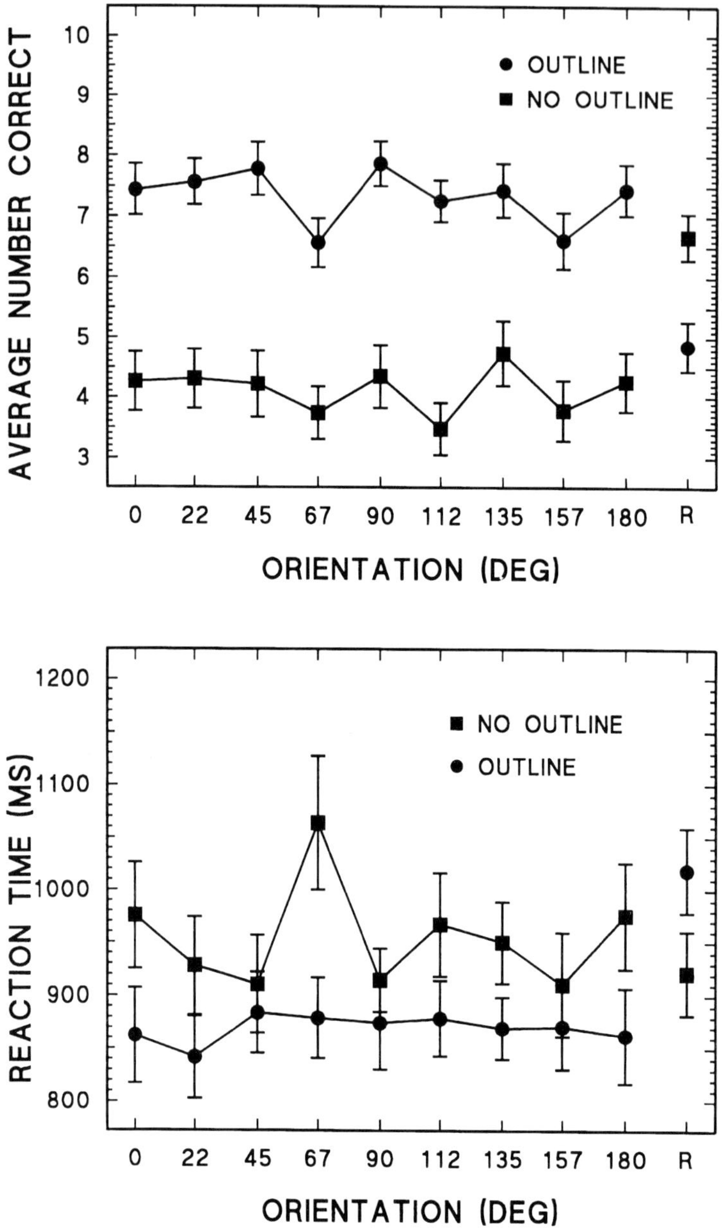

Figure 4. (a) Average number of correct responses, Experiment 2. Otherwise, as for Fig. 3. (b) Reaction times, Experiment 2. Otherwise as for (a).

3.3. Discussion

As predicted, no statistically significant effect of axis orientation was obtained in Experiment 2 and, although no statistical test was done between the two experiments, the obtained overall level of performance was worse in Experiment 2, with fewer correct responses as well as longer reaction times on symmetrical trials. These results are consistent with those of Saarinen (1988) and Barlow and Reeves (1979) in showing reduced performance when the symmetry axis is peripherally rather than centrally viewed. Nevertheless, it is clear from Fig. 4 that subjects could do the symmetry detection task when the symmetrical dots formed an annulus more than 5 deg in the periphery. There is no obvious explanation for the apparently longer reaction time at the 67 deg axis orientation in the no outline condition of Fig. 6.

4. EXPERIMENT 3

It seemed clear from the results of the first two experiments that although the shape of the outline of symmetrical dot patterns assists symmetry detection, and that symmetry can be detected well into the periphery, the presence of random masking dots hinders performance and the relative salience of different axis orientations depends upon the presence of symmetrical dot pairs close to the fixation point. In the light of the present results, it seemed surprising that Barlow and Reeves (1979) obtained such high d' values when symmetry was only present in the outer strips of their display with all other strips containing random dots. Accordingly, Experiment 3 was designed to attempt to re-examine their finding but by dividing the dot patterns into three logical circular areas rather than vertical strips and by randomly mixing trials so that the subject was never aware whether the next stimulus would contain symmetrical dots and, if it did, was also unaware of which region would contain symmetrical dots.

In this experiment all symmetry axes, when present, were vertical and not only were there trials on which symmetry was present in only the centre region (called '1'), the middle region (2) or the outer region (3), but there were also trials in which 2/3 and all three regions were symmetrical.

4.1. Methods

Again, these were generally as in Experiment 1 with 13 subjects drawn from the same population. The outer diameters of three logical annuli were 6.8, 13.7 and 20.5 deg, with each containing 28 dots so that the whole pattern had 84 dots. Subjects were again given written instructions, with illustrations of the various pattern types, and the instruction was to respond 'symmetrical' if there was *any* symmetry in the pattern at all. There were 700 trials, with 350 completely random (called 'rrr') and 350 containing some symmetry, 50 each of the remaining 7 types of symmetrical stimuli: 'vrr' (inner or '1' only symmetrical); 'rvr' (middle or '2' only symmetrical); 'rrv' (outer or '3' only symmetrical); and 'vvr', 'rvv', 'vrv' and 'vvv'.

4.2. Results

The average number correct and reaction time data are shown in Fig. 5. In Fig. 5(a), the open circles show the number correct when there was only one symmetrical area (1 or 2 or 3) and the filled circles show the number correct when there was

(a)

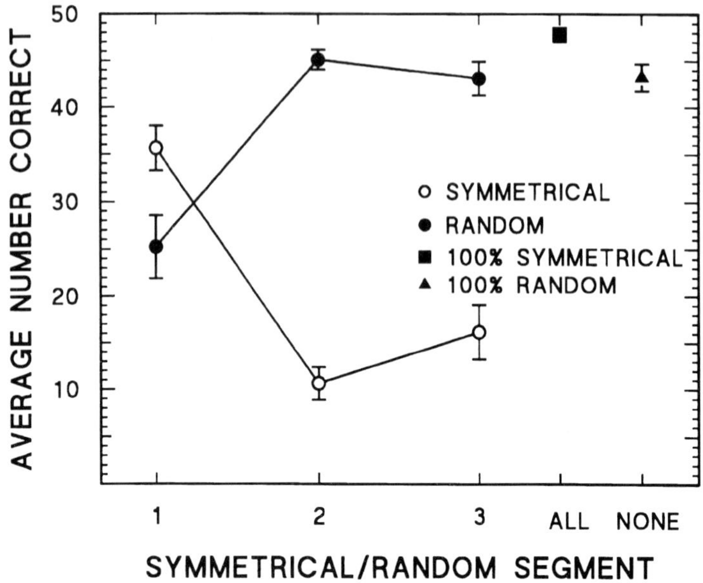

(b)

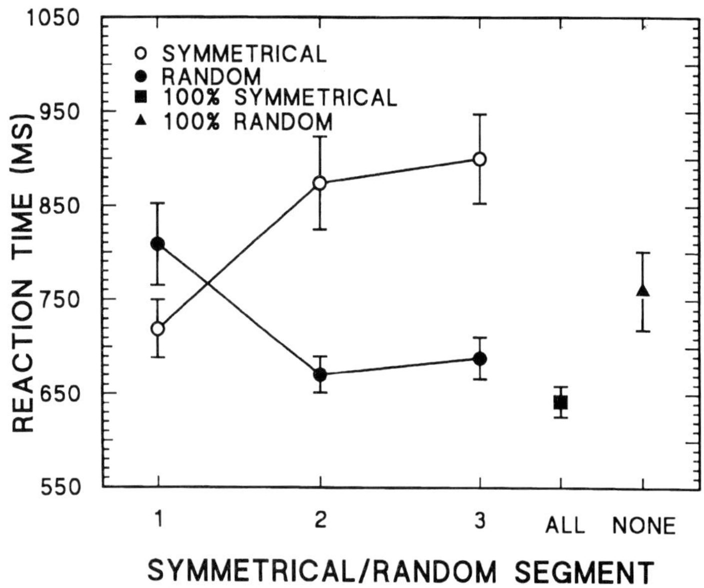

Figure 5. (a) Average number of correct responses (out of 50), Experiment 3. Segments 1, 2 and 3 refer to the inner or central, the middle and the outer parts of the dot pattern, respectively. Filled circles show number correct when only area 1, only area 2 or only area 3 was random (the other areas being vertically symmetrical). Open circles show number correct when only area 1, only area 2 or only area 3 was vertically symmetrical (the other areas being random). 100% symmetrical or random means that all three areas were symmetrical or all three were random. (b) Reaction times, Experiment 3. Otherwise as for (a).

only one random area (1, 2 or 3). Thus, when there was only one symmetrical area (open circles) subjects correctly detected the symmetry on an average of 35.62 out of 50 trials when it was in area 1 but detection fell to 10.62 for area 2 and 16.15 for area 3. In almost mirror fashion (filled circles) subjects easily detected symmetry when the sole random area was 2 (45.08 correct) or 3 (43.08) but performed much worse (25.23) when only the central area, 1, was random with the two other areas being symmetrical. The reaction time data in Fig. 5(b) show the same pattern. A repeated measures analysis of variance with planned contrasts showed that, in Fig. 5, the mean number correct with symmetry in area 2 (rvr: 10.62) was not different from the mean with symmetry in area 3 (rrv: 16.15), with $F(1,84) = 3.88$, $p = 0.052$. However, the average of these two means (13.38) did differ from the mean number correct in area 1 (vrr: 35.62), with $F(1,84) = 83.26$, $p < 0.0001$. For the random data, area 2 (vrv: 45.08) was not different from area 3 (vvr: 43.08), with $F(1,84) = 0.51$, $p = 0.48$ but the mean of these two did differ from the mean of area 1 (rvv: 25.23), $F(1,84) = 59.84$, $p < 0.0001$. Other, less interesting contrasts showed that the overall mean of the symmetrical data differed from that of the random data ($F(1,84) = 109.55$, $p < 0.0001$) and that the mean of all six of these conditions differed from the overall mean of the all symmetrical (vvv) and all random (rrr) conditions, with $F(1,84) = 99.35$, $p < 0.0001$. The results were identical in the reaction time data, with the respective F and p values as follows: $F = 0.63$, $p = 0.43$; $F = 35.08$, $p < 0.0001$; $F = 0.30$, $p = 0.59$; $F = 20.64$, $p < 0.0001$; $F = 12.95$, $p = 0.0005$; and $F = 32.89$, $p < 0.0001$.

4.3. Discussion

The results of Experiment 3 indicate that it is the area of the pattern close to the central fixation point which essentially determines whether symmetry in part of a dot pattern is detected or not. When the central area, 1, was symmetrical — as in the conditions vrr, vvr, vrv and vvv — the proportion of trials on which symmetry was correctly detected was 0.71, 0.86, 0.91 and 0.96, respectively. When that area was not symmetrical — as in the conditions rvr, rrv, rvv — the proportion of trials on which symmetry was correctly detected was 0.21, 0.32 and 0.50. These proportions also suggest that it is not just the central area which is important: adding symmetry to other areas increases the proportion of correct detections and this appears to be so even when the central area is random.

There was no evidence in this experiment that a symmetrical outline was more easily detected than a symmetrical middle region, as indexed by the contrasts comparing detectability in areas 2 and 3. Also, area 2 was very similar to the region in the periphery in which symmetry was quite easily detected in Experiment 2 *but only in the absence of central random dots*. This suggests that the poor detection of area 2 symmetry in Experiment 3 was not due merely to the difficulty of detecting symmetry in that region but rather to the interfering effect of the random dots in area 1. It would be of interest to repeat Experiment 3 with highly experienced observers like Barlow and Reeves and with blocked trials, such as those probably used by Barlow and Reeves, to see whether symmetry can be better detected in areas 2 and 3 if the subject knows where the symmetry will be, when it is present.

From the foregoing experiments it has been postulated that symmetry detection is largely determined by detection of symmetrical dot pairs near the fixation point. This is to be distinguished from the similar but different hypothesis which merely attributes symmetry detection to the existence of symmetrical dot pairs close to the axis of symmetry, regardless of how peripheral they are. These two factors — proximity to the axis and to the fixation point — were confounded in Barlow and Reeves' six-strip experiment, although they and others have performed other experiments to show that off-axis fixation also reduces detectability of symmetry (Barlow and Reeves, 1979; Saarinen, 1988). In Experiment 3, there were symmetrical dot pairs close to the axis in all 7 kinds of stimuli containing symmetry; but correct detections were only high when these pairs were in area 1.

5. EXPERIMENT 4

Let us now turn to different questions and a new paradigm. First, if symmetry is detected by judging the relationship between dot pairs near the fixation point through which the axis runs, can observers use symmetry to judge the orientation of a symmetrical dot pattern? That is, how well can observers discriminate the orientations of bilaterally symmetrical axes around vertical? Second, if they can do that task reasonably easily, what roles do dot pairs near fixation and the outline of the dot pattern play in the orientation judgments? To answer the latter question, subjects were required to judge the orientation of symmetrical patterns with and without a random dot outline mask, similar to that used in Experiment 1.

5.1. Methods

Eleven observers from the usual subject pool used the same keys as in previous experiments to make responses but now the '*' key signalled 'left of vertical' and the '-' key signalled 'right of vertical'.[3] There were 9 symmetry axis orientations: 86 deg (4 deg left of vertical at the top) through vertical (90 deg) to 94 deg in 1 deg steps. There were 100 trials at each orientation with a different random pattern on every trial. Observers completed two blocks of 450 trials. In one block, the 20-dot symmetrical patterns were presented in a 10.25 deg circular area. In the other block, the 20-dot symmetrical patterns were surrounded by 40-dot random patterns, in a fashion similar to the method used in Experiment 1. Alternate observers completed the outline masked and outline present conditions in opposite order, to control for practice effects.

5.2. Results

Figure 6(a) shows percentage of right responses plotted against axis orientation and Fig. 6(b) shows reaction times. The statistical sigmoid curves in Fig. 6(a) were fitted using a standard algorithm in a graphing package (Fig. P, Biosoft).[4] From these group data the points of subjective vertical (PSVs) taken from the fitted curves were 90.36 deg for the patterns with non-masked outline and 90.82 deg for the patterns

(a)

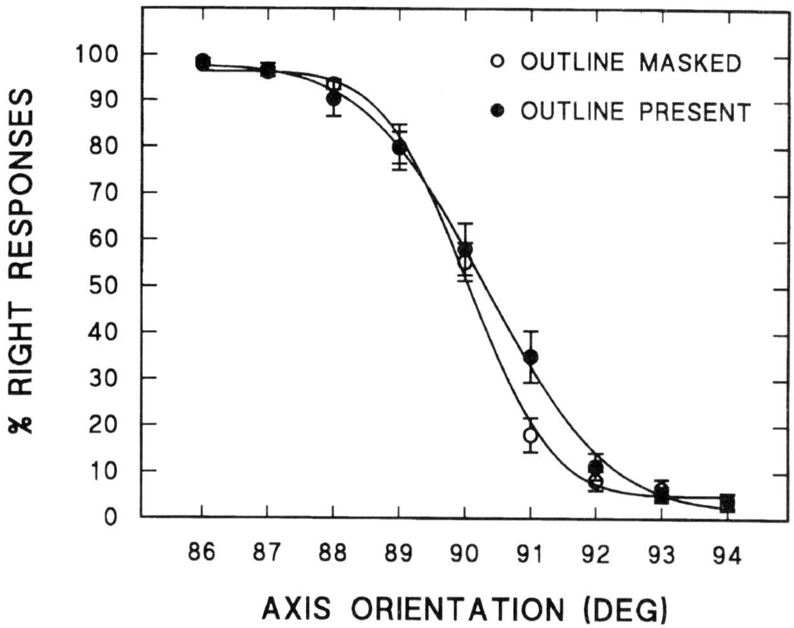

(b)

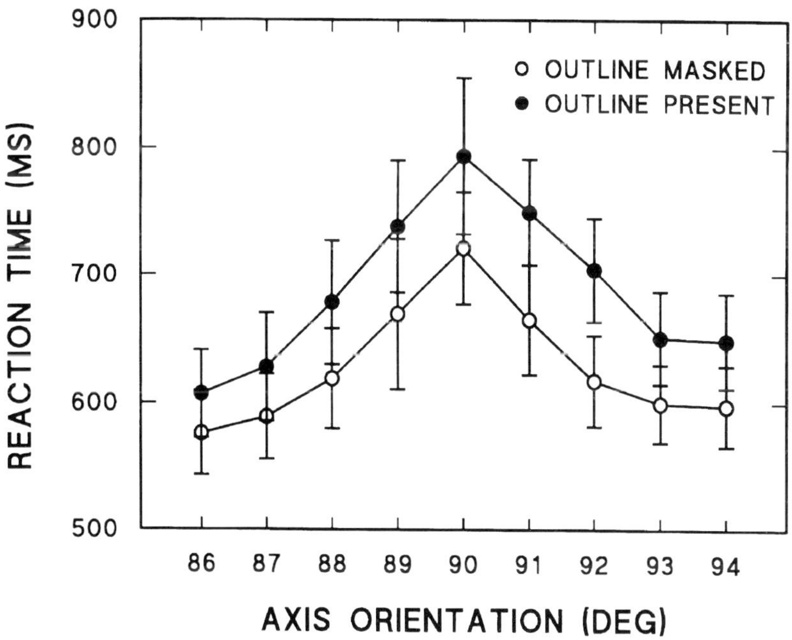

Figure 6. (a) Orientation judgments (% 'Right' responses) of bilaterally symmetrical 20-dot patterns with symmetry axes oriented 1–4 deg left or right of vertical, and with outlines masked or not masked by random dots, Experiment 4. (b) Reaction times, Experiment 4.

with the outline mask. A simple F test for repeated measures showed that these were not different, with $F(1, 10) = 0.29$ and $p = 0.61$. The mean standard deviations (slopes) of the two curves were 1.26 and 1.00, respectively, and these were also not different, with $F(1, 10) = 2.64$ and $p = 0.14$.

5.3. Discussion

These data show that even naive subjects are astonishingly adept at judging the orientation of axes of bilateral symmetry relative to subjective vertical; and that the outline of the dot pattern plays no role in that judgment. That is, the judgment is not based on the appearance of a tilted object-like pattern (insect, vase, etc.) defined by the outline but appears to be based upon the relative orientations of the virtual lines joining symmetrical dot pairs near fixation. As might be expected, the reaction time data in Fig. 10 indicate that the closer the axis is to vertical, the more difficult the judgment becomes. It is not clear why the reaction times appear slower when the outline is not masked, unless the masking of the outline merely enforces the strategy of attending to symmetrical dot pairs. A question which arises immediately is whether the task of judging the orientation of the symmetry axis is dependent upon the number of dots in the pattern and this was addressed in the final experiment. It should be noted, however, that subjects were explicitly instructed in Experiment 5 to judge the orientation of the virtual line joining the dot pair, not the orientation of the orthogonal axis of symmetry.

6. EXPERIMENT 5

It has been reported that subjects are able to judge vertical when it is defined by a pair of dots and that the mean PSV is the same whether the dots are close together or far apart (Wade, 1972). In this experiment, subjects were presented with patterns similar to those in Experiment 4, except that no random masks were used and the patterns contained just one pair of dots placed symmetrically and horizontally or near-horizontally about the fixation point (which, as always, was only present when the patterns were not). As before, the axis between the dots was oriented between 86 and 94 deg inclusive, in 1 deg steps. Ten subjects from the same pool were asked to judge whether the dots were tilted down to the left or down to the right. The dots were separated by either 3.85 or 20.5 deg.[5]

6.1. Results

The percentage of right responses are shown with fitted sigmoid functions in Fig. 7(a); and Fig. 7(b) shows reaction times. For unknown reasons, the PSVs in this experiment are biased slightly to the right of vertical, being 89.77 deg for the closer dot condition and 89.81 deg for the larger separation condition. However, in agreement with Wade (1973), a repeated measures analysis of variance showed that the difference between the two means was not significant, with $F(1, 9) = 0.54$ and $p = 0.48$. Interestingly, the standard deviations of the functions were much smaller than in Experiment 4, being 0.44 in the close condition and 0.48 in the larger separation condition. These slopes were also not different, with $F(1, 9) = 0.17$ and $p = 0.69$.

(a)

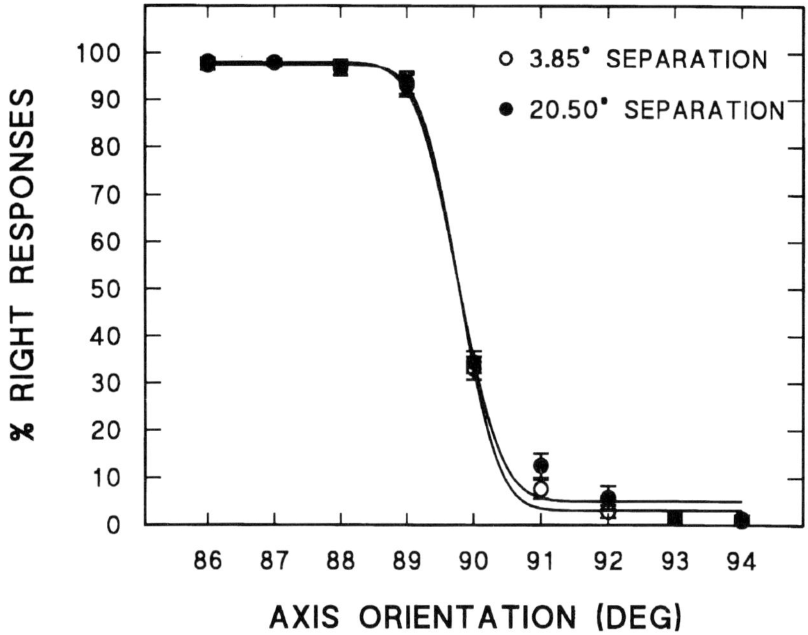

(b)

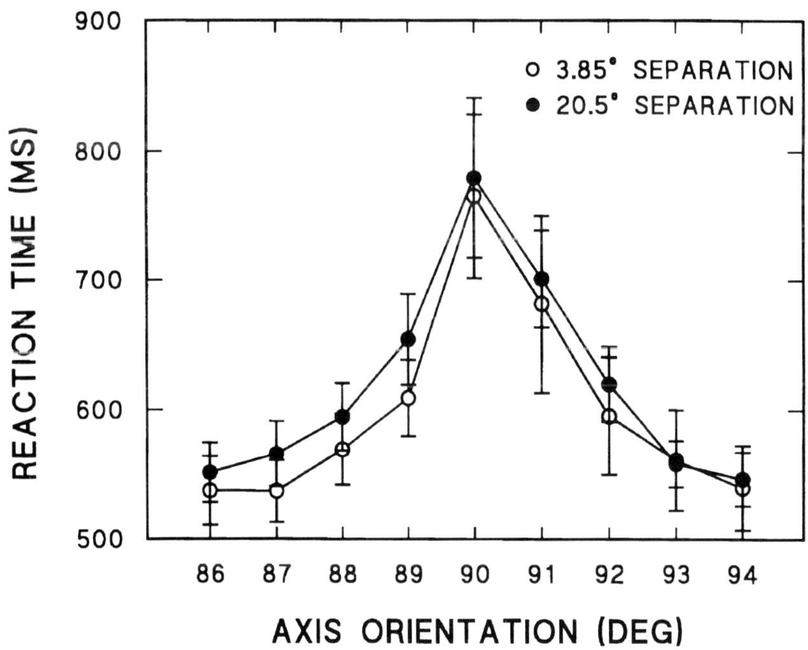

Figure 7. (a) As for Fig. 6(a), but with two-dot patterns, no masking dots, and dot pairs separated horizontally by 3.85 or 20.5 deg, Experiment 5. (b) Reaction times, Experiment 5.

7. GENERAL DISCUSSION

Taken together, the five experiments reported here indicate that the outline of symmetrical dot patterns plays some role in symmetry detection because performance is poorer, both in terms of proportion correct and reaction time, when 10.25 deg diameter 28-dot patterns are surrounded by a 28-dot random mask with outside diameter 20.5 deg (Experiment 1). However, when there is no central pattern at all but the surrounding 28-dot pattern is symmetrical (Experiment 2) performance is reasonably good: mean proportions correct and reaction times are only slightly different when the symmetrical dots are in the centre (0.806, 793.49 ms) compared with the surround (0.733, 869.08 ms). Yet when a random 28-dot central pattern is added to the 28-dot symmetrical surround (Experiment 2), performance deteriorates considerably (0.41, 955.01 ms). This suggests that while peripheral outline information can be used to detect symmetry, the presence or absence of symmetrical dot pairs near the fixation point or the pattern centre is crucial to the detection of symmetry in dot patterns. In addition, when symmetry is present only peripherally in the pattern, the usual relative salience of the different axis of symmetry orientations is not observed (Experiment 2), suggesting that symmetry axis salience in dot patterns derives from symmetry relations near the pattern centre or near fixation.[6]

Experiment 3 confirmed the latter assertion by measuring symmetry detection in the presence or absence of symmetry in one or more of three annuli. Here, symmetry was rarely detected when not present in central vision or the pattern centre, although performance improved as symmetry was added to areas other than the central one.

Experiments 4 and 5 showed, for the first time, that bilaterally symmetrical dot patterns can be used as tokens for orientation judgments, with JNDs of the order of 1 deg. That the slopes of the fitted sigmoid functions were steeper when two dots, rather than 20 dots, defined the axis and that the slopes were identical for 3.85 deg and 20.5 deg dot separations, speaks to the high sensitivity of orientation judging mechanisms, and the similarity of the functions in Experiments 4 and 5 may indicate that the mechanisms involved in judging orientation are related to the mechanisms involved in detecting a pattern's axis of symmetry. Finally, the fact that very peripheral dot pairs, in Experiment 5, could be judged just as well as much less separated pairs whereas symmetry in the periphery could not be used as effectively in the other experiments where other dots intervened, supports the idea that central dots mask the effectiveness of peripheral symmetry-detecting mechanisms.

It can be concluded that the major determinant of bilateral symmetry detection in dot patterns, such as those used here, is the presence of symmetrical dot pairs near fixation or near pattern centre when dot patterns are relatively dense. The pattern outline plays some role in symmetry detection but appears merely to make the overall task easier without playing any role in the relative salience of the various orientations of symmetrical axes.

Nevertheless, experiments have demonstrated that subjects can discriminate symmetry from asymmetry in closed forms or polygons which by definition do not have dot pairs close to pattern centre or fixation (Carmody *et al.*, 1977; Palmer and Hemenway, 1978) and there also is evidence for orientational anisotropies, including vertical salience, in that kind of task (Palmer and Hemenway, 1978). However, the forms typically used in these experiments subtended 4 deg of visual angle and had sharp, angular protuberances so that the task was most likely done by comparing small paired features of the shapes across the axis, in a fashion similar to that postulated here for detecting symmetry in dot patterns. (See Fig. 8.)

Furthermore, to generate the symmetrical stimuli, Palmer and Hemenway defined 16 radii in polar coordinates, equally spaced 22.5 deg (their Figure 13a) and, starting with vertical, placed dots at given distances from the centre on these radii. The dots were then joined to produce the required shapes (their Figure 13b). However, Palmer and Hemenway did not use random shapes as the alternative to symmetrical shapes but rather shapes which were *nearly* bilaterally symmetrical or which had rotational

(a)

(b)

(c)

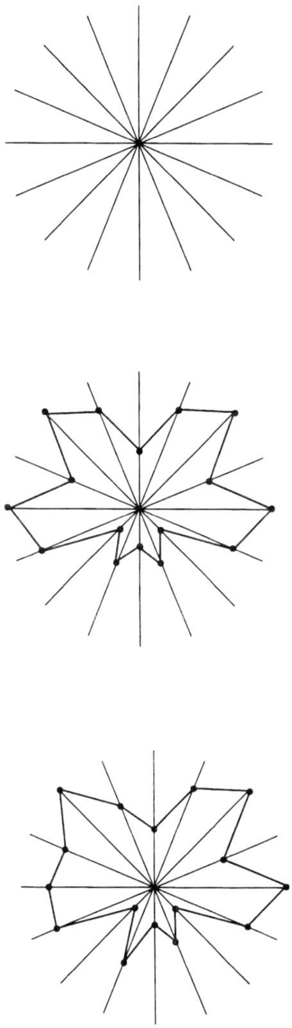

Figure 8. A representation of the polar axis method (a) used by Palmer and Hemenway (1978) to generate bilaterally symmetrical patterns (b) and near symmetrical patterns (c).

symmetry. The nearly symmetrical shapes were produced by moving 4 of the dots used to generate the figure by a small increment along the radius, two being moved out and two in (their Figure 13c). Clearly, this discrimination requires comparing pairs of vertices and it also requires scrutiny, since reaction times were of the order of 2–4 s, far greater than those obtained here. In brief, it is not clear that the Palmer and Hemenway task was very different from the discrimination of symmetry in dot patterns. At the same time, their task was not the same as discriminating symmetrical from random patterns: symmetrical shapes, near symmetrical shapes and shapes with rotational symmetry *all* have fairly regular looking outlines so that the utility of the overall outline regularity may have been minimal in their experiments. Finally, if it is reasonable to postulate that symmetry detection in dot patterns is performed by orientation analysing mechanisms, so that the task used in Experiment 5 here is actually a simplified symmetry detection task, then what might account for the restriction of the symmetry axis orientation salience effect to central vision, if that is indeed the critical variable rather than the centre of the pattern? It has been claimed that vertical and horizontal lines or objects can be detected or discriminated more easily and identified more rapidly and accurately compared with oblique objects due to a relative predominance of monkey V1 cells tuned to vertical and horizontal (Mansfield, 1974; Mansfield and Ronner, 1978) but, interestingly, these studies have provided evidence that *both* the psychophysical oblique effect *and* the preponderance of vertically and horizontally tuned cells are restricted to foveal vision. It might then be the case that the relative saliences of vertical and horizontal symmetry are determined partly by these neural factors and partly by attentional factors: a hybrid hypothesis. In these terms, the neural factors would account for the fact that vertical and horizontal are detected equally well, then non-diagonal obliques and then the diagonals. The additional mechanisms of selective attention would then account for both the relative superiority of vertical over horizontal symmetry detection and the finding (Wenderoth, 1994) that oblique symmetry detection is next best after vertical and horizontal, rather than poorest of all. Several studies have indicated recently that attentional factors can modulate the strength of what were previously considered reflections of purely low-level neural substrates (e.g. Chaudhuri, 1990).

Acknowledgements

This research was supported by an Australian Research Council project grant 1992-4. Thanks are due to Neil Hickey who assisted in testing subjects and in data entry. The symmetry program was written by Tim Cooper. I thank Darren Burke for valuable discussion.

NOTES

1. I thank the Editor for suggesting this algorithm.
2. As noted by the Editor, one way to do this, although it was not done here, would be to do an experiment which is the converse of the Fisher and Bornstein (1982) study: use dot patterns in which the dots' outline is either symmetrical or random with the internal dots always being random or forming a regular array.

3. On the PC keyboard the '*' key is to the left of the '-' key.

4. The algorithm is the Marquardt–Levenberg adaptation of the Gauss–Newton algorithm.

5. The 3.85 deg separation was chosen because at that separation, a 1 deg tilt corresponds to a shift of 1 pixel. At the larger separation, enough pixels are involved in a 1 deg tilt to ensure sufficient accuracy of orientation.

6. A reviewer pointed out, quite correctly, that the key variable may not be proximity to the fixation point but proximity to the centre of the pattern. Thus, even if all patterns were presented peripherally, performance in detecting symmetry when the outer annuli of the patterns were symmetrical might be good when the centre of the pattern also contains symmetrical dots or contains nothing, whereas performance might be poor when the pattern's centre contains random dots. These experiments cannot discriminate between these possibilities.

REFERENCES

Barlow, H. B. and Reeves, B. C. (1979). The versatility and absolute efficiency of detecting mirror symmetry in random dot displays. *Vision Res.* **19**, 783–793.

Bruce, V. G. and Morgan, M. J. (1975). Violations of symmetry and repetition in visual patterns. *Perception* **4**, 239–249.

Carmody, D. P., Nodine, C. F. and Locher, P. J. (1977). Global detection of symmetry. *Percept. Motor Skills* **45**, 1267–1273.

Chaudhuri, A. (1990). Modulation of the motion aftereffect by selective attention. *Nature* **344**, 60–62.

Corballis, M. C. and Roldan, C. E. (1975). Detection of symmetry as a function of angular orientation. *J. Exp. Psychol: Human Percept. Perform.* **104**, 221–230.

Fisher, C. B. and Bornstein, M. H. (1982). Identification of symmetry: Effects of stimulus orientation and head position. *Percept. Psychophys.* **32**, 443–448.

Jenkins, B. (1982). Redundancy in the perception of bilateral symmetry in dot textures. *Percept. Psychophys.* **32**, 171–177.

Jenkins, B. (1985). Orientational anisotropy in the human visual system. *Percept. Psychophys.* **37**, 125–134.

Locher, P. J. and Nodine, C. F. (1973). Influence of stimulus symmetry on visual scanning patterns. *Percept. Psychophys.* **13**, 408–412.

Mansfield, R. J. (1974). Neural basis of orientation perception in primate vision. *Science* **186**, 1133–1135.

Mansfield, R. J. W. and Ronner, S. F. (1978). Orientation anisotropy in monkey visual cortex. *Brain Res.* **149**, 229–234.

Palmer, S. E. and Hemenway, K. (1978). Orientation and symmetry: Effects of multiple, rotational, and near symmetries. *J. Exp. Psychol: Human Percept. Perform.* **4**, 691–702.

Pashler, H. (1990). Coordinate frame for symmetry detection and object recognition. *J. Exp. Psychol: Human Percept. Perform.* **16**, 150–163.

Royer, F. L. (1981). Detection of symmetry. *J. Exp. Psychol: Human Percept. Perform.* **7**, 1186–1210.

Saarinen, J. (1988). Detection of mirror symmetry in random dot patterns at different eccentricities. *Vision Res.* **28**, 755–759.

Wade, N. J. (1972). The influence of eye movements and eye position on perception of the visual vertical. *Psychologische Forschung* **35**, 131–142.

Wagemans, J., Van-Gool, L. and D'Ydewalle, G. (1992). Orientational effects and component processes in symmetry detection. *Q. J. Exp. Psychol: Human Exp. Psychol.* **3**, 475–508.

Wenderoth, P. (1994). The salience of vertical symmetry. *Perception* **23**, 221–236.

Human Symmetry Perception, pp. 71–83
C. W. Tyler (Ed.)

Detection and identification of mirror-image letter pairs in central and peripheral vision

KENT E. HIGGINS,[1] ARIES ARDITI[1] and KENNETH KNOBLAUCH[2]

[1] *Vision Research Laboratory, The Lighthouse Research Institute, 111 East 59th Street, New York, NY 10022, USA*

[2] *Cerveau et Vision, INSERM Unité 371, 18 avenue du Doyen Lépine, 69500 Bron, France*

Abstract—Reading performance is poorer in the peripheral than the central visual field, even after size-scaling to compensate for differences in visual acuity at the different eccentricities. Since several studies have indicated that the peripheral retina is deficient with respect to spatial phase discrimination, we compared the psychometric functions for detection (D) and identification (I) of size-scaled, mirror-symmetric letters (i.e. letters differing in the phase spectra of their odd symmetric components) at three inferior field eccentricities (0 deg, 4 deg, and 7.5 deg) using a 2-alternative, temporal, forced-choice procedure and retinal image stabilization to control retinal locus. Each subject's data were fit with Weibull functions and tested for goodness-of-fit under several hypotheses. This analysis revealed that while the psychometric functions were of constant shape across eccentricity for the respective tasks, they showed statistically significant variations in the D/I threshold ratios. However, these variations were so small that poorer reading outside the fovea is unlikely to be due to reduced letter discriminability that might occur secondary to a loss of peripheral field phase sensitivity.

1. INTRODUCTION

Reading involves two initial activities, one motor and one sensory. The motor activity consists of a series of eye movements (saccades), the direction and magnitude of which are influenced by visual information from the text (McConkie and Rayner, 1975; Rayner and McConkie, 1975; O'Regan, 1990; Vitu, 1991). Between saccades, when the eyes are relatively motionless, the normally-sighted reader must acquire the necessary visual information for detecting and recognizing the word(s) fixated (i.e. 'foveated').

Numerous everyday activities depend upon our ability to perform this task which, from a visual perspective, should be quite effortless. Legge *et al.* (1985a) have shown that the visual requirements for normal reading are quite modest. When, however, the central visual field is compromised as a consequence of ocular disorder, this simple, everyday activity can become an inefficient and frustrating task for the individual (e.g. Faye, 1984; Legge *et al.*, 1985b; Cummings *et al.*, 1985). Teaching a person with central field loss to read using peripheral vision often requires numerous hours of training and practice (Goodrich *et al.* 1977; Watson and Berg, 1983) and, even with the use of magnification to compensate for the poorer peripheral acuity, such training rarely results in restoring reading performance to the level achievable with a normal central visual field.

Two broad types of hypotheses, each corresponding to one of the initial component activities of reading, have been offered to explain the deficient performance associated with using peripheral vision. The oculomotor hypothesis attributes the problem to inadequate eye movement control (e.g. Whittaker *et al.*, 1988; Rubin and Turano, 1994). In oversimplified terms, an individual with a central field loss might persist in attempting to use saccadic eye movements to foveate words, even though this would result in the word being imaged within the non-seeing scotoma. However, measurements of reading using rapid serial visual presentation (RSVP) to eliminate the need for saccadic eye movements indicate that oculomotor factors cannot explain the entire deficit, either in patients with central field loss or in patients having other types of visual loss (Rubin and Turano, 1994).

The sensory hypothesis, in contrast, attributes the deficit to an inability of peripheral vision to perform complex pattern recognition (e.g. Rubin and Turano, 1994). In the case of patients having some type of ocular disorder, for example, peripheral retina reading rates might arguably be limited by subclinical field losses. However, it is unlikely that this version of the hypothesis would, by itself, suffice; reading performance was found to be deficient in the periphery of otherwise normal subjects (Turano and Rubin, 1988; Rubin and Turano, 1994).

Another variant of the sensory hypothesis might attribute the problem to a difference in the resolution of central vs peripheral vision. For example, declines in letter acuity, grating acuity, and contrast sensitivity with increased eccentricity are well documented (e.g. Westheimer, 1982). However, such declines are clearly not sufficient to explain peripheral reading deficits as Rubin and Turano (1994), for example, found that reading rate declined with eccentricity even though text was size-scaled to normalize peripheral and and central field spatial contrast sensitivity values.

It is possible that the Rubin and Turano approach was, in principle, appropriate, but that their spatial scaling factor was inappropriate. While many peripheral visual thresholds can be normalized to foveal levels by spatial scaling (e.g. Virsu and Rovamo, 1979), there is, in fact, no scale factor that will normalize all peripheral thresholds to central field values (e.g. Westheimer, 1987). For this reason, Levi *et al.* (1987) have suggested that different types of visual tasks may be limited by different factors, with, for example, conventional acuities being limited by retinal factors and positional acuities (e.g. vernier) being limited by cortical factors. Accordingly, reading might be better related to visual phenomena that decline faster with eccentricity than do conventional letter acuity or contrast sensitivity.

The 'crowding' effect offers one possibility. It is well known that visual acuity is better when the letters of the chart are presented in isolation than when presented with nearby contours (e.g. Stuart and Burian, 1962; Flom *et al.*, 1963). There is evidence that the magnitude of the crowding effect may be greater in peripheral vision (e.g. Loomis, 1978; Jacobs, 1979). If true, it might imply that optimal spacing differs for central and peripheral vision. However, while there is some evidence that letter spacing may affect word recognition (Whittaker *et al.*, 1989; Arditi *et al.*, 1995) and reading rate (Arditi *et al.*, 1990), there is not complete agreement on the importance of this factor (Legge *et al.*, 1985b).

Another potentially important variant of this hypothesis is that reading with peripheral retina may be constrained or limited by a loss of positional information. Evidence from several studies suggests that human observers are less certain about the spatial location of targets presented in the peripheral than central field (e.g. Pelli, 1981; Cohn and Wardlaw, 1985). Levi *et al.* (1987), for example, found a greater reduction in bisection acuity than grating acuity in the normal periphery and, also, in the central field of strabismic amblyopes. Their findings were consistent with a model that invokes cortical undersampling of stimuli as a means of producing positional uncertainty in these two visual systems. In general, the hyperacuities (e.g. vernier) decline 3–4 times faster than grating or Snellen acuities (e.g. Westheimer, 1982). Thus, the deficit observed by Rubin and Turano (1994) finds a closer parallel with the decline in position acuity than with that of resolution acuity.

The concept of position is frequently linked with the concept of phase in the Fourier domain. Moreover, several studies indicate that peripheral vision shows deficits in spatial phase discrimination relative to central vision (Braddick, 1981; Rentschler and Treutwein, 1985; Harvey *et al.*, 1985; Bennett and Banks, 1987). Further, sensitivity to spatial phase declines with eccentricity more rapidly than contrast sensitivity, i.e. at a rate similar to that of the positional acuities (e.g. Rentschler and Treutwein, 1985; Bennett and Banks, 1987). However, none of these studies examined the possible implications of phase anomalies for the generally slower reading rates characteristic of patients with central field loss (e.g. Legge *et al.*, 1985b) or for normal subjects using peripheral retina (Rubin and Turano, 1994). Phase deficits could, however, interfere with word recognition by interfering with (i) the processing of letter position information, as already noted, and/or (ii) the identification of individual letters of words.

How phase deficits would affect letter discrimination would, of course, depend on how the Fourier components of individual letters contribute to their discriminability. Gervais *et al.* (1984), for example, found that among several models for predicting the confusability of uppercase letters, a spatial frequency model based on letter amplitude spectra weighted by the normal contrast sensitivity function accounted well for the data. Moreover, a model including phase as well as filtered amplitude information was superior to a model based on only amplitude information alone. Phase information could, for example, contribute to the discrimination of letters like ‘M’ and ‘W’, ‘p’ and ‘q’, or ‘b’ and ‘d’. Members of such letter pairs have identical magnitude spectra and differ primarily in terms of the phase spectra of their odd symmetric components (horizontal components for the first pair of letters and vertical components for the latter two pairs).

We have attempted to evaluate the similarity of the mechanisms mediating detection and identification of alphabetic characters in the central and peripheral fields by using size-scaled mirror symmetric, lower case letters. Stimuli were scaled according to the procedure previously used by Rubin and Turano (1988, 1994). Specifically, this scale factor failed to normalize peripheral reading and, consequently, our objective was, in part, to determine whether this scale factor would also fail to normalize peripheral field detection/identification performance for mirror symmetric letters, i.e. differing in phase. By analyzing the psychometric functions for these tasks, we were able to distinguish between possible changes in precision, as reflected in the steepness of the psychometric function, and changes in the relative sensitivity of the processes mediating contrast detection (D) and identification (I), as reflected in the D/I threshold ratio.

2. METHODS

2.1. Subjects

The subjects were 3 normal observers, ranging in age from 26–50 yrs; two were authors, and the third was naive to the purpose of this experiment. All subjects had 20/20 or better visual acuity and normal visual fields.

2.2. Viewing conditions

Each subject was tested monocularly (right eye only) while viewing the stimulus field through the optics of a stimulus deflector unit (Crane and Clark, 1978). The latter contains two orthogonally-rotating mirrors which were driven by the horizontal and vertical eye movement signals from a Double Purkinje-image Eyetracker. These mirrors were used to produce the optical displacement of the stimulus field necessary to compensate for the subject's eye movements (Crane and Steele, 1978). To preclude accommodative fluctuations from influencing test results, each subject was administered a cycloplegic (1% cyclopentolate). Since this procedure also produced mydriasis, a 3 mm artificial pupil was used. Each subject was tested foveally and at two inferior field eccentricities, 4.0 and 7.5 deg. Head position was maintained by the use of a bite bar and forehead rest.

2.3. Stimulus generation and psychophysical procedure

A two-alternative temporal forced-choice procedure was used to measure detection and identification of the letters 'b' and 'd' at each of several contrast levels (e.g. Thomas, 1987). The subject indicated in which of two successive 50 ms time intervals a letter was presented and, in addition, identified which letter was presented.

Stimuli were presented on a Commodore Amiga 1084 RGB monitor and appeared as dark letters against a white background. The luminance of the white background of the stimulus field was 5 cd/m^2 after passing through the optics of the stimulus deflector unit. Stimulus letter height was 13′ for the foveal test distance (183 cm). Stimulus contrast values were selected to produce levels of correct detection and identification of from 50% (chance) to about 95–98% and so that identification judgments were nearly asymptotic. Stimulus size was scaled for the 4.0 and 7.5 deg eccentricities by decreasing the test distance to 81 and 54.7 cm, respectively, i.e. by an amount that Virsu and Rovamo (1979) showed was necessary to normalize peripheral spatial contrast sensitivity to foveal values (see also Rubin and Turano, 1994).

Subjects were dark-adapted for 10–12 min prior to each session. Each subject was aligned in the eyetracker, allowed to light adapt for 2–3 min, and then required to make 100 judgments using a fixed contrast and eccentricity. Subsequent to a short rest, testing proceeded in similar fashion for each contrast level. From six to nine contrast values were tested at each eccentricity, depending on the subject, with the procedure repeated two to three times for each subject and for each eccentricity. Thus each of the plotted points in figures 1–3 below was based on a minimum of 200 judgments.

In summary, for each of three eccentricities, a pair of frequency-of-seeing functions was generated for each subject. One member of each pair represented performance on the detection task as a function of contrast. The other member represented performance on the identification task over the same range of contrasts.

3. RESULTS

Figure 1 shows the proportion correct detection (○) and identification (●) for one subject at one eccentricity. Chance performance (50%) is shown by the dashed horizontal line. The identification data showed a shallower psychometric function relative to that for detection, a result similar to that reported by Knoblauch (1995) for the detection and identification of color. The higher threshold for identification was also a regular feature in our data and probably indicates that additional processing is required to make identification judgments.

The critical question to be answered in the subsequent analysis is whether similar differences in slope and threshold are obtained at two additional eccentricities tested

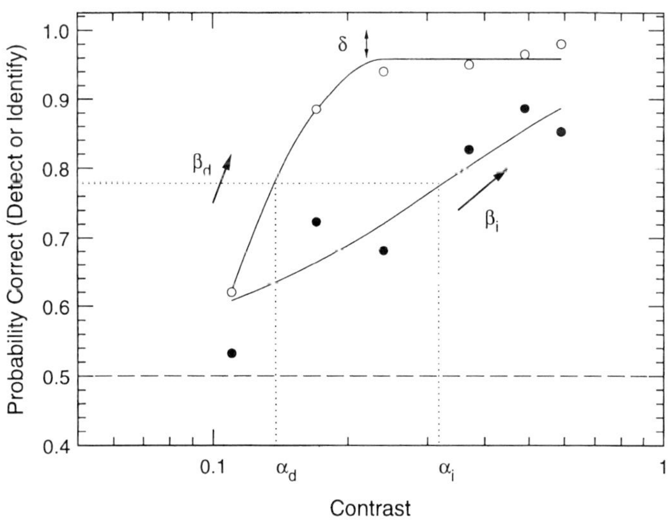

Figure 1. Probability correct detection (○) and identification (●) as a function of contrast for one subject under foveal viewing conditions. The solid curves represent Weibull functions to which the respective sets of data were fit. The 'attentional' parameter is indicated by δ and the 'steepness' parameter by β. The respective threshold contrasts for detection and identification are designated along the abscissa by α_d and α_i and represent the log contrast values (vertical dashed lines) required to produce a criterion response level after correction for the attentional factor, δ.

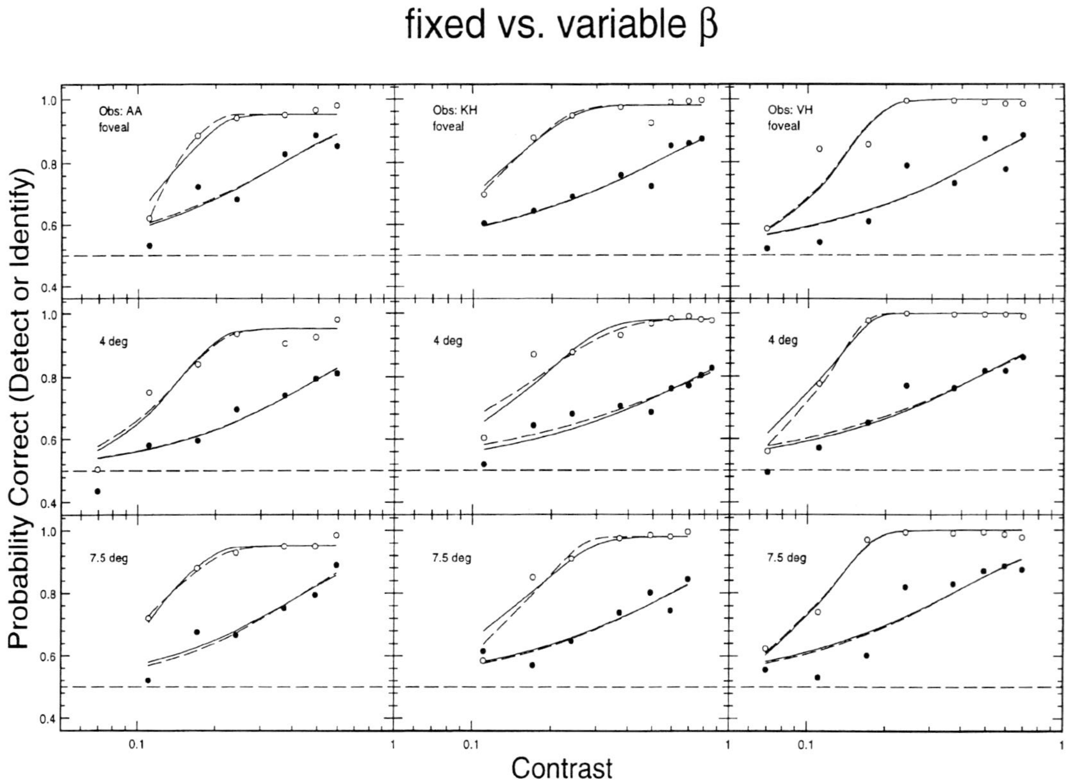

Figure 2. Comparison of fits to Weibull function across eccentricities under Hypothesis 1 (β's and α's free to vary across eccentricities and tasks) vs Hypothesis 2 (β constant within task and across eccentricities) for each of the three subjects, with each column representing a different subject. Here, as in Figs 1 and 3, ○ represent detection results and ● represent identification results. The dashed curve represents the fit under Hypothesis 1 and the solid curve represents the fit under Hypothesis 2. Values of the attentional and steepness parameters, δ and β, obtained from the fit provided by Hypothesis 2 for each subject are given in Table 2. See Table 1 for a summary of results of statistical comparison of goodness-of-fit illustrated in this figure as well as Fig. 3.

for each subject and using the size-scaled stimuli. The frequency-of-seeing data for each subject at the three eccentricities were fit with Weibull functions and then tested for goodness-of-fit under three hypotheses described below. These fits were carried out using a maximum likelihood procedure described previously by Watson (1979) and a minimization routine (Chandler, 1965). The solid lines in Fig. 1 represent one such fit. Since this was a two alternative forced-choice procedure, a function of the following general form was used:

$$f(\alpha) = 0.5 + 0.5 \times \exp\left[-\ (\alpha/\alpha_0)^{\beta} \right] \times (1 - \delta),$$

where α is the contrast; α_0 is the contrast at which f is approximately $(1-\delta) \times 0.81$, i.e. threshold for detection or identification respectively; β is a steepness parameter related to the slope; δ = an 'attentional' factor.

We define the D/I threshold ratio as α_d/α_i. The attentional factor was introduced to account for the fact that the psychometric functions did not reach 100% at the highest contrast levels used. There were a number of reasons for this. Occasionally, for example, a subject would indicate 'interval 1' (or 'b') when (s)he meant to indicate 'interval 2' (or 'd'), or vice-versa. For each subject, however, the value of δ was held constant across tasks and eccentricities. Holding δ constant produced satisfactory fits with values of delta that were always less than 10% (see Table 2 below). Since δ was held constant across eccentricity and task for a given subject, the primary interest in this analysis concerns possible changes across eccentricity in the two parameters that are free to vary, i.e. the β's and the α's for detection and identification.

Figure 2 shows the results of fitting the data at the three eccentricities for each of the three subjects under two hypotheses. Columns represent subjects and rows represent eccentricities. Under Hypothesis 1 (dashed curves), all of the parameters of the psychometric function were free to vary across tasks and eccentricity, except for the attentional parameter, δ. Under Hypothesis 2 (solid curves), the slopes of the psychometric functions were constrained to be constant within tasks (D or I), while the D/I threshold ratio was allowed to vary. Although visual inspection of Fig. 2 appears to show slight differences between the fits provided by the two hypotheses, Table 1 indicates no significant differences in the goodness-of-fit for any of the three subjects. These results are thus consistent with the view that the precision of judgments was constant across the range of eccentricities tested for these size-scaled stimuli. Table 2 gives values of β and δ for each subject, as derived from the fit carried out under Hypothesis 2.

Figure 3 compares the fit under Hypothesis 2 with that of a third hypothesis having the added constraint that the threshold ratio, α_d/α_i, was constant across eccentricity. As in Fig. 2, the fits provided by Hypothesis 2 (constant β's) are illustrated by the solid curves. The fits provided by Hypothesis 3 (constant β's and constant α_d/α_i) are shown by the dashed curves. We compared the goodness-of-fit under Hypothesis 3 with that under Hypothesis 2; Hypothesis 3 was rejected for all 3 subjects (see Table 1). Thus, while the slopes of the functions for the respective tasks were constant across eccentricity, the D/I threshold ratios varied significantly.

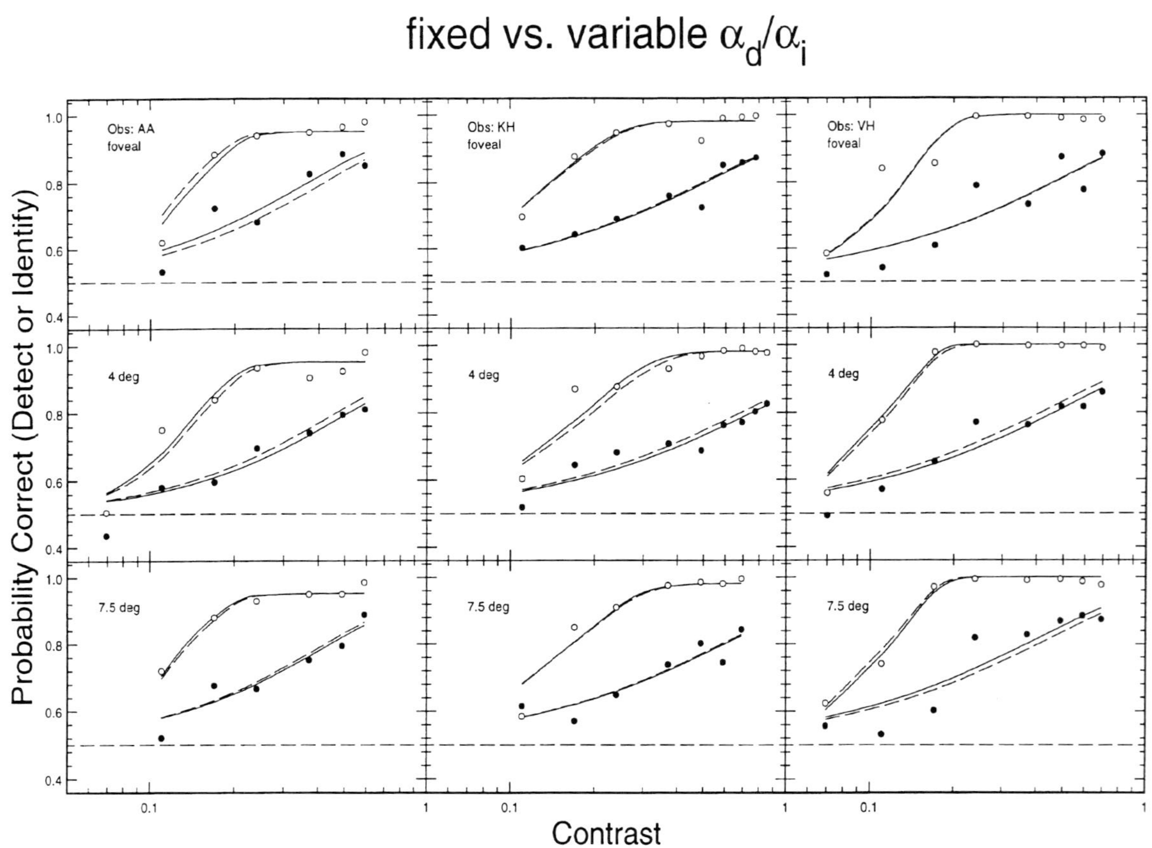

Figure 3. Comparison of fits to Weibull function across eccentricities under Hypothesis 2 (β constant within task) vs Hypothesis 3 (β constant within task and across eccentricity and, in addition, D/I threshold ratio constant across eccentricity). As in Fig. 2, the solid curve represents the fit under Hypothesis 2; the dashed smooth curve represents the fit under Hypothesis 3. See Table 1 for summary of statistical results.

Table 1.
Results of statistical comparison of hypotheses by subject

Hypotheses (H)		Subject		
		VH	AA	KH
Hypothesis 2 vs Hypothesis 1				
	χ^2	5.62	6.72	7.045
	d.f.*	4	4	4
	p	0.23	0.15	0.134
Hypothesis 3 vs Hypothesis 2				
	χ^2	10.302	9.02	14.3
	d.f.*	2	2	2
	p	0.006	0.00078	0.01

*The indicated d.f. correspond to the difference in d.f. (in number of free parameters) under the two hypotheses being tested. For Hypothesis 1, 13 values are free to vary (6 α's, 6 β's, and 1 δ for each subject). For Hypothesis 2, only 9 of these values are free to vary as there are only 2 β's under this hypothesis. This corresponds to a difference of 4 d.f., as indicated above. For Hypothesis 3, seven parameters are free to vary (3 α's, 2 β's, 1 δ, and 1 D/I threshold ratio).

Table 2.
Weibull parameters, β and δ, from fit using Hypothesis 2

Parameter	Subject		
	VH	AA	KH
$\beta_{\text{detection}}$	2.58	2.54	1.78
$\beta_{\text{identification}}$	0.99	1.25	0.98
$\delta \times 100\%$	1.7%	9.5%	4.0%

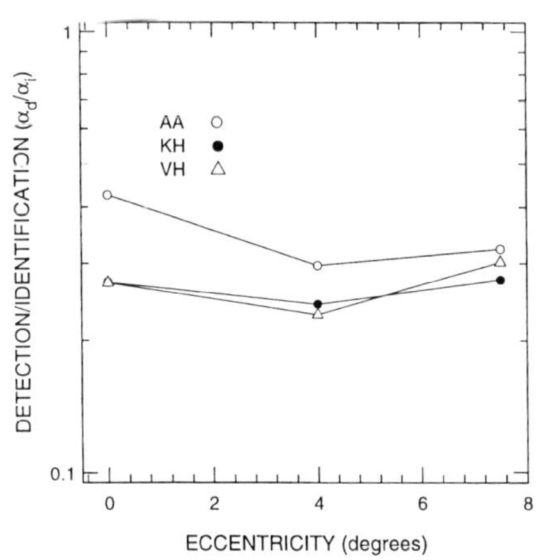

Figure 4. Change in D/I threshold ratio across eccentricity. The D/I threshold ratio was calculated for each subject from the fit provided by Hypothesis 2. The magnitude of the ratio for each subject and eccentricity corresponds to the lateral separation of the respective solid curves in Figs 2 and 3.

Figure 4 shows the magnitude of the variation in the D/I threshold ratio across eccentricities for each of the 3 subjects. These ratios were computed for each subject from the fit provided by Hypothesis 2. It is important to note that while the data do not fall on perfect horizontal lines, as would have been predicted by Hypothesis 3, there did not appear to be any systematic trends across subjects. It should also be noted that the largest magnitude of within-subject change in the D/I ratio was, in absolute terms, small.

4. DISCUSSION AND CONCLUSION

These results are generally consistent with an accumulating body of evidence indicating that many of the differences between peripheral and central visual thresholds can be resolved by the application of an appropriate spatial scaling factor (e.g. Virsu and Rovamo, 1979; Thomas, 1987; Saarinen *et al.*, 1989; Farrell and Desmarais, 1990). In the present study, we used a scaling factor that previous researchers had shown was sufficient to normalize peripheral contrast sensitivity to central field values. This same scale factor was sufficient, on average, to normalize peripheral field detection and identification of mirror-symmetric, lower case letters to central field performance levels.

In a previous investigation of spatial frequency and orientation discrimination across eccentricity, Thomas (1987) reported trends analogous to those described here. Thus, while he obtained results that were, on average, consistent with the view that the major change from central to peripheral vision was a change in spatial scale, he also found that results for individual subjects showed statistically reliable, but unsystematic, changes with eccentricity. This latter result finds a ready parallel in the rejection of Hypothesis 3 which was predicated on constancy of the D/I ratio as well as constancy of the slope factor, β. However, it should also be noted that the magnitude of within-subject variation in the D/I ratios was small in absolute terms, with the largest magnitude in Fig. 4 corresponding to the smallest magnitude of within-subject variation observed by Thomas (1987, Fig. 3). Whether the observed departures from strict constancy of the D/I ratio we observed were due to the use of a single 'average' spatial scaling factor for all subjects is not known. More importantly, there was no consistent trend in the D/I threshold ratio that would offer a viable quantitative parallel for the marked decline (approaching 1 log unit) in peripheral retina RSVP reading rate in normal subjects over the same range of eccentricities (cf. Rubin and Turano, 1994, Fig. 8).

These results might seem at odds with previous studies such as that of Bennett and Banks (1987). The latter, for example, reported a selective loss in the discriminability of mirror-symmetric compound gratings (i.e. compound gratings differing by 90–270 deg of relative phase), a finding that they concluded was consistent with a selective loss in sensitivity of odd-symmetric mechanisms with increasing eccentricity. More recently, however, Morrone *et al.* (1989) used stimuli which, while one-dimensional, were more complex, and found no evidence of a selective decline in phase sensitivity. Their stimuli consisted of the sum of 256 vertically-oriented cosine harmonics. The harmonics were added in different phases to produce pairs of contrast-reversed patterns having 'edge-like' (90 and 270 deg) or 'line-like' (0 and 180 deg) features (Burr *et al.*, 1989). Morrone *et al.* found that the discriminability of the pairs was as good in the periphery as it was in the fovea. Thus, the use of a spatial scaling factor that was sufficient to equate peripheral contrast sensitivity and grating acuity to central-field values, was also sufficient to equate peripheral and central phase sensitivity for these more complex stimuli.

Bennett (1992) has also suggested a possible means of resolving the discrepancy between results of studies using the narrow- vs the broad-band stimuli, with only the former suggesting a selective loss in phase sensitivity with eccentricity. Specifically, Bennett argued that both types of stimuli can be thought of as an array of spatial features such as lines and edges. In the case of gratings, the size and separation of the features capable of mediating discrimination are correlated such that the features typically abut spatially. Thus, it is possible that closer proximity of features capable of mediating the 90–270 phase shift discrimination may have rendered it selectively more vulnerable to the deleterious effects of lateral masking or crowding. This would not necessarily be the case for the Morrone *et al.* type of stimulus pattern where the feature size and separation can be independently varied. Using the latter type of stimuli, Bennett (1992) found that peripheral thresholds for the 90–270 and the 0–180 stimuli rose at the same rate as feature separation was decreased.

The stimuli used in the present study are more similar to those of Morrone *et al.* than to those in studies using compound gratings. Mirror-symmetric letters of the same contrast polarity contain an identical and broad spatial frequency representation where the phase of only the vertical odd components differ by 180 deg (e.g. Bracewell, 1986). Moreover, for these stimuli, like those of Morrone *et al.*, the use of a spatial scale factor sufficient to equate central and peripheral contrast and phase sensitivity was sufficient to produce generally similar detection/identification performance at the different eccentricities for mirror-symmetric letters.

In conclusion, these results argue against the hypothesis that poorer reading performance outside the fovea is due to reduced letter discriminability secondary to peripheral-field phase sensitivity loss. Instead they suggest that other factors are more likely responsible for the decline in peripheral retina reading rate described by Rubin and Turano (1994). The crowding phenomenon may represent a more important source of confusion in letter discriminability in the periphery. Rubin and Turano have shown that reading rate appears to decline at a faster rate with eccentricity than does conventional acuity and, as noted in the introduction, there is evidence to suggest that the magnitude of the crowding phenomenon may be disproportionately greater in peripheral vision. However, the significance of the crowding phenomenon to the deficient reading performance characteristic of peripheral retina is, as yet, unknown.

Acknowledgment

This research was supported by NIH grants EY07886 and EY07747 and NASA-Ames NCC2–541.

REFERENCES

Arditi, A., Cagenello, R. and Jacobs, B. (1995). Letter, strokewidth, spacing, and legibility. In: *Vision Science and Its Applications, Technical Digest Series*, Vol. 1. Optical Society of America, Washington, DC, pp. 324–327.

Arditi, A., Knoblauch, K. and Grunwald, I. (1990). Reading with fixed and variable character pitch. *J. Opt. Soc. Am.* **A7**, 2011–2015.

Bennett, P. J. (1992). The effects of lateral masking on the discrimination of lines and edges in the periphery. *Invest. Ophthalmol. Visual Sci.* **33** (Suppl.), 1349.

Bennett, P. J. and Banks, M. S. (1987). Sensitivity loss in odd-symmetric mechanisms and phase anomalies in peripheral vision. *Nature* **326**, 873–876.

Bennett, P. J. and Banks, M. S. (1991). The effects of contrast, spatial scale, and orientation on foveal and peripheral phase discrimination. *Vision Res.* **31**, 1759–1786.

Burr, D. C., Morrone, M. C. and Spinelli, D. (1989). Evidence for edge and bar detectors in human vision. *Vision Res.* **29**, 419–431.

Bracewell, R. N. (1986). *The Fourier Transform and its Applications* (2nd edn). McGraw-Hill, New York.

Braddick, O. (1981). Is spatial phase degraded in peripheral vision? *Documenta Ophthalmologica, Proc. Ser.* **30**, 255–262.

Chandler, J. P. (1965). STEPIT — Direct Search Optimization; Solutions of Least Squares Problems. Quantum Chemistry Program Exchange. QCEP Program 307, Chemistry Department, Indiana University, Bloomington, IN.

Cohn, T. E. and Wardlaw, J. C. (1985). Effect of large spatial uncertainty on foveal luminance increment detectability. *J. Opt. Soc. Am.* **A2**, 820–825.

Crane, H. D. and Clark, M. R. (1978). Three-dimensional visual stimulus deflector. *Appl. Opt.* **17**, 706–714.

Crane, H. D. and Steele, C. A. (1978). Accurate three-dimensional eyetracker. *App. Opt.* **17**, 691–705.

Cummings, R. A. W., Whittaker, S. A., Watson, G. R. and Budd, J. A. M. (1985). Scanning characters and reading with a central scotoma. *Am. J. Optom. Physiol. Opt.* **62**, 833–843.

Farrell, J. E. and Desmarais, M. (1990). Equating character-identification performance across the visual field. *J. Opt. Soc. Am.* **A7**, 152–159.

Faye, E. (1984). *Clinical Low Vision*. (2nd edn). Little, Brown & Co, Boston.

Flom, M. C., Weymouth, F. W. and Kahneman, D. (1963). Visual resolution and contour interaction. *J. Opt. Soc. Am.* **53**, 1026–1032.

Gervais, M. J., Harvey, L. D., Jr. and Roberts, J. D. (1984). Identification confusions among letters of the alphabet. *J. Exp. Psychol.: Hum. Percept. Perform.* **10**, 655–666.

Goodrich, G., Mehr, E. B., Quillman, R. D., Shaw, H. K. and Wiley, J. K. (1977). Training and practice effects in performance with low-vision aids: A preliminary study. *Am. J. Optom. Physiol. Opt.* **54**, 312–318.

Harvey, L. O., Rentschler, I. and Weiss, C. (1985). Sensitivity to phase distortions in central and peripheral vision. *Percept. Psychophys.* **38**, 392–396.

Jacobs, R. J. (1979). Visual resolution and contour interaction in the fovea and periphery. *Vision Res.* **19**, 1187–1195.

Knoblauch, K. (1995). Dual bases in dichromatic color space. In: *Colour Vision Deficiencies XII*. B. Drum (Ed.). Kluwer, Dordrecht, pp. 165–176.

Legge, G. E., Pelli, D. G., Rubin, G. S. and Schleske, M. M. (1985a). Psychophysics of reading I. Normal vision. *Vision Res.* **25**, 239–252.

Legge, G. E., Rubin, G. S., Pelli, D. G. and Schleske, M. M. (1985b). Psychophysics of reading II. Low Vision. *Vision Res.* **25**, 253–266.

Levi, D. M., Klein, S. A. and Yap, Y. L. (1987). Positional uncertainty in peripheral and amblyopic vision. *Vision Res.* **27**, 581–591.

Loomis, J. M. (1978). Lateral masking in foveal and eccentric vision. *Vision Res.* **18**, 335–338.

McConkie, G. W. and Rayner, K. (1975). The span of the effective stimulus during a fixation in reading. *Percept. Psychophys.* **17**, 578–586.

Morrone, M. C., Burr, D. C. and Spinelli, D. (1989). Discrimination of spatial phase in central and peripheral vision. *Vision Res.* **29**, 433–445.

O'Regan, J. K. (1990). Eye movements and reading. In: *Eye Movements and their Role is Visual and Cognitive Processes*. E. Kowler (Ed.). Elsevier, Amsterdam, pp. 395–453.

Pelli, D. G. (1981). The effect of uncertainty: detecting a signal at one of ten-thousand possible times and places. *Invest. Ophthalmol. Visual Sci.* **20** (Suppl.), 178.

Rayner, K. and McConkie, G. W. (1975). What guides a reader's eye movements. *Vision Res.* **15**, 829–837.

Rentschler, I. and Treutwein, B. (1985). Loss of spatial phase relationship in extrafoveal vision. *Nature* **313**, 308–310.

Rubin, G. and Turano, K. (1994). Low vision reading with sequential word presentation. *Vision Res.* **34**, 1723–1733.

Saarinen, J., Rovamo, J. and Virsu, V. (1987). Texture discrimination at different eccentricities. *J. Opt. Soc. Am.* **A4**, 1699–1703.

Saarinen, J., Rovamo, J. and Virsu, V. (1989). Analysis of spatial structure in eccentric vision. *Invest. Ophthalmol. Visual Sci.* **30**, 293–296.

Stuart, J. A. and Burian, H. M. (1962). A study of separation difficulty. *Am. J. Ophthalmol.* **53**, 471–477.

Thomas, J. P. (1987). Effect of eccentricity on the relationship between detection and identification. *J. Opt. Soc. Am.* **A4**, 1599–1605.

Turano, K. and Rubin, G. (1988). Reading performance with peripheral viewing using rapid serial visual presentation. In: *Noninvasive Assessment of the Visual System, Technical Digest Series*, Vol. 3. Optical Society of America, Washington, DC, pp. 192–195.

Virsu, V. and Rovamo, J. (1979). Visual resolution, contrast sensitivity, and the cortical magnification factor. *Exp. Brain Res.* **37**, 475–494.

Vitu, F. (1991). The influence of parafoveal processing and linguistic context on the optimal landing position effect. *Percept. Psychophys.* **50**, 58–75.

Watson, A. B. (1979). Probability summation over time. *Vision Res.* **19**, 515–522.

Watson, G. R. and Berg, R. V. (1983). Near training techniques. In: *Understanding Low Vision*. R. Jose (Ed.). American Foundation for the Blind, New York, pp. 317–362.

Watt, R. J. and Hess, R. F. (1987). Spatial information and uncertainty in anisometropic amblyopia. *Vision Res.* **27**, 661–674.

Westheimer, G. (1982). The spatial grain of the perifoveal visual field. *Vision Res.* **22**, 157–162.

Westheimer, G. (1987). Visual acuity and hyperacuity: resolution, localization, form. *Am. J. Optom. Physiol. Opt.* **64**, 567–574.

Whittaker, S., Budd, J. and Cummings, R. (1988). Eccentric fixation with macular scotoma. *Invest. Ophthalmol. Visual Sci.* **29**, 268–278.

Whittaker, S., Rohrkaste, F. and Higgins, K. (1989). Optimum letter spacing for word recognition in central and eccentric fields. In: *Noninvasive Assessment of the Visual System, Technical Digest Series*, Vol. 7. Optical Society of America, Washington, DC, pp. 56–59.

Human Symmetry Perception, pp. 85–109
C. W. Tyler (Ed.)

Evidence for the use of scene-based frames of reference in two-dimensional shape recognition

PHILIP T. QUINLAN
Department of Psychology, University of York, Heslington, York YO1 5DD, UK

Abstract—A series of reaction time (RT) experiments are reported in which subjects had to match pairs of two-dimensional shapes. The shapes were either symmetrical or asymmetrical and were arranged into either symmetrical or asymmetrical displays. In the initial experiments the shapes straddled the vertical midline of the display. It was found that Same RTs to symmetrical displays were shorter than Same RTs to asymmetrical displays, and that Same RTs to symmetrical shapes were shorter than Same RTs to asymmetrical shapes. Moreover, display symmetry and shape symmetry produced an additive effect on responses. This pattern of results obtained when both shapes were presented simultaneously and when a delay was introduced between the presentation of one shape and the presentation of the other. However, display and shape symmetry interacted when the displays were rotated 90 deg within the plane so that the shapes straddled the horizontal meridian. The implications of these results are discussed in terms of both a transformational account of shape recognition and an account concerning the imposition and use of scene-based frames of reference.

1. INTRODUCTION

There is a strong tradition in perceptual psychology of studying the recognition of two-dimensional (2-D) shapes (see Quinlan, 1991 for a review). Recent work in this area has investigated the imposition and use of perceptual frames of reference for the purposes of shape recognition (Quinlan and Humphreys, 1993). From the available evidence it seems that objects appear as they do because the visual system recovers a structural description that specifies both salient parts and relations inherent to those objects (Quinlan, 1991). Such a description, in coding the spatial disposition of salient parts, takes into account some form of frame of reference because the spatial relations must be coded relative to the origin of a corresponding co-ordinate system: The frame of reference fixes the origin of this co-ordinate system.

Although many sorts of perceptual reference frames have been discussed (Hinton and Parsons, 1988), only three will be considered here, namely, a retinal-based frame, an object-based frame and a scene-based frame. The description of an object relative to each of these frames is quite different. Whereas the retinal description of an object specifies the spatial disposition of the parts of the object in retinal co-ordinates, the object-based description specifies the spatial disposition of the parts of the object in terms of a frame of reference centred on itself. In turn the disposition of the object in the scene can be described relative to a scene-based frame of reference.

Hinton (1981) developed a working model of single letter and shape recognition based on the distinctions between these three sorts of object representations. Figure 1 shows the components of the model in a schematic form. At an early stage in its development (i.e. boxes (i)–(iii) in the figure), the model was an attempt to show how the perceptual system takes a viewer-centred and retinotopic representation of a shape and then derives a viewpoint-independent representation. In developing the account, Hinton addressed the problem of how the perceptual system deals with shapes in the context of more global scene information. He therefore added two more sorts of processing units to the model, namely, higher-order mapping units and scene-based units (boxes (iv) and (v) in the figure). In the model the scene-based units encode the relationships between the parts of the scene in a coordinate system abstracted away from the scene itself. For instance, a scene-based description of a room remains constant regardless of whether the observer remains stationary or moves around (Hinton and Parsons, 1988). As a consequence the viewer may retain a record of the relative dispositions of the objects in the room independently of their own position. The scene-based units establish a scene-independent description of parts of the scene analogous to the way in which the object-based units establish a viewpoint-independent description of the parts of an object. According to the model such a scene-based representation is derived by imposing a scene-based frame of reference. The derivation of a scene-based frame is essentially due to the operation of the higher-order mapping units. They establish which of the many candidate frames are used to code the spatial disposition of the objects in the scene.

Consideration of this model and the general ideas about imposing and using scene-based frames of reference provided the initial motivation for the present study. The experimental methodology was adapted from other studies in which the subject's task was to decide whether or not two simultaneously presented letters were physically identical (see Krueger, 1978, for a review). It seemed that such a task might be adapted to provide insights into possible inter-relationships between scene and item information during the normal course of shape recognition.

Historically, Fox (1975) was the first to use the letter matching task in order to assess the effects of item and display symmetry on matching performance. In his initial experiment he used two kinds of Same displays: those in which the letters were vertically symmetrical (e.g. TT) and those in which the letters were not symmetrical about a vertical axis (e.g. QQ). Particular interest was with the displays comprising vertically symmetrical letters, because these displays were globally symmetrical. In these cases, the symmetry of the whole display was perfectly correlated with the Same response. Fox hypothesised that subjects might therefore use display symmetry as a 'diagnostic' for responding Same and responses to the vertically symmetrical displays should have been faster than those to the vertically asymmetrical displays. Unfortunately, the results of his first experiment did not provide compelling support for this hypothesis because the RTs to the vertically symmetrical displays were no

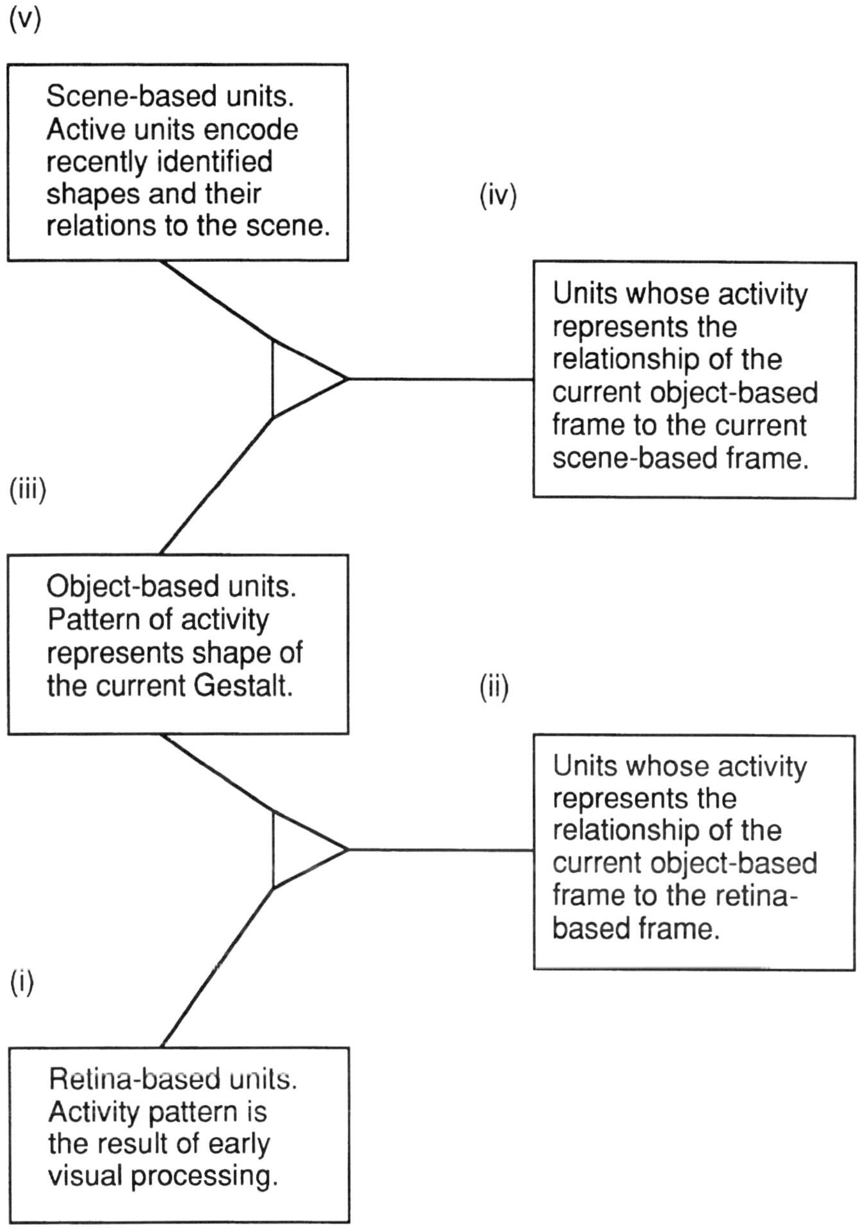

Figure 1. Schematic representation of Hinton's (1981) connectionist model of shape recognition. Each rectangle stands for a different set of processing units. (Copyright © 1981 by Morgan Kaufman. Adapted with permission. Copies of the Proceedings of the Third Cognitive Science Conference are available from Morgan Kaufman Publishers, 2929 Campus Drive, San Mateo, California 94403.)

faster than those to the vertically asymmetrical displays. Nevertheless, there was a provocative pattern of errors, with fewer errors made to vertically symmetrical Same displays (e.g. TT) than to Different displays, where two different vertically symmetrical letters were present (e.g. TA).

Although, the effects of display symmetry on letter matching performance are not therefore as clear-cut as they might be, this could be because letter familiarity overrides any facilitatory effects due to display symmetry. That is, subjects' Same responses may be based on the familiarity of the items in the displays irrespective of organisational factors. Indeed the strongest display effects have arisen when the task is to match relatively unfamiliar items. For example, Wolff (1971) obtained a substantial facilitatory effect on matching pairs of horse-shoe figures when they were arranged into vertically symmetrical displays as compared to when they were arranged into asymmetrical displays. Also Richards (1978, Experiment 4) found a significant 70 ms advantage in RTs to symmetrical displays, relative to asymmetrical counterparts, when the displays comprised nonsense dot patterns.

Previous evidence with relatively unfamiliar stimuli suggests therefore that the processing of item and display symmetries might fruitfully be examined using unfamiliar 2-D shapes. If the shapes are relatively unfamiliar, then strong effects of display symmetry should obtain when subjects have to match shapes instead of letters. Also, apart from testing the generality of already-established effects, other manipulations were introduced to allow further examination of possible interactions between item and display symmetries. For example, a condition was introduced in which Same displays comprised two instances of a symmetrical shape organised into an asymmetrical display. This type of configuration has never before been examined in the item matching task.

In summary the tradition of using Same-Different matching with pairs of simultaneously presented items has provided many insights into the perceptual inter-relationships between item and display information. Thus it seemed that this kind of matching task provided an ideal tool to examine the more specific ideas about the imposition and use of shape-based and scene-based frames of reference.

2. EXPERIMENT 1. SIMULTANEOUS PRESENTATION

In the following experiments only two shapes were used, the combination of which into different spatial arrangements generates the various displays. The complete set of displays used in the experiment is shown in Fig. 2. The symmetrical shape possesses a major axis that is an axis of both symmetry and elongation, whereas the asymmetrical shape possesses an axis of elongation alone. Here the axis of elongation was defined as the longest possible line to be drawn through the shape. Previous work with these shapes (Quinlan and Humphreys, 1993) has been taken to suggest that, whereas the symmetrical shape appears to be described relative to an axis-based frame of reference centred on its axis of symmetry, the designated axis of elongation of the asymmetrical shape does not fulfil the role of a frame of reference for the description of that shape. The symmetrical shape therefore provides an example of a shape that possesses an unambiguous principal axis which is perceptually important. In contrast, the asymmetrical shape possesses no such axis. Given that the perceptual processing of these particular shapes has been studied extensively, the idea was that a better understanding of any possible configural effects would arise if the displays contained these shapes.

In the present series of studies, the first experiment was carried out to see if a display symmetry effect would occur in a simultaneous Same-Different item matching task. Independently of the possible display effects, it was also of interest to see if any item effects obtained. Previous experiments revealed that subjects were faster in matching successive versions of the symmetrical shape than successive versions of the asymmetrical shape (see Quinlan and Humphreys, 1993).

2.1. Method

2.1.1. Subjects. 12 subjects were selected from the Birkbeck College Subject Panel and all reported having normal or corrected-to-normal vision.

2.1.2. Design. The experimental task was Same-Different matching of pairs of simultaneously presented shapes. Half of the displays comprised two instances of the symmetrical shape and were locally symmetric (Loc Symm), while half comprised two instances of the asymmetrical shape and were locally asymmetric (Loc Asymm). In the Same displays half the time the two shapes were in the same orientation; i.e. they both 'pointed' downwards (see Fig. 2(a) and (b)) or they both 'pointed' upwards (see Fig. 2(c) and (d)) so as to be globally symmetric about the vertical midline (Glob Symm). In the remaining displays the shapes were in different orientations; i.e. one shape 'pointed' upwards and the other 'pointed' downwards (see Fig. 2(e)–(h)) so as to be globally asymmetric (Glob Asymm). Across the eight Same displays, global and local symmetries were combined orthogonally to give four basic types of displays: (i) the Glob Symm–Loc Symm displays contained two instances of the symmetrical shape in the same orientation, (ii) the Glob Symm–Loc Asymm displays contained two instances of the asymmetrical shape in the same orientation, (iii) the Glob Asymm–Loc Symm displays contained two instances of the symmetrical shape, but they 'pointed' in different directions, and, (iv) the Glob Asymm–Loc Asymm displays contained two instances of the asymmetrical shape, which 'pointed' in different directions.

With the Different displays, because each includes one instance of each shape, all were globally asymmetrical. However, in four of the Different displays the shapes 'pointed' in the same direction (Same orientation displays) and in the remaining half the shapes 'pointed' in different directions (Different orientation displays).

In the experiment, the computer generated 12 random multiples of the 16 basic displays to give a total of 192 experimental trials. Half the trials comprised Same displays and half comprised Different displays. Overall there were 24 trials for each of the four types of Same displays labelled above and 48 trials for the two general types of Different trials. Half the subjects responded Same with their preferred hand and half responded Different with their preferred hand.

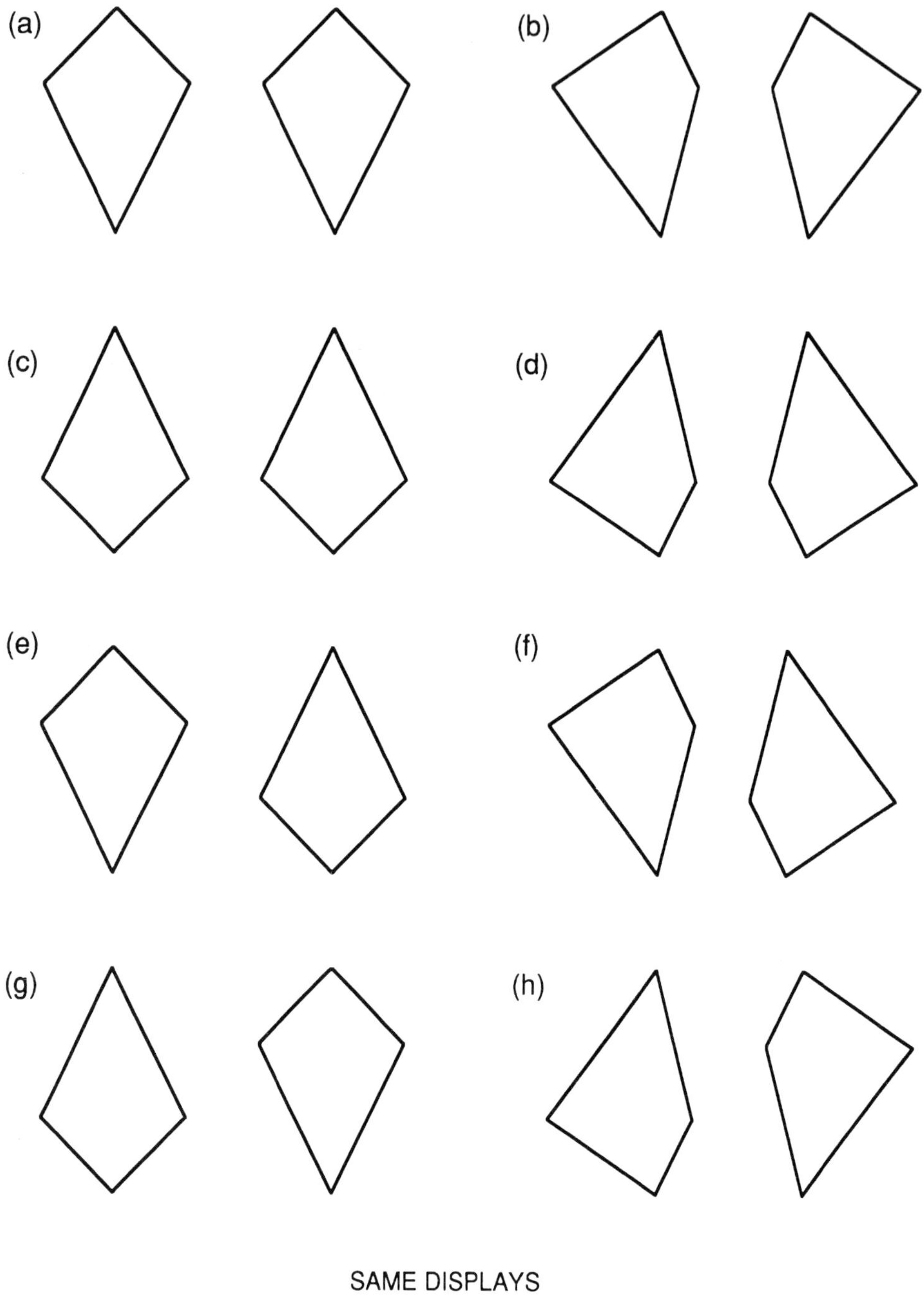

Figure 2. The complete set of displays used in Experiment 1. (a)–(h) are Same displays: (a) and (c), Glob Symm–Loc Symm; (b) and (d), Glob Symm–Loc Asymm; (e) and (g), Glob Asymm–Loc Symm; (f) and (h), Glob Asymm–Loc Asymm.

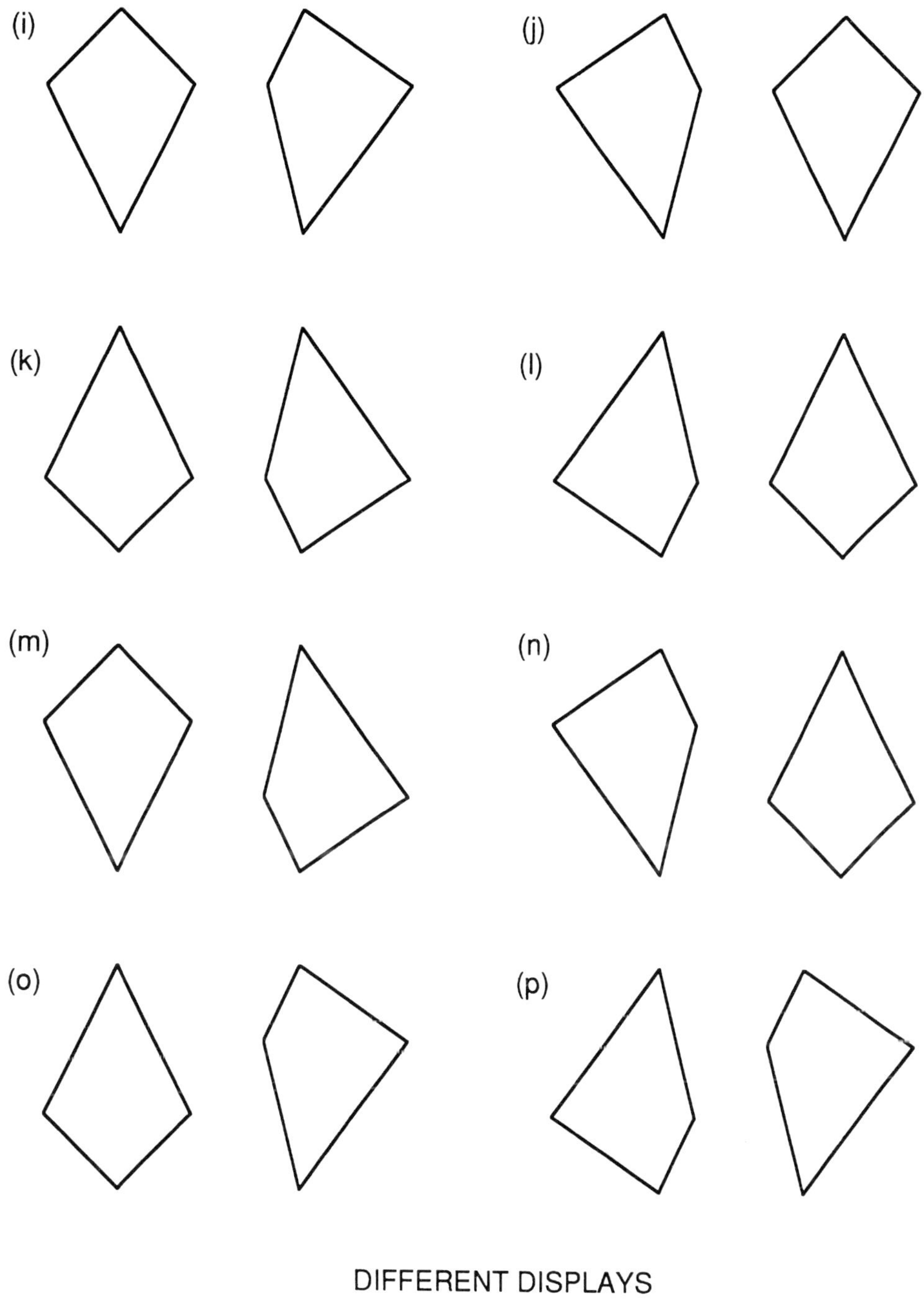

Figure 2. (Continued). (i)–(p) are Different displays: (i)–(l) 'point' in the same direction; (m)–(p) 'point' in different directions.

2.1.3. Apparatus and procedure. The shapes were plotted on an Advance Instrument oscilloscope (05 250) possessing a P-31 rapid decay phosphor. Removing a stimulus led to a drop to 1% of maximum brightness in 0.25 ms. All stimulus presentations, timing and data collection were under control of a PDP-11/10 computer. The oscilloscope and a pair of response keys were interfaced to the computer and were situated in a quiet room separate from the computer.

Subjects were run individually in a single session lasting approximately 40 min. They were shown examples of the shapes and the displays and were told that on each trial a pair of shapes would appear for a brief interval (i.e. 200 ms). When a shape was plotted on the oscilloscope it measured 3×2 deg of visual angle. The shapes were separated by 1 deg measured between their nearest corners. Subjects were told to judge if the two shapes were the same or not irrespective of overall orientation. They were also told to respond as quickly and as accurately as possible given the time constraint of a 2 s response deadline.

The sequence of events on a trial was as follows. Initially, a fixation point was plotted for 200 ms. At the offset of the cross a pair of shapes appeared and response timing was initiated. The shapes were presented simultaneously for 200 ms and feedback contingent on a button press, or elapsed time until the deadline, was presented. The presentation of the fixation point signalled the start of the next trial after an ITI of 600 ms.

Subjects were given an initial block of 32 practice trials comprising two random orders of the basic 16 displays, following which they ran through two blocks of 96 experimental trials.

2.2. Results and discussion

The same methods of analyses were adopted throughout. In the analyses of the RTs, means per condition per subject were analysed with appropriate parametric tests. Only correct RTs were considered; errors and RTs 3 standard deviations (SDs) away from each subject's mean per condition were discarded from these analyses so that the analysed data were uncontaminated by either anticipatory 'fast' responses or extremely long responses. The corresponding error data were analysed in an analogous fashion to the RTs. Percentage errors per condition per subject were computed and were arcsin-transformed prior to being analysed.

Unless otherwise stated, the same pattern of results arose in the analyses of RTs and errors. There were no signs of any speed-error trade-offs in any of the experiments. Discussion centres on the analyses of the data on Same RTs. Reference to performance on Different trials is limited to picking out cases where a statistically reliable effect obtained. In general, there were no effects on Different trials in either the RT or error analyses. Summaries of the statistical results are included in Table 1 and mean performance on Same trials is depicted in Fig. 3.

A 2×2 repeated measures ANOVA was carried out on the RT data. Global symmetry (Glob Symm vs Glob Asymm) and local symmetry (Loc Symm vs Loc Asymm) acted as fixed factors. Subjects acted as a random factor. The analysis revealed statistically significant main effects of global symmetry and local symmetry, but the global $\times$ local

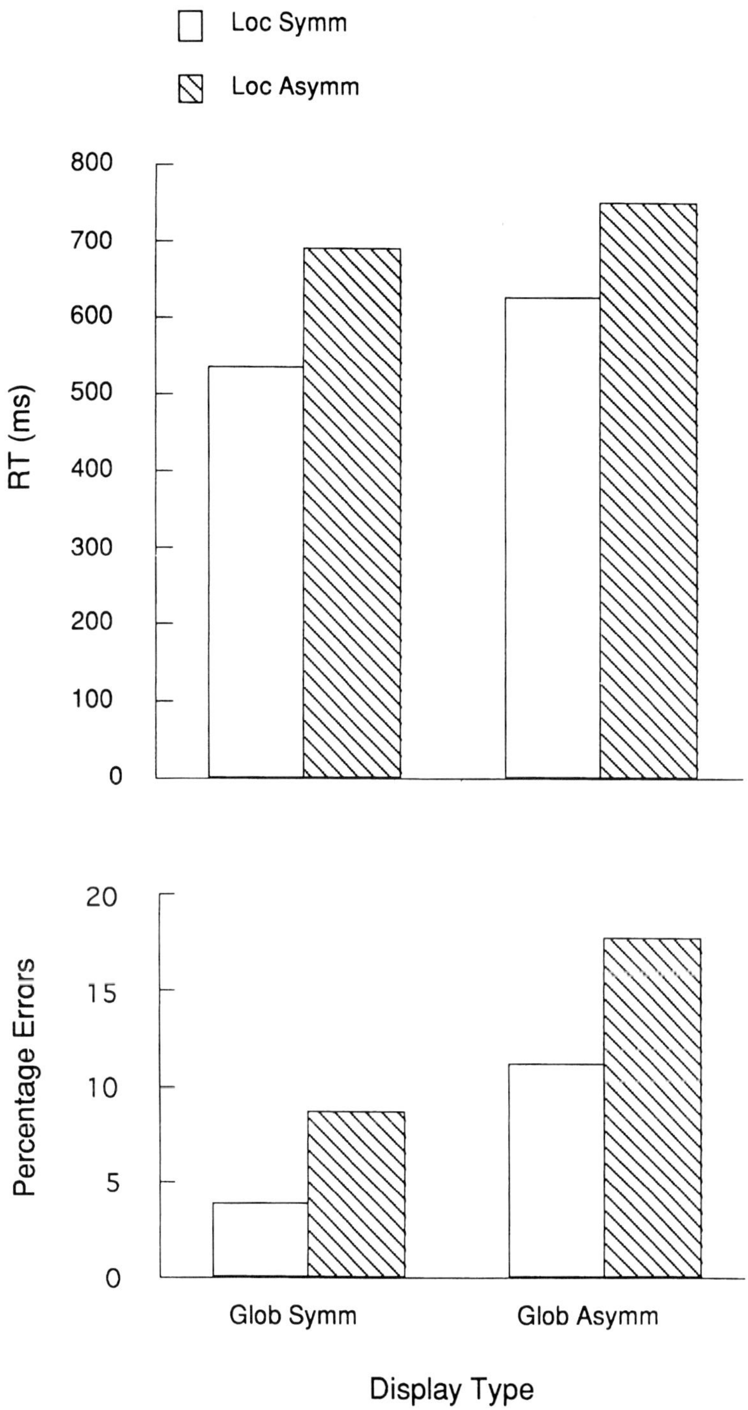

Figure 3. Mean RTs and mean percentage errors on Same trials in Experiment 1.

Table 1.
Summary of the statistical analyses for Experiment 1

Same responses			
	F	df	Significance
RTs			
Main effect of global symmetry	13.38	1,11	$p < 0.01$
Main effect of local symmetry	17.11	1,11	$p < 0.01$
Global × local symmetry interaction	0.49	1,11	n.s.
Errors			
Main effect of global symmetry	11.09	1,11	$p < 0.01$
Main effect of local symmetry	5.68	1,11	$p < 0.01$
Global × local symmetry interaction	0.52	1,11	n.s.
Different responses			
	t	df	Significance
RTs			
Orientation of shapes	−0.03	11	n.s.
Errors			
Orientation of shapes	1.48	11	n.s.

n.s. indicates that the effect did not reach statistical significance.

symmetry interaction was not reliable. The main effects are clear in showing that RTs to Glob Symm displays were faster than to Glob Asymm displays and that RTs to Loc Symm displays were faster than RTs to Loc Asymm displays. Overall therefore the RT data for Same trials reveal strong facilitatory effects to both Glob Symm displays (a 78 ms advantage) and to Loc Symm displays (a 140 ms advantage) as compared to their corresponding asymmetrical counterparts. For Different trials the data were analysed according to whether the two shapes 'pointed' in the same direction or not (see Fig. 2). Neither the difference in RTs or errors reached statistical significance (see Table 1).

For Same trials, the display symmetry effect replicates the findings of earlier studies with non-letter stimuli where subjects responded Same faster to symmetrical displays than to asymmetrical displays. In addition, subjects responded faster to displays comprising symmetrical shapes than asymmetrical shapes. However, it is interesting to look at these effects in more detail. For instance, it is not the case that all symmetrical displays were responded to faster than all asymmetrical displays. RTs to the Glob Asymm–Loc Symm displays and RTs to the Glob Symm–Loc Asymm displays were equivalent ($p > 0.05$, Newman–Keuls test). This result appears to be damaging to the diagnostic account of shape matching (Fox, 1975) which maintains that Same responses to symmetrical displays will invariably be faster than to asymmetrical displays. By this account, Same responses to symmetrical displays will be executed on the basis of a successful and preliminary 'fast test' for global symmetry. The present results disconfirm this prediction of the diagnostic account.

Of perhaps more importance is the finding of an additive effect of display symmetry and item symmetry. The size in the difference between the RTs to Loc Symm and Loc Asymm for the Glob Symm displays was 156 ms and the corresponding difference for Glob Asymm displays was 125 ms. By a straightforward application of additive factors assumptions (Sternberg, 1975), the additive result may be interpreted in terms of the independence of two sorts of processing: one being concerned with item information and the other with display information. The following experiments examine this simple account of performance in more detail.

3. EXPERIMENT 2. SUCCESSIVE PRESENTATION

The display symmetry effect obtained in Experiment 1 replicates the findings of Richards and Wolff in which the stimuli were relatively unfamiliar items. An aim now was to further examine this effect by introducing a delay between the presentation of one item of a pair and the other (i.e. the introduction of a stimulus onset asynchrony — SOA). Richards (1978) used this manipulation to test the hypothesis that response facilitation with the symmetrical displays was critically dependent on the simultaneous presentation of both items. According to Richards, simultaneous presentation was special because it made explicit the symmetric inter-item relationship in the display. By this view, the introduction of an SOA between the presentation of the two items should abolish the RT advantage to the symmetrical displays. In his experiment, Richards increased the SOA across separate blocks of trials to see if the size of the display-symmetry effect would vary indirectly with the length of the SOA. He found that the size of the RT facilitation to the symmetrical displays as compared to the asymmetrical displays did diminish as a function of increasing SOA and that the display symmetry effect had been abolished with an SOA of 200 ms.

Following from this result, an initial experiment (Experiment 2a) was run with one group of subjects and the SOA was set at 400 ms (an inter-stimulus interval, ISI, of 200 ms). This value seemed appropriate given the disruption to the display symmetry effect witnessed in Richards' results with a similar interval between the presentation of the two items. As it turned out, the present display symmetry effect was robust over this interval. Therefore a second experiment (Experiment 2b) was run with a different group of subjects and the ISI was set at 500 ms.

3.1. Method

Both Experiments 2a and b stand in contrast to Experiment 1 through using successive rather than simultaneous presentation of the shapes to be matched. In both experiments, on each trial each shape was presented for 200 ms. However, in Experiment 2a (the Short ISI experiment) the ISI was 200 ms and in Experiment 2b (the Long ISI experiment) it was 500 ms. Apart from this procedural difference across the two experiments the apparatus, procedure and design were identical to those used in Experiment 1.

3.1.1. Subjects. Across the experiments two groups of 12 subjects from the Birkbeck College Subject Panel were used. All reported having normal or corrected-to-normal vision.

3.1.2. Design. The same design used in Experiment 1 was used here. In this case, though, the term 'display' refers to the successive presentation of a pair of shapes. The displays used map onto those shown in Fig. 2. An added factor taken into account in balancing the different instances of the shapes across the displays was the order in which they were presented. Balancing was achieved by ensuring that on half the trials the left shape was presented first, whilst on the remaining trials the right shape was presented first. Overall there was a total of 192 experimental trials and 32 practice trials.

3.1.3. Apparatus and procedure. Following the offset of the central fixation point, one shape (henceforth, S1) was presented for 200 ms. There was then an appropriate ISI, followed by the second shape (henceforth, S2) for 200 ms. In all other respects the events on a trial were the same as before.

3.2. Results and discussion

Summaries of the statistical results are included in Tables 2 and 3 and mean performance on Same trials is depicted in Fig. 4.

The same pattern of results arose in both experiments. In the RT data there were statistically significant main effects of global symmetry and local symmetry, but the global $\times$ local symmetry interaction was not statistically reliable. A similar pattern of results obtained with the analysis of the errors, although the main effect of global symmetry just failed to reach statistical significance for both experiments.

Overall, and as in Experiment 1, subjects were faster to respond Same to symmetrical displays than to asymmetrical displays; they were also overall faster to respond Same to symmetrical shape displays than to asymmetrical shape displays.

To enhance comparisons across the three experiments, a mixed design, 3-way ANOVA was carried out on the complete set of RT data from Same trials. Experiments (1, 2a, and 2b) acted as a between-groups factor, global symmetry (Glob Symm vs Glob Asymm) and local symmetry (Loc Symm vs Loc Asymm) acted as fixed factors and subjects were entered as a random factor. The results of the analysis revealed statistically significant main effects of experiment, global symmetry, and local symmetry (see Table 4). None of the interactions reached statistical significance. The main effect of the experiment revealed that RTs were overall slower with simultaneous presentation (in Experiment 1) than with presentations involving a long (500 ms) ISI. Although this main effect failed to reach significance in the analysis of the errors, the pattern of significance was otherwise the same as the analysis of the RTs.

It is interesting to note that the overall difference in speed of responding with simultaneous and successive displays is reminiscent of a finding attributed to Podgorny (1975; reported by Shepard and Cooper, 1982). Although Podgorny examined Same-Different performance in a 3-D block figure comparison task, he argued that,

Table 2.
Summary of the statistical analyses for Experiment 2a

Same responses			
	F	df	Significance
RTs			
Main effect of global symmetry	13.42	1,11	$p < 0.01$
Main effect of local symmetry	19.91	1,11	$p < 0.01$
Global × local symmetry interaction	0.58	1,11	n.s.
Errors			
Main effect of global symmetry	4.07	1,11	n.s.
Main effect of local symmetry	5.38	1,11	$p < 0.05$
Global × local symmetry interaction	0.00	1,11	n.s.
Different responses			
	t	df	Significance
RTs			
Orientation of shapes	1.89	11	n.s.
Errors			
Orientation of shapes	−0.52	11	n.s.

Table 3.
Summary of the statistical analyses for Experiment 2b

Same responses			
	F	df	Significance
RTs			
Main effect of global symmetry	12.53	1,11	$p < 0.01$
Main effect of local symmetry	13.89	1,11	$p < 0.01$
Global × local symmetry interaction	2.53	1,11	n.s.
Errors			
Main effect of global symmetry	3.49	1,11	n.s.
Main effect of local symmetry	6.99	1,11	$p < 0.05$
Global × local symmetry interaction	2.81	1,11	n.s.
Different responses			
	t	df	Significance
RTs			
Orientation of shapes	0.42	11	n.s.
Errors			
Orientation of shapes	0.58	11	n.s.

Table 4.
Summary of the statistical analyses of the cross experiment comparisons between Experiments 1 and 2

Same responses			
	F	df	Significance
RTs			
Main effect of experiment	3.35	2,33	$p < 0.05$
Main effect of global symmetry	36.48	1,33	$p < 0.05$
Main effect of local symmetry	49.29	1,33	$p < 0.01$
Experiment × global symmetry interaction	1.52	2,33	n.s.
Experiment × local symmetry interaction	0.45	2,33	n.s.
Global × local symmetry interaction	0.08	1,33	n.s.
Experiment × global symmetry × local symmetry interaction	1.14	2,33	n.s.
Errors			
Main effect of experiment	1.72	2,33	n.s.
Main effect of global symmetry	18.32	1,33	$p < 0.01$
Main effect of local symmetry	18.00	1,33	$p < 0.01$
Experiment × global symmetry interaction	1.57	2,33	n.s.
Experiment × local symmetry interaction	0.23	2,33	n.s.
Global × local symmetry interaction	1.99	1,33	n.s.
Experiment × global symmetry × local symmetry interaction	0.72	2,33	n.s.

with simultaneous presentation of the figure pairs, the intercept value of the function relating RT to angular difference between the two block figures reflected the time to encode two items. In contrast, the intercept value for successive presentation reflected the time to encode a single item (namely, the second item). A similar account seems to provide a reasonable explanation of the overall RT differences observed here between simultaneous and successive forms of presentation.

The strong display symmetry effect found with both simultaneous and successive displays provides a striking contrast with the data reported by Richards (1978). His display symmetry effect was abolished when he introduced a 200 ms SOA between the presentation of the letters in his displays. Consequently, he argued that the display symmetry effect was critically dependent on the presence and registration of the inter-item spatial relationships. In contrast, the present results show that the display symmetry effect is robust over relatively long intervals between the two items. The present robust effects suggest that the same processes are operating in the simultaneous and successive presentation conditions.

One possible reason for this difference may again be linked to the relative familiarity of the two sets of stimuli: Richards used letters, whereas relatively unfamiliar 2-D shapes were used here. It is possible that when highly familiar stimuli are used, subjects are more likely to describe the items as being separate objects rather than as parts of a composite display. By this account, high familiarity dictates that the items be treated as single objects. In Richards' study, the introduction of an SOA

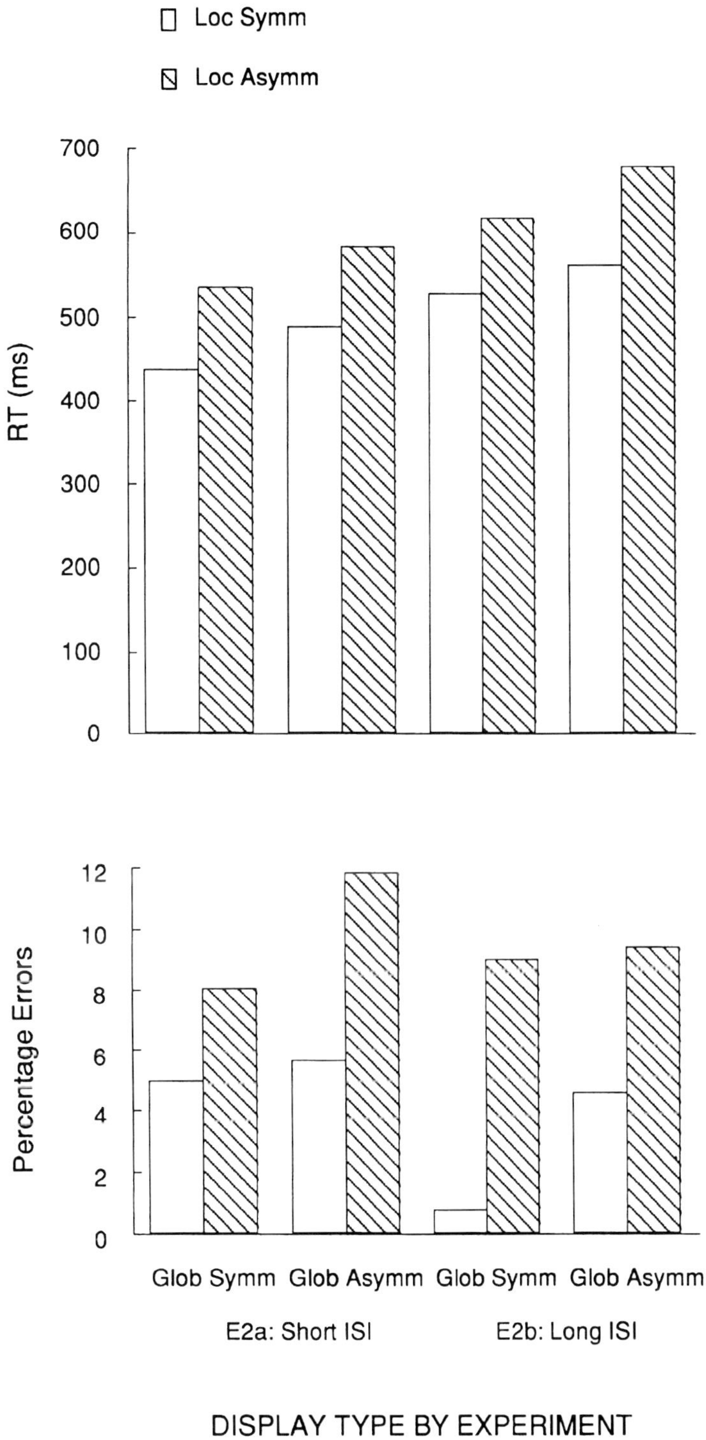

Figure 4. Mean RTs and mean percentage errors on Same trials in Experiments 2a and b.

minimized any processing benefits that may accrue because of the diagnosticity of display symmetry. Even though the letters were symmetrical about the point of fixation, subjects may have treated them as being distinct objects and not as members of a composite display. In this respect, item familiarity and organizational factors may produce opposing effects.

Before speculating further about the nature of the processes involved in the matching task, a third study will be described that is another attempt to manipulate the display symmetry effect. Experiment 3 examines whether the display symmetry effect is robust when symmetry detection is slowed by rotating the whole display 90 deg within-the-plane. This experiment provides a test of an alternative account of shape matching performance advocated by Shepard (1984). By Shepard's account, performance in the matching task is intimately tied up with the nature of the rigid transformations which separate the two instances of the shapes in 3-D space (see also Kahn and Foster, 1986, for an alternative transformational account). On this view, subjects perform the task by attempting to 'superimpose' the two instances via the application of rigid transformations, the ultimate aim being to affect something akin to a template match. Such an account can be applied to the present data.

First consider the Glob Symm–Loc Symm displays. Two separate transformations will bring the shapes into correspondence: (i) a translation along the horizontal meridian or, (ii) a 180 deg through-the-plane reflection effected about the vertical meridian. With the Glob Symm–Loc Asymm displays, however, only the reflection is the appropriate transformation. The translation does not bring the two instances into correspondence. With the Glob Asymm–Loc Symm and the Glob Asymm–Loc Asymm displays, there are a number of different transformations that will bring the two items into correspondence. Importantly, though, only a particular pattern of transformations with the various displays can account for the present results, if additional assumptions concerning the different amounts of time needed to affect different transformations are provided. The account is as follows. Subjects match the Glob Symm–Loc Symm displays via a translation and they match the Glob Symm–Loc Asymm displays via a 180 deg reflection with respect to the vertical meridian. According to the data, the difference in the speed of responding to the two symmetrical displays mirrors the difference in the time to enact a translation (t) as compared to a reflection (r) (i.e. $r - t$ ms).

Different transformations apply with the two asymmetrical displays. Here the assumption is that subjects match the Glob Asymm–Loc Symm displays via an inversion effected about the centre of the shape followed by a translation. They match the Glob Asymm–Loc Asymm displays via a similar inversion followed by a 180 deg reflection about the centre of the display. Consequently, the difference in the speed of responding to the two types of asymmetrical displays again reduces to $r - t$ ms. The overall difference in speed of responding between the Glob Symm and Glob Asymm displays is due to the additional time needed to effect the inversion with the latter displays. The elegance of this account is that it offers a straightforward explanation of the additive results found here between display and item symmetry.

The transformational account also predicts that an RT difference should arise between the Same and Different Orientation conditions on Different trials (cf. Shepard and Cooper, 1982; Chapter 3). Such a difference would reflect the time taken to apply the necessary transformations to bring the shapes into partial alignment. The Different orientation displays should take longer to process than Same orientation displays because of the need to enact the extra operation of an inversion with the Different orientation displays. Unfortunately, though, this prediction has no support in the present data as no orientation effects have so far been observed.

In summary, Experiment 3 provides a test of the transformational account in that it predicts that performance should be the same as in Experiment 1. By this account there is no reason to suppose that performance should be any different if the shapes are vertically as opposed to horizontally aligned because essentially the same set of rigid transformations apply in the two cases. Indeed, the transformation account predicts that the effects of global and local symmetry should continue to be additive.

4. EXPERIMENT 3. EFFECTS OF ROTATING THE WHOLE DISPLAY

4.1. Method

The design, apparatus procedure and the sequence of events on each trial were identical to those used in Experiment 1. In this case, the displays used were rotated within-the-plane by 90 deg relative to those used previously. Examples of Same displays are shown in Fig. 5.

In pilot work for this experiment, it became obvious that some subjects found the task to be impossible. This stemmed from the fact that as in Experiment 1, the display time of the shapes was kept at 200 ms (in order that direct comparisons could be made between the results of the two experiments). So although 11 subjects were run in total, only the data from eight of them are reported. A deliberate attempt was made to re-contact the people who had been tested in the first experiment and to run them again in this experiment. Only eight such individuals were available and all eight were able to perform the task competently.

4.2. Results and discussion

Summaries of the statistical results are included in Table 5 and mean performance on Same trials is depicted in Fig. 6.

Initially, the eight subjects' data from Experiment 1 (the Vertical presentation experiment) were re-analysed so as to ensure that their pattern of performance fitted that of the whole group. This, in the main, turned out to be true: the pattern of significance for their RTs was identical to that previously reported (Table 1). The analysis of their errors revealed a very similar pattern of significance although the main effect of local symmetry only approached statistical significance ($F(1, 7) = 4.15$, $p = 0.08$). Overall therefore the sample of eight subjects appears to be representative of the original complete sample.

Next, the data from the Horizontal presentation experiment (Experiment 3) were analysed. In the analysis of the RT data, the main effect of global symmetry was

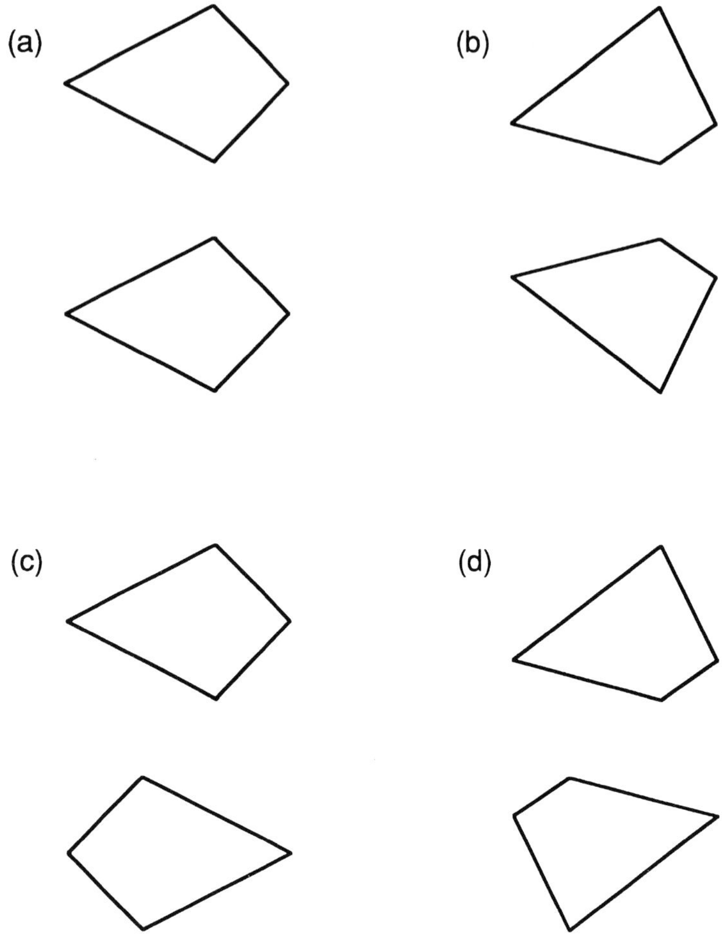

Figure 5. A sample of the Same displays used in Experiment 3. (a) Glob Symm–Loc Symm, (b) Glob Symm–Loc Asymm, (c) Glob Asymm–Loc Symm, (d) Glob Asymm–Loc Asymm.

not statistically significant. However, both the main effect of local symmetry and the global $\times$ local symmetry interaction were statistically reliable. The main effect of local symmetry showed that subjects were generally faster in responding to the symmetrical shape displays than they were in responding to the asymmetrical shape displays. However, further analysis of the significant interaction showed that, whereas there was a display-symmetry effect in the responses to symmetrical shape displays, there was no such effect in the responses to the asymmetrical shape displays (a similar interaction was witnessed in the analysis of the errors).

In order to make direct comparisons between the Horizontal and the Vertical presentation experiments, the complete set of data was entered into a 3-way, $2 \times 2 \times 2$ repeated-measures ANOVA. Experiment (Vertical presentation vs Horizontal presentation), global symmetry (symmetrical vs asymmetrical) and local symmetry (symmetrical vs asymmetrical) acted as fixed factors, and subjects acted as a random factor (see Table 6). The analysis of the RT data revealed statistically significant main ef-

Table 5.
Summary of the statistical analyses for Experiment 3

Same responses			
	F	df	Significance
RTs			
Main effect of global symmetry	1.34	1,7	n.s.
Main effect of local symmetry	8.21	1,7	$p < 0.05$
Global × local symmetry interaction	15.75	1,7	$p < 0.01$
Errors			
Main effect of global symmetry	1.16	1,7	n.s.
Main effect of local symmetry	0.24	1,7	n.s.
Global × local symmetry interaction	23.03	1,7	$p < 0.01$
Different responses			
	t	df	Significance
RTs			
Orientation of shapes	−1.02	7	n.s.
Errors			
Orientation of shapes	0.09	7	n.s.

fects of experiment, global symmetry, local symmetry, and a statistically significant global × local symmetry interaction. There were no statistically significant interactions involving the experiment factor. Overall, subjects were slower in responding to the horizontal presentation displays than they were to respond to the vertical presentation displays. A corresponding analysis of the error data revealed a similar pattern of differences.

The RT data from Different trials were subjected to a 2-way, 2×2 repeated-measures ANOVA in which experiment (Vertical presentation vs Horizontal presentation) and direction of pointing (Same vs Different) acted as fixed factors and subjects acted as a random factor (see Table 7). This analysis revealed only a statistical main effect of experiment: subjects responded quicker in the Vertical than in the Horizontal presentation experiment. A comparable analysis of the error data revealed no significant results although the main effect of experiment showed a trend towards significance ($p = 0.07$).

The main result of interest in the Horizontal presentation experiment was that, whereas there was a strong display symmetry effect for the symmetrical shape displays, there was no such effect for the asymmetrical shape displays. Indeed, performance with the asymmetrical shape displays was surprisingly slow and inaccurate. In contrast, the data from the Vertical presentation experiment revealed strong display symmetry effects for both the symmetrical and asymmetrical shape displays with the effects of global and local symmetries being additive. Unfortunately, the omnibus analysis of these two sets of data failed to substantiate the difference in per-

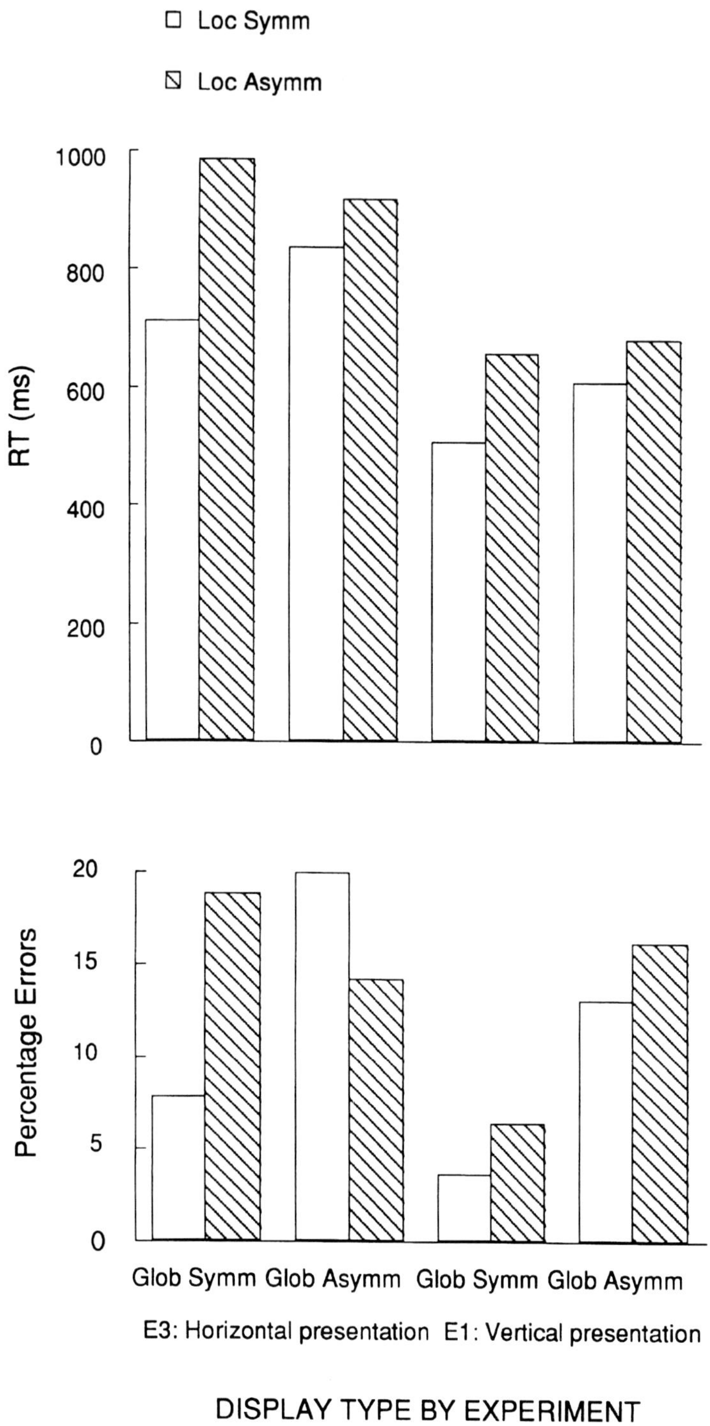

Figure 6. Mean RTs and mean percentage errors on Same trials for the eight subjects used in Experiments 3 and 1.

Table 6.
Summary of the statistical analyses of the cross experiment comparisons involving Same responses for the eight subjects that took part in Experiments 3 and 1

	F	df	Significance
RTs			
Main effect of experiment	26.51	1,7	$p < 0.01$
Main effect of global symmetry	11.62	1,7	$p < 0.05$
Main effect of local symmetry	16.76	1,7	$p < 0.01$
Experiment × global symmetry interaction	1.02	1,7	n.s.
Experiment × local symmetry interaction	0.76	1,7	n.s.
Global × local symmetry interaction	19.30	1,7	$p < 0.01$
Experiment × global symmetry × local symmetry interaction	2.10	1,7	n.s.
Errors			
Main effect of experiment	4.29	1,7	n.s.
Main effect of global symmetry	6.40	1,7	$p < 0.05$
Main effect of local symmetry	2.41	1,7	n.s.
Experiment × global symmetry interaction	1.27	1,7	n.s.
Experiment × local symmetry interaction	0.45	1,7	n.s.
Global × local symmetry interaction	10.20	1,7	$p < 0.05$
Experiment × global symmetry × local symmetry interaction	3.19	1,7	n.s.

Table 7.
Summary of the statistical analyses of the cross-experiment comparisons involving Different responses for the eight subjects that took part in Experiments 3 and 1

	F	df	Significance
RTs			
Main effect of experiment	22.34	1,7	$p < 0.01$
Main effect of direction of 'pointing'	1.32	1,7	n.s.
Experiment × direction of 'pointing' interaction	0.60	1,7	n.s.
Errors			
Main effect of experiment	4.63	1,7	n.s.
Main effect of direction of 'pointing'	0.35	1,7	n.s.
Experiment × direction of 'pointing' interaction	0.33	1,7	n.s.

formance revealed by the separate analyses, presumably because of the small number of subjects. All this analysis revealed was that RTs to the displays in the Vertical presentation experiment (i.e. Experiment 1) were faster than were RTs to the displays in the Horizontal presentation experiment (i.e. Experiment 3): a significant mean RT difference of 251 ms. As a consequence, the data show that both shape symmetry and display symmetry are less salient when the corresponding axes of symmetry are aligned horizontally than when they are aligned vertically.

Nevertheless, the most striking aspect of the results of the Horizontal presentation experiment is that subjects experienced great difficulty in responding Same to the Glob Symm–Loc Asymm displays. In this respect, performance runs against the predictions of the transformational account offered to explain the results of Experiments 1 and 2. Essentially, the transformational account predicted the same additive relationship between display and shape symmetries in both the Horizontal and Vertical presentation experiments. This is simply because the identical set of transformations apply to the displays in both cases: all that differs across the experiments is the direction of the transformations. Clearly, this prediction was not borne out by the data. In contrast, the data show that subjects experienced difficulty in detecting the equivalence of the two instances of the asymmetrical shape when the instances were arranged into a horizontally symmetrical configuration. It is almost as if the horizontally symmetrical arrangement made the two instances of the asymmetrical shape appear to be instances of different shapes.

As the transformational account has not been supported by the present data, it is appropriate to consider an explanation based on perceptual frames of reference. From Experiments 1 and 2 the evidence suggests there are two different sets of processes being tapped by the tasks. One set of processes operate to impose a display frame of reference and another set of processes operate at the level of individual shapes. From consideration of Hinton's (1981) model, it is assumed that the imposition of display frames is governed, in part, by the geometrical characteristics of the display. Therefore there will be a tendency to impose a frame that coincides with an axis of symmetry or elongation of the display. In particular, if the shapes are arranged symmetrically then the corresponding global axis of symmetry will be selected as the frame of reference for the display. An additional assumption is that, if the perceptual system establishes a particular display frame quickly, then this will aid in the selection of frames for the individual shapes: hence Same RTs will be shorter for symmetrical displays than for asymmetrical displays. On this view, symmetrical displays are processed faster than are asymmetrical displays simply because the time to impose a display frame is shorter in the former than the latter case. The local symmetry effect, also witnessed in the first two experiments, is consistent with the finding that, even when processed in isolation, the symmetrical shape is processed more quickly than the asymmetrical shape (Quinlan and Humphreys, 1993).

In addition to the intrinsic factors governing the imposition of a display frame, it is also assumed that the bias to impose a frame that is coincident with the extrinsic vertical also operates here (Rock, 1973). So again, if the system settles on a vertical frame relatively quickly (as in the Vertical presentation experiment) then this will bias the selection of vertical frames at the level of the individual shapes. This interpretation provides something of a framework for understanding why subjects were overall slower to respond to the displays in the Horizontal presentation experiment than in the Vertical presentation experiment. By the frame of reference account, in the Vertical presentation experiment, the shapes are described relative to vertical axes, yet in the Horizontal presentation experiment this normal mode of processing is unlikely to best serve the demands of the task. For subjects to perceive the symmetry of the display, they must impose a horizontal frame of reference on the display. Similarly, to perceive the symmetry of the shapes, subjects must impose horizontally aligned frames coincident with each shape's axis of elongation. In this respect the imposition of a vertical frame of reference would impede the recovery of a description of the shapes relative to the horizontal direction. Thus the gross difference in performance across the Horizontal and Vertical presentation experiments is taken to reflect the tendency to impose vertical frames of reference in both cases.

In order to overcome this tendency, it seems that information about the orientation of the shapes plays a role. In particular, if a shape frame is derived relatively quickly, then this may influence those processes concerning the derivation of a display frame. In other words, shape frames can bias the selection of a display frame (a point also made by Palmer, 1985). Moreover, this behaviour will tend to occur with the symmetrical shape displays and not with the asymmetrical shape displays because the shape frame for the symmetrical shape is derived quickly (Quinlan and Humphreys, 1993). Given these assumptions, the frame of reference account can plausibly explain why performance with the Glob Symm–Loc Asymm displays was so poor in the Horizontal presentation experiment. First, the bias to impose a vertical display frame will endure because there will be no information about the orientation of the shape frames to help overcome this bias. Second the bias to impose a vertical frame will result in the shapes being described in terms of the top-bottom (vertical) direction of the whole display (Rock, 1973). This will prove to be particularly problematic for processing the Glob Symm–Loc Asymm displays. Whereas the two instances of the symmetrical shape are identical with respect to the top-bottom direction (exactly half of the shape is above and half is below the horizontal midline) with the asymmetrical shape this is not the case. Describing the asymmetrical shapes in terms of the top-bottom direction will give rise to quite different shape descriptions. Consider Fig. 5b; here the upper shape appears top-heavy with a sharp angle uppermost. In contrast, the lower shape appears bottom-heavy with a shallow angle at the top. So, if it assumed that the bias is to describe these instances in terms of the top-bottom directions of the display as defined relative to the vertical midline, quite different shape descriptions will be derived. Even though the instances are identical they will be treated as being instances of two different shapes.

5. CONCLUSION

In summary, three experiments have been reported which have examined the influence that shape and display symmetries produce on 2-D shape-matching performance. The typical findings were that subjects were quicker to respond to symmetrical displays than to asymmetrical displays, i.e. there was a display symmetry advantage, and that subjects were quicker to respond to symmetrical shapes than to asymmetrical shapes, i.e. there was a shape symmetry advantage. Moreover, the display symmetry advantage was typically the same size for both symmetrical and asymmetrical shape displays. This pattern of results when the shapes were presented simultaneously was replicated when they were presented successively (i.e. when the ISI was 200 or 500 ms). The robustness of the results seems to indicate that the experiments have tapped a fundamental characteristic of shape processing in composite displays. Nevertheless when the shapes were arranged symmetrically about the horizontal midline of the display a different pattern emerged. The results showed that there was essentially no display-symmetry effect for displays comprising the asymmetrical shapes. Moreover, comparisons between performance in this horizontal symmetry task and that in a vertical symmetry task revealed that subjects were overall less efficient in processing the shapes in the horizontal than in the vertical conditions, a result that is consistent with the idea that subjects naturally code the visual world primarily in terms of the gravitational vertical (Rock, 1973).

Overall, the experiments provide strong evidence showing that both shape symmetry and display organization affect shape identification. Shape identification is intimately tied-up with imposing frames of reference and generating descriptions relative to such frames. In turn, reference frames can be determined by the intrinsic geometry of shapes or the intrinsic geometry of the scene of which the shapes form parts. A similar conclusion has been reached previously by Kahn and Foster (1986) and Palmer (1985) using quite different stimuli and methods to those adopted here.

Acknowledgements

This work was completed at Birkbeck College, University of London and was funded by the SERC of Great Britain. The work benefited from discussion with Glyn Humphreys.

REFERENCES

Fox, J. (1975). The use of structural diagnostics in recognition. *J. Exp. Psychol: Human Percept. Perform.* **1**, 57–67.

Hinton, G. E. (1981). The role of spatial working memory in shape perception. In: *Proc. Third Cognitive Sci. Conf.* Morgan Kaufman, San Mateo, CA, pp. 56–60.

Hinton, G. E. and Parsons, L. M. (1988). Scene-based and viewer-centered representations for comparing shapes. *Cognition* **30**, 1–35.

Kahn, J. I. and Foster, D. H. (1986). Horizontal-vertical structure in the visual comparison of rigidly transformed patterns. *J. Exp. Psychol: Human Percept. Perform.* **12**, 422–433.

Krueger, L. E. (1978). A theory of perceptual matching. *Psychol. Rev.* **85**, 278–304.

Palmer, S. E. (1985). The role of symmetry in shape perception. *Acta Psychologica* **59**, 67–90.

Podgorny, P. (1975). *Mental Rotation and the Third Dimension.* Unpublished senior honors thesis, Stanford University.

Quinlan, P. T. (1991). Differing approaches to two-dimensional shape recognition. *Psychol. Bull.* **109**, 224–241.

Quinlan, P. T. and Humphreys, G. W. (1993). Perceptual frames of reference and two-dimensional shape recognition: Further examination of internal axes. *Perception* **22**, 1343–1364.

Richards, J. T. (1978). Interitem structure and the facilitation of simultaneous comparison. *J. Exp. Psychol: Human Percept. Perform.* **4**, 72–87.

Rock, I. (1973). *Orientation and Form.* Academic Press, New York.

Shepard, R. N. (1984). Ecological constraints on internal representation: Resonant kinematics of perceiving, imagining, thinking and dreaming. *Psychol. Rev.* **91**, 417–447.

Shepard, R. N. and Cooper, L. A. (1982). *Mental Images and Their Transformations*. MIT Press, Cambridge, MA.
Sternberg, S. (1975). Memory scanning: New findings and current controversies. *Q. J. Exp. Psychol.* **27**, 1–32.
Wolff, P. (1971). Mirror-image confusability in adults. *J. Exp. Psychol.* **91**, 268–272.

Human Symmetry Perception, pp. 111–121
C. W. Tyler (Ed.)

Independence of bilateral symmetry detection from a gravitational reference frame

G. LEONE,[1,*] M. LIPSHITS,[2] J. McINTYRE[1] and V. GURFINKEL[2]

[1]*Laboratoire de Physiologie de la Perception et de l'Action, CNRS Collège de France, 15 rue de l'Ecole de Médecine, 75270 Paris Cedex 06, France*

[2]*Institute for Problems of Information Transmission, Russian Academy of Sciences, Moscow, Russia*

Abstract—Data were collected during three orbital flights aboard the Russian MIR space station from eight cosmonauts tested on a bilateral symmetry detection task. It is known that on earth subjects' performance is significantly superior for a vertical or horizontal than for an obliquely oriented axis of symmetry, giving a so-called oblique effect. The present results show that this oblique effect did not disappear in microgravity. They confirm that the detection of visual symmetry is not tied to a gravitational reference frame. An unexpected result of the experiments was that practice reduced the reaction time for detection of asymmetrical patterns below that for symmetrical patterns. This result suggests the presence of two separate detection processes. The detection of symmetry, being useful for the efficient encoding of visual information, is well developed in naive subjects. Extensive practice on a symmetry-detection task, however, increases the efficacy of the asymmetry-detection process.

1. INTRODUCTION

An axis of symmetry is one of the major intrinsic axes of a shape and could be important in the recognition of a disoriented form (Howard, 1982), and a vertical axis of symmetry is much more salient than any other (Mach, 1897/1959). Corballis and Roldan (1975) investigated the effect of stimulus orientation in a symmetry-detection task. They presented subjects with stimuli composed of 12 points divided in two equal groups by a line. The left part of a stimulus could be the reflection of the right part (symmetrical stimulus) or could be identical to the right part (repeated stimulus). The orientation of the stimulus line was varied from trial to trial (horizontal, vertical, oblique left and oblique right). The task for the subjects was to classify the stimuli into symmetrical and asymmetrical patterns. Under these conditions, response time (RT) increased linearly for stimulus orientations from vertical to horizontal. In contrast, the authors found an oblique effect

*To whom correspondence should be addressed.

for error rates. (The term oblique effect describes a general pattern of responses in a task where superior performance is reported for vertically and horizontally oriented stimuli as compared to obliquely oriented stimuli; Essock, 1980.) More importantly, they demonstrated that the effect on RT of symmetry-axis orientation with respect to the head was independent of head tilt. This suggested that the perceptual reference frame was tied to the head-retinal vertical rather than to the gravitational vertical. The authors proposed that subjects mentally rotated the patterns to a retinally vertical orientation prior to testing for symmetry, and therefore that visual operators to detect symmetry are aligned with the vertical axis of the eyes.

Note, however, that the linear increase of RT with axis orientation seen by Corballis and Roldan has not been reproduced in any other experiments on symmetry detection. Indeed, when the axis of symmetry was not explicitly displayed within the stimuli, the pattern of discriminability (Barlow and Reeves, 1979) as well as the pattern of error rates or RT as a function of axis orientation showed an oblique effect (Palmer and Hemenway, 1978; Royer, 1981). It is supposed that the apparent 'mental rotation' observed by Corballis and Roldan is due to the explicit presence of the possible axis of symmetry in the image.

Contrary to the results obtained by Corballis and Roldan, Rock's studies (Rock and Leaman, 1963; Rock, 1974) indicate that in symmetry detection the most salient orientation of the shape is what is 'perceived' as being the environmental vertical and not the head-retinal vertical (in the head tilted position, for example). This discrepancy can be partly attributed to the fact that the sides of the screen were visible in Rock's experiments and could have encouraged subjects to use an environmental visual reference frame.

In addition to this apparent discrepancy with Rock's observations, the head-retinal fixed processing supported by Corballis and Roldan is in conflict with other visual oblique effects. For tasks of identification and discriminability Attneave and his colleagues (Attneave and Olson, 1967; Attneave and Reid, 1968; Olson and Attneave, 1970) reported response-time oblique effects that were related to a gravitational or phenomenal reference frame adopted by the subject. In the light of these conflicting results, the question of a visual vs gravitational reference frame for symmetry detection remains open.

We studied the effect of prolonged weightlessness on the performance of symmetry detection. On earth, one can measure only relative effects of gravity with head-tilt paradigms, as otoliths will always detect the gravitational acceleration. Furthermore, neck muscle activities during head tilt could interfere with the processing of information. Weightlessness is the only environment in which otoliths do not indicate to the brain the orientation of the head referred to a stable exocentric direction. We tested eight cosmonauts, living in orbit from 14 days to 7 months, to assess the potential role of weightlessness on the symmetry-detection process. If gravity plays an important role in symmetry detection by assigning coordinate reference frames to visual shapes, prolonged exposure to weightlessness should change the patterns of response times and accuracy. Conversely, if gravity is not involved in the process of symmetry detection, we should find no change of performance.

2. MATERIALS AND METHODS

2.1. Subjects

Eight male cosmonauts (6 Russian, 2 French) with normal vision, aged between 35 and 48 years, participated in this experiment. They stayed in a state of weightlessness between 14 days and 7 months, during three consecutive missions aboard the space station MIR: the French–Russian mission Antares (1992), the Russian mission Post-Antares (1993) and the French–Russian mission Altaïr (1993–94).

2.2. Hardware

The equipment included a body-restraint system to ensure similar postural configurations in all experimental situations. The subjects were seated in this body-restraint system in front of a computer screen. During sessions in weightlessness, the subjects were attached to the seat with belts in order to prevent them from floating freely and to create similar tactile surface contacts as during 1 g sessions. The screen was viewed through a cylindrical black optical tube to avoid any external visual reference frames and to maintain a constant distance of 60 cm between the eyes and screen. Response times (RTs) were recorded with a resolution of 80 μs. The computer screen on which the stimuli were displayed had a 600×600 pixel resolution. The uniform white (dark) areas of the screen had a luminance of approximately 80 cd m^{-2} (0.004 cd m^{-2}).

2.3. Stimuli

Forty 2-D random shapes were generated using method 4 of Attneave and Arnoult (1956). The shapes differed in the following form parameters: symmetry (symmetrical, asymmetrical, or mixed) and complexity as measured by the numbers of sides (12, 14, 16, 18, or 20 sides). Symmetrical shapes (20 in the set) were symmetrical about only one axis. Asymmetrical shapes (10 in the set) were not symmetrical about any axis and thus contained no element of symmetry. Mixed shapes (10 in the set) contained symmetrical and asymmetrical components in a single shape; the numbers of symmetrical corners within one of these shapes was fixed at 50%. In half of the set (5 polygons) the symmetrical corners were the vertices closest to, and in the other half those farthest from, the local axis of symmetry. The locations of the non-symmetrical corners were randomly determined. Shapes subtended 8 deg of visual angle in approximately all directions. The range of number of sides per polygons (12 to 20) was chosen following a control experiment in which it was demonstrated that this factor did not influence reaction times or accuracy. One shape from each category is shown in Fig. 1.

Each shape could be presented in 12 different orientations of the generation axis from 0 to 330 deg in 30 deg steps. Nevertheless, in the limited time available for in-flight experiments it was not possible to test all these orientations for each shape and each subject. We demonstrated (unpublished results) that the relative orientation with respect to vertical is the only important factor. Mirror image transformations around a vertical or horizontal axis have no effect on the response for a given shape. Thus,

Figure 1. Examples of 2D polygons used as stimuli in this experiment. There were three categories of polygons: symmetrical (top left), locally symmetrical (top right) and asymmetrical (bottom centre), with varying number of sides (12, 14, 16, 18 or 20), resulting in a set of 40 figures.

such shapes have no up or down and no left or right. RT and accuracy are similar if the shape is presented upright at 0 deg or upside down at ± 180 deg and similarly if the axis of generation is rotated by ± 90 deg. By extension, we pooled orientations of ± 30 and ± 150 deg into one group, and ± 60 and ± 120 deg into another. Thus, each shape was presented to the subject in only four different orientations (0, 30, 60, and 90 deg), called the relative orientation in the following. Finally, the combination of all the factors (relative orientation $\times$ sides $\times$ shapes) resulted in a total of 160 stimuli each being shown at least once during each session to each subject.

The stimuli were displayed during 50 ms in synchrony with the screen refresh. This rate of presentation was chosen to prevent eye movements. Shapes were white figures (80 $\mathrm{cd\,m^{-2}}$) on a black background (0.04 $\mathrm{cd\,m^{-2}}$). To prevent retinal after-images, the pre- and post-exposure visual fields were masked with a pattern of 1-cm black stripes placed vertically across the field at 1-cm intervals. A white fixation square in the centre of the screen was added to the mask to help the subject fixate the centre of the screen, where stimuli were displayed.

2.4. Procedure

Subjects were asked to determine if the shape presented was symmetrical or not. Subjects pressed the right button to indicate that the shape was symmetrical or the left button to indicate a mixed or asymmetrical shape. They were instructed to do so as quickly as possible but without sacrificing accuracy. The subject's response triggered the presentation of the next trial with a random delay ranging from 120 to 220 ms. When the subject did not give a response within 15 s, a message 'Press a button to continue' was displayed on the screen and the trial was marked in order to be rejected in subsequent analysis. All experimental sessions for all subjects were performed in the second half of the day in order to decrease the influence of circadian rhythms on performance. Each experimental session lasted 16 min with 10-s pauses every 4 min. Only one set of stimuli (presented above) was used for all mission sessions (except for training sessions), and trials were randomly ordered. Their number depended on the subject's RTs, but each shape was presented approximately equally. The complete mission consisted of a pre-flight session, an in-flight session and a post-flight session.

2.5. Practice

Each subject was trained on the experiment from three to six times prior to the pre-flight sessions, with a catalogue of stimuli different from the flight catalogue. Such extensive training is a way, first, to check that subjects understand the task correctly; second, to ensure that their performance reaches a kind of asymptote and is stable from one session to the other; and, third, to prevent differences in performance related to refinements of strategies used to solve the task after several sessions. At the end of the practice period, we had achieved these aims, as the performance was stable from one session to the next.

2.6. Statistical methods

We rejected from raw data the trials without response (i.e. RT > 15 s). Furthermore, subjects indicated that sometimes the buttons did not respond, or that they missed the presentation of the stimuli. In these cases, RTs are rather large. We deleted large RTs by the following procedure. Any trials with RT more than two standard deviations away from the mean for that cell was deleted and the mean was recalculated from the remaining raw data.

The statistical analyses were not performed on error rates but, as advised by Snedecor and Cochran (1989) for experiments that produce small amount of errors (less than 20%), by the arcsine of their square root, in order to improve the equality of variance between cells. The values resulting from this transformation are called transformed error in the following. All the statistical tests were performed on the transformed error, but the means and standard deviations are given for error rates. We used STATISTICA for Windows (version 4.3, Statsoft Inc.) for all statistical analyses.

As all subjects did not perform the same number of sessions in each different condition, we decided to group data according to the mission phase during a flight, that is, all the sessions performed pre-flight were grouped together as were in-flight and post-flight sessions. This design allowed us to test the effects of mission phase (3), type of polygon (3) and orientation of generation axis (4) by a within-subject analysis of variance on transformed error and mean RT for correct responses. When not explicitly specified, all the comparisons between multiple pairs of means were achieved with conservative Scheffé tests (Snedecor and Cochran, 1989).

In addition to the flight subjects, we tested four backup subjects as a control group. Since we will not report here differences between flight and ground conditions, the results for these subjects are not presented. However, the ANOVA statistical analysis included these data as a between-subjects factor, and thus the F values reported below take into account these data.

3. RESULTS

3.1. Error rate

Overall, the subjects erred on 6.3% (±5.8%) of trials (means ± SD). There was a significant effect of type of shape ($F(2, 20) = 15.6$, $p < 0.0001$). Specifically, the subjects made significantly more errors on symmetrical polygons (13.3% on average) than on mixed shapes (3.7% on average) and much more than on asymmetrical shapes (1.3% on average).

Furthermore, there was a significant effect of orientation of generation axis ($F(3, 30) = 23.7$, $p < 0.0001$) as well as a significant interaction between this factor and the type of polygon ($F(6, 60) = 10.5$, $p < 0.001$). Indeed, both groups showed a significant oblique effect in accuracy for symmetrical shapes. The error rate for a vertical symmetry axis (6.4%) was significantly lower than for any other orientation (all $p < 0.05$). Similarly, the error rate for a horizontal symmetry axis (11.4%) was significantly lower than for oblique orientations ($p < 0.05$), but there was no significant difference between errors for 30 and 60 deg tilts of the symmetry axis (17.8%).

For mixed shapes, there was also a significant effect of orientation of generation-axis ($F(3, 30) = 4.6$, $p < 0.01$). Indeed, error rate increased as orientation departed from vertical and ranged from 2.1% for a vertical generation axis to 6.1% for a horizontal generation axis. The effect of orientation of generation axis was not significant for asymmetrical shapes ($p > 0.1$). Figure 2 displays the error rate for mission phase and type of polygon separately as a function of generation-axis orientation. There was no other significant effect or interaction between factors.

3.2. Response times

Subjects responded on average in 563 ms (±43 ms). There was a significant effect of polygon type ($F(2, 20) = 4.22$; $p < 0.05$). Subjects in both groups responded more slowly to symmetrical polygons (580 ms on average) than to mixed (553 ms) or asymmetrical polygons (545 ms). The difference in RT between mixed and asymmetrical shapes was not significant ($p > 0.5$), but the differences in RT between symmetrical and either mixed or asymmetrical polygons were significant (both $p < 0.001$).

There was also a significant effect of generation-axis orientation on RT ($F(3, 30) = 27.8$; $p < 0.0001$) as well as a significant interaction between this factor and the type of polygon ($F(6, 60) = 8.0$; $p < 0.0001$). Thus, subjects exhibited an oblique-effect pattern of RT for symmetrical shapes. They responded in 557 ms for a vertical symmetry axis, in 577 ms for a horizontal one and finally in 597 ms (595 ms) for

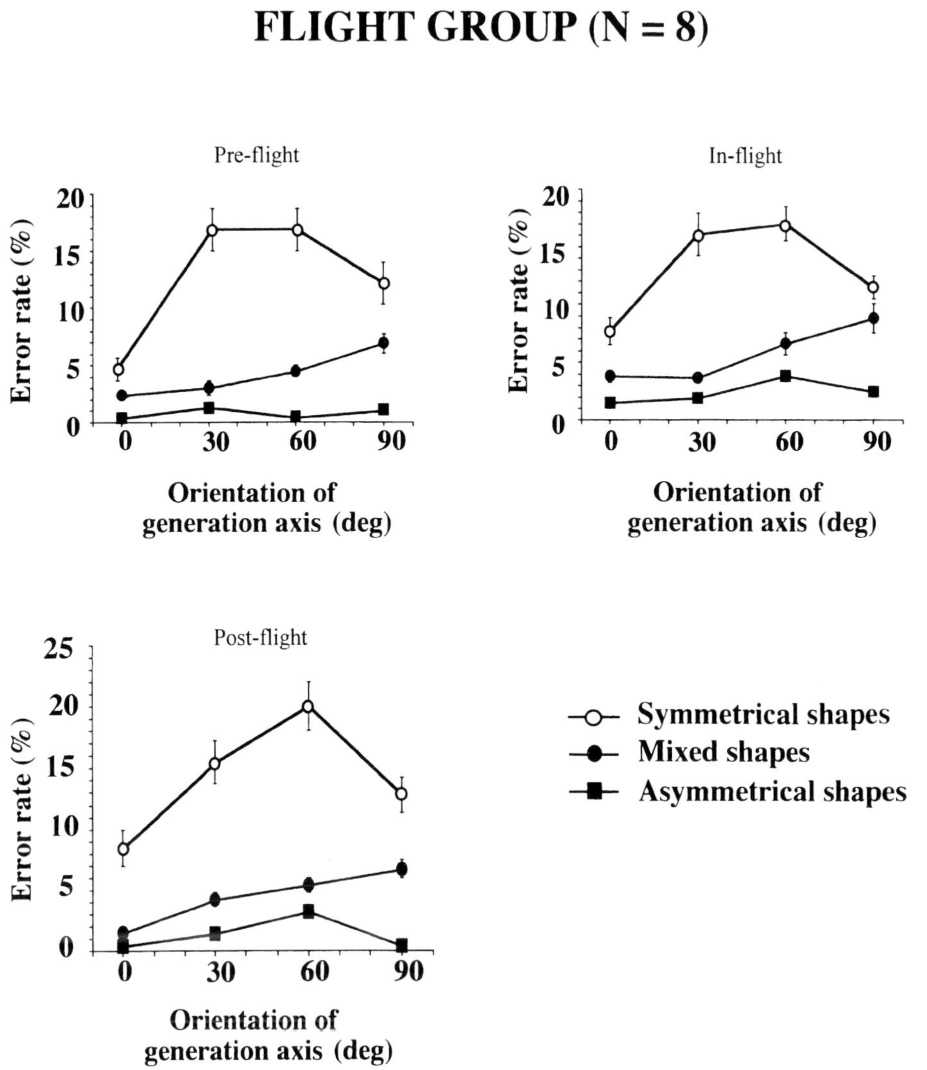

Figure 2. Mean error rates as a function of the orientation of the generation axis (0 deg vertical) for each type of polygon (symmetrical, mixed, and asymmetrical) and each mission period (pre-flight, in-flight, and post-flight). The bars represent the standard error of the mean.

a 30 deg (60 deg) tilt of the symmetry axis. Post-hoc comparisons indicate that the RT for vertical orientation of symmetry axis was significantly lower than for any other orientation (all $p < 0.001$); horizontal orientation RT was significantly lower than for oblique orientations (all $p < 0.001$); and that oblique-orientation RTs were not significantly different ($p > 0.95$). Furthermore, the difference in RTs between vertical or horizontal symmetry-axis orientations versus oblique ones (mean $\pm$ SD: -30 ± 18 ms) was significantly lower than 0 (paired t-test, $p < 0.05$) and remained constant during the different conditions, clearly indicating that the oblique effect for symmetrical shapes was not modified by weightlessness or training.

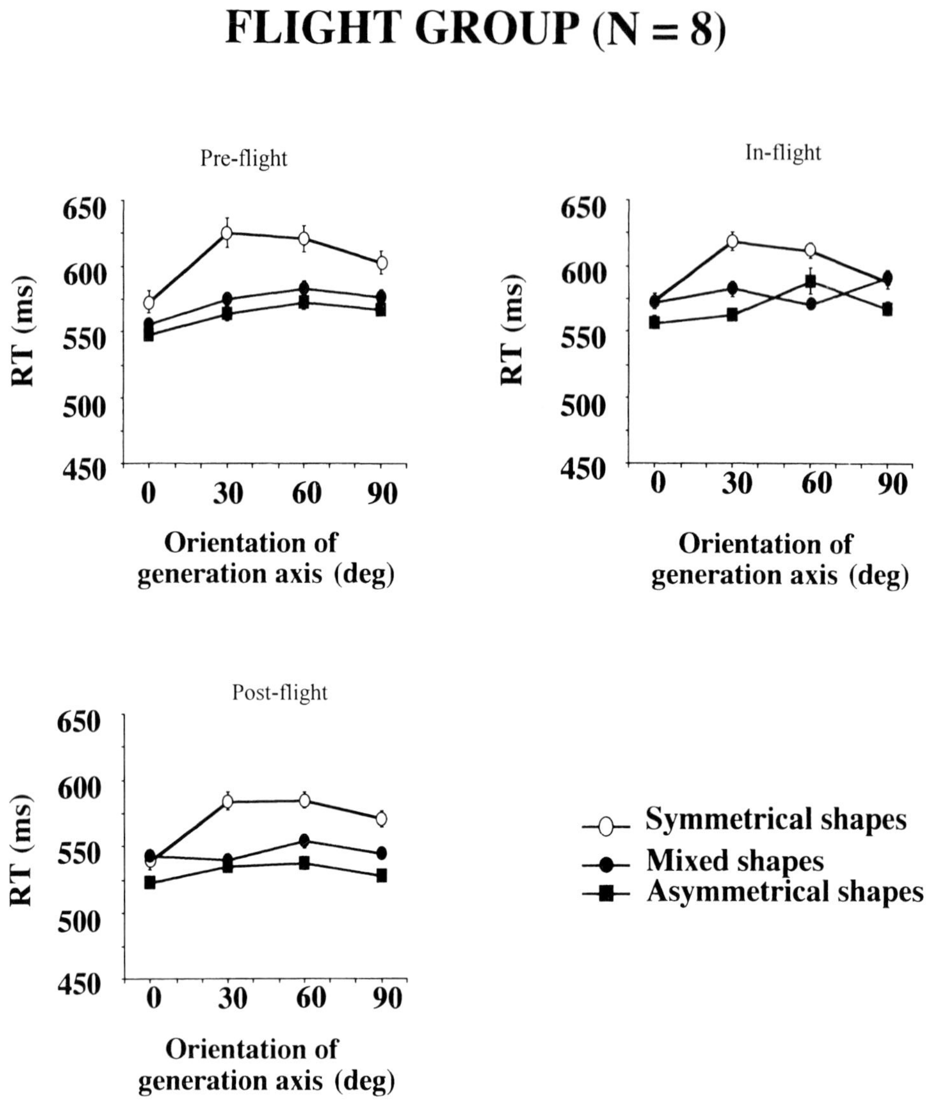

Figure 3. Mean correct response times as a function of orientation of generation axis (0 deg vertical) for each type of polygon (symmetrical, mixed, and asymmetrical) and each mission period (pre-flight, in-flight, and post-flight). The bars represent the standard error of the mean.

For mixed shapes, there was an effect of orientation of generation axis ($F(3, 30) = 7.5$; $p < 0.001$). RT increased linearly from a vertical (545 ms) to a horizontal orientation (559 ms). For asymmetrical shapes, there was no effect of orientation of generation axis ($p > 0.05$).

To sum up, the oblique effect for both response time and error rates for symmetry detection persists in-flight. Furthermore, the detection of non-symmetry was more rapid than the detection of symmetry for these subjects.

4. DISCUSSION

Our results confirm that symmetry detection is an efficient mechanism. Indeed, with a short presentation time (50 ms) of complex polygons, subjects responded quickly but without sacrificing accuracy.

The fact that subjects responded more quickly to asymmetrical than to symmetrical shapes is an intriguing result. Many authors (see Royer, 1981; Pashler, 1990) have found that response times as well as error rates were lower for symmetrical than for asymmetrical shapes. This discrepancy is rather important, as some models of symmetry detection suppose that subjects are predisposed to detect symmetry and respond 'asymmetrical' only when they fail to find a symmetry axis. In these conditions, subjects should detect symmetrical shapes more quickly than they detect asymmetrical ones. For example, the Royer (1981) model of a serial decomposition of shape code, terminating in a judgment of non-symmetry, would not agree with our results. It might be argued that the requirement to explicitly detect the presence of an axis of symmetry biased the subjects' response times. However, Corballis and Roldan (1974) have demonstrated that, even when the instructions are not biased in favour of symmetry, the response times to detect symmetry were also lower than the response times to detect its absence. Nevertheless, in the studies cited above, the subjects were naive or, at least, not highly trained on the task. If we suppose that symmetry detection involves two parallel processes, one detecting the presence of an axis of symmetry and a second detecting the failure in shape symmetry, then it can be supposed that the first process is efficient even in naive subjects, due to the likely economical role that symmetry can play in shape visual computation (Barlow and Reeves, 1979). The second process has no 'ecological meaning' for naive subjects, as detecting no axis of symmetry in a shape will not reduce the total amount of visual information to process. But, with practice on this particular task, this second process can become more efficient, significantly decreasing the RT for asymmetrical shapes.

Our results indicate also that the patterns of RT and accuracy for symmetrical stimuli exhibited an oblique effect in the sense that the performance with a vertical or horizontal orientation of symmetry axis was significantly better than with an oblique orientation. This contradicts the results of Corballis and Roldan (1975) supporting the use of a mental rotation process in the detection of bilateral symmetry. In contrast, the patterns of RT and accuracy obtained with mixed shapes showed a linear increase from vertical to horizontal orientations. This pattern could be accounted for by the use of a mental rotation process for mixed shapes, but the average mental rotation velocity would reach 6000 $\deg s^{-1}$. Mental rotation velocities usually obtained in another task with simple 2D polygons reached only 450 $\deg s^{-1}$, even with practised subjects (Shepard and Cooper, 1982). Thus, it is doubtful that such a mental rotation process can be involved in detection of non-symmetry. As suggested by Palmer and Hemenway (1978), we suspect that the results of Corballis and Roldan (1975) stem from the explicit presence of an axis of symmetry in their stimuli.

More important than the rejection of a mental rotation process, we found that the oblique effect did not disappear in weightlessness for either dependent variables (RT or error rates). The results for in-flight tests exhibited an overall pattern of RT or accuracy similar to that obtained on earth. If symmetry detection required a gravitational reference frame, the oblique effect should have disappeared with the disappearance of gravity. The pattern of RT as a function of orientation of symmetry axis should have become virtually flat. In fact, the oblique effect remains largely unchanged in weightlessness, with a preference for vertical or horizontal orientations over oblique orientations. Corballis and Roldan (1975) also found that the pattern of RT was unchanged in a symmetry-detection task when the subject's head was tilted. Our results reinforce the conclusions of these authors that the directional anisotropy of performance in symmetry is not absolutely dependent on a gravitational reference frame.

On the other hand, it can be argued that subjects effectively used a gravitational reference frame on earth for symmetry detection, but quickly substitute another perceptual reference frame when gravity information is missing, mainly the body reference frame. Friederici and Levelt (1990) studied explicitly the existence of such a substitution of perceptual reference frame on two cosmonauts. Their subjects were required to adjust the spatial position of a black ball compared to a white ball, surrounded by a schematic context. By tilting the subject's head, they investigated whether the body reference frame or head-retinal reference frame was used in the absence of gravity. Their main result was that subjects used on ground a gravitational-body reference frame, whereas in microgravity, the head-retinal reference frame became dominant. Moreover, the substitution took place within one day of exposure to microgravity.

Some of our subjects also acknowledged that the substitution of a head-retinal reference frame for a gravitational reference frame in microgravity was very important for them, especially during the first three weeks of their mission aboard MIR (informal after-flight reports). These remarks cast doubt on the idea that subjects use a body reference frame to detect bilateral visual symmetry in microgravity, as it is known that they use a head-retinal reference frame in normal gravity conditions (Corballis and Roldan, 1975). Our results emphasize the tie between the RT oblique effect for symmetry detection and a retinal (as opposed to gravitational) reference frame. Consequently, our results support the hypothesis that visual symmetry detection is a 'class 1' oblique effect, as defined by Essock (1980). The properties of symmetry detection should be an emergent property of the basic functioning of the visual system.

Acknowledgements

We thank Y. Matsakis (Institut de Médecine Spatiale), A. Shoulenin (Institute of Medical and Biological Problems), A. Koulev (Centre for Preparing Cosmonauts), Olivier Charade (Laboratoire de Physiologie de la Perception et de l'Action) and A.E.T.A for technical support. We particularly acknowledge Alain Berthoz and Jacques Droulez (Laboratoire de Physiologie de la Perception et de l'Action) for their helpful comments and their rewriting work. This research was in part supported by a grant from C.N.E.S (Centre National d'Etudes Spatiales) (92/CNES/0247), a grant from Russian fund for fundamental research (93-04-20520), and a grant from Ecole Polytechnique (DRET 91-815/10). We gratefully thank the cosmonauts (C. André-Deshays,

S. Avdeiev, J. P. Haignère, G. Manakov, A. Polestchuk, A. Serebrov, A. Soloviov, M. Tognini, V. Tsibliev) for their fruitful participation.

REFERENCES

Attneave, F. and Arnoult, M. D. (1956). The quantitative study of shape and pattern perception. *Psychol. Bull.* **53**, 452–471.

Attneave, F. and Olson, R. K. (1967). Discriminability of stimuli varying in physical or retinal orientation. *J. Exp. Psychol.* **74**, 149–157.

Attneave, F. and Reid, K. W. (1968). Voluntary control of frame of reference and slope equivalence under head rotation. *J. Exp. Psychol.* **78**, 153–159.

Barlow, H. B. and Reeves, B. C. (1979). The versatility and absolute efficiency of detecting mirror symmetry in random dot displays. *Vision Res.* **19**, 783–793.

Corballis, M. C. and Roldan, C. E. (1974). On the perception of symmetrical and repeated patterns. *Percept. Psychophys.* **16**, 136–142.

Corballis, M. C. and Roldan, C. E. (1975). Detection of symmetry as a function of angular orientation. *J. Exp. Psychol: Human Percept. Perform.* **1**, 221–230.

Essock, E. A. (1980). The oblique effect of stimulus identification considered with respect to two classes of oblique effects. *Perception* **9**, 37–46.

Friederici, A. D. and Levelt, W. J. M. (1990). Spatial reference in weightlessness: perceptual factors and mental representations. *Percept. Psychophys.* **47**, 253–266.

Howard, I. (1982). *Human Visual Orientation.* John Wiley, New York.

Mach, E. (1897). *The Analysis of Sensations.* Republished 1959 by Dover, New York.

Olson, R. K. and Attneave, F. (1970). What variables produce similarity grouping? *Am. J. Psychol.* **83**, 1–21.

Palmer, S. E. and Hemenway, K. (1978). Orientation and symmetry: effect of multiple, rotational and near symmetries. *J. Exp. Psychol: Human Percept. Perform.* **4**, 691–702.

Pashler, H. (1990). Coordinate frame for symmetry detection and object recognition. *J. Exp. Psychol: Human Percept. Perform.* **16**, 150–163.

Rock, I. (1973). *Orientation and Form.* Academic Press, New York.

Rock, I. and Leaman, R. (1963). An experimental analysis of visual symmetry. *Acta Psychol.* **21**, 171–183.

Royer, F. L. (1981). Detection of symmetry. *J. Exp. Psychol: Human Percept. Perform.* **7**, 1186–1210.

Shepard, R. N. and Cooper, L. A. (1982). *Mental Images and their Transformations.* MIT Press, Massachussets.

Snedecor, G. W. and Cochran, W. G. (1989). *Statistical Methods.* Iowa State Press University

Human Symmetry Perception, pp. 123–134
C. W. Tyler (Ed.)

Level of processing in the perception of symmetrical forms viewed from different angles

JANET P. SZLYK,[1,2,*] IRVIN ROCK[3] and CELIA B. FISHER[4]

[1]*The Research and Development Service of The West Side Veterans Administration Medical Center, Chicago, Illinois 60612, USA*

[2]*Department of Ophthalmology and Visual Sciences, UIC Eye Center, University of Illinois at Chicago College of Medicine, Chicago, Illinois 60612, USA*

[3]*Department of Psychology, University of California, Berkeley, California 94720, USA*

[4]*Department of Psychology, Fordham University, Bronx, NY 10458, USA*

Abstract—This study is concerned with the level of processing underlying the perception of symmetry about a vertical axis. Specifically, it asks whether the equality of the two sides of a symmetrical pattern must be present retinally or perceptually for the impression of symmetry to be realized. Sixty-four subjects were assigned to groups that viewed symmetrical and asymmetrical figures when the figures were in either a frontoparallel plane or one slanted by 65 deg from the line of sight. In the 65 deg condition, the objectively symmetrical stimuli projected an asymmetrical shape on the retina, and conversely the objectively asymmetrical stimuli produced a symmetrical retinal projection. From each viewing angle, patterns were observed under either full or reduced (monocular) depth cue conditions. Like their counterparts in the frontoparallel condition, observers in the 65 deg, full depth-cue condition identified the objective symmetry of the figures. By contrast, 65 deg, reduced depth-cue observers responded primarily to the retinal structure of the stimuli. The same pattern of responding was observed for both holistic and multi-element patterns and for both 100 ms and 1 s exposures. These findings and the significant relationship obtained between phenomenal judgment of symmetry and a measure of shape constancy suggest that the perception of symmetrical figures depends upon the perception of the equality of their two halves, and is thus another example of perceptual causality.

1. INTRODUCTION

It is now a well-established fact that bilateral or mirror symmetry is spontaneously perceived when the axis of such symmetry is vertical (e.g. Mach, 1897/1959, 1898; Goldmeier, 1936, 1972; Attneave, 1955; Rock and Leaman, 1963; Arnheim, 1974; Garner and Sutliff, 1974). When the axis is in some other orientation, such as

*Requests for reprints should be sent to Janet P. Szlyk, PhD, Department of Ophthalmology and Visual Sciences, University of Illinois at Chicago College of Medicine, 1855 West Taylor St., Chicago, Illinois.

horizontal or oblique, we are not apt to perceive symmetry or, if we do, the impression of equality on two sides that is characteristic of such perception is not very striking.

A related aspect of the vertical symmetry effect that concerns us here bears on the equality of the two halves of the figure on either side of the axis of symmetry. Regardless of the figure's orientation in a frontal plane, these halves are equal in their retinal and thus in their cortical projection. But suppose the figure is slanted in the third dimension so it is no longer the case that the two halves will project equally because the image of the half on the far side of the axis will be foreshortened more than the half on the near side of the axis? If the symmetry effect is based on a low level of processing closely correlated with the proximal input, one might predict that it will be eliminated under these conditions. But if it is based on the perception of the equality of the two halves of the figure, then, given shape constancy operations, the two halves will appear to be equal and, if so, the impression of symmetry will be maintained. In that event, the effect would be based on a higher, post-constancy, level of processing.

If the perception of symmetry depends upon constancy operations that lead to the perception of an axis orientation as vertical or to the veridical perception of the width of the halves of a figure slanted in depth, then it can be regarded as an example of perceptual causality or perceptual interdependency, i.e. a state of affairs in which one perception depends upon or is linked to another (see Gogel, 1973; Hochberg, 1974; Epstein, 1982; Rock, 1983). The question we ask is this: does the spontaneous impression of symmetry about a vertical axis depend upon the perception of the slant of the figure and the achievement of veridical perception of the size of its two halves?

In a recent study, Locher and Smets (1992) compared symmetry detection in two- and three-dimensional volumetric dot patterns. The patterns were displayed orthogonally and non-orthogonally to the viewer's line-of-sight (in the investigation the forms were rotated) with full depth cues available. Because there were no differences in detection accuracy between the two- and three-dimensional patterns, they concluded that depth does not provide perceptually useful information about a pattern's symmetry. However, their subjects performed more poorly in the non-orthogonal condition.

In our present study, we assume that if shape constancy governs symmetry perception then the availability of depth cues should be critical. Therefore, a full depth-cue and a reduced depth-cue condition were included in this study, as was a measure of shape constancy. By manipulating these depth conditions, and having conditions where the subjects view the display screen at an angle, rather than the patterns being rotated, our intent was to determine the role of depth information in symmetry discrimination, and ultimately test the perceptual causality hypothesis.

2. METHODS

2.1 Subjects

Sixty-four normally sighted subjects or subjects with corrected vision of at least 20/20 (Snellen visual acuity) between the ages of 18 and 27 yr were recruited.

2.2. Stimuli

There were 8 pairs of closed polygon holistic stimuli and 8 pairs of multi-element stimuli used for testing. Two additional sets of each pattern type were used to train subjects. Examples of these stimuli are illustrated in Fig. 1. By 'pair' we only mean that for each particular pattern there was a version of it that was symmetrical and a version of it that was asymmetrical. However, the figures were seen one at a time and the symmetrical and asymmetrical versions of each pattern were randomly presented and, therefore, rarely shown one after the other but rather separated by many other figures. When viewed at a 65 deg angle, objectively symmetrical stimuli projected an asymmetrical retinal image and objectively asymmetrical stimuli projected a symmetrical retinal image. To maximize the retinal asymmetry while maintaining the

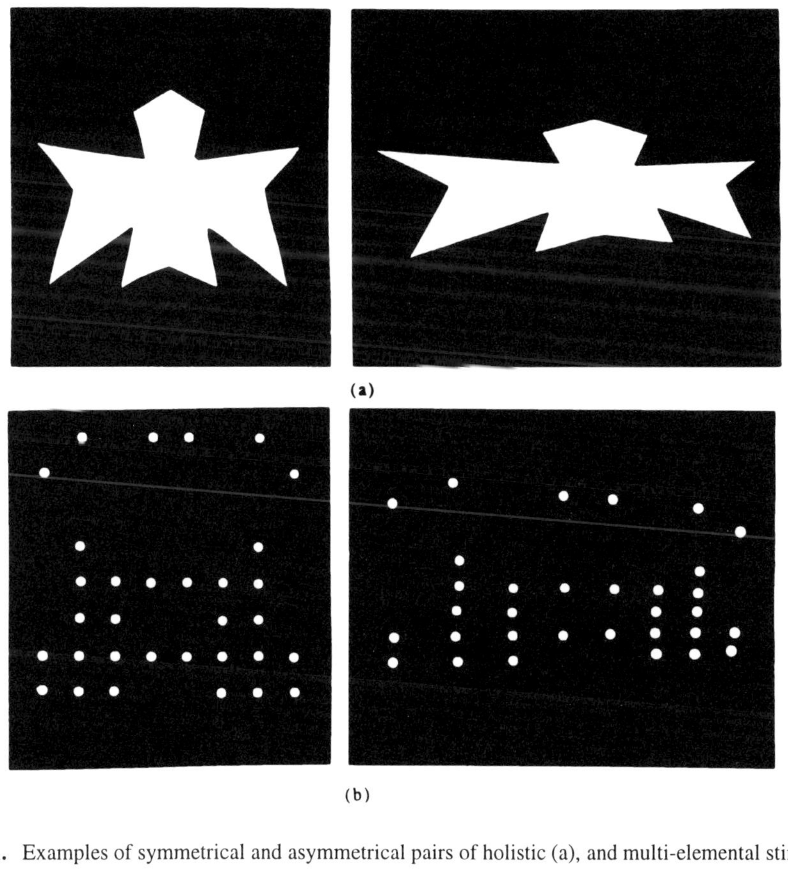

(a)

(b)

Figure 1. Examples of symmetrical and asymmetrical pairs of holistic (a), and multi-elemental stimuli (b).

slant within the range of shape constancy, we chose a slant angle of 65 deg, which has been found to be adequate for shape constancy under full cue conditions (King *et al.*, 1976).

The holistic figures were adapted from stimuli originally employed by Rock and Leaman (1963), Palmer and Hemenway (1978), and Fisher *et al.* (1981). The multi-elemental dot patterns were adapted from Royer (1981) and Fisher and Bornstein (1982). All figures were photographed with Kodalith film, and developed into slides. Two Kodak 4600 Carousel Slide Projectors back-projected slides of all stimuli onto a translucent screen. Both the projectors and the screen were hidden from the view of the observer behind a black partition. When projected onto the screen, the figures appeared white on an opaque black background. The figures subtended an average visual angle of approximately 4 deg. The average luminance of the figures was 6.1 $\mathrm{cd\,m^{-2}}$. The projector luminance was increased by approximately 0.33 in the reduced-cue conditions to compensate for the decreased light entering the eye through the pinholes that were used in these conditions.

2.3. *Design and procedure*

The 64 subjects, both men and women, were randomly assigned to one of four groups of 16 subjects each. The four groups included: the full-cue, 0 deg group; the full-cue, 65 deg group; the reduced-cue, 0 deg group; and the reduced-cue, 65 deg group.

2.3.1. Symmetry identification. Prior to each experimental session, all subjects were instructed to label the figures presented to them as simply 'symmetrical' or 'asymmetrical'. To insure that subjects understood the terms symmetrical and asymmetrical, two sets of asymmetrical and symmetrical forms (not used later in actual test trials) representing examples of the multi-element and closed polygonal patterns were shown in the vertical plane, outside the testing room prior to testing. All subjects identified all 4 training forms correctly.

In the full-cue conditions, the figures were viewed through either of two 15.3-cm apertures, both at 1.5 m from the screen. The screen was stationary and the depth tilt was produced by viewing the screen through the aperture straight ahead or the other one set at a 65 deg angle to the center of the screen. Through the apertures, subjects had full depth information concerning the orientation of the projector screen. In the reduced-cue conditions, both apertures were covered with black plastic and equipped with 1 mm pinholes. Subjects viewing the stimuli through the pinhole monocularly (with an eye patch over the non-dominant eye) saw only the white stimulus figure on a black background; they did not see the edges of the projector screen. Subjects in all conditions had their heads held in place by a chin-rest. To avoid dark adaptation, subjects were exposed to a low-luminance, diffuse light between trials that did not cause glare or localized light-adaptation effects.

Without binocular cues to depth or differential blur from far or near regions of the slide, reduced-cue subjects had only the retinal image on which to base their judgments of symmetry. By contrast, in the full-cue (binocular) condition, the subject had full view of the screen with both eyes and he or she could see its edges. Thus both binocular depth information and cues from the screen edge of the orientation of the screen were available to the full-cue subject.

Each subject saw 8 vertically symmetrical and 8 vertically asymmetrical holistic figures along with 8 vertically symmetrical and 8 vertically asymmetrical multi-element forms, presented in random order. All 32 stimuli were presented in blocks of two stimulus durations: 100 ms and 1 s. The order of the two exposure durations was counterbalanced across subjects, totaling 64 trials. A fixation point, projected from a third projector, appeared in the center of the screen immediately preceding each stimulus presentation so that the subject's gaze was localized.

2.3.2. Shape constancy. Subjects from both the full-cue, 65 deg group and the reduced-cue, 65 deg group were tested for shape constancy following the standard 64 trial session. Subjects from these two groups received an additional 32 trials: eight trials (4 holistic and 4 multi-element figures) at each of two durations (1 s and 100 ms) in each of two viewing positions (0 deg and 65 deg) for a total of 32 trials. Only symmetrical slides were used. Each subject made 16 judgments of the extent of the figures viewing them in the 0 deg frontoparallel plane and 16 in the 65 deg plane. Shape constancy was defined by the comparison of height-to-width ratios between judgments made during frontoparallel (0 deg) viewing and judgments made during 65 deg viewing. An illuminated apparatus containing two glow-in-the-dark pegs was used to measure subjects' perceptions of vertical and horizontal figural extents in both frontoparallel and rotated presentations.

3. RESULTS

3.1. Symmetry detection

Subject performance is presented in Table 1 in terms of mean proportion of correct objective shape identifications (with their corresponding standard deviations). A five-way ANOVA was performed with the factors of viewing condition (full-cue, reduced-cue) $\times$ type of figure (symmetrical, asymmetrical) $\times$ viewing angle (0 deg, 65 deg) $\times$ pattern type (holistic, multi-element) $\times$ exposure duration (100 ms, 1 s). Viewing condition and viewing angle were between-subjects factors; the remaining were within-subjects factors. The proportion of judgments (out of eight) that were objectively correct was the dependent measure.

3.1.1. Main effects. For the between-subjects factors, significant main effects were found for viewing condition [$F(1, 60) = 96.21$, $p < 0.001$] and viewing angle [$F(1, 60) = 117.58$, $p < 0.001$]. An advantage for symmetry emerged [$F(1, 60) = 44.85$, $p < 0.001$], with more correct judgments made for symmetrical patterns (mean proportion M = 0.74) than for asymmetrical patterns (M = 0.59). Consistent with expectations based on information processing models (Attneave, 1955; Chipman, 1977; Royer, 1981), the holistic patterns were judged correctly (M = 0.71) more often than

Table 1.
Mean proportions and standard deviations of correct objective shape identifications

	Holistic				Multi-elemental				
	100 ms		1 s		100 ms		1 s		
Overall	Sym	Asym	Sym	Asym	Sym	Asym	Sym	Asym	Means
Full-cue, 0 deg	0.89 (0.08)	0.81 (0.19)	0.89 (0.13)	0.93 (0.13)	0.82 (0.19)	0.66 (0.19)	0.91 (0.14)	0.79 (0.20)	0.84
Full-cue, 65 deg	0.76 (0.16)	0.66 (0.21)	0.82 (0.15)	0.79 (0.21)	0.75 (0.18)	0.50 (0.21)	0.77 (0.20)	0.56 (0.20)	0.70
Reduced-cue, 0 deg	0.87 (0.14)	0.66 (0.30)	0.90 (0.14)	0.79 (0.23)	0.72 (0.15)	0.49 (0.19)	0.83 (0.23)	0.55 (0.24)	0.73
Reduced-cue, 65 deg 65 deg	0.58 (0.25)	0.29 (0.19)	0.47 (0.21)	0.31 (0.19)	0.44 (0.19)	0.34 (0.20)	0.41 (0.19)	0.27 (0.16)	0.39

Note. The maximum possible proportion correct for each condition was 1.00. The number of subjects in each cell was 16. Standard deviations for each cell appear in parentheses.

Table 2.
Mean proportions for between-subjects factors

	Angle	
Viewing condition	0 deg	65 deg
Full-cue	0.84a	0.70a
Reduced-cue	0.73a	0.39b

Note. Means with different subscripts are significantly different from each other at the 0.05 level (Newman–Keuls).

multi-element patterns (M = 0.61), $[F(1,60) = 44.73,\ p < 0.001]$, and 1 s exposure durations resulted in more correct judgments (M = 0.69) than 100 ms exposures (M = 0.64), $[F(1,60) = 10.29,\ p < 0.002]$.

If perceptual equality is responsible for the perception of symmetry when shapes are viewed from a 65 deg angle, then the number of objective judgments of symmetry and asymmetry reported by subjects in the full-cue, 65 deg condition should not differ from responses of subjects in the full-cue, 0 deg condition, while objective judgments of reduced-cue, 65 deg subjects should drop significantly. Post-hoc analysis of the significant viewing condition by viewing angle interaction $[F(1,60) = 21.48,\ p < 0.001]$ yields such findings. Table 2 presents the means (combining correct reports of symmetry and asymmetry) involved in the interaction. Newman–Keuls tests ($p < 0.05$) on these means yielded significantly fewer objectively correct responses for the reduced-cue, 65 deg condition than for any of the other three conditions, which in turn did not differ from one another. Moreover, subjects in these three other groups identified the objective shape of the stimuli above chance levels [$t(15) = 10.65$, $p < 0.01$; $t(15) = 6.39$, $p < 0.01$; $t(15) = 7.10$, $p < 0.01$; for the full-cue, 0 deg, the full-cue, 65 deg, and the reduced-cue, 0 deg conditions, respectively]. By contrast, the reduced-cue, 65 deg condition produced responding below chance level [$t(15) = -3.51$, $p < 0.01$], indicating judgments influenced by, but not exclusively dependent upon, the retinal projection of the stimuli (see below).

The focus of this investigation is on performance of the subjects when they view the figures at a slant but have full depth cues available, the full-cue, 65 deg condition. Do they respond in terms of the symmetry or asymmetry, as these properties may be perceived, or in terms of how they are given within the retinal image? The mean of 0.70 correct responses clearly is in the direction of the perceptual interpretation, since responding on the basis of retinal symmetry or asymmetry would lead to the expectation of a mean close to zero (because in the latter case the subjects would always be incorrect). Moreover, the mean of 0.70 should not be compared to the value of 1.00, the maximum proportion that could be correct, because in the full-cue, 0 deg condition the mean is only 0.84. For whatever reason, that value is the level of correct responses achieved when both perceptual and retinal symmetry is present.

3.2. Shape constancy

The baseline ratio (of height in cm to width in cm) for the two frontoparallel plane conditions turned out to fall roughly between 0.94 and 1.14, depending upon whether full or reduced-cues were available and upon the type of figure and duration of exposure. The means were 1.0 with full cues available and 1.07 in the reduced-cue condition. This result corresponds rather well with the objective height-to-width ratio, which approximates 1.0.

Were the subjects able to achieve full constancy for the shapes of the figures despite their rather extreme slant of 65 deg from the frontal plane? At the 65 deg angle in the full-cue condition, the mean ratio was 1.26 across exposure duration and type of figure, which significantly differed from the full-cue, 0 deg condition mean of 1.0 [$t(15) = 5.04$, $p < 0.01$]. This increase reflects the decrease in perceived width of the stimuli relative to the judgment of width in the frontoparallel position and represents a loss of constancy.

In the reduced-cue, 65 deg condition, where of course constancy is not expected to be obtained, the ratio rises sharply to a mean of 2.03 for the two kinds of figure and two exposure interval conditions combined. Using the full-cue, 0 deg mean height-to-width ratio of 1.0 as the value for complete constancy and the value of this ratio when the figures are viewed through the artificial pupil from an angle of 65 deg of 2.03 as the value for zero constancy, we can derive a measure for the degree of constancy obtained for the mean ratio in the full-cue, 65 deg condition of 1.26. Following the formula that is generally used, we have a Constancy Ratio $= (2.03 - 1.26)/(2.03 - 1.0) = 0.77$, which is, of course, an appreciable but not complete degree of constancy.

Furthermore, the mean value of the height-to-width ratio obtained in the reduced-cue, 65 deg condition, of 2.03, is lower than the value to be expected in this condition if the perceived width of the figure were a function of the compression of the projected figure images relative to the unchanged image of the height of the figures. We computed this value trigonometrically to be 2.35. The departure of 2.03 from 2.35 (which we take to be the objective value indicating zero constancy) is statistically significant [$t(15) = 5.99$, $p < 0.01$]. The Constancy Ratio $(2.35 - 2.03)/(2.35 - 1.0) = 0.24$. Thus, it can be concluded that the reduced-cue, 65 deg condition did not succeed in completely eliminating cues to depth. This in turn undoubtedly explains why symmetry and asymmetry responses in that condition were not what would have been expected if depth perception had been entirely eliminated. For, if it had, subjects should have responded incorrectly as often as they responded correctly in the 0 deg condition. That is because symmetrical figures yielded asymmetrical retinal images and asymmetrical figures yielded symmetrical retinal images in the reduced-cue, 65 deg condition and, without depth cues, these image relations should have determined responses. If so, the expected mean proportion correct should have been $1.0 - 0.73$ (using the mean proportion correct in the reduced-cue, 0 deg condition as baseline) or 0.27. But the mean proportion of correct responses in the reduced-cue, 65 deg condition was 0.39.

The influence of exposure duration and pattern type on shape constancy was assessed by a two-way repeated measures ANOVA. Neither the main effects nor the interaction were significant [$F(1, 31) < 1.0$]. Thus, neither exposure duration, nor pattern type significantly affected shape constancy in our experiment.

4. DISCUSSION

The results give a clear answer to the main question posed in this investigation. The perception and response to symmetry or the lack of it in patterns symmetrical about their vertical axis is determined by whether or not the two halves of the pattern appear to be equal. Thus, although a figure is rotated about its vertical axis quite appreciably away from the frontoparallel plane, it will generally continue to appear symmetrical if it is symmetrical and asymmetrical if it is not. This is so despite the fact that the symmetrical figures then project an image to the retina that is asymmetrical and the asymmetrical figures project an image that is symmetrical about the vertical axis.

The data that support this claim are the results of the condition in which the subject views the figures in a plane slanted away from him or her by 65 deg with full cues to depth available. In comparison to the results when the figures are seen in the frontoparallel plane, where the mean proportion of correct responses was 0.84, the mean proportion correct in the slanted-plane condition was 0.70. But, for responses based on the retinal state of affairs, the predicted mean ought to be close to 0, since responses based on retinal symmetry or asymmetry would always be incorrect. This prediction assumes that subjects would always be 'correct.' However, a more likely expectation would be the difference between the proportion correct in the frontoparallel plane condition and the perfectly correct performance of 1.00, namely $1.00 - 0.84$, or 0.16. So the closeness of the obtained mean of 0.70 to 0.84 and its distance from 0.16 tells the story.

Moreover the measure of shape constancy tells us that the perceived shape of the figures seen at a slant was not fully veridical. This is what might be expected for a slant as extreme as 65 deg. Thus a symmetrical figure might be expected to appear to be slightly less than perfectly symmetrical and an asymmetrical one to appear not quite as asymmetrical as it is. Given that expectation, the slight departure of the results in the slanted-plane condition from those in the frontal-plane condition become fully understandable.

The results in the condition in which subjects viewed the figures in the slanted plane, but presumably without any cues to depth, round out the picture. Here the mean proportion correct plummets to 0.39. Thus with the only difference between the two slanted-plane conditions being the presence or absence of depth information, the responses shift from those governed predominantly by the objective state of affairs to those governed predominantly by the retinal state of affairs. This is not to imply that subjects are perceiving their retinal images in the reduced-cue condition. A better formulation would be to say that in this condition subjects tend to perceive the figures as lying in a plane orthogonal to their line of sight and thus retinal symmetry or the lack of it would have to signify objective symmetry or the lack of it.

But the responses in these conditions are not fully in accord with a 'retinal' prediction. The mean of 0.39 is significantly greater than the value suggested above of 0.16, which is what we should expect in this condition were retinal symmetry or asymmetry to govern the outcome. However, a further correction in this prediction is required. The use of the artificial pupil not only reduces depth information, but also appreciably lowers the luminance level of the figures. The greater intensity of light we introduced in the two reduced-cue conditions would not have totally compensated for the loss of luminance at the eye. Thus the results of the other reduced-cue condition, in which the figures were seen in a frontoparallel plane, become relevant and useful. Here the mean proportion correct was 0.73. Therefore it is the difference between 1.00 and 0.73, or 0.27, that is the best prediction of what to expect in the reduced-cue slanted-plane condition if the retinal state of affairs governs the outcome. The obtained mean of 0.39 is still significantly greater than 0.27 [$t(15) = 3.60$, $p < 0.01$]. From this we can conclude that the artificial pupil did not succeed in completely eliminating cues to depth in this condition. This conclusion jibes with the fact that a slight tendency toward shape constancy was found to occur in this condition. It is possible that the asymmetrical retinal projection of the symmetrical slides served as a kind of perspective cue, suggesting a symmetrical pattern viewed at a slant.

4.1. *Symmetry bias*

Subjects consistently did better with the symmetrical than with the asymmetrical patterns. They were more often correct on these trials. We believe that the finding reveals a bias. It must be borne in mind that our asymmetrical figures are deliberate distortions away from what is otherwise a symmetrical figure. Therefore, one might say that all the figures look somewhat symmetrical. The question the subject has to answer for him- or herself is rather whether a given figure looks *perfectly* symmetrical. Given some uncertainty, the subject is more likely to respond 'symmetrical'. A signal detection analysis supports this interpretation of a symmetrical bias. Across all conditions and stimuli, the average proportion of hits (symmetrical response when a symmetrical stimulus was presented) was 0.74, misses (asymmetrical response/symmetrical stimulus) 0.26, false alarms (symmetrical response/asymmetrical stimulus) 0.41, and correct rejections (asymmetrical response/asymmetrical stimulus) 0.59, yielding a d' of 0.87. Saying 'symmetrical' when the stimulus was asymmetrical (false alarm) should have been as likely as saying 'asymmetrical' when the stimulus was symmetrical (miss), if no bias were evident. However, false alarms were considerably more likely (0.41) than were misses (0.26), offering further support for a symmetrical bias, with beta = 0.82. This interpretation is supported by the further finding that the symmetrical bias increases with the difficulty of the condition.

4.2. Effects of figure type and exposure duration

We have already noted the expected finding that subjects do better with the holistic figures than with the multi-element figures. Similarly, it hardly needs to be explained that subjects would do better with the longer 1 s exposure duration than the 100 ms one. The fact is that subjects do perform better in the slanted-plane full-cue condition with the holistic figures and with the longer exposure duration. However, such figural and duration differences occur for all the other conditions as well, so that, in the absence of a significant interaction effect, we cannot say that these factors specifically affect the constancy operation. Although it is known that constancy is achievable in brief durations, there is no evidence of which we are aware that shape constancy is a function of the particular shape or pattern investigated (except perhaps some bias that improves performance when the shape is a regular one such as a circle (King *et al.*, 1976) or one with vertical symmetry as in our experiments here). Hence our failure to find specific effects of figural characteristics and duration of exposure do not violate existing knowledge on this subject.

4.3. A test of perceptual causality

Some investigators might be inclined to regard our finding of the dependency of symmetry perception on constancy operations as surprising (Locher and Smets, 1992). That is because symmetry can be detected in exposure durations of as little as 50 ms followed by a mask. In fact, for this reason, the detection of symmetry has been thought to be preattentive as determined by studies using 2-D dot patterns rotated clockwise and counter clockwise in 2-D space (Wagemans *et al.*, 1991, 1992), whereas it would seem unlikely to many that a constancy process of taking account of depth information would be preattentive. Some might even be inclined to reverse our claim about the direction of cause and effect of perception of the plane and perception of symmetry. The argument would be that the skewed symmetry projected to the retina in our slanted-plane conditions is preattentively interpreted as bilateral symmetry and this in turn allows the correct interpretation that the plane of the figure is slanted with respect to the viewer.

Our answer to this line of reasoning is as follows. First, none of all the preattentive research of which we are aware succeeds in testing perception without attention. Either a divided attention paradigm is employed or it is assumed that texture segregation or pop-out based on brief presentations of an array of multiple items must be preattentive because the outcome entails parallel processing (Treisman and Gelade, 1980;

Julesz, 1981; Beck, 1982). But the fact remains that subjects in these paradigms are attending to the array. When this is prevented by a method in which the subject is not attending to the stimulus or stimulus array to be tested, then neither texture segregation, perceptual groupings, pop-out, nor shape is perceived (Mack *et al.*, 1992; Rock *et al.*, 1992). Therefore it is unlikely that any overall property of shape such as symmetry would be detected preattentively. Second, it does not at all follow that, *with* attention as in our experiments and others on symmetry, depth perception and shape constancy would not be achieved in brief durations. We are inclined to believe they would because there is evidence that processes such as depth from retinal disparity and constancy are indeed achieved in very brief durations (Dove, 1841; King *et al.*, 1976). In our experiment, depth and constancy were achieved in 100 ms presentations although the figures were not masked.

Third, there is a test of the cause-effect reversal hypothesis contained in our experiment. We refer to the condition in which the pattern on the screen is viewed through a pinhole at a 65 deg slant thus reducing depth information. For in this case, the skewed symmetry image is available and should, according to the hypothesis, yield good symmetry perception of symmetrical figures. However, the other cues allowed in the 65 deg, full-cue condition are not available in this condition, but, according to the hypothesis, they are not relevant. The result in this condition is a marked decline in correct reports of symmetry and, going along with this, a marked decline in constancy. However, there still is a small residual tendency toward constancy and there are more correct symmetry responses than would be expected from the asymmetrical image of the symmetrical figures. Therefore, as we acknowledge above, it may well be that a skewed symmetrical image is one kind of cue that the object producing it is slanted away from the frontal plane.

There is another relevant fact to consider. It concerns the subjects' responses to asymmetrical figures presented in the 65 deg full depth-cue condition. These yield a symmetrical retinal image so that there is no skewed symmetry image present. If the outcome were not based on taking account of the slant of the screen, subjects should respond 'symmetrical' given the symmetrical image. However, subjects tend to respond 'asymmetrical' to these stimuli seen at a slant; whereas in the reduced-cue 65 deg condition subjects tend to respond 'symmetrical' to these stimuli. This finding is no doubt the result of information from retinal disparity and the interaction of appropriate accommodation and convergence. The trapezoidal image of the slanted rectangular screen may also serve as a pictorial cue to the slant of the screen.

It can be argued that our findings are not exclusively concerned with symmetrical figures. While this is undoubtedly true, it does not imply that our findings have nothing to do with symmetry *per se*. One of the main purposes of this study was to demonstrate that there is nothing special about the perception of symmetry beyond the perception of the equality of the halves of a figure on each side of its vertical bisector. If such perception of equality is indeed crucial, as our data suggest, then we believe that the vertical-symmetry effect is one more example of perceptual causality.

REFERENCES

Arnheim, R. (1974). *Art and Visual Perception.* University of California Press, Berkeley, CA.

Attneave, F. (1955). Symmetry, information, and memory for patterns. *Am. J. Psychol.* **68**, 209–222.

Beck, J. (1982). Textural segmentation. In: *Organization and Representation in Perception.* J. Beck (Ed.). Lawrence Erlbaum Assoc., Hillsdale, New Jersey, pp. 285–317.

Chipman, S. F. (1977). Complexity and structure in visual patterns. *J. Exp. Psychol.* **106**, 269–301.

Dove, H. W. (1841). Über stereoskopie. *Ann. Phys., Series 2* **110**, 494–498.

Epstein, W. (1982). Percept-percept couplings. *Perception* **11**, 75–83.

Fisher, C. B. and Bornstein, M. H. (1982). Identification of symmetry: Effects of stimulus orientation and head position. *Percept. Psychophys.* **32**, 443–448.

Fisher, C. B., Ferdinandsen, K. and Bornstein, M. H. (1981). The role of symmetry in infant form discrimination. *Child Develop.* **52**, 457–462.

Garner, W. R. and Sutliff, D. (1974). The effect of goodness of encoding time in visual pattern discrimination. *Percept. Psychophys.* **16**, 426–430.

Gogel, W. (1973). The organization of perceived space: I. Perceptual interactions. *Psychologische Forschung* **36**, 195–221.

Goldmeier, E. (1936). Uber Ahnlichkeit bei gesehen Figuren. *Psychologische Forschung* **21**, 146–208.

Goldmeier, E. (1972). Similarity in visually perceived forms. *Psycholog. Issues* **8**, 1–136.

Hochberg, J. (1974). Higher-order stimuli and inter-response coupling in the perception of the visual world. In: *Studies in Perception: Essays in Honor of JJ Gibson.* R. MacLeod and H. L. Pick, Jr (Eds). Cornell University Press, Ithaca, New York.

Julesz, B. (1981). Figure and ground perception in briefly presented isodipole textures. In: *Perceptual Organization.* M. Kubovy and J. Pomerantz (Eds). Lawrence Erlbaum Assoc., Hillsdale, New Jersey, pp. 27–54.

Locher, P. and Smets, G. (1992). The influence of stimulus dimensionality and viewing orientation on detection of symmetry in dot patterns. *Bull. Psychonom. Soc.* **30**, 43–46.

King, M., Meyer, G. E., Tangney, J. and Biederman, I. (1976). Shape constancy and a perceptual bias towards symmetry. *Percept. Psychophys.* **19**, 129–136.

Mach, E. (1898). *Popular Scientific Lectures*, 3rd edition. T. J. McCormack (translator). Open Court Publishing House, Chicago, Illinois.

Mach, E. (1899/1955). *The Analysis of Sensations*, Revised Edition. C. M. Williams (translator). Dover Publishing, New York, NY.

Mack, A., Tang, B., Tuma, R., Kahn, S. and Rock, I. (1992). Perceptual organization and attention. *Cognit. Psychol.* **24**, 475–501.

Palmer, S. E. and Hemenway, K. (1978). Orientation and symmetry: Effects of multiple, rotational, and near symmetries. *J. Exp. Psychol: Human Percept. Perform.* **4**, 691–702.

Rock, I. (1983). Perceptual interdependencies. In: *The Logic of Perception.* Bradford Books/MIT Press, Cambridge, Massachusetts, pp. 283–299.

Rock, I. and Leaman, R. (1963). An experimental analysis of visual symmetry. *Acta Psychol.* **21**, 171–183.

Rock, I., Linnett, C. M., Grant, P. and Mack, A. (1992). Perception without attention: Results of a new method. *Cognit. Psychol.* **24**, 502–534.

Royer, F. L. (1981). Detection of symmetry. *J. Exp. Psychol.* **7**, 1186–1210.

Treisman, A. and Gelade, G. (1980). A feature integration theory of attention. *Cognit. Psychol.* **12**, 97–136.

Wagemans, J., Van Gool, L. and d'Ydewalle, G. (1991). Detection of symmetry in tachistoscopically presented dot patterns: effects of multiple axes and skewing. *Percept. Psychophys.* **50**, 413–427.

Wagemans, J., Van Gool, L. and d'Ydewalle, G. (1992). Orientation effects and component processes in symmetry detection. *Q. J. Exp. Psychol.* **44A**, 475–508.

Human Symmetry Perception, pp. 135–155
C. W. Tyler (Ed.)

Determinants of symmetry perception

SEONGHEE HONG[1,*] and M. PAVEL[2]

[1]*Department of Psychology, Yonsei University, Seoul, Korea*
[2]*Computer Science and Engineering, Oregon Graduate Institute, PO Box 91000, Portland OR 97291-1000, E-mail: pavel@cse.ogi.edu*

Abstract—We investigated perception of symmetry by measuring the ability of the visual system to detect small amounts of asymmetry. Our vertically symmetric stimulus prototypes were perturbed by scaling one of the half planes relative to the other. Using stimuli composed of small dots, we determined that the visual system detects symmetries using multiple spatial scales. In particular, symmetry can be computed from the position of individual dots, but the performance is improved if the dots form larger features such as lines and corners. Our data suggest that curved lines might be less effective than straight lines. We also found that symmetry of 1 deg patterns can be perceived in the parafovea more accurately than would be expected from the results of prior bisection and hyperacuity experiments. Finally, we determined that the luminance distribution may affect estimates of the axis of symmetry. We discuss a feasible model of symmetry perception based on a multiresolution representation, such as the cortex transform proposed by Watson (1987).

1. INTRODUCTION

Symmetry has been considered an important characteristic influencing human perception For example, Dinnerstein and Wertheimer (1957) argued that symmetry, along with other Gestalt principles, is used to interpret ambiguous figures. Thus, an image of a partially occluded object that can be interpreted by the visual system in many different ways is often perceived uniquely. Even for impoverished images, such as the one shown in Fig. 1 observers usually agree on a small number of potential interpretations. The visual system appears to use constraints, such as potential or partial symmetry, to eliminate many otherwise possible interpretations.

This theoretical account, however, implies that a potential regularity, e.g. symmetry, is used as a constraint. The visual system must, therefore, be able to determine partially symmetric patterns in images. In the present study we investigated the ability of the visual system to perceive symmetry as a continuous rather than an all-or none phenomenon. We were particularly interested in assessing the role of spatial patterns (simple features), and the location of individual points on symmetry perception. In this chapter we review briefly prior attempts to characterize perception of symmetry, describe our four pilot experiments and a subset of their theoretical implications.

---

*Deceased. Address correspondence to Professor Pavel.

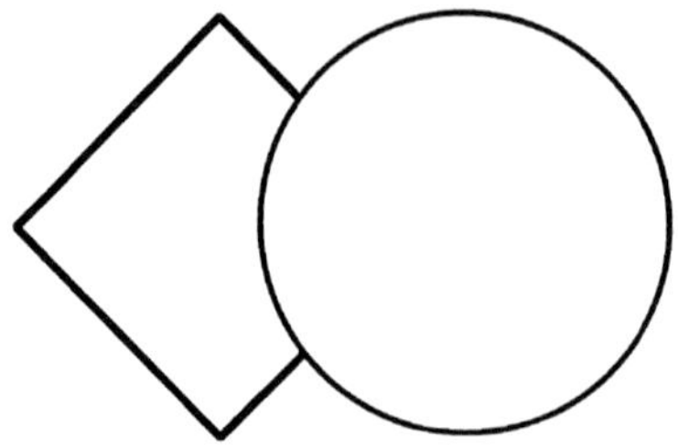

Figure 1. An example of a symmetric completion of occluded objects. People usually see a diamond behind a circle.

2. BACKGROUND

Preference for seeing symmetric objects is a property that has been considered together with other Gestalt principles to constrain possible interpretations of ambiguous percepts. A typical interpretation of Gestalt principles is based on the theoretical notion of (computational) complexity, or equivalently, on the amount of information necessary to represent an object. Properties such as symmetry can be used by the visual system to construct economical (short) codes.

Although information content for finite objects cannot be defined and measured independently of the representation (e.g. Kanizsa, 1979), psychologists tend to associate information content with the notion of geometric regularity, especially symmetry (e.g. Attneave, 1955). When confronted with partial information, the human visual system, which is assumed to prefer the simplest interpretation, would tend to see highly redundant or patterned objects with low information content (Attneave, 1954). Thus, symmetry is important because it allows the image to be represented efficiently (Barlow and Reeves, 1979).

Buffart *et al.* (1981) proposed a particular coding scheme to model the recognition of occluded polygonal objects. In their theory, each edge, corner, angle, and sequences of edges and angles can be represented by separate symbols. A visual object is then described by a sequence of these symbols. Because a redundant sequence arising from object symmetry would require only a single symbol, symmetry would reduce the length of the code, and thereby the information load. Because the human visual system is assumed to select the interpretation that minimizes the length of the code, the coding theory of Buffart *et al.* (1981) predicts that partially occluded objects would be perceived as maximally symmetric. This theory is partially supported by the data of Locher and Nodine (1973), who demonstrated that subjects tend to look at one half of symmetrical shapes.

Because of its deterministic and algebraic nature, the theory of Buffart *et al.* (1981) is difficult to apply to real images and perceptual phenomena. A more realistic approach was taken by Brady (1982) who proposed a theory of perception of occluded objects based on smoothed local symmetries. According to this theory, a visible portion of an object is first segmented, and then, using an iterative approach, recombined into largest symmetric basic shapes.

These theories illustrate how symmetry of objects might be involved in occluded object perception. In the following discussion we describe an account of how the visual system might compute symmetry from visual images. Although there are many kinds of symmetry, we focus our discussion on a single-fold mirror symmetry along the vertical axis in a two-dimensional plane. Most of the discussed issues generalize to other kinds of symmetries.

One of the first researchers to recognize the saliency of symmetry in perception was Mach (1897). He showed that people can easily perceive bilateral and centric (or radial) symmetries of simple amorphic shapes. One implication of his results is that symmetry can be detected before a pattern is recognized. Mach, therefore, concluded that symmetry computation is performed on a very low level of image representation. In a more recent study, Julesz (1971) used random dot figures and found that symmetry can be perceived with stimulus duration less than 50 ms when the center of symmetry coincides with the fixation point. However, unlike Mach's study, subjects could not detect symmetry when the center of symmetry was shifted a few degrees peripherally. To explain the discrepancy between his and Mach's results, Julesz suggested two different mechanisms for symmetry perception. One process, based on the symmetric organization of neural anatomy in the fovea, performs a high resolution point-by-point comparison (a more formal definition of the point-by-point comparison will be given in the discussion Section 7). The other process, perhaps mediated by a more central mechanism, computes symmetry of larger low-frequency patterns. The former process is restricted to the fovea, but the latter process might be responsible for detecting symmetries in the periphery.

Corballis and Beale (1976) described the point-by-point comparison mechanism from a neurological perspective. They suggested that the point-by-point comparison mechanism proposed by Julesz involves homotopic comparison between symmetrically opposite corresponding points in the brain. Mirror-image points in the two hemispheres are compared when the retinal image maps onto the visual cortex topologically (Cumming, 1970). When the figure is fixated on the axis of symmetry, each half of the symmetrical figure projects to the opposite side of the brain.

This theory and the supporting evidence are consistent with Julesz's (1971) finding that bilateral symmetry is more easily detected than repeated patterns. Fitts *et al.* (1956) also found that symmetrical figures were recognized more rapidly than repeated figures. Furthermore, their finding that vertically symmetrical figures were recognized more rapidly than horizontally symmetrical ones was consistent with the notion of a homotopic comparison mechanism.

The point-by-point comparison process has been criticized (e.g. Jenkins, 1983) for several reasons. In particular, this approach to the computation of symmetry is inefficient and requires a complete image of the object, which is not in the case of occlusions. Moreover, Jenkins (1983) argued that there is no neurophysiological evidence of point-by-point organization in the human visual system. Instead, Jenkins proposed that symmetry can be perceived by the detection of collinearity of midpoints of hypothetical line segments connecting each pair of points. He assumed that this computation can be performed by a group of similarly oriented receptive fields of different sizes (Hubel and Wiesel, 1977). To support his theory, Jenkins (1983) demonstrated that his subjects could detect the symmetry of pairs of points even when the pairs were not all located in exactly symmetric positions. In fact, this approximate symmetry was detected if the midpoints were randomly distributed over less than 1 deg.

Using similar stimuli, Barlow and Reeves (1979) demonstrated that subjects could detect symmetry even when the proportion of symmetric points in random dot textures was less than 50%, and when the axes of symmetry were located up to 3 deg from fovea. Moreover, their subjects could perceive symmetry when dot pairs were not placed at the exactly symmetric positions, but at randomly chosen positions within a limited area. They concluded that the visual system processes information with a certain tolerance range, and therefore, than symmetry is perceived by comparisons of small areas rather than by comparison of individual points.

The point-by-point hypothesis was also questioned by Pizlo (1988) who argued that symmetry perception may be aided by higher level features, such as lines or angles of corners. Pizlo asked subjects to discriminate between an isosoceles and a scalene triangle, and recorded the loci of their eye fixations during recognition. The resulting eye fixations were clustered around a line between the apex of a triangle and the midpoint of its base, which suggests that subjects verified whether the apex was located at the bisector of the triangle in order to make symmetry judgments.

The advantage of higher level features for symmetry assessment has been also recognized in research on machine pattern recognition. For example, Biswas and Majumder (1988) proposed to measure the degree of symmetry by approximating a closed curve by a polygon with two parameters: the lengths of sides and the angles between them. These two parameters are used to index symmetry. Others used edges and the angles between them (Biswas and Majumder, 1988), curvatures (Chaudhuri and Majumder, 1980), and orientation and lengths of sides (Parui and Majumder, 1983).

The work summarized thus far indicates that humans can perceive symmetry with stimuli ranging in complexity from several dots, via various patterns, to random dot textures. Since a triplet of points is sufficient stimulus for symmetry judgment, the visual system must have the capability to compare distances of individual points. However, in order to characterize the processes underlying symmetry perception, it is necessary to determine to what extent the visual system can take advantage of patterns of points. We carried out several experiments designed to assess the relative contribution of individual points and simple patterns of points to symmetry judgments. In Experiments 1 and 2 we investigated the need for corresponding points on opposite sides of an axis of symmetry. Experiment 3 was carried out to compare the effectiveness of curved and straight lines as well as the effects of eccentricity. Finally, Experiment 4 was designed to assess the contribution of asymmetry in luminance on the judgments of symmetry in location. In order to investigate the sensitivity of the visual system to symmetry, the basic task in all our experiments is to detect small asymmetries in nearly symmetric patterns. The details of the underlying theoretical issues will be considered in the discussion section.

3. EXPERIMENT 1

According to the point-by-point comparison assumption, symmetry is perceived best if, for every point on one side of the axis of symmetry, there is a corresponding point on the other side. In this experiment we compared the effectiveness of symmetric features to the pointwise correspondence. We used three types of stimuli. The first type of stimulus consisted of symmetric features (lines) defined by symmetric (matched) points. The second type of stimulus consisted symmetric lines, but without corresponding points. The third type of stimulus consisted of matched points in general position, i.e. no lines.

3.1. Method

Subjects. Two subjects participated in Experiment 1. Both had normal or corrected vision.

Apparatus. The stimuli were presented on a color graphics display monitor with 60 Hz refresh rate, controlled by a Silicon graphics IRIS-2400 workstation. The resolution of the screen was 767 pixels (height) $\times$ 1023 pixels (width). The observer's head was partially stabilized by a chin rest at a distance of 136 cm from the screen.

Stimuli. Three types of stimuli used in this experiment are shown in Fig. 2. They consisted 16 blue disks or dots, each subtending 3.3 min in diameter displayed on a white background.[1] In the first type of stimulus, the dots were arranged to form diamond shown in Fig. 2(a). Each dot on the left side of the diamond was matched, in the vertical dimension, by a corresponding dot on the right. The diamond stimulus, therefore, consisted of pairs of corresponding points forming corresponding lines. Each side of the diamond consisted of dots at both ends and two more dots that equally divided the side. The second type of stimulus, illustrated in Fig. 2(b), was also a diamond, but the dots on one side were each randomly shifted along the edge of the diamond to eliminate any spatial correspondence. The third type of stimulus, shown, in Fig. 2(c), consisted of all the pairs in the first type, except that the vertical dimension of each pair was randomized. The randomization was performed to

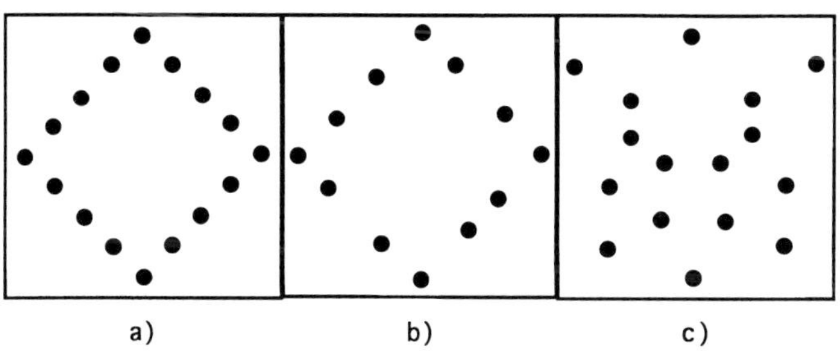

Figure 2. Examples of the three types of symmetric stimuli in Experiment 1. a) is a sample stimulus of the corresponding dotted diamond condition; b) is a sample of the non-corresponding dotted diamond condition; and c) is a sample of the random dot figure condition.

eliminate collinearity so that the dots in each hemifield were general position. Each of the three type of stimuli subtended 1.5 deg. A small cross subtending 8.4 min visual angle was used as a fixation point. The fixation point and the stimuli were presented at the center of the rectangular CRT screen.

Procedure. The stimuli were presented in blocks of 60 trials for each condition. The first block of each session was used for practice. At the beginning of each block, a vertical bar was presented to inform the observer to judge the symmetry about the vertical line. On each trial, the fixation point was displayed for 1 s on the center of the screen. 80 ms after the fixation point was extinguished, the stimulus was displayed for 50 ms. Because of the bright background, no mask was used. The symmetric stimuli were perturbed by a linear scaling of either the right or the left half of the horizontal axis. For each point P' on the perturbed side of the diamond, the horizontal coordinate was multiplied by a constant $0 < \lambda$, $P'(x, y) = P(\lambda x, y)$. This transformation preserves straight lines, but affects their orientations. The amount of of asymmetry specified by λ was controlled by randomly interleaved up-down staircases.

The observer's task was to decide whether the pattern was protruding to the right or to the left. For example, if $\lambda > 1$, and was applied to the right side $x > 0$, the pattern would protrude to the right. The same response would be correct if $\lambda < 1$, but applied to the left side of the image, $x < 0$. The observers were asked to indicate their judgment by pressing one of two buttons on a mouse device. The fixation point then reappeared and the next trial began.

3.2. Results and discussion

Observers' ability to perceive symmetry was characterized by their ability to judge the direction of asymmetry. The amount of physical asymmetry was expressed in terms of the distance of the furthest point in a pattern from its nominal symmetric position. For each condition, observers' responses were summarized by logistic fit to the empirical psychometric function; i.e. the proportions of right-responses for each level of asymmetry. These raw estimates of the psychometric functions were used to find the maximum likelihood fit of a logistic function. A logistic function is an ogival, monotonically increasing function of the form,

$$f(x) = \frac{1}{1 + \varepsilon^{-(a+bx)}},$$

where a determines the bias and b determines the slope. The sensitivity to asymmetry was summarized by just noticeable differences (JND) defined as the difference between the 50% and 75% points on the logistic psychometric function.

The resulting JNDs for both observers and all three conditions are shown in Fig. 3. The JNDs for non-corresponding and corresponding dotted diamonds were virtually the same, about 1.5 min. This corresponds to approximately 1.5% of the extent of the stimuli. This Weber fraction is comparable to that obtained in bisection experiments

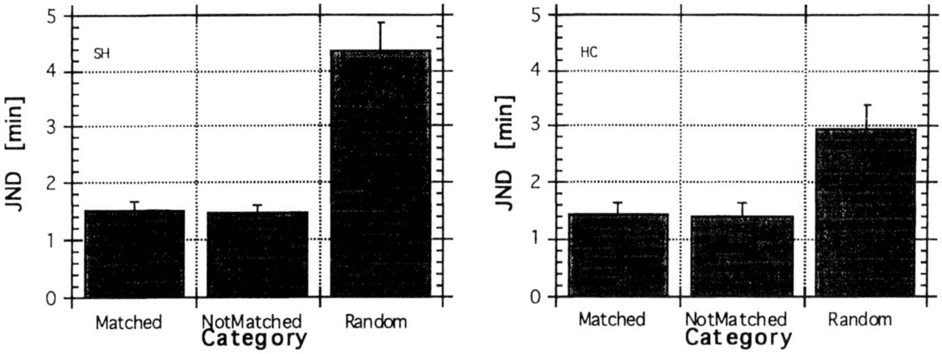

Figure 3. Resulting just noticeable differences for the three conditions and two subjects in Experiment 1. The conditions are: Matched (matched points and lines), Not Matched (not matched points but matching lines), and Random (matched points but no matched lines).

(e.g. Yap *et al.*, 1987). The JNDs for the random dot patterns were approximately twice as large, approximately 3 min. Both subjects showed the same results.

The fact that the performance with corresponding points was much better when they were arranged in a diamond shape suggests that it is not simply the presence of corresponding points that allows the visual system to detect asymmetry about the vertical axis. If the visual system depended solely on a comparison of corresponding points then the detectability of asymmetry for the random dot patterns should be the same as that of the diamond with corresponding points. Our results, therefore, contradict the hypothesis that the visual system determines symmetry only by a point-by-point comparison process. The fact that performance was similar for diamond patterns with and without full point symmetry suggests that the visual system is capable of using simple patterns of dots to determine the symmetry or asymmetry of images.

There are, however, two drawbacks of the stimuli used in this experiment. The first is due to the fact that the symmetric diamond consisted of four 90 deg angles. The amount of asymmetry was directly related to the deviations of the angles from 90 deg, and the observers might use this cue to make their judgment. Although this explanation is unlikely, it is possible that this absolute information aided in the performance of the task. The second drawback of the stimuli used in this experiment was the fact that the noncorresponding stimuli had two corresponding dots marking the most extreme points of the pattern in the horizontal direction. The observers could, theoretically, ignore the remaining non-corresponding points and make their judgments on the basis of these two points. We carried out Experiment 2 with stimuli that did not provide these cues.

4. EXPERIMENT 2

Experiment 2 was identical to Experiment 1 except for small modifications in the stimuli.

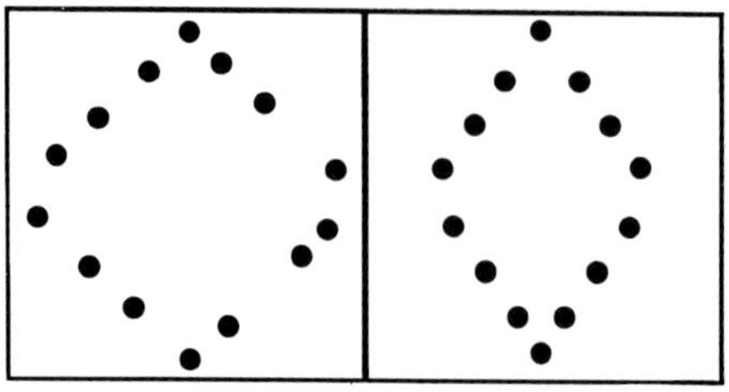

Figure 4. Examples of symmetric stimuli in Experiment 2, where a) is an example of the non-corresponding dotted diamonds with an angle of 82 deg, b) is a sample stimulus of the corresponding dotted diamonds with an angle of 98 deg. Unlike the stimuli used in Experiment 1 (see Fig. 2), these patterns have no dots at the extreme left and right corners.

4.1. Method

Subjects. Two naive subjects participated in Experiment 2. Both had normal or corrected vision.

Apparatus. The apparatus was the same as used in Experiment 1.

Stimuli. The stimuli were diamonds constructed in a similar manner as those in Experiment 1, except that the right and the left corners deviated from 90 deg. Moreover, the there were no dots at the lateral corners of the diamonds. Examples of two of these stimuli are shown in Fig. 4. We used five different symmetric prototype diamonds with angles 82 deg, 86 deg, 90 deg, 94 deg, 98 deg at both the right and the left corners. For each angle, there were two types of stimuli: with corresponding and without corresponding points. The placement of the dots was chosen to permit a direct comparison between the corresponding and non-corresponding stimuli. For the corresponding diamonds, one dot was placed at the top and bottom corners, and four more dots were spaced along the perimeter dividing the length from the top corner to the bottom in accordance with the length ratios 2.5 : 2 : 2.5 : 2 : 2. For the non-corresponding diamonds, one side of the diamond was generated exactly the same way as the corresponding one, and the other side was generated using the same ratios in the reverse order. The asymmetric stimuli were generated by scaling the left or the right half of the field along the horizontal axis as in Experiment 1.

Procedure. The procedure was the same as that of Experiment 1, except that we used mixed conditions design. Subjects ran one block of 240 trials after a block of practice trials for each condition. The stimuli with different angles were randomly selected within a block of trials with equal probability. The amount of symmetry of the stimuli was controlled using three interleaved up-down staircase methods.

4.2. Results and discussion

Observers' responses in Experiment 2 were analyzed in the same way as in Experiment 1. Observers' ability to detect asymmetry was summarized in terms of JNDs.

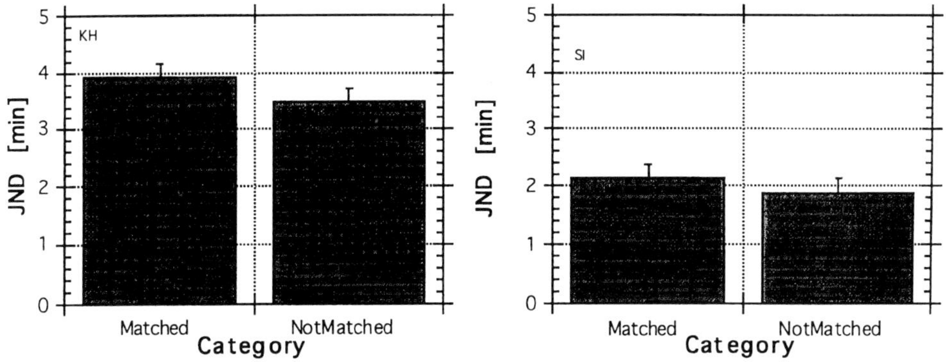

Figure 5. Results of Experiment 2 JNDs of performance on two conditions from two subjects averaged over all angles. The categories are Matched (orresponding dotted diamond condition), and Not Matched (non-corresponding dotted diamond condition).

The resulting JNDs for the corresponding and the non-corresponding diamonds are shown in Fig. 5. The most important result of this experiment is that the performance with the corresponding stimuli was no better than that with the non-corresponding ones. Both subjects showed similar data. This implies that the correspondence of the corner dots and the right angles used in Experiment 1 were not the reasons for the equivalence of the corresponding and the non-corresponding stimuli.

The average performance with the stimuli in Experiment 2 was slightly worse than with those in Experiment 1. This is consistent with the possibility that the judgments were aided by the presence of right corners. On the other hand, this difference may, in part, be due to the fact that the observers in this experiment were less experienced psychophysical subjects. Due to the circumstances beyond our control, we could not replicate this experiment with the original subjects.

In Experiments 1 and 2 we demonstrated that the visual system can use simple features, such as lines, in determining potential symmetry of images. Experiment 3 was designed to investigate whether curved lines are as effective as straight ones and whether sensitivity to asymmetry depends on eccentricity.

5. EXPERIMENT 3

This experiment was designed to investigate two aspects of symmetry perception. First, we compared the effectiveness of straight and curved lines in symmetry judgments. The effectiveness of straight lines in symmetry perception is consistent with the hypothesis that the visual system is using local orientation information. In this experiment we compared the ability of the visual system to detect symmetry in circular and diamond stimuli. If only local information is used, then making symmetry/asymmetry judgments for circles should be at least as easy as for symmetric diamonds. On the other hand if orientation information is integrated over larger regions, or if corner angles are used to assess symmetry, then performance with diamond stimuli should be better.

The second aspect of symmetry perception addressed in this experiment concerns the effect of eccentricity. In particular, we compared the ability to detect asymmetry in near periphery to that in the fovea. Our goal was to examine the hypothesis that symmetry judgments are based on a fixed neural network that is symmetric with respect to the fovea (e.g. Julesz, 1971; Corballis and Beale, 1976). Under this hypothesis, the ability to perceive symmetry would deteriorate rapidly with the eccentricity of the axis of symmetry. We acknowledge that most spatial judgments deteriorate as a function of eccentricity, and therefore, any reasonable theory would expect worse performance in the periphery. If, however, the symmetry judgments are based on a fixed neural network that is symmetric with respect to the fovea (e.g. Julesz, 1971; Corballis and Beale, 1976), we would expect much worse performance in the periphery than that observed with spatial tasks (e.g. vernier acuity) that do not depend on the network symmetry.

5.1. Method

Subjects. Two subjects participated in Experiment 3. Both had normal or corrected vision.

Apparatus. The apparatus and the experimental setup was the same as used in Experiment 1.

Stimuli. The symmetric prototype stimuli used in Experiment 3 were a diamond and a circle illustrated in Fig. 6. At a viewing distance of 136 cm, the symmetrical stimulus figure subtended 1.1 deg both horizontally and vertically. Both diamond-shaped and circle-shaped figures were outlined in blue against a white background. As in Experiments 1 and 2, the asymmetry generated by scaling the left or right half plane defined by the vertical axis was varied over trials using several randomly interleaved up-down staircases.

A small cross of size 8.4 min visual angle was used as a fixation point. The stimuli were presented in one of three locations. In the foveal condition, the pattern was centered on the fixation point. In the peripheral condition, the center of the pattern was located either 4 deg to the right or 4 deg to the left of the fixation point.

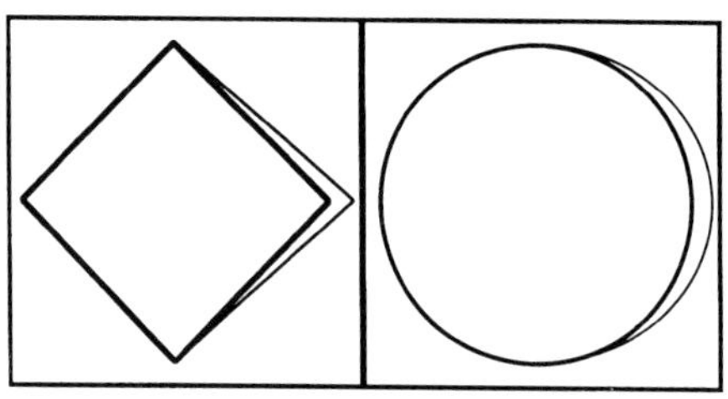

a) b)

Figure 6. Examples of the stimuli for Experiment 3. The thick lines represent symmetric stimuli, and the thin lines show the effect of asymmetric scaling.

Procedure. The procedure was identical to that of Experiment 1. Following one block of practice, each subject ran four blocks of 60 trials. The conditions (location of the stimuli) were randomized within the blocks.

5.2. Results and discussion

Observers' responses from Experiment 3 were analyzed in the same way as that in Experiments 1 and 2. The results summarized in terms of the JNDs are shown in Fig. 7. The size of the JNDs, expressed as a Weber fraction of the stimulus diameter, ranged between 0.75 and 3%. These results are consistent with previously published data. For example, Burbeck (1987) reported JNDs between 1 and 3% for judgments of distance defined by two bars.

The most important result is that the JNDs for the diamond stimuli were smaller than those of the circle stimulus. This was true for both the foveal and the peripheral conditions and for both subjects. The advantage of the diamond stimulus suggests that the visual system is using orientation information for straight lines or angular information at corners. This result also tends to contradict the hypothesis that symmetry is assessed using point-by-point comparisons. In fact, many theories based on point-by-point comparisons would predict that the symmetry of a circle would be easier to judge than that of a diamond because the perimeter of the circle is 11% longer than that of a diamond. In contrast to the diamond, however, a circle has no singular points to establish unique correspondences.

As expected, the ability to detect asymmetry is worse in the periphery than in the fovea. An important question is whether this decrement in performance is greater than that expected based on other spatial tasks such as vernier acuity. A number of experimental studies (e.g. Levi *et al.*, 1985; Cunningham and Pavel, 1986; Yap *et al.*, 1987) demonstrated that the JND for such spatial tasks decreases linearly with eccentricity. This effect, frequently attributed to the anatomical structure of the striate cortex, is typically summarized by

$$\mathrm{JND}(e) = \mathrm{JND}(0)\left(1 + \frac{e}{e_0}\right), \tag{1}$$

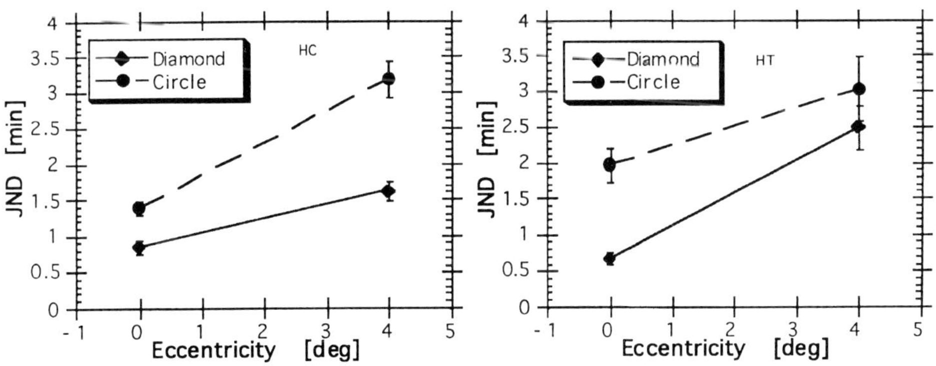

Figure 7. JNDs of performance on diamond and circle shapes in fovea and periphery conditions from two subjects (Experiment 3).

Table 1.
JNDs and obtained scaling factors (Experiment 3)

Subject	Shape	JND in fovea [min]	JND at 4 deg [min]	Scaling factor
HC	diamond	0.88	1.61	4.82
	circle	1.38	3.15	3.19
HT	diamond	0.64	2.49	1.39
	circle	1.97	3.05	7.28

where JND(e) is the threshold at a given eccentricity e, JND(0) is the threshold at the fovea, and the eccentricity scaling factor e_0 can be interpreted as the eccentricity at which JND is twice the foveal, JND(0). The greater values of the scaling factor correspond to slower decrease in performance with eccentricity. The value of the scaling factor e_0 for a task similar to ours (e.g. three-dot bisection) was estimated to be 0.6 deg (Yap *et al.*, 1987).

The scaling factors calculated using the JND data of each subject in this experiment are shown in Table 1. The scaling factors for both stimuli are larger than those obtained in the three-dot bisection studies. This implies that the performance with the peripheral stimuli was better than expected on the basis of the three-dot bisection task. This result contradicts the assumption that a symmetrical retinal projection is required for symmetry detection. The relatively good performance suggests that spatial uncertainty in the periphery might be smaller for more complex stimuli than that for a three point bisection. When the stimuli were presented in the fovea, it was possible for the visual system to assume that the axis of symmetry is located in the fovea. Since the observers were able to make symmetry judgments in the periphery, the visual system must determine the potential axis of symmetry from the image. This could be accomplished by a computation of the center of 'mass' or principal components from the image. The following experiment was carried out to investigate a prediction generated by such a model.

6. EXPERIMENT 4

If the visual system computes the axis of symmetry from the center of mass or from the principal components using a luminance distribution, we should be able to affect this computation by manipulating the intensity of, or the number of, points. In Experiment 4 we presented dotted-diamonds with different number of dots on each side to test a prediction that the perceived axis of symmetry might be shifted toward a side with more points.

6.1. Method

Subjects. The subjects were the same as in Experiment 1.

Apparatus. The apparatus was the same as used in the previous experiments.

Stimuli. The prototype stimuli were diamonds similar those used in Experiment 1 except that in some stimuli one side of the pattern had more dots than the other. An example of such stimulus is shown in Fig. 8. The number of dots on either side was between 3 and 5 and they were uniformly distributed over the perimeter of the pattern. These stimuli were presented randomly 4.2 min to the right or to the left of the fixation point. This prevented the observers from using the fixation point to locate the axis of symmetry.

Procedure. The procedure was the same as that of Experiment 1.

6.2. Results and discussion

Observers' responses in Experiment 4 were analyzed in the same way as in the previous experiments. Figure 9 shows JNDs of the unequal and the equal number of dots conditions. Both observers could detect asymmetry more easily when the pattern had an equal number of dots on each side. This suggests that the visual system uses the distribution of luminance to determine the possible location of the axis of symmetry.

If, in fact, the axis of symmetry is computed by such processes, then the stimuli used in this experiment should cause systematic biases in the responses. For example,

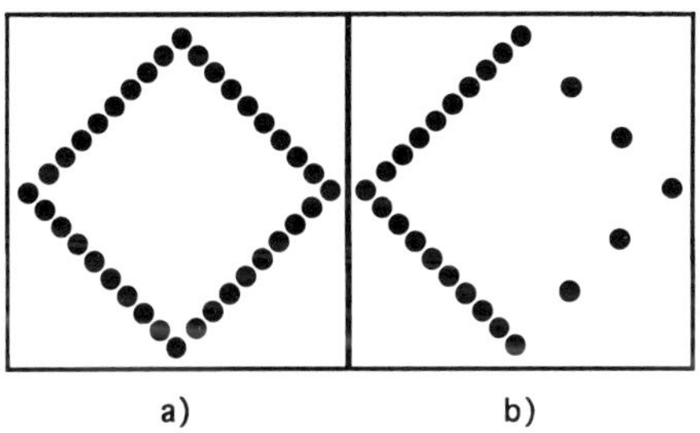

Figure 8. Examples of symmetric stimuli in Experiment 4. a) is a sample stimulus of the equal number of points condition, and b) is a sample stimulus with unequal number of points.

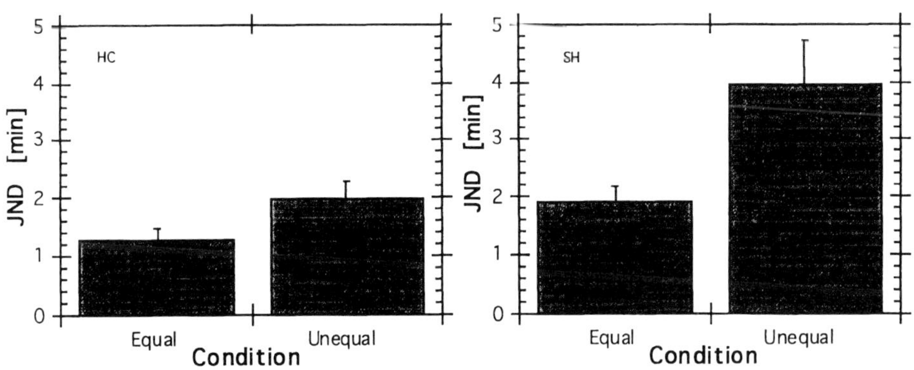

Figure 9. JNDs of performance on two conditions from two subjects in Experiment 4. (EQUAL = equal number of points condition; UNEQUAL = unequal number of points condition.)

a larger number of dots on the left side of a pattern should result in an illusory perception that the stimulus is protruding to the right. This analysis predicts that the psychometric functions should be shifted by the bias to make fewer left judgments. The analysis of the psychometric functions using the logistic functional form yielded estimates of the biases. The bias was significantly greater than zero with individual values ranging from 0.2 to 1.5 min of visual angle. Both subjects showed similar results.

7. THEORETICAL IMPLICATIONS

The four experiments described in this paper were designed to study perception of symmetry by assessing the ability of the visual system to detect small amounts of spatial asymmetry. Our empirical results suggest that the visual system can use a hierarchy of features ranging from small dots to at least lines and corners. We have also demonstrated that symmetry is a continuous property of images rather than all-or-none. In the following discussion we will outline an model of the processes that can account for our observations.

The ability to exploit a two-dimensional symmetry requires at least the following components (Tyler, 1994):

1. An image representation that embodies similarity metric of symmetric features.
2. A method of combining individual measurements of symmetry over space.
3. A procedure for detecting different types of symmetries and their parameters, e.g. axes.

We discuss a possible approach to each of these in some detail.

Representation. Our focus is the tolerance of the representation to spatial and luminance uncertainties that permit the detection of potential symmetries. For the purpose of this discussion we assumed initially that the axis of symmetry is known, as was the case in our experiments.

For the sake of simplicity we will initially restrict our discussion to the case when an image can be described by a one-dimensional luminance function $L(x, y) = L(x)$. The contrast in such image is then given by

$$f(x) = \frac{L(x) - \langle L \rangle}{\langle L \rangle}, \tag{2}$$

where $\langle L \rangle$ is an average luminance. We also assume that the location of the axis of symmetry is known to be at $x = 0$.

A function f is symmetric with respect to the origin if $f(x) = f(-x)$ everywhere in the range of x. A measure of the deviation of the difference $f(x) - f(-x)$ from zero, evaluated over the range of x, is then indicative of the degree of asymmetry. To specify an image representation requires a definition of a metric for the deviation that is consistent with perceptual results. There are a variety of potential metrics for the deviation, such as maximum deviation, sum of absolute values, various moments, etc. One example of a potential measure of asymmetry is a measure based on moments, such as the third central moment,

$$\int_{-a}^{a} x^3\left[f(x) - f(-x)\right] \mathrm{d}x$$

in which the difference $f(x) - f(-x)$ is weighted by the third power of the distance from the axis of symmetry. This measure of asymmetry, used in probability and statistics, is unfortunately inappropriate as a model of perception because the points further away from the axis of symmetry are given larger weight than the close ones. This contradicts many empirical observations that the symmetry near the axis is easier to detect (Bruce and Morgan, 1975; Barlow and Reeves, 1979).

An alternate, mathematically convenient way to define a measure of asymmetry that does not have this problem is based on squared differences of the form

$$A = c\int_{0}^{a} w(x)\left[f(x) - f(-x)\right]^2 \mathrm{d}x,$$

where c is a normalizing constant that depends on the image. The function w is a real, positive function that depends on eccentricity and accounts for the effects of eccentricity scaling given by Eqn (1). The degree of asymmetry in this formulation is proportional to the energy in the odd-symmetric component of $f(x)$. It follows, therefore, that the degree of symmetry is proportional to the energy in the even-symmetric component of the image,

$$J_{\mathrm{e}} = c\int_{0}^{a} w(x)\left[f(x) + f(-x)\right]^2 \mathrm{d}x.$$

The normalization constant,

$$c = \frac{1}{\int_{-a}^{a} w(x) f^2(x)\,\mathrm{d}x} \tag{3}$$

is be defined such that J_{e} vanishes for asymmetric images and has the value of unity for bilaterally symmetric images with identical left and right sides. With this normalization, the measure of symmetry is proportional to the correlation of the positive and negative half planes,

$$J_{\mathrm{e}} = 1 + 2c\int_{0}^{a} w(x) f(x) f(-x)\,\mathrm{d}x. \tag{4}$$

In this correlation-based measure of symmetry, each pair of points $(x, -x)$ contributes independently of other points and, therefore, the J_{e} can be taken as a mathematical definition of a point-by-point procedure (Tyler and Miller, 1994).

This convenient point-by-point procedure is, unfortunately, inconsistent with a subset of prior observations including our empirical results. This is particularly true for black-and-white (0,1) images, in which the correlation depends on the exact correspondence in the position of symmetric points. To reduce this critical dependency of the symmetry measure on the exact positions of corresponding points, Barlow and Reeves (1979) proposed that the visual system examines small regions of dot images in the neighborhood of x, and counts the number of dots in each region, N_x. Symmetry is then assessed by computing an χ^2 of the differences between corresponding regions,

$$\chi^2 = \sum \frac{(N_x - N_{-x})^2}{(N_x + N_{-x})}.$$

This measure based on counting is applicable to images consisting of black-and-white pixels. A similar effect incorporating spatial summation can be obtained by computing the symmetry measure J_e after filtering the image f by a filter h as proposed by Hong (1991) (see also Dakin and Watt, 1994). To simplify the notation we use g_h to denote the filtered image,

$$g_h(x) = [f \star h](x) = \int_{-\infty}^{\infty} f(x')h(x - x')\,\mathrm{d}x',$$

where the $\star$ operation represents a convolution, and h is an even-symmetric spatial filter. Then the measure of symmetry can be defined as

$$J = c\int_{0}^{+a} w(x)g_h(x)g_h(-x)\,\mathrm{d}x. \tag{5}$$

The spatial width of the filter h represents the tolerance or sensitivity of the symmetry mechanism to spatial deviations. According to this model, an image is judged to be symmetric in a Yes/No experiment if the measure J exceeds a fixed threshold.

A drawback of the filter-based symmetry measure is its dependence on the size of the filter, i.e. the area over which information is integrated. Finding a single, optimal size of the filter is not possible. For a filter with a large spatial extent, symmetry information in high spatial frequencies would be difficult to detect. A small-extent filter will fail to detect symmetry contained in low frequencies. One biologically plausible representation relies on a multiresolution approach in which the visual system analyzes images with different filter sizes simultaneously. This can be accomplished easily if images are represented by multiresolution pyramids (e.g. Adelson *et al.*, 1984). A multiresolution representation of an image is a set of filtered images containing different spatial frequency bands. Each filtered image corresponds to a level of a pyramid. Variations over larger regions are represented by low frequency levels, and changes over small areas are represented by high frequency levels. This approach to symmetry perception was previously suggested by Julesz and Chang (1979), who explained their results of superposition of different symmetries using an argument based on a multiresolution representation.

We propose that the symmetry measure is computed at each level of a multiresolution representation. A multiresolution-based representation with radially symmetric filters can potentially explain the continuity of symmetry perception, but it does not account for the superiority of edges and corners found in our experiments. In order to account for these results, we assert that the multiresolution pyramid represents orientation information at each level. Such a pyramidal scheme, e.g. cortex transform (Watson, 1987), is generated by filters sensitive to several orientations.

The degree of symmetry is evaluated for each neighborhood using images convolved with oriented filters $h_{\omega,\theta}(x)$, where ω is the center of the spatial frequency band, and θ is the orientation. Corresponding to each filter $h_{\omega,\theta}(x)$ we must use its even-symmetric mirror image $\bar{h}_{\omega,\theta}(x)$.

Let us denote by $g_{\omega,\theta}$ and $\bar{g}_{\omega,\theta}$ the results of a convolution of f with $h_{\omega,\theta}$, and its mirror image $\bar{h}_{\omega,\theta}$, respectively. A partial measure of symmetry with respect to an arbitrary point x is then

$$j_{\omega,\theta}(x) = c_{\omega,\theta} \int_{0}^{+a} w(x') g_{\omega,\theta}(x' - x) \bar{g}_{\omega,\theta}(-x' - x)\,\mathrm{d}x', \tag{6}$$

where $c_{\omega,\theta}$ is a normalizing factor defined in a similar manner to that in Eqn (3). The resulting measure represents a partial degree of symmetry with respect to the point x along the horizontal axis in the frequency band specified by ω.

This one-dimensional measure can be generalized to a two-dimensional case by perfoming the computation specified by Eqn (6) in all possible directions. For each direction γ and a point $\vec{x}$ we compute the contribution to symmetry of each pair of points equidistant from $\vec{x}$ along a line oriented along γ. This approach is equivalent to computing horizontal symmetry in a coordinate system rotated by γ. The partial symmetry measure then becomes

$$j_{\omega,\theta}(\vec{x}) = c_{\omega,\theta} \int_{0}^{\pi} \int_{0}^{+a} w(x') g_{\omega,\theta-\gamma}\big(\vec{z}(\tau) - \vec{x}\big) \bar{g}_{\omega,\theta-\gamma}\big(\vec{z}(-\tau) - \vec{x}\big)\,\mathrm{d}x'\,\mathrm{d}\gamma, \tag{7}$$

where τ is the distance along the rotated x axis, and $\vec{z}(\tau)$ is the point at the distance τ from $\vec{x}$. Note that the orientation of the filters in the computation of $j_{\omega,\theta}(\vec{x})$ is specified with respect to the rotated coordinate system.

Combination rules. A model of human judgment of symmetry requires that the partial measurements $J_{\omega,\theta}(x_0, y_0, \Delta x, \Delta y)$ be combined into a single decision variable. Because this decision variable must have many properties of a distance metric, we propose an additive combination of partial symmetry measures. In particular, to judge horizontal symmetry with respect to x is

$$S(x) = \sum_{\omega,\theta,x} a_{\omega,\theta}(x) j^{r}_{\omega,\theta}(x), \tag{8}$$

where $0 > r$ is the exponent of a Minkowski r-metric, and a is a weight that controls the relative contribution of each frequency and orientation. By choosing the value of r in the model, it is possible to modify the contribution of the partial measures. For example, setting $r = 1$ leads to the city-block metric, but increasing $r \to \infty$ leads to a metric that selects the maximum partial symmetry. Thus a large value of r a corresponds to the probability summation.

Parameters of symmetry. Given a representation, it is necessary to determine the type of symmetry, and the location of the axis (point) of symmetry. There are several ways to perform such a computation. The algorithm described in this paper is based on an approach motivated by the Hough transform. In this approach, the task of detecting a symmetric relation is converted to a task of detecting simple features such as lines and blobs in the parameter space. For each pair of points that could be a part of a symmetric pattern we compute the center of symmetry. We accumulate these points and then determine the existence of any clusters. A cluster is indicative of a set of points related by a symmetry determined by the type of cluster. For example, a line in this parameter space represents an axis of symmetry.

An implementation of this approach begins with a clear two-dimensional accumulator array $M(x, y)$. For the following description we assume only symmetry with respect to the vertical axis. For each pair of points (x', y) and (x'', y) such that $x' < x''$ the potential center of symmetry (x, y) is given by $x = (x' - x'')/2$. The contribution of the two points (x', y) and (x'', y) to the symmetry measure is given by the product of filtered images contributing to the correlation measure j in Eqn (7). Thus the accumulator is updated by

$$M(x, y) = M(x, y) + \left[h_{\omega,\theta} \star f\right](x' - x)\left[\bar{h}_{\omega,\theta} \star f\right](-x' - x).$$

Following an exhaustive analysis of all pairs of points, the values in the accumulator array will be related to the proportion of points that are symmetric relative to each accumulator location. An axis of symmetry will result in a ridge, and the center of a radially symmetric pattern will be represented by a point. An example of a result of this computation is shown in Fig. 10. The image on the left was obtained by thinning an inverted letter 'V'. In this example, a single axis of symmetry is found even without oriented filters. If an image consists of lines more than one pixel wide, however, even simple patterns will have complex set of symmetries. An example of this computation on a diamond pattern is shown in Fig. 11.

The computational complexity of this algorithm is on the order of n^2. Because the computation can be performed in parallel, it is not biologically infeasible. This model has a sufficient number of degrees of freedom to account for all our data. Much more empirical work is required in order to test the details of this model.

This model will not, however, not account for all symmetry judgments. For example, an odd-symmetric image with a reversed polarity as shown in Fig. 12 would be predicted to be asymmetric. Similarly, perception symmetric patterns constructed using various texture gradients would not be predicted correctly. This type of stimuli must be investigated empirically to determine whether its symmetry is judged by the same mechanism as the stimuli used in this study.

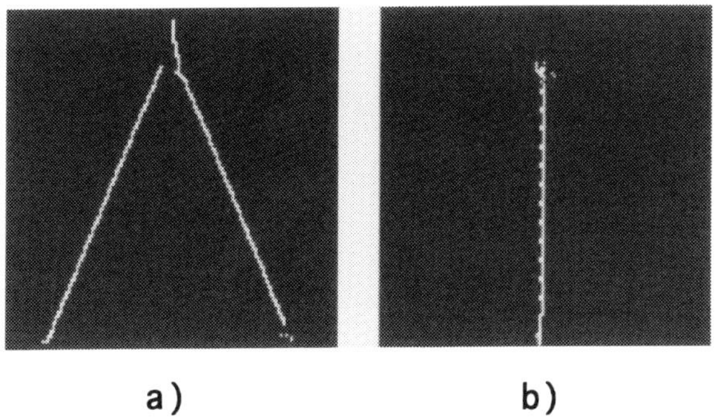

Figure 10. The result of a symmetry computation performed on the left image. The right shows the accumulator array when the computation is restricted to the horizontal axis.

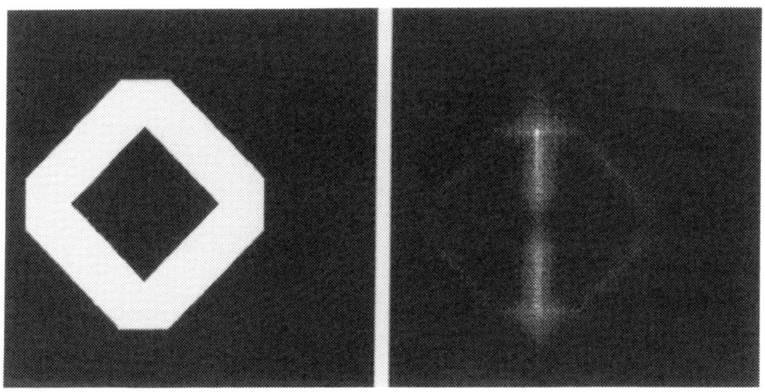

Figure 11. The result of a symmetry computation performed on the diamond pattern shown in the left image. The right image shows the accumulator array when the computation is restricted to the horizontal axis. The intensity in the accumulator array indicates the strength of the resuting symmetry.

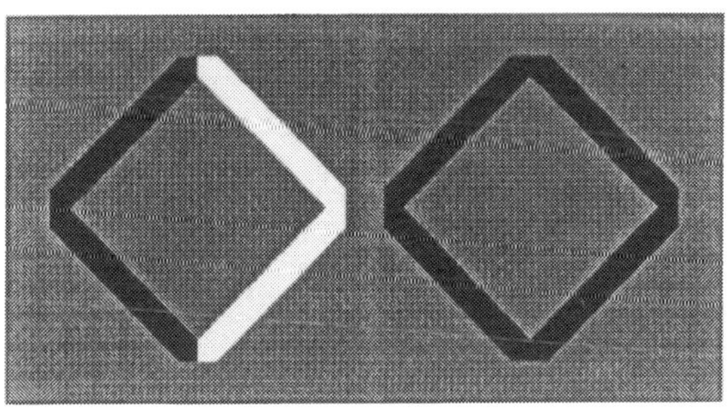

Figure 12. A example of an object with an odd and even symmetries. Both are perceived to be symmetric by the human visual system.

Acknowledgements

We thank Helen Cunningham for many suggestions regarding experimental procedures, Al Ahumada for his comments on theoretical developments, and Holly Jimison for her editorial suggestions. This work was supported in part by a grants from NASA No NAG-2-931 and NCC-2-811 to Oregon Graduate Institute.

NOTES

1. Disks were selected because they contain the same amount of energy at each orientation.

REFERENCES

Adelson, E. H., Bergen, J. R., Burt, P. J. and Ogden, J. M. (1984). Pyramid methods in image processing. *RCA Eng.* **29**, 31–41.

Attneave, F. (1954). Some informational aspects of visual perception. *Psychol. Rev.* **3**, 183–193.

Attneave, F. (1955). Symmetry, information and memory for patterns. *Am. J. Psychol.* **68**, 209–222.

Barlow, H. B. (1961). Possible principles underlying the transformations of sensory messages. In: *Sensory Communication.* W. Rosenblith (Ed.). Wiley, New York.

Barlow, H. B. and Reeves, B. C. (1979). The versatility and absolute efficiency of detecting mirror symmetry in random dot displays. *Vision Res.* **19**, 783–793.

Biswas, S. N. and Majumder, D. D. (1988). Region symmetry analysis through cantor number strings. In: *Proc. of 9th International Conference on Pattern Recognition*, pp. 427–429.

Brady, M. (1982). Smoothed local symmetries and local frame propagation. In: *Proc. Pattern Recognition and Image Processing.* Las Vegas, pp. 629–633.

Buffart, H., Leeuwenberg, E. and Restle, F. (1981). Coding theory of visual pattern completion. *J. Exp. Psychol.: Human Percept. Perform.* **7**, 241–274.

Burbeck, C. A. (1987). Position and spatial frequency in large-scale localization judgments. *Vision Res.* **27**, 417–427.

Campbell, F. W. and Robson, J. (1968). Application of Fourier analysis to the visibility of gratings. *J. Physiol.* **197**, 551–556.

Chaudhuri, B. B. and Majumder, D. D. (1980). Recognition and fuzzy description of sides and symmetries of figures by computer. *Int. J. Systems Sci.* **11**, 1435–1445.

Corballis, M. C. and Beale, I. L. (1976). *The Psychology of Left and Right. Hillsdale.* Erlbaum, NJ.

Cumming, W. J. K. (1970). An anatomical review of the corpus callosum. *Cortex* **6**, 1–18.

Cunningham, H. A. and Pavel, M. (1986). Judgments of position in near and far peripheral visual fields. *Invest. Ophthalmol. Visual Sci.* **27** (Suppl.) (3), 95.

Deregowski, J. B. (1971). Symmetry, gestalt, and information theory. *Quarterly J. Exp. Psychol.* **23**, 381–385.

Dinnersten, D. and Wertheimer, M. (1957). Some determinants of phenomenal overlapping. *Am. J. Psychol.* **70**, 21–37.

Fitts, P. M., Weinstein, M., Rappaport, M., Anderson, N. and Leonard, J. A. (1956). Stimulus correlations of visual pattern recognition: a probability approach. *J. Exp. Psychol.* **51**, 1–11.

Gibson, J. J. (1979). *The Ecological Approach to Visual Perception.* Houghton Mifflin, Boston.

Graham, N. and Nachmias, J. (1971). Detection of grating patterns containing two spatial frequencies: a comparison of single-channel and multiple-channel models. *Vision Res.* **11**, 251–259.

Hong, S. (1991). Modeling of rapid symmetry detection, Ph.D. Dissertation, Stanford University.

Hubel, D. and Wiesel, T. N. (1977). Ferrier lecture: Functional architecture of macaque monkey visual cortex. *Proc. R. Soc., London* **B198**, 1–59.

Jenkins, B. (1982). Redundancy in the perception of bilaterally symmetric dot textures. *Percept. Psychophys.* **32**, 171–177.

Jenkins, B. (1983). Component process in the perception of bilaterally symmetric dot textures. *Percept. Psychophys.* **34**, 433–440.

Julesz, B. (1971). *Foundations of Cyclopean Perception.* University of Chicago Press, Chicago.

Kanizsa, G. (1979). *Organization in Vision: Essays on Gestalt Perception.* Praeger, New York.

Levi, D. M., Klein, S. A. and Aitsebaomo, A. P. (1985). Vernier acuity, crowding and cortical magnification. *Vision Res.* **25**, 963–977.

Locher, P. J. and Nodine, C. G. (1973). Influence of stimulus symmetry on visual scanning patterns. *Percept. Psychophys.* **13**, 408–412.

Mach, E. (1897). *Analysis of the Sensations.* Open Court Publishing House, Chicago.

Parui, S. K. and Majumder, D. D. (1983). Symmetry analysis by computer. *Pattern Recogn.* **16**, 63–67.

Pizlo, Z. (1988). Physiology based simulation model of triangle shape recognition. *Biol. Cybernet.* **58**, 51–62.

Rock, I. (1975). *An Introduction to Perception.* Macmillan, New York.

Tyler, C. W. and Miller, R. T. (1994). Pattern identification by trajectory analysis in autocorrelation hyperspace. In: *Proc. of the World Congress on Neural Networks* **III**, pp. 312–316.

Watson, A. B. (1987). The cortex transform: Rapid computation of simulated neural images. *Computer Vision, Graphics, and Image Processing* **39**, 311–327.

Wilson, H. R. and Bergen, J. R. (1979). A four mechanism model for threshold spatial vision. *Vision Res.* **19**, 19–32.

Yap, Y. L., Levi, D. M. and Klein, S. A. (1987). Peripheral hyperacuity: three-dot bisection scales to a single factor from 0 to 10 degrees. *J. Opt. Soc. Am.* **4**, 1554–1561.

Human Symmetry Perception, pp. 157–171
C. W. Tyler (Ed.)

Mirror symmetry detection: predominance of second-order pattern processing throughout the visual field

CHRISTOPHER W. TYLER and LANI HARDAGE
Smith-Kettlewell Eye Research Institute, 2232 Webster St., San Francisco, CA 94115, USA

Abstract—Size-scaled random Gaussian blobs were used to determine the range of pattern matching involved in human perception of mirror symmetry and to evaluate the relative sensitivity of local (linear receptive field) or long-range (polarity-insensitive pattern-matching) mechanisms as a function of retinal eccentricity. Observers were able to perform symmetric pattern matching with presentations shorter than 100 ms across as much as 64 deg of visual angle. Sensitivities to opposite-polarity symmetry and low-density patterns suggest that this performance was mediated predominantly by polarity-insensitive mechanisms at all eccentricities. The results militate against a pattern-matching process based solely on local processing and implicate the involvement of long-range connections spanning the visual cortex. Outside the fovea, eccentricity scaling for all symmetry tasks was a close match to the self-scaling used for the stimuli in our experiments.

1. INTRODUCTION

Visual symmetry is of obvious importance to humans, as evidenced by the ubiquitous use of design symmetries in craftsmanship and fabricated environment, from the palaces, formal gardens and diadems of the past to the cars, computers and high-rise structures of today. Detection of mirror-symmetric objects in the visual field may be adaptive to survival in the animal world, where recognition of predators and prey could be based in part on discrimination of an animal's bilateral symmetry from the generally asymmetric background flora. In particular, when an animal turns to face the observing organism, seeing a meal or a mate, it displays its symmetry and becomes perceptually salient. Human symmetry detection has been well studied in the fovea (Julesz, 1971; Corballis and Roldan, 1974; Bruce and Morten, 1975; Barlow and Reeves, 1979; Jenkins, 1982), where it can be discriminated from noise in less than 100 ms (Carmody *et al.*, 1977; Barlow and Reeves, 1979; Tyler *et al.*, 1995). The rapidity of this processing suggests a local cortical mechanism in an area such as V1, supported by recent neurophysiological evidence that complex cells show specialized responses to the medial axis of textured stimuli placed symmetrically in the receptive field (Lee *et al.*, 1995).

In other types of visual processing, notably motion perception, separate mechanisms for short-range and long-range pattern matching have been identified (Braddick,

1974; Larsen *et al.*, 1983; Ledgeway and Smith, 1994; Smith, 1994; Solomon and Sperling, 1995). The short-range mechanisms are typically identified as first-order or pattern-specific (possibly mediated by the approximately *linear* or first-order response properties of simple-cell signals), whereas the long-range mechanisms typically show the second-order property of a relative independence from pattern detail and contrast polarity in the stimulus (possibly mediated by the *rectified* response properties of complex-cell signals). Most studies of symmetry detection in both the fovea and periphery have investigated its dependence on short-range mechanisms only (Mach, 1886; Julesz, 1971; Barlow and Reeves, 1979; Kahn and Foster, 1981; Jenkins, 1982; Saarinen, 1988; Locher and Nodine, 1989; Herbert and Humphrey, 1993; Wenderoth, 1995), implicating local, pattern-specific mechanisms. The discovery of a complex-cell signal at the symmetry axis (Lee *et al.*, 1995) makes it important to evaluate the role and extent of long-range mechanisms in symmetry detection.

We therefore designed experiments to determine the range of pattern matching involved in human perception of mirror symmetry and to evaluate the extent to which local (linear receptive field) or long-range (second-order pattern-matching) mechanisms were involved as a function of retinal eccentricity. The role of first- and second-order pattern-processing mechanisms in symmetry detection performance for these stimuli was evaluated in two ways: by varying contrast polarity of symmetric elements and by varying density of patterns. First-order pattern matching is, by definition, polarity-specific; a comparison between similar elements of opposite polarity will produce a negative response in a first-order mechanism and be rejected as a match. A good example is provided by cortical binocular cells, which show enhanced response only for stimuli of the same polarity in both eyes. Strictly, therefore, any ability to detect symmetry for opposite-polarity stimuli implies the presence of a polarity-insensitive or second-order mechanism.

However, this strong interpretation of first-order exclusion relies on the assumption of absolute position specification. If the position of a modulated pattern has some uncertainty, an adequate first-order match may be obtained from adjacent components of the response that happen to be of opposite polarity. When the opposite-polarity manipulation of the stimulus switches the components to the same polarity, the components in roughly corresponding positions in the image may be sufficient to produce a match in a first-order mechanism. To evaluate this possibility, we included a manipulation of the pattern density within the uniform grey background. When the density of black and white blobs around uniform grey is manipulated from 100% to 10% (Fig. 1(a) and (b)), first-order (linear) and second-order (phase-insensitive) mechanisms would be expected to respond differently. A high-density pattern (Fig. 1(a), left) should stimulate more first-order processing units than a low-density pattern (Fig. 1(a), right), by virtue of the increased number of visible elements. If first-order processing predominates, the predicted result (Fig. 1(c), left) is a greater sensitivity to the high-density pattern even if there were positional uncertainty in the first-order matching process.

The response of second-order mechanisms, on the other hand, is reflected in the (rectified) contrast envelope of the blob pattern. The contrast envelope is modulated only a little in a high-density pattern (Fig. 1(b), left) whereas its amplitude goes all

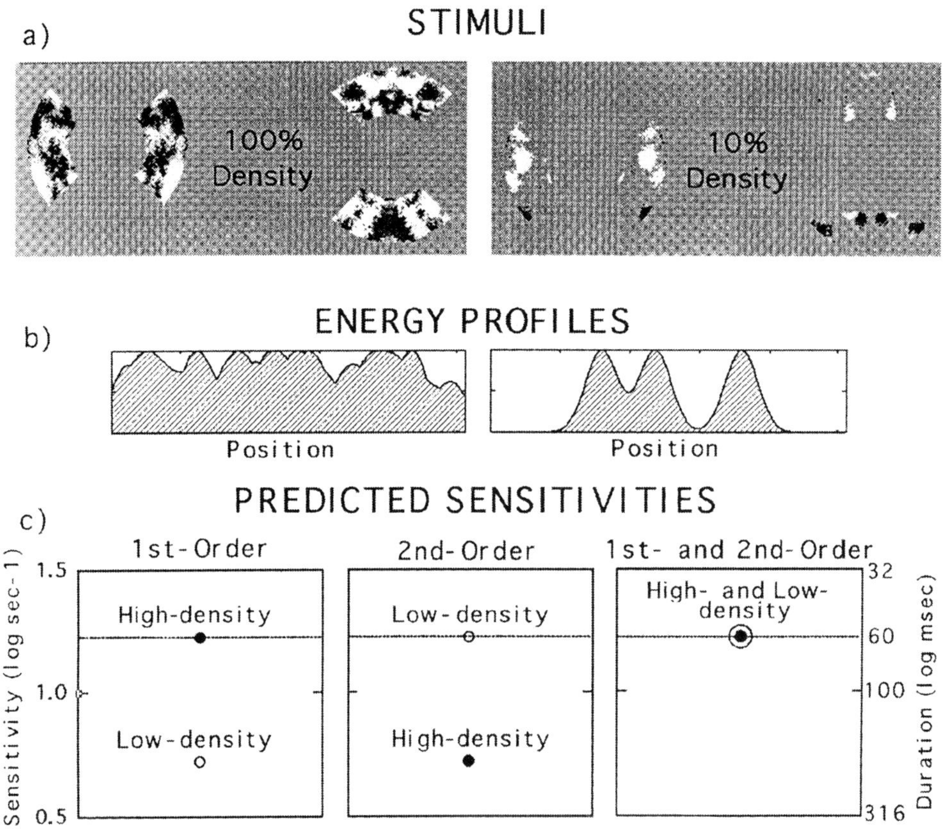

Figure 1. Predictions of first- and second-order mechanism responses to high- and low-density patterns. (a) Examples of the horizontal (left icons) and vertical (right icons) quadrant annuli stimuli at 100% (left panel) and 10% (right panel) blob densities. (b) Typical rectified energy profiles for 100% (left) and 10% (right) blob densities. (c) Qualitative predictions for the relative sensitivity relationships for high- and low-density stimuli for predominantly first-order (left panel), predominantly second-order (centre panel) and combined (right panel) symmetry-processing mechanisms.

the way from maximum to zero in a low-density pattern (Fig. 1(b), right). The counterintuitive prediction for predominant second-order processing (c, center), therefore, would be a greater sensitivity to the low-density pattern due to its greater modulation of the overall contrast envelope compared to that in a high-density pattern. If both types of mechanism are present, the most sensitive mechanism is expected to predominate (Fig. 1(c), right). Thus, the relationship between sensitivities for the high- and low-density pattern provides a second assay of the relative role of the first- and second-order mechanisms in the pattern matching processes of symmetry perception.

2. METHODS

The test patterns consisted of Gaussian blobs coloured randomly either black or white, reflected bilaterally, with the same or opposite color, about a vertical axis. The blobs were generated by random modulation of pixel density in a Gaussian distribution. To compensate for the rapid fall-off of spatial frequency resolution in the periphery, the test images were scaled in both element size and aperture size with eccentricity. To avoid an *a priori* judgment of the retinocortical scaling factor involved in this task, we employed a self-scaling image configuration (Tyler and Apkarian, 1982; Watson, 1987) in which the image was scaled in direct proportion to its eccentricity simply by changing the viewing distance of the observer from the display. The applicable retinocortical scaling factor may then be evaluated from the residual sensitivity variations with stimulus eccentricity. The resulting stimulus consisted of bilaterally reflected, randomly-coloured blobs scaled in proportion to distance from the fovea, with their diameter doubled from the inner to the outer border of a sector, as shown in exploded view in Fig. 2 (offset and overlaid). For each doubling of eccentricity, the blobs were scaled to cover twice the distance on the retina, permitting a comparison of local scale at points across the retina. The annuli were windowed into opposing 90 deg quadrants of either a vertical (Fig. 2(a)) or horizontal (Fig. 2(b)) pair. The vertical quadrant annulus presented patterns with a local vertical symmetry axis visible, whereas horizontal sectors presented widely separated symmetric regions with no information near the symmetry axis. Random dynamic noise blobs were presented during pre- and post-test periods of the 2 s stimulus interval to obliterate processing of the symmetry based on afterimages of the test stimuli.

The pairs of quadrant annuli were presented at 6 viewing distances to cover eccentricities on the retina of 1–2 deg, 2–4 deg, 4–8 deg, 8–16 deg, 16–32 deg and 32–64 deg; at the greatest eccentricity, pairs spanned 128 deg of visual angle or a mean eccentricity of 90 deg. Additional eccentricities of 0.25–0.5 deg and 0.5–1 deg

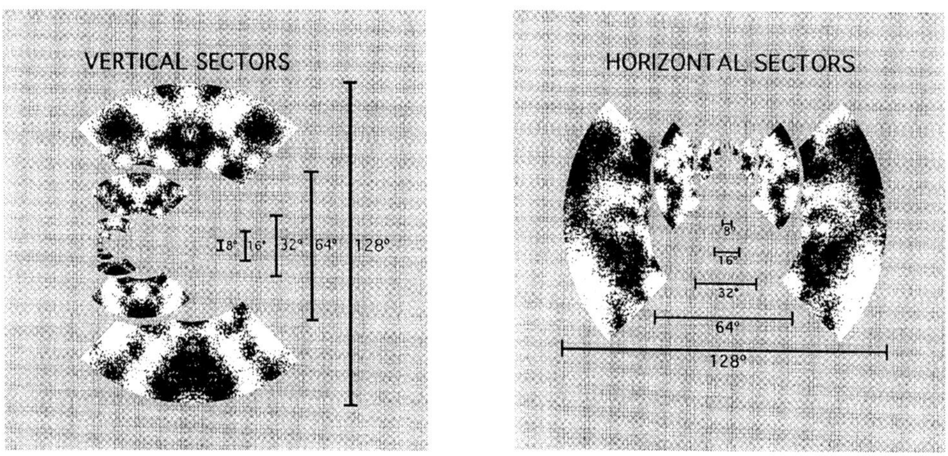

(a) (b)

Figure 2. Pairs of quadrant annular sectors presenting the random-blob stimuli used for testing symmetry discrimination with size-scaling as a function of eccentricity. (a) Vertical sectors containing the symmetry axis. (b) Horizontal sectors requiring pattern matching between the pairs to extract the symmetry information. Each pair of sectors is offset laterally or vertically from its neighbors to provide a clearer view of the individual shapes.

were presented to one observer to evaluate foveal sensitivity. In each testing condition, either vertical (Fig. 2(a)) or horizontal (Fig. 2(b)) sector pairs were selected, with the remaining quadrants masked to match the mean-luminance background. Patterns of 10% density were also generated (Fig. 1(a) right), in which 90% of the blobs were set to the mean-luminance background. The stimuli were generated on a Macintosh IIfx with a Motorola 68030 CPU and presented on a monochrome monitor for durations varying in increments of 15 ms, from 30 ms to 1800 ms.

Measurements were taken by the method of constant stimuli with durations separated by intervals of about 0.5 octaves chosen to span a range of performance levels from a level of below 60% to above 90% correct. There were blocks of 150 trials per duration, and observers were instructed to identify symmetry with a 'yes' response and asymmetry (random blobs) with a 'no' response.

3. RESULTS

3.1. Psychometric functions

Psychometric functions for same-polarity symmetry detection (Fig. 3) for the two observers were unusual and did not resemble the psychometric functions typical of contrast sensitivity studies (Nachmias and Kocher, 1970; Stromeyer and Klein, 1975; Foley and Legge, 1981). Vertical or horizontal sector pairs were presented at 100% density; data for 3 eccentricities are plotted. For both vertical sectors (upper panel) and horizontal sectors (lower panel), observers' sensitivity increased rapidly with duration up to about 75 ms, and then often increased at a markedly slower rate. The segmented properties of these psychometric functions are significant in revealing the contributions of discrete mechanisms at shorter and longer durations (Tyler *et al.*, 1995). With the scaling depicted in Fig. 2, the psychometric functions are similar across the visual field, suggesting little change in the cortical processing properties with eccentricity beyond that implied by the self-scaling of the stimuli.

Opposite-polarity symmetry elicited similar psychometric functions (Fig. 4) from both observers. Vertical (top panels) or horizontal (bottom panels) were presented at 100% density, and data for 3 representative eccentricities are shown. For both vertical and horizontal sectors, observers' sensitivity again increased rapidly for short durations, then increased at a much lower rate. As with same-polarity symmetry detection, the psychometric functions suggest the presence of discrete mechanisms at shorter and longer durations and little change in properties with eccentricity.

3.2. Eccentricity variation for same-polarity symmetry

Sensitivity was estimated from the psychometric functions as the reciprocal of duration required to reach a threshold criterion of $d' = 0.5$ (well above the theoretical standard deviation of 0.2 for 150 trials per datum point). The criterion is shown as the dashed line in Figs 3 and 4. This low criterion for threshold was selected in order to stay within the steep portion of the psychometric function (see Fig. 5), resulting in

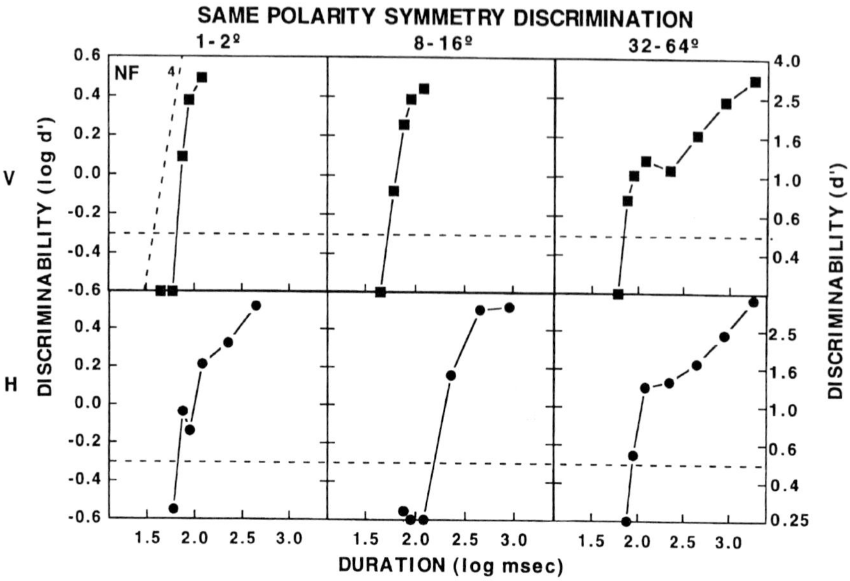

(a)

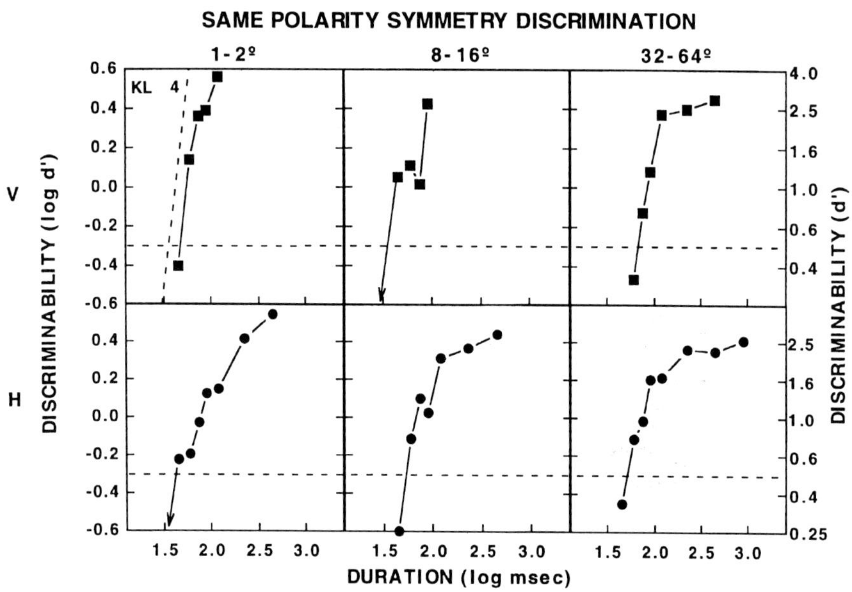

(b)

Figure 3. Psychometric functions for same-polarity symmetry discrimination across eccentricity. The psychometric functions from two naive observers show $\log d'$ as a function of duration in log ms. Data from three representative eccentricities are presented: 1–2 deg, 4–8 deg and 32–64 deg. For durations below about 75 ms, functions show a steep slope of about 4 (dashed line in first panel). The slope becomes shallower at longer durations. Shapes of the psychometric functions are quite similar across eccentricity, with minor differences seen at higher d' values. A threshold criterion of $d' = 0.5$ (horizontal dashed line in all panels) was chosen on the steepest portion of the psychometric function to minimize variability in the detection analysis.

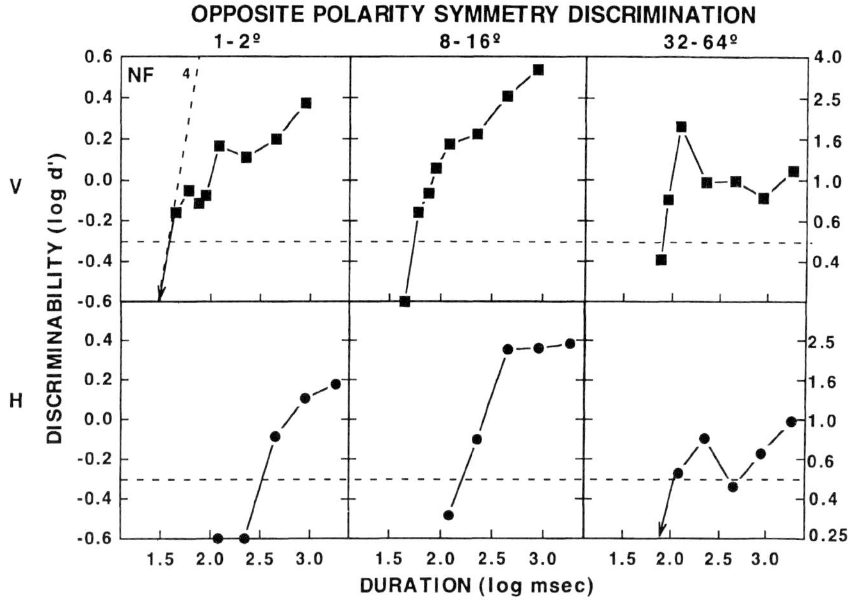

(a)

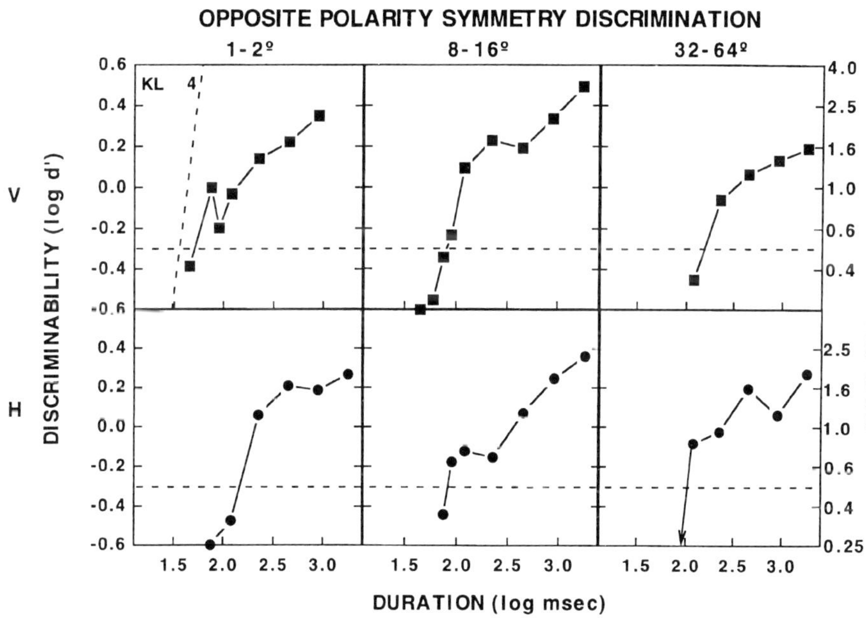

(b)

Figure 4. Psychometric functions for opposite-polarity symmetry discrimination across eccentricity in the same format as Fig. 3. For durations below about 75 ms for the vertical conditions, longer for the horizontal conditions, functions show a steep slope of about 4 (dashed line in first panel) becoming shallower at longer durations. Shapes of the psychometric functions are similar across eccentricity, with minor differences seen at higher d' values. The threshold criterion of $d' = 0.5$ (horizontal dashed line in all panels) targets the steepest portion of the psychometric functions, as in Fig. 3.

substantially less variability in estimate of sensitivity than if the criterion encroached into the shallow portion, at higher d' values.

Sensitivities for detection of same-polarity symmetry at two pattern densities in vertical and horizontal sector pairs are shown in Fig. 5 for eccentricities ranging from fovea to extreme periphery. The principal result for both observers is that, under the self-scaling regime, duration sensitivity for both horizontal and vertical sectors is essentially uniform across eccentricity for stimuli outside the fovea, implying that the self-similar scaling matched within a factor of about $\pm\sqrt{2}$ the cortical scaling factor for this task. The slopes of the eccentricity functions are not significantly different from zero for any condition (at the criterion of $p < 0.08$ after correction for multiple tests).

Within the fovea (eccentricities less than 2 deg), sensitivity was measured for vertical same-polarity stimuli for the 100% density condition for observer NF only; sensitivity dropped by about a factor of two toward the central fovea. To evaluate how this drop compared with the prediction from cortical magnification studies, we plot the sensitivity predictions for cortical magnification from Klein and Levi (1987) (Fig. 5, observer NF, solid curves). The two curves represent the steepest average scaling for a positional acuity task (upper curve) and the shallowest average prediction for a resolution task (lower curve) that they reported. (Note that the steeper scaling falls closest to the complete self-scaling employed for our stimuli.) The data fall closer to self-scaling even than the resolution-task prediction, implying that cortical magnification for symmetry detection is at least as steep as for the steepest magnification for any positional acuity task (Klein and Levi, 1987), i.e. symmetry scaling is essentially self-scaling beyond about 2 deg eccentricity.

For both vertical sectors, which contain local symmetry-axis information, and horizontal sectors, which do not, sensitivity was significantly higher for the low-density patterns than for the high-density ones ($p < 0.01$, combined across observers). According to the prediction of Fig. 1(c), this relationship implies that the long-range symmetrical pattern matching, which requires comparisons over approximately 90 deg of visual angle at the extreme eccentricity, is mediated predominantly by second-order (polarity-insensitive) mechanisms of pattern-matching. Thus, even when the symmetry axis was present (vertical sectors), the observers appeared to rely on second-order mechanisms to detect the symmetry. The similarity of sensitivities for low- and high-density patterns nevertheless suggests substantial involvement of first-order mechanisms, although our current understanding of symmetry processing is insufficient to allow quantitative predictions.

It is noteworthy that, in the low-density condition, the observers were able to discriminate symmetric from non-symmetric patterns for presentations of 75 ms or less at any eccentricity. The same was true for many high-density stimuli, although one observer required up to 200 ms for stimuli away from the symmetry axis (horizontal condition, Fig. 5, filled symbols). Sample low- and high-density stimuli are reproduced in Fig. 6 for the reader to compare the visibility of peripheral symmetry in the two cases.

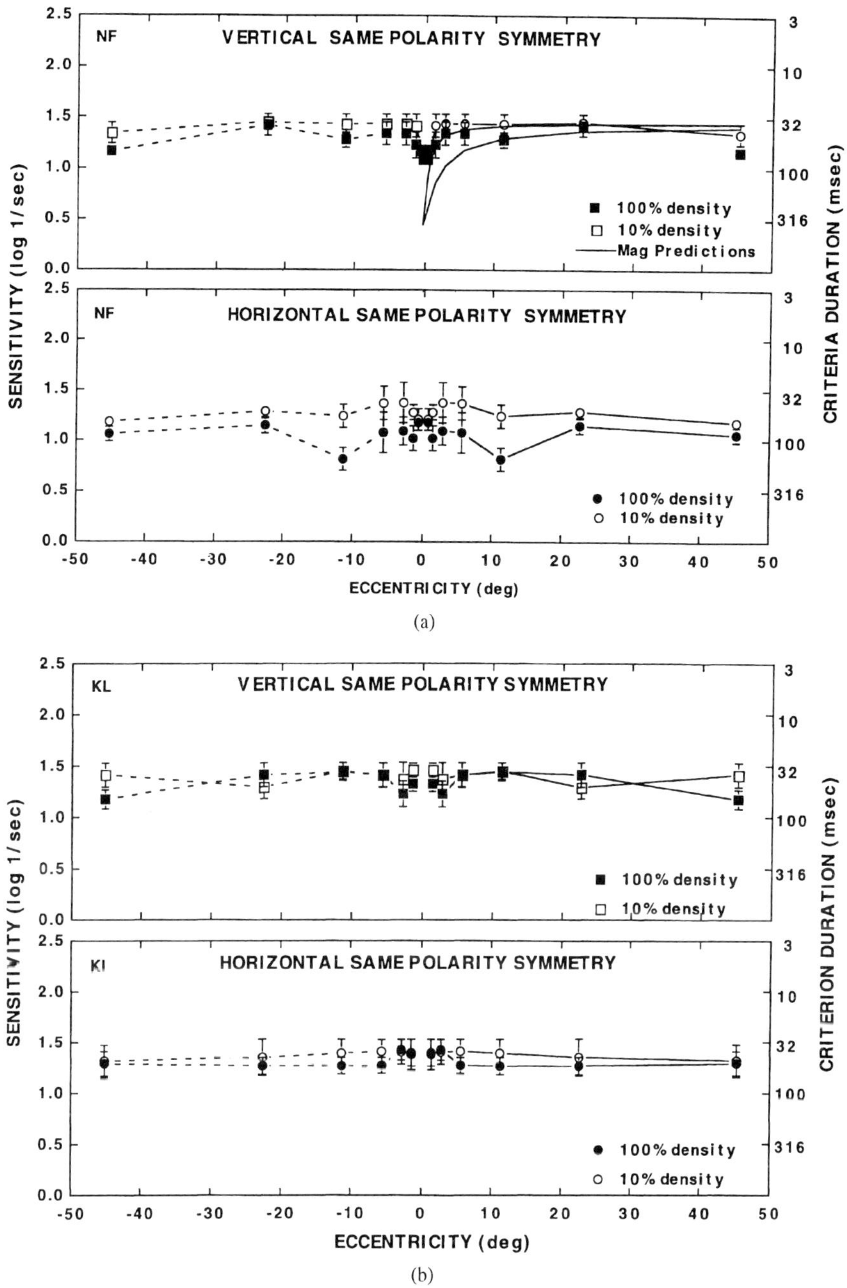

Figure 5. Sensitivities for same-polarity symmetry at two pattern densities in sector pairs at mean eccentricities ranging from 0.3–45 deg, for two observers. Upper panels, discrimination of symmetry in vertical sectors; lower panels, symmetry discrimination in horizontal sectors. Filled symbols: 100% blob density; open symbols, 10% blob density with grey background. With the self-scaling depicted in Fig. 1, duration sensitivity for both observers is uniform across eccentricity (±0.1 log unit in each condition). The slopes of the eccentricity functions from 3 deg to 45 deg show no change in sensitivity even in the extreme periphery.

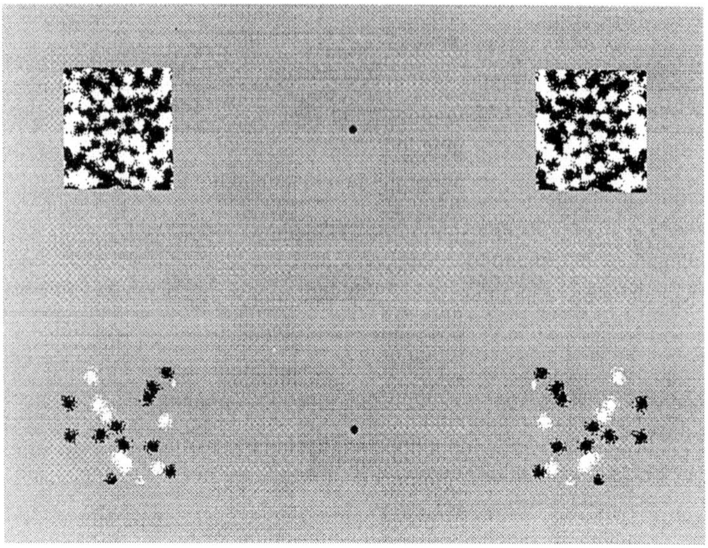

Figure 6. Examples of same-polarity symmetry. Note that, with fixation on the upper fixation dot, it is hard to detect symmetry in the high-density pair, even with prolonged viewing. The symmetry of low-density pair, on the other hand, is immediately obvious on central fixation of the lower dot, illustrating the ready long-range association between the laterally-flipped patterns.

3.3. Eccentricity variation for opposite-polarity symmetry

Since the density variation was designed to test for the presence of a polarity-independent processing for symmetry, we ask whether switching the stimuli to have opposite contrast polarity on either side of the symmetry axis has any effect on the detection the symmetry at any of the eccentricities tested. This manipulation should abolish processing by a polarity-specific mechanism Sensitivities for detection of opposite-polarity symmetry were therefore evaluated at the same criterion of $d' = 0.5$ as for the same-polarity stimuli. Data for the two pattern densities in the vertical and horizontal sector pairs are shown in Fig. 7 for eccentricities ranging from fovea to extreme periphery. Remarkably, the observers were again able to discriminate symmetric from non-symmetric patterns of opposite polarity for presentations of 75 ms or less at any eccentricity when the patterns were of low density. For this density, there was no significant loss in sensitivity relative to the same-polarity condition, supporting the implication that symmetry detection in both conditions was mediated by a second-order mechanism.

The same conclusion is implied by the effect of density variation on sensitivities for the opposite-polarity stimuli (Fig. 7). In general, duration sensitivity for high-density patterns was reduced by about a factor of two relative to the low-density conditions (significant at $p < 0.01$, with the exception of the vertical condition for observer NF). Thus, the sensitivity relationships corresponded to that predicted for a predominantly second-order task (center panel of Fig. 1, lower). (Readers may compare for themselves the visibility of the opposite-polarity stimuli provided in Fig. 8 with that of the same-polarity stimuli of Fig. 6.)

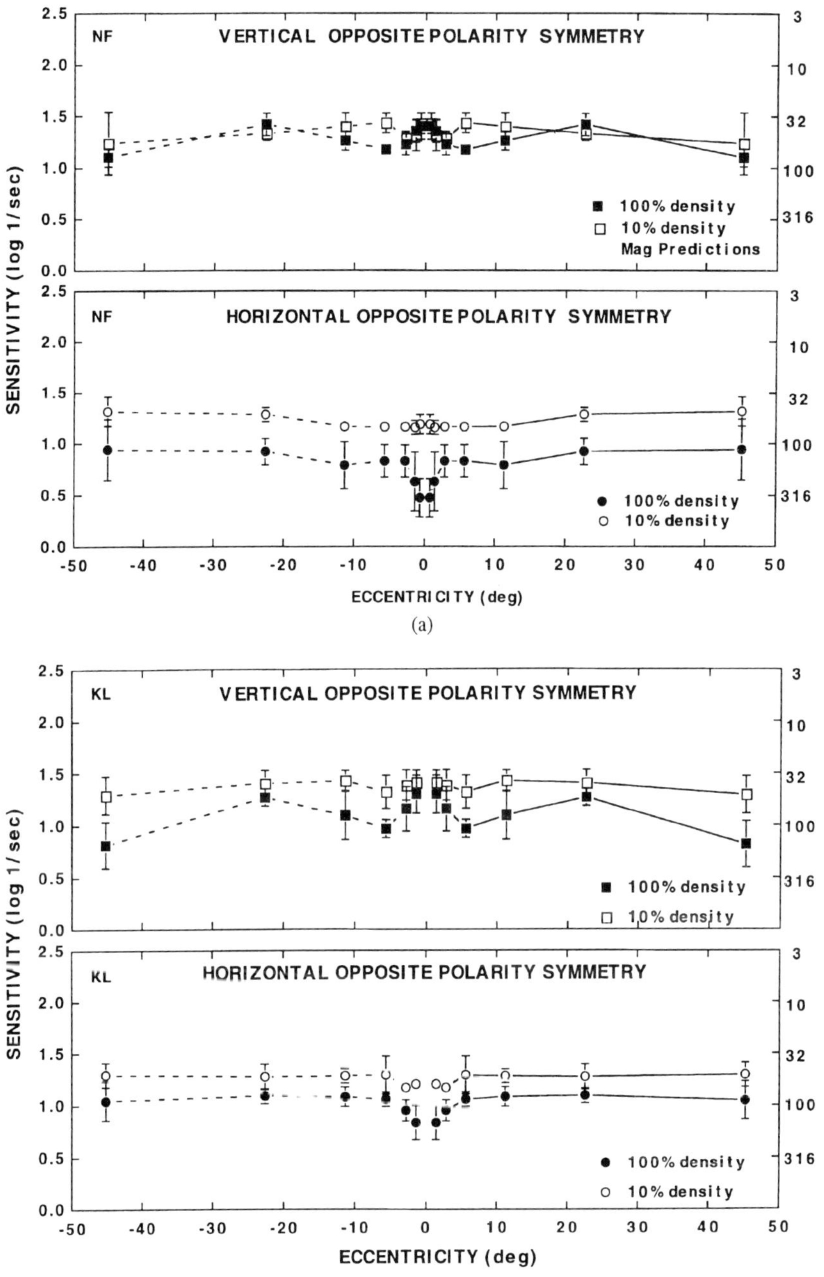

Figure 7. Sensitivities for opposite-polarity symmetry at two pattern densities in sector pairs at eccentricities ranging from 1.4–45 deg, in same format as Fig. 5. Except in one condition, sensitivity is significantly greater for low-density patterns (open symbols), at all eccentricities. High-density patterns (filled symbols) reveal a foveal specialization for opposite-polarity symmetry near the symmetry axis: for both observers, there is a peak in sensitivity within 4 deg of the fovea for vertical annuli and a reciprocal reduction in sensitivity for horizontal annuli.

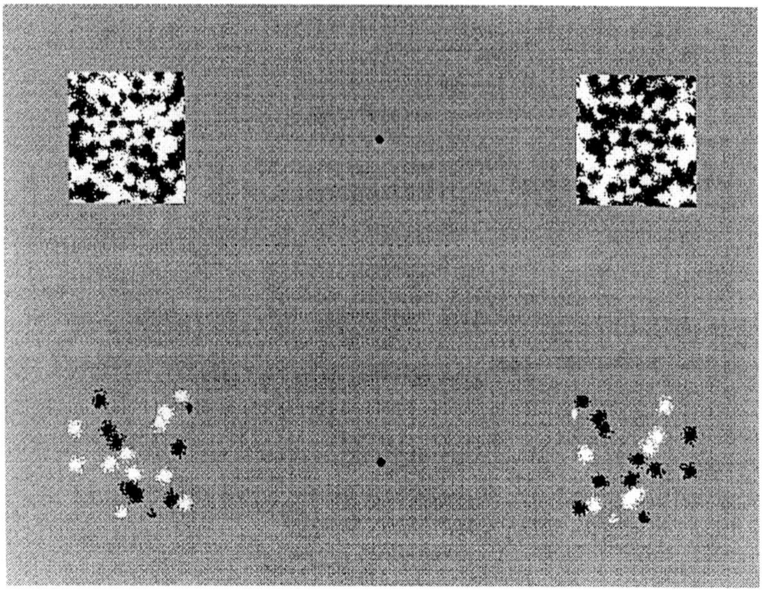

Figure 8. Examples of opposite-polarity symmetry in the same format as Fig. 6. Note that, with fixation on the upper fixation dot, it is again quite difficult to detect symmetry in the high-density pair. The symmetry of the low-density pair, on the other hand, sings out strongly during central fixation despite the polarity switch, illustrating the ready long-range association between the doubly-inverted patterns. With alternating foveal inspection, on the other hand, one tends to associate blobs of the same polarity and pick out the approximate same-polarity match that is present by chance. Note also that, in the low-density pair, the polarity inversion of the symmetric elements is perceptually evident, supporting the interpretation that both second- and first-order mechanisms are operating in this long-range pattern-matching process.

As seen in results for same-polarity stimuli, duration sensitivity for both observers for both horizontal and vertical sectors is essentially uniform across eccentricity for stimuli outside the fovea, again implying that the self-similar scaling closely matched the cortical scaling factor for this task. Between 1 deg and 64 deg in the periphery, the slopes of the eccentricity functions are not significantly different from zero for any condition (at the criterion of $p > 0.1$ after correction for multiple tests).

In addition, high-density patterns (filled symbols) reveal a perifoveal specialization for opposite-polarity symmetry near the symmetry axis. For both observers, there is a significant peak in sensitivity for vertical sectors within 4 deg of the fovea. (Curiously, there is a second statistically significant peak for both observers at 35 deg eccentricity, but this is harder to interpret.) A reciprocal *reduction* in sensitivity appears near the fovea for horizontal sectors. There is, conversely, no significant variation with eccentricity for low-density stimuli. The perifoveal improvement for dense vertical sectors is not unexpected, given the degree of foveal specialization that has often been reported (e.g. Rovamo and Virsu, 1979; Tyler, 1986) but the relative loss in sensitivity for dense horizontal sectors suggests that some more specific processing change is occurring, as will be developed in Discussion.

4. DISCUSSION

The results show that observers were able to perform symmetric pattern matching with presentations shorter than 75 ms across as much as 64 deg of visual angle. The effects of density manipulation suggest that this performance was mediated predominantly by polarity-insensitive mechanisms at all eccentricities. The results militate against a pattern-matching process based solely on local processing, such as classical receptive fields, and implicate the involvement of long-range connections in the cortex. Some neural system must be able to compare the form of unexpected stimulus patterns presented at as much as a 90 deg separation for temporally masked presentations of 75 ms or less. Its precise nature remains a puzzle for future research, however.

The foveal specialization seen in the data for 100% density, opposite-polarity stimuli (Fig. 7) suggests an interesting hypothesis about early processing mechanisms. It is well known that the periphery has poor phase discrimination (Rentschler and Treutwein, 1985; Klein and Tyler, 1986; Bennett and Banks, 1987; Bennett and Banks, 1991). Bennett and Banks (1991), in particular, provide evidence that the periphery is limited to the representation of even-symmetric phase. Suppose, then, that the fovea alone had independent access to signals from odd-symmetric cellular mechanisms. Such signals would provide a first-order representation of the opposite-polarity symmetry around the symmetry axis, explaining the improvement for the 100% density stimuli specific to vertical sectors when presented near the fovea.

Horizontal sectors, on the other hand, always remain too far apart to provide a basis for processing by local first-order mechanisms. They were designed as a probe of long-range second-order processing for symmetry. The consistent advantage of the 10% density stimuli for detection in horizontal sectors (Figs 5 and 7) bears out the second-order hypothesis. In this regard, it is remarkable that the detectability for opposite-polarity symmetry at low density over a long range is comparable to that of any other condition. Clearly, long-range processing is the dominant pattern-recognition process throughout much of the retina (as demonstrated in Fig. 8). However, just where the short-range (paraxial, high-density) processing seems to be enhanced for vertical sectors, the long-range processing is degraded (Fig. 7). Perhaps the same mechanisms that are segregated to provide the first-order phase information are removed from service in second-order (phase-insensitive) mechanisms. The low-density stimuli may be a poor probe for this change because the patches are sufficiently widely separated to provide an adequate second-order match even without the odd-symmetric contribution to a phase-invariant signal. High-density stimuli, on the other hand, are more demanding, in that loss of the odd-symmetric component will produce a local variation in the second-order envelope signal that can disrupt its match to that in the opposite patch. Thus, the reciprocal perifoveal effects for high-density vertical and horizontal sectors are consistent with a restriction of odd-symmetric mechanisms from contributing to second-order processing near the fovea.

One issue that should be addressed is the difference between the form of the present results for high-density same-polarity stimuli and those for uniform random-dot stimuli in Tyler *et al.* (1995), in which the sensitivities showed a head, shoulders and skirts appearance as information was progressively removed from around the symmetry axis. The main comparison, accordingly, would be with the horizontal (axis omitted) conditions near the fovea in the present study. The present stimuli differ in four important respects. First, they are made of Gaussian blobs, so they constitute a low-pass filtered version of the fine random dot stimuli of the earlier study. Second, they are restricted to small, roughly homogeneous, retinal regions. Third, both the blob size and the region size is scaled with eccentricity. Finally, they go out to much greater eccentricities.

It is certainly not surprising that the eccentricity scaling would eliminate the dramatic decline with eccentricity seen for the random-dot stimuli. The present data thus support and amplify the earlier conclusion that sensitivity for static symmetry extends far farther into the periphery than previously suggested. There is, however, little evidence of any local specialization from the presence of discrete mechanisms across eccentricity in the high-density same-polarity conditions of Fig. 5 (except, perhaps for observer NF at 10 deg). By limiting the stimulus elements to relatively low-spatial-frequency blobs and constraining the area to a small patch, the present stimuli may be expected to have focused on the sensitivity of one type of neural mechanism, eliminating the possibility of symmetry detection from either fine, local details or elongated features of fine filaments, for example. Thus, the two types of stimuli are not commensurate because they were designed to address different aspects of the neural processing of symmetry.

5. CONCLUSION

The remarkable pattern-processing abilities revealed by symmetry discrimination across the entire range of visual eccentricity pose a challenge for models of cortical pattern recognition. Both first- and second-order mechanisms must be available to match up the symmetric information from one cortical hemisphere to the other on the basis of an extremely short stimulus presentation followed by a masking stimulus. This interpretation is not intended to imply that all symmetry processing is interhemispheric, merely that it happens to be interhemispheric for our particular stimulus configuration. The peak response evoked when the median axis of a textured strip is placed within the receptive field of some complex cells in area V1 (Lee *et al.*, 1995) could subserve a limited form of this long-range pattern-matching role.

Acknowledgment

Supported by NIMH grant #49044.

REFERENCES

Barlow, H. B. and Reeves, B. C. (1979). The versatility and absolute efficiency of detecting mirror symmetry in random dot displays. *Vision Res.* **19**, 783–793.

Bennett, P. J. and Banks, M. S. (1987). Sensitivity loss in odd-symmetric mechanisms and phase anomalies in peripheral vision. *Nature* **326**, 873–878.

Bennett, P. J. and Banks, M. S. (1991). The effects of contrast, spatial scale, and orientation on foveal and peripheral phase discrimination. *Vision Res.* **31**, 1759–1786.

Braddick, O. (1974). A short-range process in apparent motion. *Vision Res.* **14**, 519–527.
Bruce, V. G. and Morgan, M. J. (1975). Violations of symmetry and repetition in visual patterns. *Perception* **4**, 239–249.
Carmody, D. P., Nodine, C. F. and Locher, P. J. (1977). Global detection of symmetry. *Percept. Motor Skills* **45**, 1267–1273.
Corballis, M. C. and Roldan, C. E. (1974). On the perception of symmetrical and repeated patterns. *Percept. Psychophys.* **16**, 136–142.
Foley, J. M. and Legge, G. E. (1981). Contrast detection and near-threshold discrimination in human vision. *Vision Res.* **21**, 1041–1053.
Herbert, A. M. and Humphrey, G. (1993). Bilateral symmetry detection: detectability as a function of axis orientation and eccentricity. *Invest. Ophthalmol. Visual Sci.* **34**(4), 1866.
Jenkins, B. (1982). Redundancy in the perception of bilateral symmetry in dot textures. *Percept. Psychophys.* **37**, 125–134.
Julesz, B. (1971). *Foundations of Cyclopean Perception.* U. Chicago Press, Chicago, IL.
Kahn, J. I. and Foster, D. H. (1981). Visual comparison of rotated and reflected random-dot patterns as a function of their positional symmetry and separation in the field. *Quarterly J. Exp. Psychol.* **33A**, 155–166.
Klein, S. A. and Levi, D. M. (1987). Position sense of the peripheral retina. *J. Opt. Soc. Am.* **A4**, 1543–1553.
Klein, S. A. and Tyler, C. W. (1986). Phase discrimination of compound gratings: generalized autocorrelation analysis. *J. Opt. Soc. Am.* **A3**, 868–879.
Larsen, A., Farrell, J. E. and Bundesen, C. (1983). Short- and long-range processes in visual apparent movement. *Psychol. Res.* **45**, 11–18.
Ledgeway, T. and Smith, A. T. (1994). Evidence for separate motion-detecting mechanisms for first- and second-order motion in human vision. *Vision Res.* **34**, 2727–2740.
Lee, T. S., Mumford, D. and Schiller, P. H. (1995). Neural correlates of boundary and medial axis representations in primate striate cortex. *Invest. Ophthalmol. Visual Sci.* **36**, S477.
Locher, P. and Nodine, C. (1989). The perceptual value of symmetry. *Comput. Math. Appl.* **17**, 475–484.
Mach, E. (1886). *The Analysis of Sensations* (1959 edn.). Dover, New York.
Nachmias, J. and Kocher, E. C. (1970). Visual detection and discrimination of luminance increments. *J. Opt. Soc. Am.* **60**(3), 382–389.
Rentschler, I. and Treutwein, B. (1985). Loss of spatial phase relationships in extrafoveal vision. *Nature* **313**, 308–310.
Rovamo, J., Virsu, V. and Nasanen, R. (1978). Cortical magnification factor predicts the photopic contrast sensitivity of peripheral vision. *Nature* **271**, 54–56.
Saarinen, J. (1988). Detection of mirror symmetry in random dot patterns at different eccentricities. *Vision Res.* **28**, 755–759.
Smith, A. T. (1994). Correspondence-based and energy-based detection of second-order motion in human vision. *J. Opt. Soc. Am.* **A11**, 1940–1948.
Solomon, J. A. and Sperling, G. (1995). 1st- and 2nd-order motion and texture resolution in central and peripheral vision. *Vision Res.* **35**, 59–64.
Stromeyer, C. F. and Klein, S. (1975). Evidence against narrow-band spatial frequency channels on human vision: The detectability of frequency-modulated gratings. *Vision Res.* **15**, 899–910.
Tyler, C. W. and Apkarian, P. A. (1982). Properties of localized pattern evoked potentials. *Annals N.Y. Acad. Sci.* **388**, 662–670.
Tyler, C. W. and Gorea, A. (1986). Different encoding mechanisms for phase and contrast. *Vision Res.* **26**, 1073–1082.
Tyler, C. W., Hardage, L. and Miller, R. (1995). Multiple mechanisms in the detection of mirror symmetry. *Spatial Vision* **9**, 79–100.
Watson, A. B. (1987). Estimation of local spatial scale. *J. Opt. Soc. Am.* **A418**, 1579–1582.
Wenderoth, P. (1995). The role of pattern outline in bilateral symmetry detection with briefly flashed dot patterns. *Spatial Vision* **9**, 57–77.

Human Symmetry Perception, pp. 173–183
C. W. Tyler (Ed.)

Human discrimination of surface slant in fractal and related textured images

P. J. PASSMORE* and A. JOHNSTON
Department of Psychology, University College London, Gower St., London, UK

Abstract—Slant-discrimination thresholds were measured for textures with the property that their power spectra, when log transformed, are inversely proportional to log spatial frequency: $P(f) \propto f^{-\beta}$. As the exponent β changes from high values to low values, the slope of the power spectrum of the image decreases. As the parameter passes through values in the fractal range the resulting texture changes from having the appearance of a cloud-like surface through to a granite-like surface. Exponents below the fractal range produce textures that converge towards the appearance of a random grey-level noise pattern. Since fractal patterns are self-similar at a range of scales, one might think it would be difficult to recover changes in depth in fractal images; however, slant-discrimination thresholds did not differ substantially as a function of the slope of the power spectrum. Reducing the size of the viewing aperture increased thresholds significantly, suggesting that slant discrimination benefits from a global analysis. The effect of texture regularity on perceived slant was investigated using bandpassed fractal textures. As the bandwidth of a bandpass filter is reduced, the bandpassed texture was perceived to be increasingly more slanted than its fractal counterpart.

1. INTRODUCTION

Symmetries involve transformations of an object that leave the appearance of the object unchanged. The mathematical ideal can be relaxed to provide the concept of a perceptual symmetry, which we can define as a transformation of a pattern that, to the human observer, leaves the pattern indistinguishable from the original, in the sense that they cannot be distinguished without detailed scrutiny. For example, two patches of the same texture may appear interchangeable or two images of the same fractal surface, produced at different scales, may appear to be self-similar.

Surface texture has, since Gibson (1950), been seen as an important cue to the orientation of surfaces. The two assumptions mostly commonly made about textures in shape-from-texture algorithms are that the surface texture is either isotropic

*To whom correspondence should be addressed. Present address: School of Computing, University of North London, 166-220 Holloway Rd., London N7, UK.

(Witkin, 1981; Blake and Marinos, 1990; Brown and Shvaytser, 1990), or homogeneous (Aliomonos, 1988; Kanatani and Chou, 1989). Isotropy implies that there is no orientational bias in the texture while homogeneity embodies the constraint proposed by Gibson (1979) that there are equal amounts of texture per unit of surface area. Shape-from-texture algorithms typically estimate slant by attempting to apply an inverse perspective or orthographic transform to the image texture to maximize isotropy or homogeneity in the resulting pattern.

Psychophysical studies of shape-from-texture have demonstrated the importance of a number of cues affecting perceived slant. Gibson (1950) found, using wallpaper textures and a cross-modal matching task, that perception of slant from texture is improved by increasing the regularity of textures. More recently, investigation has focused on the relative importance of three texture gradients: perspective, compression, and density. If we consider a uniform texture in the ground plane, the perspective gradient is the x-axis width of texture elements (texels) projected onto the image plane orthogonal to the line of sight, the compression gradient is the ratio of y/x axis measures on the projected plane, and the density gradient corresponds to the number of texels per unit visual angle.

A number of studies have suggested that the compression gradient is the most important for shape-from-texture (Cutting and Millard, 1984; Todd and Akerstrom, 1987; Cumming *et al.*, 1993). Cumming *et al.* proposed that shape-from-texture involved assumptions of isotropy rather than homogeneity as they found that disrupting isotropy reduced the relative influence of texture in judgments of surface shape in stereoscopic displays. Stone (1993) suggests that the critical factor is that the distribution of the orientation of the local tangent vectors to the brightness contours in the image is invariant with position — a more constrained form of anisotropy. There has also been some debate whether shape-from-texture is a local or global process, with Stevens (1981) suggesting that the process may be considered local under the assumption of approximately circularly symmetric texture elements, whilst Cumming *et al.* (1993) find evidence for global processes. Elongated texture elements, which formed elliptic contours on the test surfaces, were not interpreted as regular circular contours in an alternative depth plane.

Psychophysical investigations of shape-from-texture often use idealized textures which are typically produced by tiling a planar or curved surface with standardized elements such as lines or circles (e.g. Cutting and Millard, 1984) or carving from volumetric textures containing ideal shapes such as spheres and ellipsoids (Cumming *et al.*, 1993). Fractal images, on the other hand, have the appearance of naturally occurring textures (such as those produced by clouds or granite) and many images of different natural scenes have been shown to have amplitude spectra in the range of the fractal model (Field, 1987). A fractal surface is characterized by its fractal dimension, which corresponds to the roughness of the surface, a property that is invariant over scale.

When a textured surface is slanted, there is a change in scale in the image with distance as the more distant surface patches are projected smaller in the image. As a fractal texture has spatial structure which is invariant with scale, it has been assumed that surface orientation would be difficult to recover from fractal textures. However, as Stevens (1981) points out, although in the direction orthogonal to the tilt of the surface perspective projection simply results in a spatial scaling of the image features, in the direction in which the texture is slanted there is an anisotropic transformation (foreshortening) which varies as a function of spatial position. If subjects are sensitive to this property of perspective projection then it should be possible to see depth in slanted fractal textures. We decided to test this idea directly by measuring slant-discrimination thresholds for fractal textures.

2. GENERAL METHODS

2.1. Stimulus generation

Many natural images have a $1/f^2$ power spectrum (Field, 1987). Note that a $1/f^2$ power spectrum corresponds to a $1/f$ amplitude spectrum. We generated synthetic textures with the property that their power spectra, when log transformed, are inversely proportional to log spatial frequency:

$$P(f) \propto f^{-\beta}.$$

As the exponent β changes from high values to low values, the slope of the power spectrum of the image decreases. The resulting texture changes from having the appearance of a cloud-like surface through a granite-like texture to a random grey-level noise pattern. Knill *et al.* (1990) explain that, for a fractal model based on fractional Brownian motion, there is a relationship between the fractal dimension and the slope of the power spectrum. For brightness surfaces the exponent β can be shown to be related to the fractal dimension D by:

$$\beta = 8 - 2D,$$

where D falls in the range 2 to 3 (Knill *et al.*, 1990). Thus a fractal image would give rise to values of the exponent β between 2 and 4. We can of course design images with power spectra which have values of β outside this range. These patterns would not be representative of the class of images that can be generated by a process based on fractional Brownian motion.

The spectral synthesis or Fourier filtering method was used to generate stimuli (Saupe, 1988). The stimuli were constructed in frequency space with a given slope specified by the square root of the β exponent and a random phase spectrum. An inverse Fourier transform was then applied to produce 2048 by 2048 pixel images. The images were rotated in 3D space to produce 512 by 512 pixel texture gradients. Aliasing was reduced by supersampling and perspective averaging. The intensity of a pixel in the frontal plane was determined by averaging over an elliptic area of the rotated surface that projects onto the neighbourhood of that pixel. The average value was computed by projecting an elliptic grid of points onto the rotated surface and summing over the intensity values of those points, which are calculated by nearest

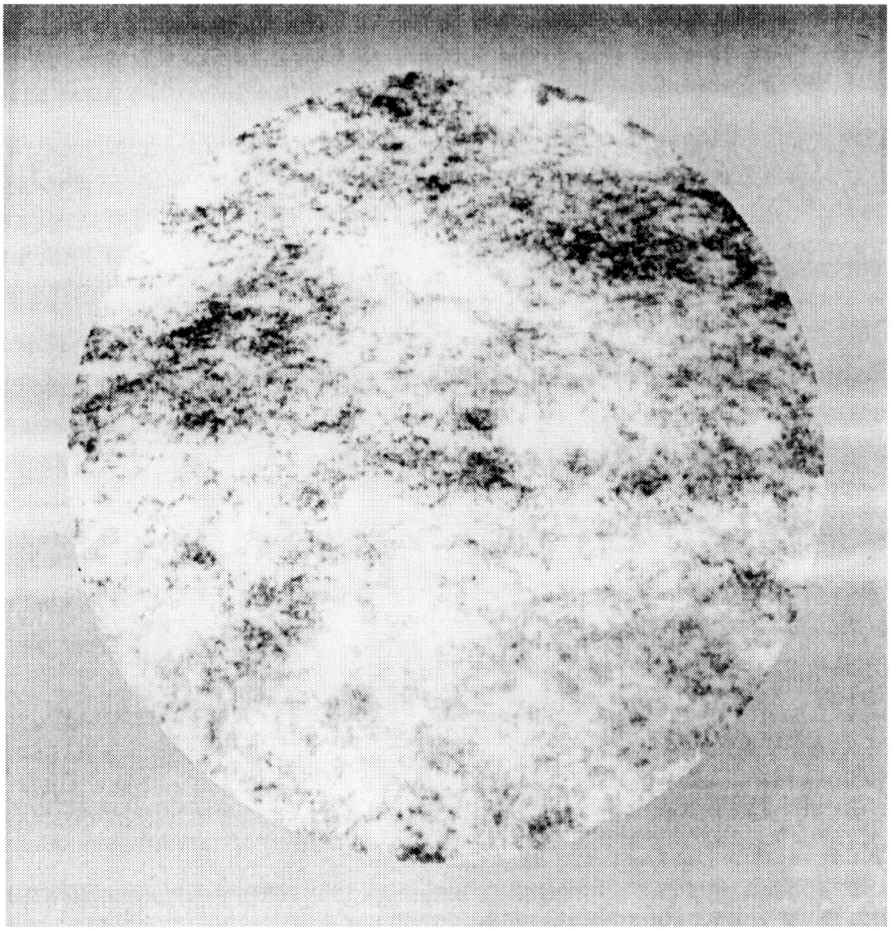

Figure 1. This is an example fractal stimulus ($\beta = 2.2$) slanted by 50 deg towards the ground plane.

neighbour grey-level bilinear interpolation. The resulting stimuli were scaled to have the same Michelson contrast. The images were displayed with eight-bit precision on a nineteen-inch Sony Trinitron monitor under the control of a Sun SPARCstation 330. In order to linearize the display, a lookup table of luminance values was determined with a micro-photometer and used to control stimulus brightness. Figure 1 shows an example fractal stimulus.

2.2. General procedure

Slant-discrimination thresholds were measured by a method of constants for stimuli that depicted textures that had been rotated around a central horizontal axis; subjects were required to decide whether the stimuli were slanted towards the ground plane or the sky plane. No feedback was given during the experiments but subjects were required to conduct a few practice runs. The textures were occluded by a circular window, which subtended 27 deg of visual angle. The size of the aperture was randomly jittered by up to ±12% between presentations, and the position of the aperture in the window was randomly jittered by up to ±1.7 deg in a random direction between trials. These randomizations were employed to prohibit subjects using simple comparisons of texture size over trials as a cue, thus forcing them to rely on estimates of the slant of the surface. Stimuli were viewed monocularly from 25 cm. The head was kept still using a headrest, and subjects were required to fixate a central point embedded in the image. Due to the technical difficulty of generating completely novel stimuli from trial to trial, stimuli were chosen from a pool of 36 images grouped into 9 deg of slant. The stimuli in each group had identical amplitude spectra but different random phase spectra. Thresholds were calculated by probit analysis (Finney, 1971) and each threshold was the result of at least 280 judgments. Discrimination threshold was defined as the standard deviation of the error distribution and corresponds to the 84% point on the psychometric function.

3. EXPERIMENT 1. SLANT DISCRIMINATION THRESHOLDS AS A FUNCTION OF THE SLOPE OF THE POWER SPECTRUM

Slant discrimination was measured as a function of the slope of the power spectrum with the exponent β ranging from 0.2 to 3.8. This includes stimuli whose β exponent fell within the fractal range and extends outside it to include patterns whose power spectrums have shallower slopes. The slants used were in the range of 0 to 50 deg in either direction with respect to the image plane. The two authors and two naive subjects, one of whom (WC) is stereo blind due to a late-corrected strabismus, participated in the experiment. It should be noted that WC's results did not differ substantially from those of other subjects in these experiments.

Figure 2 shows the slant discrimination thresholds averaged over the four subjects as a function of the exponent β. We found that subjects could see surface slant in these displays. Thresholds were slightly higher for the fractal range than for the textures

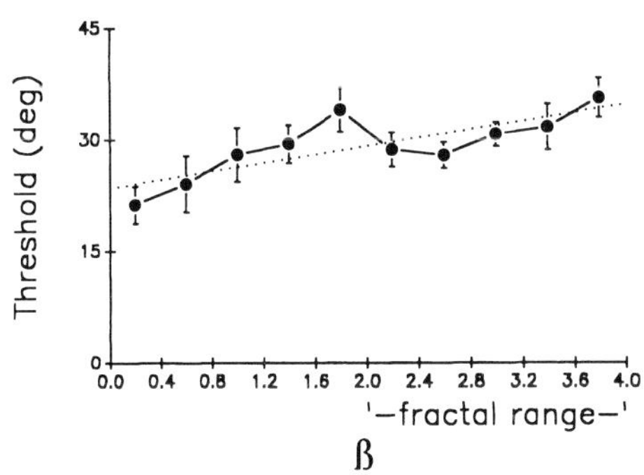

Figure 2. Averaged slant discrimination thresholds are plotted as a function of the slope of the power spectrum (β) for four subjects. Although there is a slight decrease in threshold with decrease in the slope the effect is small and demonstrates a substantial degree of scale invariance for slant discrimination with these textures.

with flatter power spectra but there were no substantial differences. Thresholds ranged from around 20 to 35 deg increasing slightly as the slope of the amplitude spectrum increased. There was some variation between individuals but no radical effect of the slope of the amplitude spectrum on discrimination thresholds was found.

4. EXPERIMENT 2. SLANT DISCRIMINATION THRESHOLDS: EFFECTS OF FILTERING GAUSSIAN NOISE AND FLAT SPECTRA WITH A $f^{-\beta}$ FILTER

In Experiment 1 we used the spectral synthesis method to generate fractal textures. In essence this technique involves filtering a Gaussian random variable by a $f^{-\beta}$ spectrum. The amplitude of each spectral component is drawn from a Gaussian distribution and then multiplied by $f^{-\beta}$. However, the expected power spectrum for an ensemble of fractal images with a particular fractal dimension would be a smooth sloping line, the slope of which is related to the fractal dimension of the ensemble. In order to quantify the effects of the random variable employed in Experiment 1 we compared slant discrimination thresholds for Gaussian noise and flat spectra both of which were modified by a $f^{-\beta}$ filter. The experimental procedure was the same as in Experiment 1. Two subjects, the authors, participated in this experiment.

The results are shown in Figs 3a and b. While thresholds for subject PP are generally lower than for subject AJ, both subjects have similar results for both the Gaussian and flat spectra conditions, demonstrating that the random variation in the amplitude of the power spectra in Experiment 1 had no radical effects on slant perception.

5. EXPERIMENT 3. EFFECT OF APERTURE SIZE ON SLANT DISCRIMINATION THRESHOLDS

It is widely accepted that there exists a range of spatial channels at each retinal position in the human visual system, and that the range of spatial analysis shifts progressively to a coarser spatial scale moving towards the periphery of the visual field. Increasing the retinal image size of a slanted stimulus will increase the number

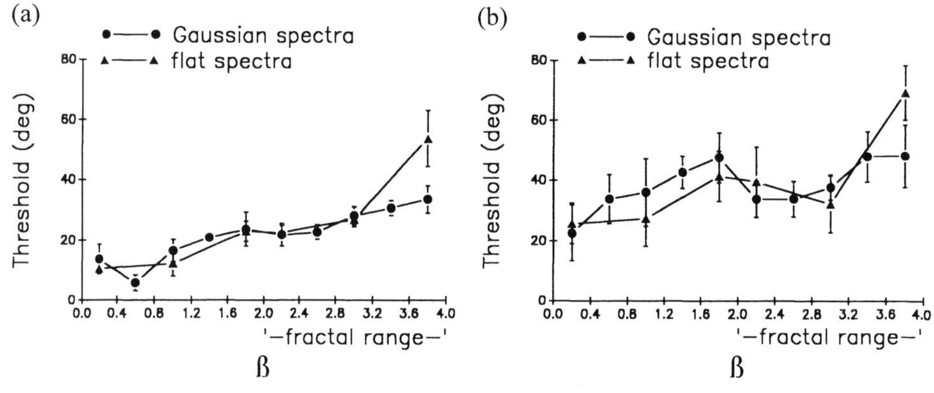

Figure 3. Slant discrimination thresholds plotted as a function of the slope of the power spectrum for Gaussian noise and flat power spectra. (a) subject PP, (b) subject AJ.

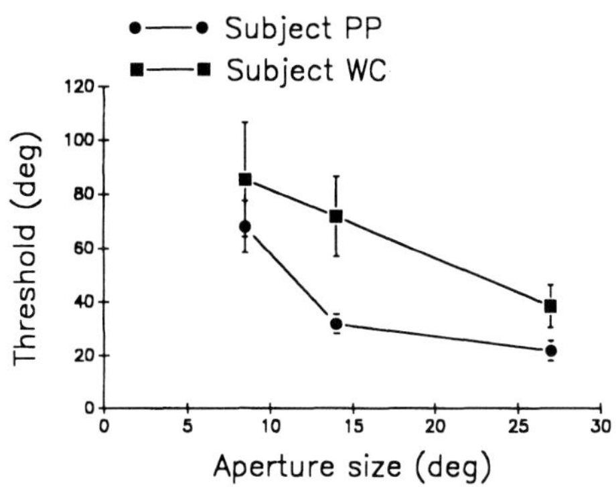

Figure 4. Slant discrimination thresholds are plotted as a function of aperture size. Thresholds increase with a decrease in aperture size suggesting that slant discrimination is made on the basis of global information.

and scale of spatial filters that the visual system may use to process a stimulus. This may improve performance. If the visual system can determine slant on the basis of local information then we would expect that the reduction of stimulus aperture size would not greatly affect slant discrimination thresholds. To investigate whether slant discrimination from texture is achieved by local or global processes, we used the same experimental procedures as in Experiment 1 but varied the aperture size. The power spectra of the stimuli were fixed ($\beta = 2.2$). The apertures subtended 27, 14, and 8.5 deg of visual angle. One author (PP) and a naive subject (WC) participated in the experiment.

Figure 4 shows slant discrimination thresholds as a function of aperture size. For both subjects thresholds increased significantly with a decrease in aperture size. Thus, for these textures, slant discrimination benefits from access to global information.

6. EXPERIMENT 4. SLANT DISCRIMINATION OF BANDPASSED FRACTAL TEXTURES

For fractal textures, bandpass filtering tends to increase regularity. The regularity increases systematically as bandwidth is reduced. For example, when the exponent β has the value 2.2, the texture changes from being granite-like to being more mottled in appearance with local blob-like texels becoming discriminable in the bandpassed texture. Increasing regularity can also be considered to introduce a greater degree of perceptual symmetry in a pattern. In the one-dimensional case, filtering a noise function with a sufficiently narrow bandpass filter will produce a single sine wave which has reflective, translational, and rotational symmetries.

In this experiment, the effect of texture regularity on perceived slant was investigated by bandpass filtering fractal textures prior to 3D rotation. We used a range of bandwidths centred on a frequency of 512 cycles per image. The image size was 2048 by 2048 pixels. Perceived difference in slant between the original fractal stimuli and bandpassed fractal stimuli was measured using a two-alternative forced-choice task. The fractal stimuli were randomly chosen from a pool of four images that had the same exponent ($\beta = 2.2$) but different random phase spectra, and were slanted at

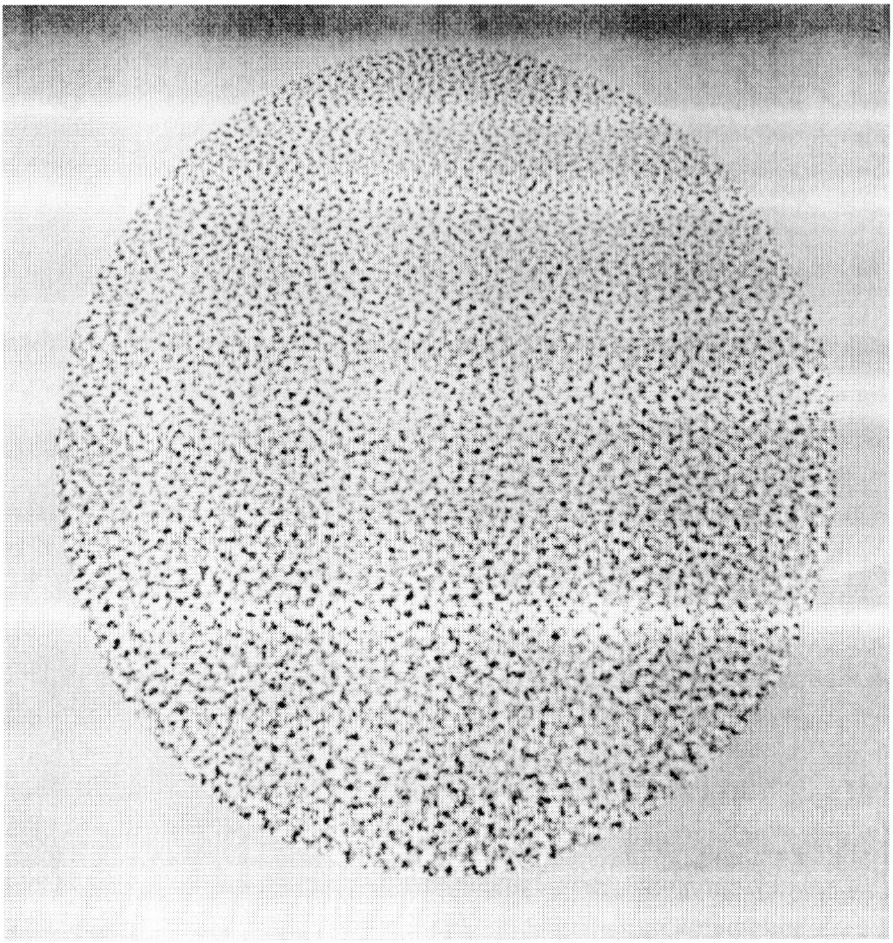

Figure 5. This is an example stimulus used in Experiment 4. It was generated by bandpass filtering a fractal image prior to rotation using a filter with a bandwidth of 796 cycles per image. Note that this texture, like that of Fig. 1, is also slanted by 50 deg towards the ground plane but appears to most observers to be more slanted owing to the effects of texture regularity. We may consider the bandpass process as increasing the symmetry of the texture.

50 deg towards ground plane. The exponent value of $\beta = 2.2$ was arbitrarily chosen, being near the middle of the range of values used but in the fractal range. Bandpassed stimuli were chosen from a pool of images in which nine levels of slant (from 25 to 50 deg) and four random phase spectra were combined. The experimental procedure followed that used in Experiment 1 except that the fractal and bandpassed texture were presented consecutively. For each trial the order of presentation was randomized. The subjects' task was to indicate which texture appeared more slanted. Stimuli were displayed for approximately 2 sec and an inter-stimulus grey-level image was displayed for approximately 3 sec between the two textures. An example bandpassed stimulus is shown in Fig. 5. Two subjects, the authors, participated in the experiment.

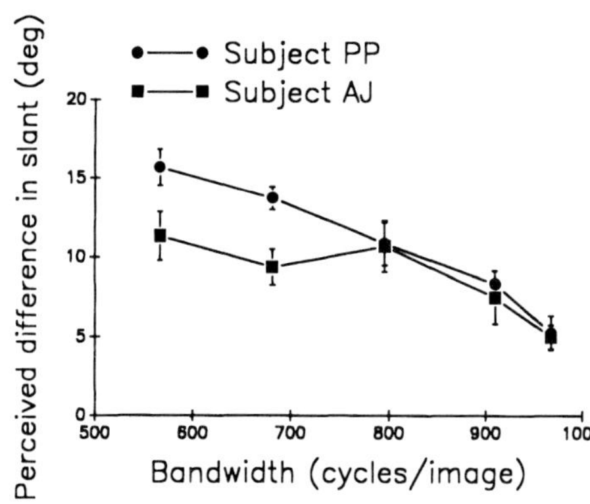

Figure 6. Perceived difference in slant is plotted as a function of the bandwidth of the bandpass filter. The increase in perceived difference in slant for both subjects with decrease in bandwidth size is due to the effects of texture regularity.

Figure 6 shows the perceived difference in slant as a function of bandwidth. For both subjects, perceived difference in slant increases with the decrease of bandwidth. This replicates, for fractal textures, the finding that as texture regularity increases slanted surfaces appear more slanted. This result agrees with the findings of Gibson (1950), Flock and Moscatelli (1964) and Phillips (1970) that increases in the regularity of texture increases perceived slant. Applying a bandpass filter to the texture makes it visually segmentable. Although self-similarity over changes in scale is reduced, there is an increase in self-similarity with respect to spatial position, since translation involving multiples of the spatial period of the pattern would not be easily discriminable.

7. DISCUSSION

Because fractal surfaces display self-similarity under changes in scale, it is often thought that it would be difficult to see a change of distance in a fractal surface since changing distance simply results in a change of scale in the retinal image. However, we have shown that subjects can make judgments of surface slant for surfaces defined by fractal textures and the slant discrimination thresholds are similar to those for textures with flatter power spectra, which appear rather like random grey-level textures. The results of Experiment 1 show that there is little effect of the slope of the amplitude spectrum (and consequently fractal dimension) on slant discrimination thresholds. Although we might expect difficulties in recovering changes in local scale brought about by changes in slant, there are a number of possible sources of information available in slanted fractal textures that might be used to recover the slant. Any effects must be related to perspective projection since our subjects made decisions about slant for a range of stimuli that included surface rotations in both directions with respect to the vertical. The scaling effects resulting from orthographic projection are identical for surfaces slanted by the same amount in either direction. Subjects therefore must be using perspective cues. One possible source of information is in the distribution of local orientations along isophote contours, or lines of equal reflectance (Witkin, 1981; Stone, 1993). When an isotropic textured surface is slanted, there are local changes in the distribution of orientations that can be used to recover slant. This effect results from the same geometry that gives rise to linear perspective cues, which also signal depth through the orientation of image contours.

For an ideal texture produced by tiling a plane with circles, it may be possible to base slant discrimination on local estimates as proposed by Stevens (1981). However, a characteristic of the textures used in the first three experiments is that they are not visually segmentable into local texture elements. Thus the experiments provide further evidence for the role of global processes as against local processes in slant perception, since slant discrimination thresholds are markedly increased for small aperture sizes. This effect may also provide an explanation for the increases in thresholds for the reduced aperture conditions if we accept that reducing aperture size reduced the ability of subjects to make use of foreshortening and convergence cues.

The techniques used in Experiment 4 allowed us to systematically manipulate texture regularity by varying the pass-band of the fractal grey-level texture and we have replicated earlier findings that regular textures appear to be more slanted. We can characterize this manipulation in terms of increasing the symmetry in the image; as an image is bandpass filtered with increasing narrow spatial filters it becomes increasingly more cyclical, visually segmentable, and regular. Thus, for a fractal image, bandpass filtering decreases self-similarity over scale and increases self-similarity over spatial position. Both of these manipulations can be expected to act together to improve slant discrimination.

Acknowledgement

This work was supported by a grant from the SERC Image Interpretation Initiative.

REFERENCES

Aloimonos, J. (1988). Shape from texture. *Biol. Cybernet.* **58**, 345–360.

Blake, A. and Marinos, C. (1990). Shape from texture: estimation, isotropy and moments. *Artif. Intel.* **45**, 323–380.

Brown, L. G. and Shvaytser, H. (1990). Surface orientation from projective foreshortening of isotropic texture autocorrelation. *IEEE Trans. PAMI* **12** (6), 584–588.

Cumming, B. G., Johnston, E. B. and Parker, A. J. (1993). Effects of different texture cues on curved surfaces viewed stereoscopically. *Vision Res.* **33**, 827–838.

Cutting, J. E. and Millard, R. T. (1984). Three gradients and the perception of flat and curved surfaces. *J. Exp. Psychol. Gen.* **113**, 198–216.

Field, D. J. (1987). Relations between the statistics of natural images and the response properties of cortical cells. *J. Opt. Soc. Am. A* **4**, 2379–2394.

Finney, D. J. (1971). *Probit Analysis*, 3rd edn. Cambridge University Press, Cambridge, UK.

Flock, H. R. and Moscatelli, A. (1964). Variables of surface texture and accuracy of space perceptions. *Percept. Motor Skills* **19**, 327–334.

Gibson, J. J. (1950). *The Perception of the Visual World.* Houghton Mifflin, Boston.

Gibson, J. J. (1979). *The Ecological Approach to Visual Perception.* Houghton Mifflin, Boston.

Kanatani, K. and Chou, T. (1989). Shape from texture: general principle. *Artif. Intel.* **38**, 1–48.

Knill, D. C., Field, D. and Kersten, D. (1990). Human discrimination of fractal images. *J. Opt. Soc. Am. A* **7** (6), 1113–1123.

Phillips, R. J. (1970). Stationary visual texture and the estimation of slant angle. *Q. J. Exp. Psychol.* **22**, 389–397.

Saupe, D. (1988). Algorithms for random fractals. In: *The Science of Fractal Images*. H. Peitgen and D. Saupe (Eds). Springer-Verlag, New York, pp. 71–136.

Stevens, K. (1981). The information content of texture gradients. *Biol. Cybernet.* **42**, 95–105.

Stone, J. V. (1993). Shape from local and global analysis of texture. *Phil. Trans. Roy. Soc. Lond. B* **339**, 53–65.

Todd, J. T. and Akerstrom, R. A. (1987). Perception of three-dimensional form from patterns of optical texture. *J. Exp. Psychol: Human Percept. Perform.* **13**, 242–255.

Witkin, A. P. (1981). Recovering surface shape and orientation from texture. *Artif. Intel.* **17**, 17–47.

Part Two

THEORETICAL ISSUES IN SYMMETRY ANALYSIS

Human Symmetry Perception, pp. 187–207
C. W. Tyler (Ed.)

Detection of bilateral symmetry using spatial filters

S. C. DAKIN and R. J. WATT*
Department of Psychology, University of Stirling, Scotland FK9 4LA, UK

Abstract—When bilaterally symmetric images are spatially filtered and thresholded, a subset of the resultant 'blobs' cluster around the axis of symmetry. Consequently, a quantitative measure of blob alignment can be used to code the degree of symmetry and to locate the axis of symmetry. Four alternative models were tested to examine which components of this scheme might be involved in human detection of symmetry. Two used a blob-alignment measure, operating on the output of either isotropic or oriented filters. The other two used similar filtering schemes, but measured symmetry by calculating the correlation of one half of the pattern with a reflection of the other. Simulations compared the effect of spatial jitter, proportion of matched to unmatched dots and width or location of embedded symmetrical regions, on models' detection of symmetry. Only the performance of the oriented filter + blob-alignment model was consistent with human performance in all conditions. It is concluded that the degree of feature co-alignment in the output of oriented filters is the cue used by human vision to perform these tasks. The broader computational role that feature alignment detection could play in early vision is discussed, particularly for object detection and image segmentation. In this framework, symmetry is a consequence of a more general-purpose grouping scheme.

1. INTRODUCTION

The identification of bilaterally symmetric figures is thought to be a component of a number of basic visual tasks. It has been proposed that symmetry plays a major role in object recognition by providing a canonical axis for the representation of shape (Marr, 1982). When a symmetrical object is tilted, the degree of *skew* of the resulting symmetry is a cue to deriving three-dimensional structure (Kanade and Kender, 1987). It is also possible that, by identifying symmetric regions of an image, the presence of salient objects can be detected. This is reflected by the use of symmetry, within machine-vision systems, as a cue for orienting attention (e.g. Yeshurun *et al.*, 1992).

Barlow and Reeves (1979) demonstrated that human identification of symmetry is not an 'all-or-nothing' process but is *graded* with respect to a number of attributes

*To whom correspondence should be addressed.

of symmetrical patterns. They showed that discrimination of symmetrical from non-symmetrical dot textures smoothly deteriorates as the proportion of paired to unpaired dots in the symmetrical pattern decreases. Barlow and Reeves (1979) also showed that symmetry detection smoothly deteriorates with perturbations in dot location. They propose that human symmetry detection operates using a symmetrical distribution of feature detectors sensitive to local dot densities. Assuming that the orientation and position of the axis of symmetry are known in advance, this organisation may be practically realised by performing a simple correlation of one half of the image with a reflected version of the other. In order to account for our tolerance to spatial jitter of features, Barlow and Reeves (1979) do not correlate individual image locations, but calculate the local density of features and correlate these (coarse scale) estimates.

Jenkins (1983) noted that because a reflection is used to generate symmetrical stimuli this does not imply that a reversed mapping must be used to detect structure, especially given that this has no obvious interpretation in terms of the known structure of the human visual system. Instead he proposed a three-stage model of symmetry detection: detection of orientation uniformity, fusion of pairs into features, and the detection of symmetry of the resultant feature. Symmetrical dot pairs are detected directly by the output of receptive fields, of various sizes, oriented orthogonal to the axis of symmetry.

Symmetry detection seems to operate at a low spatial resolution. It is resistant to local perturbation of the position of the constituent elements of symmetrical textures (Barlow and Reeves, 1979). If symmetric textures are composed of short oriented lines, the orientation of lines does not greatly affect detection of symmetrically *positioned* features (Koeppl, 1993; Locher and Wagemans, 1993). These findings implicate mechanisms sensitive to *low spatial frequencies*. This is supported by two findings which demonstrate that the degree to which features activate filters determines their contribution to the percept of symmetry. Firstly that discrimination of symmetry from noise is not possible, at 150 ms exposure duration, when elements on either side of the axis of symmetry are of opposite contrast (Zhang, 1991). Secondly that the *proximity* of symmetrical features to one another is critical: the area around the mid-line of bilaterally symmetric texture has been found to contribute more to the perception of symmetry than other areas (Bruce and Morgan, 1975; Julesz, 1975; Barlow and Reeves, 1979; Jenkins, 1982). Violations of symmetry are more easily detected near the mid-line (Bruce and Morgan, 1975) and detection of symmetry is less affected by random displacement of dots that are nearer the outskirts than the mid-line (Barlow and Reeves, 1979; Jenkins, 1982).

These results suggest that spatial filtering mechanisms are involved in the perception of symmetry. Recently, it has been proposed that isotropic, Laplacian-of-Gaussian (LoG) filtering might suffice for the grouping of symmetrical elements (Locher and Wagemans, 1993). However, no specific scheme has been described for measuring symmetry from the output of LoGs, or from any other pre-processing system.

The starting point for our study is the observation, illustrated in Fig. 1, that when images of bilaterally symmetric objects are filtered with anisotropic filters that are oriented orthogonal to the axis of symmetry, a striking and simple pattern of response

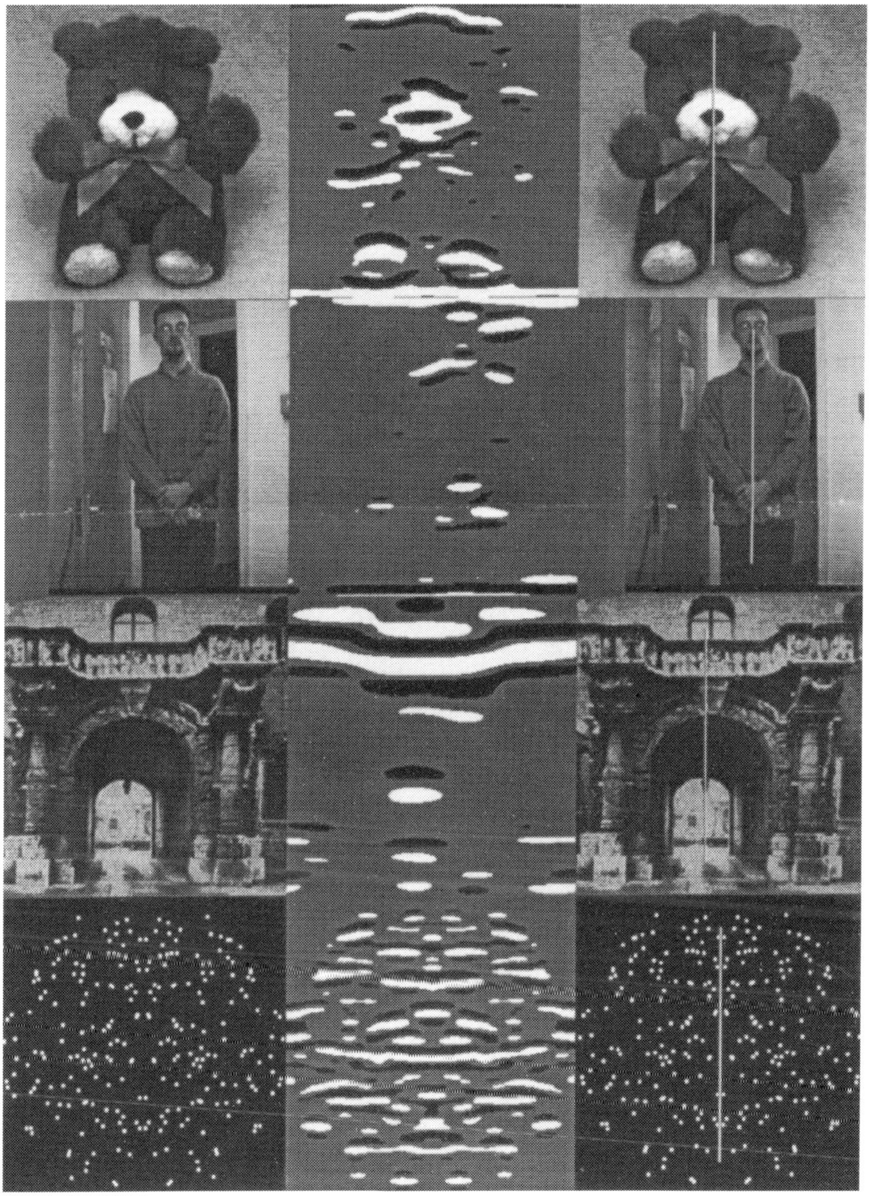

Figure 1. This figure shows (left column) natural images, (middle column) the output of a horizontal DoG at the optimal scale for symmetry detection, and (right column) the original image with the x-location which maximizes $A(x)$ marked.

is found. In each of the filtered images in Fig. 1, there is a pattern of parallel aligned stripes in the response, each stripe centred very close to the axis of symmetry. Such alignment of stripes is not likely to occur by chance, and could be used as an indication of the presence of an object in the image. We have examined the hypothesis that such a pattern of response could be used by the visual system in detecting bilateral symmetry.

The proposed mechanism does not in fact detect symmetry. Bilateral symmetry is defined by the relationship:

$$I(x_a, y_a) = I(-x_a, y_a),$$

where x_a and y_a are positions orthogonal and parallel to the axis of symmetry, respectively. Any pattern that conforms to this relationship is symmetric. To measure whether a pattern is symmetric, it is strictly necessary to assess whether this relationship holds for all (x_a, y_a). A simple numerical way to perform this computation is to note that the correlation measure:

$$4\frac{I(x_a, y_a)I(-x_a, y_a)}{\left(I(x_a, y_a) + I(-x_a, y_a)\right)^2} \tag{1}$$

reaches a maximum value of 1.0 when $I(x_a, y_a)$ equals $I(-x_a, y_a)$. This provides a means for calculating whether an image region is symmetric about a particular axis by averaging this measure for all values of $I(x_a, y_a)$. Since the symmetric pattern generally would not fill the entire image, in principle this measure would need to be assessed for all possible axes and regions. This would be achieved by applying a family of spatial weighting functions of varying size, orientation and perhaps shape, to correlations performed at every point in the image. Such a proposition would place very heavy computational demands on any neural implementation.

A much simpler mechanism, but one that does not exactly calculate symmetry, is proposed. First, we make the simplifying assumption that image values can be represented by just two values, which for convenience, we will take to be -1 and 1. We can then note that the measure:

$$I(x_a, y_a) + I(-x_a, y_a)$$

will reach an extremum value (± 2) when $I(x_a, y_a)$ equals $I(-x_a, y_a)$, and will average to the value zero otherwise. We can further note that summing along a particular row (y_a) orthogonal to the axis of symmetry:

$$\sum^{x_a} I(x_a, y_a)$$

produces a measure that will be more extreme the more similar all the values are to each other along that row. The second simplifying assumption is that where bilateral symmetry in natural images exists, it will frequently be of a form that allows this type of summation to be used. The second assumption is not as severe as it appears — given the spatial correlations that exist in natural images.

This proposal is equivalent to the starting observation, that parallel stripes occur in images that have been filtered with oriented filters. The filtering process is equivalent to some form of differentiation operation followed by an anisotropic summation. The first component will not alter bilateral symmetry, and the second component effectively calculates the appropriate summation. Hence, any place in an image where the filtered response is high could potentially be an axis of symmetry. In practice there are many reasons why the filter response should be high. However, where several peaks in response become aligned on a common axis that is orthogonal to the preferred orientation of the filter, the likelihood is high that this identifies a region of local bilateral symmetry. Note that the computational problems associated with the correlation measure have been avoided by using the combinatorics of filtering itself.

In the study reported in this paper, implementations of the correlation measure and the alignment measure in image processing simulations of psychophysical tasks are compared with the performance of human subjects. All simulations were of the same basic form. For each condition of each task, we created a sequence of pairs of test images to be discriminated, exactly as in the real task: one member of each pair containing more symmetric structure than the other. For each pair in turn, a computational procedure was applied, resulting in a measure of degree of symmetry for each of the two stimuli. A 2AFC psychophysical response could then be generated according to the task. Four computational procedures were employed: two different types of filter (isotropic and oriented) and the two different types of symmetry measure (correlation and alignment).

2. GENERAL METHODS

2.1. Filtering

There is accumulating evidence that the early visual system consists of mechanisms selective for certain spatial frequencies in the input (e.g. Campbell and Robson, 1968; Sachs *et al.*, 1971). Models that have been proposed to describe the exact point-spread function of cells in visual cortex include Gabors (e.g. Daugman, 1985) and Difference-of-Gaussians (DoGs) (e.g. Wilson and Gelb, 1984). Similarly, it has been proposed that retinal ganglion or LGN cells can be described by the Laplacian-of-Gaussian (e.g. Marr, 1976). The principal functional difference between these models is whether filters are oriented or isotropic. In the present study both isotropic and oriented filters are examined.

The isotropic filter used was the Laplacian-of-Gaussian with a point-spread function defined as:

$$\mathrm{f}(x, y, s) = \frac{1}{s^2}\left(1 - \frac{x^2 + y^2}{s^2}\right)\mathrm{e}^{-(x^2+y^2)/2s^2},$$

where s is the space constant of the filter.

For the oriented filter the point-spread function used was an elongated, horizontal DoG filter:

$$\mathrm{f}(x, y, s) = \left(\mathrm{e}^{-y^2/2s^2} - \frac{1}{2.23}\mathrm{e}^{-y^2/2(2.23s)^2}\right)\mathrm{e}^{-x^2/2(3s)^2},$$

where s is the space constant of the filter. The ratio of the amplitudes of the positive and negative parts of the DoG and the aspect ratio of the filter are based on those derived by Wilson and co-workers (Phillips and Wilson, 1983; Wilson and Gelb, 1984). Because only vertical bilateral symmetry is under consideration, only horizontally oriented filters are considered.

Seven sizes of LoGs were used: $s = 2.00–16.00$ pixels, in multiplicative steps of $\sqrt{2}$. Seven sizes of horizontal DoG filter were used: $s = \sqrt{2}–8\sqrt{2}$ pixels, in multiplicative steps of $\sqrt{2}$. Both sets of filters had peak spatial-frequency sensitivities of between 28.8 and 3.5 cycles per image, respectively.

2.2. *Primitive extraction*

The result of the filtering is divided into positive and negative signals, by thresholding the image. Grey levels less than one standard deviation greater or less than the mean grey level are set to zero. This non-linearity is introduced as a way of delineating individual image features.

Filter output images are then converted into a symbolic or 'primal sketch' type representation (Marr, 1976). Blobs from the filtering/thresholding stage are described using measurements of their centroids, lengths, etc. Such schemes have previously been proposed as practical methods for deriving texture statistics (e.g. Voorhees and Poggio, 1987). Watt's (1991) image description scheme was used to describe each zero-bounded region blob in the form:

$$(\mathrm{cx}, \mathrm{cy}, \mu, \lambda, \theta),$$

where (cx, cy) is the centroid, μ is the mass, λ is the length, and θ is the orientation of the blob. (Details of how these parameters are derived are described in Watt (1991), pp. 114–120.)

2.3. *Measuring alignment*

Consider the alignment, $A(x)$, of all blobs which intersect a particular image column, x. A simple measure of blob alignment is to calculate the distance from x to the centroid of the blob:

$$A(x) = \frac{1}{M} \sum_{i=1}^{N_x} \exp\left[- \left(\frac{x^2 - cx_i^2}{2\lambda_i^2} \right) \right] \mu_i, \tag{2}$$

where N_x is the number of blobs intersecting the xth image column, and M is their total mass. Deviation of the blob centroid from x is weighted in inverse proportion to the length of the blob, λ_i, and in direct proportion to the mass of the blob, μ_i. Normalization by M means the alignment measure falls in the range 0.0–1.0. The maximum value of A for the image, $\max_x(A)$, is used as a measure of the symmetry of the pattern. The location at which the maximum occurs $x_{\max}$ is used to indicate the axis of symmetry. Figure 1 demonstrates the application of this technique to axis location in natural and artificial symmetrical images.

2.4. Correlation

Equation (1) was used to calculate an image correlation for each image.

2.5. Estimating discriminability

Given two sets of measurements from each model, one of the reference set and one of a particular level of the cue, these sets will have means of μ_{ref} and μ_{cue}, respectively, and standard deviations of σ_{ref} and σ_{cue}. Typically, the cue and reference sets had unequal variance, suggesting an appropriate criterion as:

$$c = \mu_{\text{ref}} + \frac{\sigma_{\text{ref}}}{\sigma_{\text{ref}} + \sigma_{\text{cue}}} \left(\mu_{\text{cue}} - \mu_{\text{ref}}\right). \tag{3}$$

Values from the reference and cue files were randomly selected and compared to the criterion to establish probability of detecting a cued stimulus (hit), versus probability of falsely identifying a reference stimulus as the cue (false alarm). From these two values d' was calculated as:

$$d' = P^{-1}\,(\text{hit}) - P^{-1}\,(\text{false alarm}),$$

where $P^{-1}\,(y)$ is the inverse function of the Gaussian probability function:

$$P(y) = \frac{1}{\sqrt{2\pi}} \int_{\infty}^{y} \mathrm{e}^{(-x^2/2)}\,\mathrm{d}x.$$

This function can be calculated to an arbitrary level of precision using an approximation to the incomplete Gamma function (Press *et al.*, 1992).

3. SIMULATION PROCEDURE

For each image, symmetry measures were derived using four models. The first two models used the alignment measure in conjunction with oriented or isotropic filters (the 'I + A' and 'O + A' models, respectively). Alignment estimates were made, as described above, and these values placed in files.

The third and fourth models were correlators, operating on the output of similar oriented and isotropic filters. The convolved images were divided at the axis of symmetry, correlated with a reflected version of the other half, and the degree of correlation recorded. These are referred to as the 'I + C' and the 'O + C' models, according to the type of pre-processing used.

All models had prior knowledge of the location of the axis of symmetry. This meant that the degree of alignment or correlation was calculated only around the central image column. Uncertainty on axis location/orientation is known to reduce performance of human subjects on this task (Barlow and Reeves, 1979) and similarly affects the output of the models.

4. SIMULATIONS 1 AND 2: EFFECT OF SIGNAL-TO-NOISE RATIO AND POSITIONAL JITTER

Barlow and Reeves (1979) measured the discriminability of symmetric from non-symmetric random-dot textures as a function of the ratio of paired to unpaired dots in the symmetric texture. The ratio of paired to unpaired dots (the signal-to-noise ratio, SNR) was varied and discrimination from pure noise and pure symmetry was measured. As the number of paired dots in the stimulus decreases (a) discrimination from noise deteriorates and (b) discrimination from pure symmetry improves.

Barlow and Reeves (1979) also showed that discrimination of symmetry from noise deteriorates as jitter on the position of symmetrical elements increases. The second simulation in this section is to investigate if operation of the models at coarse spatial scales can explain subjects' performance in this condition.

4.1. Stimuli

The stimuli were textures composed of 100 dots spatially randomly distributed in a centrally-positioned circular region, with radius 128 pixels. Each dot was composed of a square of four individual pixels and appeared black on a white background.

In generating noise stimuli, dots were allowed to fall randomly in the circular field. The symmetrical patterns were generated by dropping a fixed proportion of dots randomly in the circular field and then reflecting each of their positions around a vertical, centrally positioned axis of symmetry and placing a corresponding dot there. In the second simulation, dot locations were generated as above and then each was subjected to a random shift.

4.2. Procedure

For Simulation 1, 128 stimuli were generated at each of eleven ratios of paired to unpaired elements (0.0–1.0, in steps of 0.1). For Simulation 2, 128 stimuli were generated with between ± 4.06 and ± 65.02 pixels of positional jitter in multiplicative steps of $\sqrt{2}$ pixels (corresponding to shifts of ± 2 to ± 32 arcmin, at the viewing distance quoted in the original paper). For each stimulus level, and for each model, d' was measured in a procedure exactly analogous to that employed in the original experiments.

4.3. Simulation results and discussion

Results from Simulation 1 are presented in Fig. 2. The left half of the figure shows results for discrimination of an increasing proportion of symmetry from noise. In accord with human data, all four models predict a gradual improvement in performance with increasing proportion of symmetry. As the spatial scale is increased, the predictions of the O + C model match human data more closely. Note that both of the alignment models are less sensitive to the scale of filter selected.

The right half of the figure shows the predicted discriminability of pure symmetry from increasing proportions of symmetry. The correlation models overestimate performance, typically because their estimates of *pure* symmetry show little or no variation

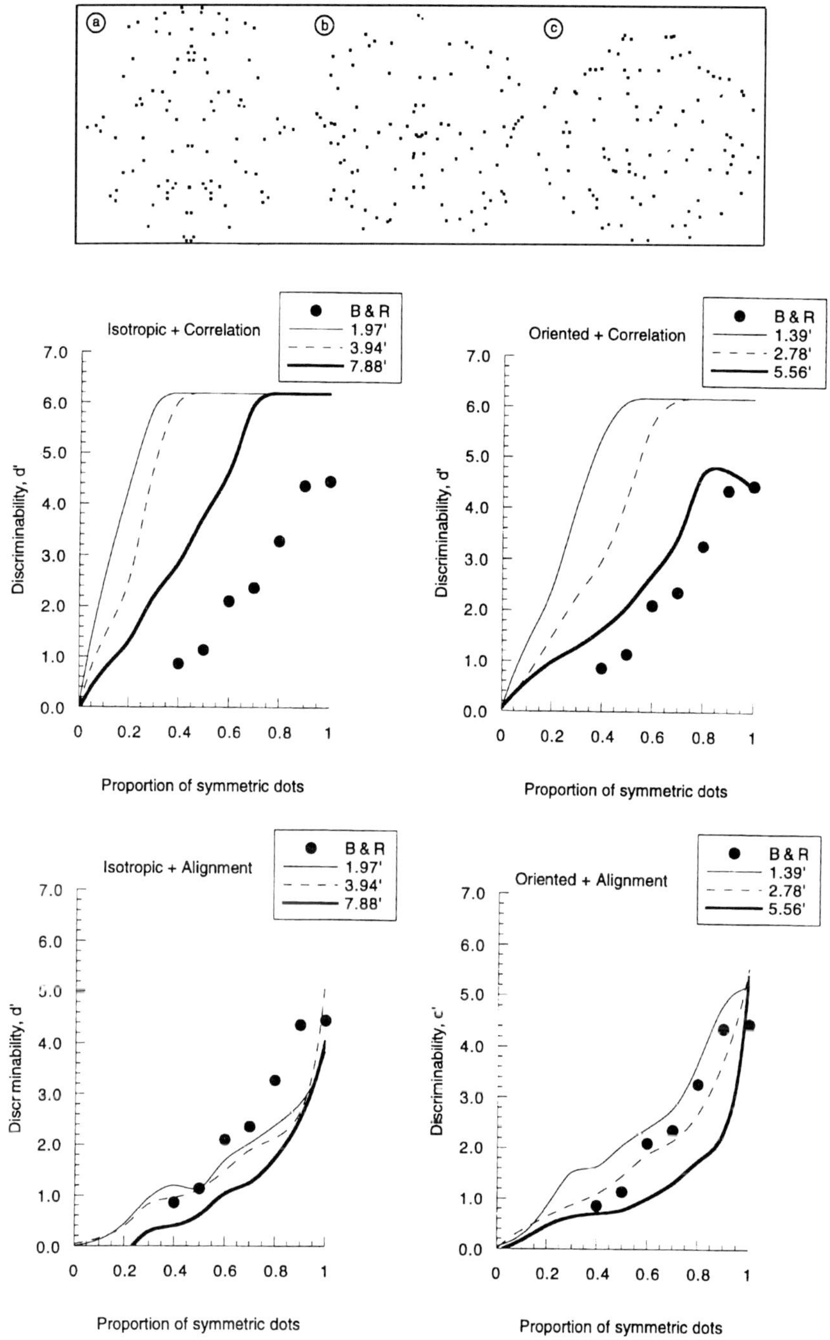

Figure 2. (Top part) Examples of the stimuli used in Simulation 1. Patterns have the following proportions of paired dots: (a) 1.0, (b) 0.75, and (c) 0.5. (Bottom part) Solid symbols show data from Barlow and Reeves (1979), for discrimination from (this page) noise and (next page) pure symmetry patterns, as a function of the proportion of paired to unpaired dots in the stimulus. Lines show predictions from models using filters with space constants given in the legend (in arcmin).

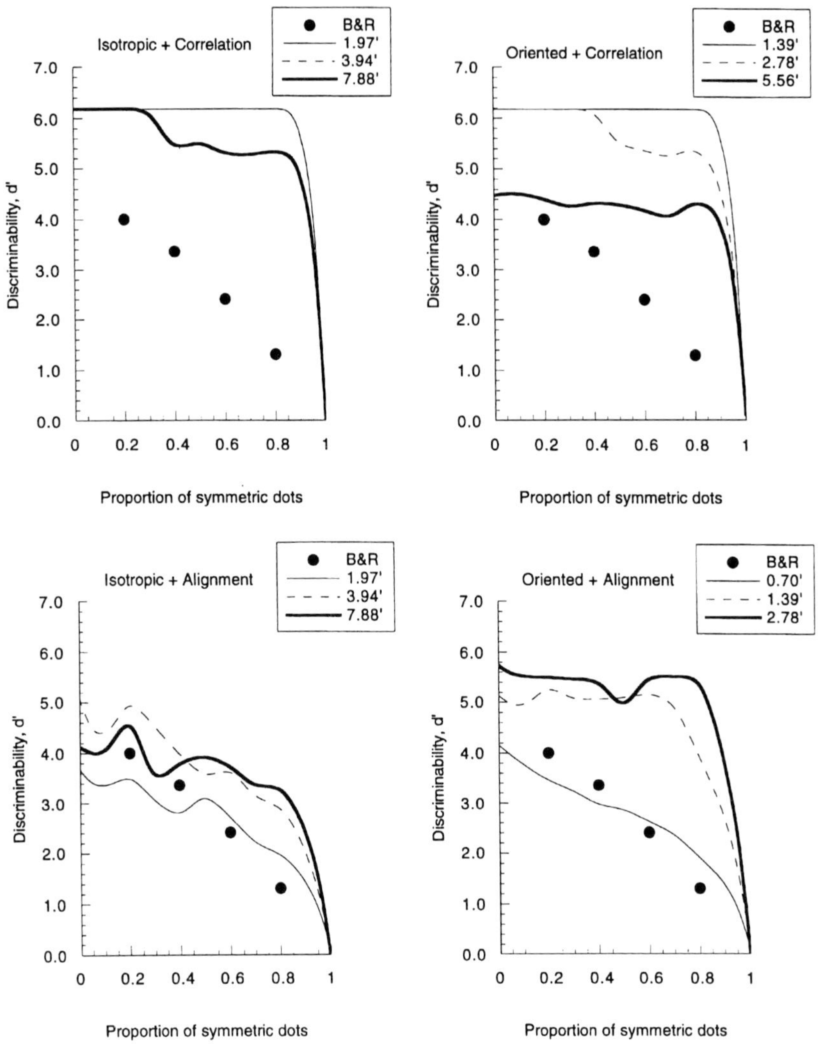

Figure 2. (Continued).

(around 1.0), so d' measures tend to be very high. Predictions from the alignment models, however, do show the same gradual deterioration as human subjects. These models introduce variability in the representation of pure symmetry that is consistent with human performance on this task.

Simulation results for structure discrimination in the presence of positional jitter are plotted alongside human data in Fig. 3. All models produce a gradual drop-off in discrimination of symmetry from noise, as positional jitter increases. However, both correlation models overestimate performance at very low levels of jitter. The I + A model fails to achieve human levels of performance in all but the extremely perturbed patterns. A failure to fit because of overly high performance can be accounted for by the assumption of additional noise in a system, but a failure to achieve human

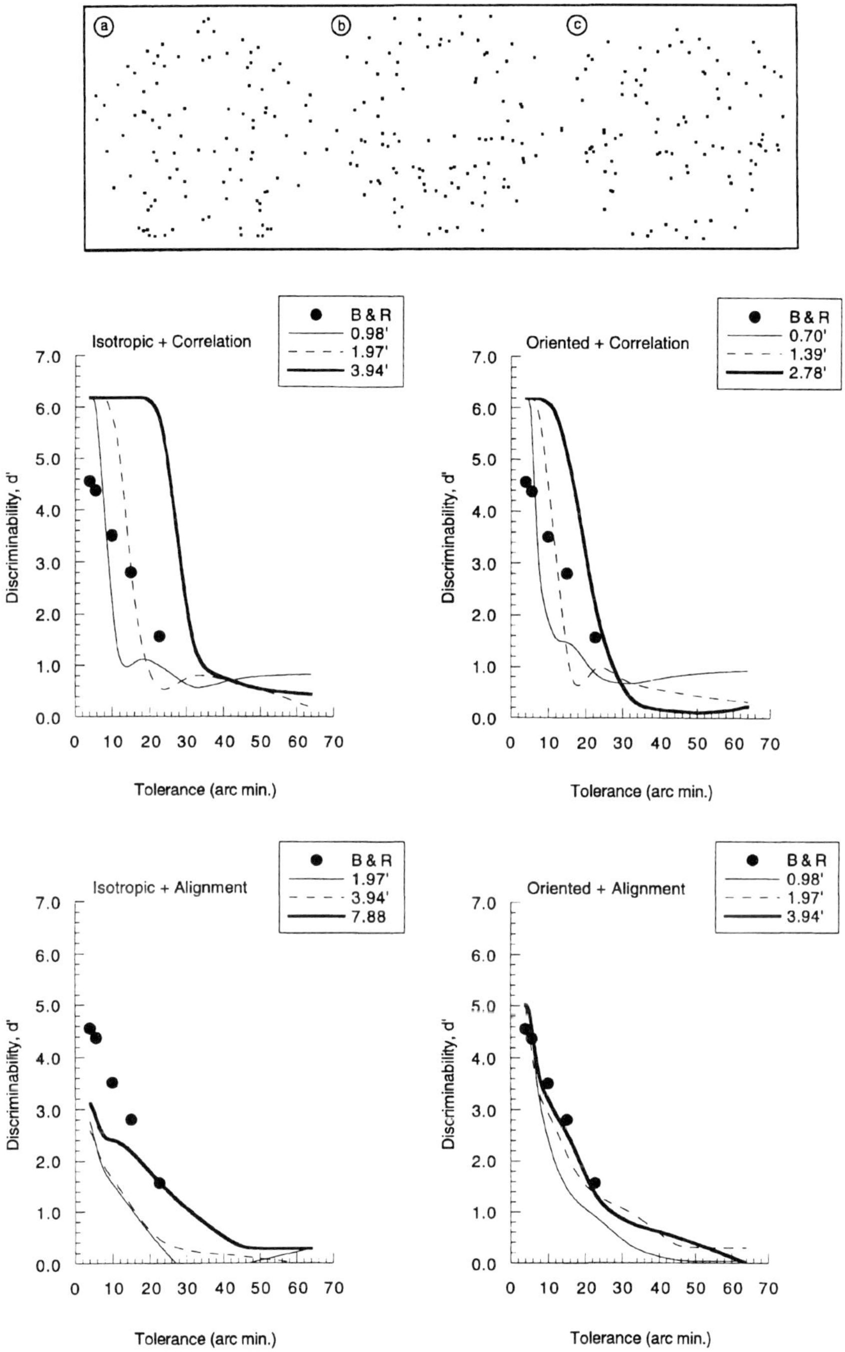

Figure 3. (Top part) Examples of the stimuli used in Simulation 2. Patterns have positional jitter of (a) ±4.0, (b) ±8.0, and (c) ±16 pixels. (Bottom part) Solid symbols are psychophysical data from Barlow and Reeves (1979), for discriminating symmetrical patterns from noise in the presence of spatial jitter. Lines represent the predictions from the models at a number of spatial scales.

levels of performance must be taken as very strong evidence against the model. The agreement of the O + A with human data models is extremely good, especially given that no fitting of the model to the data has been used. Note that the O + A model predicts similar performance across two octaves of filter sizes.

All of the proposed models produce a graded response to symmetry in the presence of unpaired dots, but the I + C model, consistently overestimates performance. In the presence of spatial perturbation of elements, the I + A model fails to perform as well as humans and for that reason can be rejected. The correlators' behaviour is broadly in accord with human data, but the fits of the O + A model are consistently better.

5. SIMULATIONS 3–5: EFFECT OF LOCATION OF SYMMETRY WITHIN A TEXTURE

A pattern containing a set of symmetrical dots embedded in a background composed of noise dots is most effectively discriminated from noise patterns when the symmetrical region is located around the axis of symmetry (Barlow and Reeves, 1979). However, the next most discriminable region is at the outer boundaries of the pattern. The region producing lowest discriminability is located between the axis and boundary. If any model using filtering predicts that symmetry is simply related to the proximity of the closest matched pairs then there should be no such advantage for structure around the boundary. In this section we simulate three psychophysical tasks examining the role of the *location* of the symmetrical features within the pattern.

Jenkins (1983) measured the effect of the spatial location of a symmetrical region embedded in noise. The tasks measured discriminability of patterns (from noise or pure symmetry) as a function of the width of the embedded symmetrical region. The tasks were discrimination of: (a) noise patterns from noise containing symmetry around the axis, (b) perfect symmetry from symmetrical patterns with noise encroaching from the boundary, and (c) noise patterns from symmetrical patterns containing noise around the axis of symmetry.

5.1. Stimuli

Stimuli were generated in approximate accord with those described in Jenkins (1983).[1] Textures contained 650 dots, where each dot was an individual pixel, appearing white on a black background.

The stimuli for Simulations 3 and 4 were noise textures with a strip of symmetrical dots around the axis of symmetry. Dots were randomly positioned throughout the image, except for a region around the axis of symmetry. A proportion of the total number of dots was then placed in the central strip, and reflected around the axis. The number of dots in the central region was proportional to the area of the strip, so that there was no difference in density across the pattern. The reference stimuli for Simulation 3 were noise textures (i.e. the task was to detect the pattern which contained a strip of symmetry), and the references for Simulation 4 were purely symmetrical patterns (i.e. the task was to detect which pattern had noise encroaching from the boundaries).

The stimuli for Simulation 5 were generated in a similar way to above, except that textures were symmetrical but contained a strip of randomly positioned elements around the axis of symmetry. The reference stimuli used were pure noise patterns (i.e. the task was to detect which patterns had symmetry at the boundaries).

5.2. Procedure

128 stimuli were generated at each width of symmetrical/noise strip. The width of strips used were: for Simulation 3, 0–24 pixels, in steps of 4 pixels (i.e. from 0–0.468 deg, assuming a 5 deg wide display), for Simulation 4, 0–72 pixels, in steps of 8 pixels (i.e. from 0–1.41 deg), and for Simulation 5, 0–24 pixels, in steps of 4 pixels. Measurements of symmetry were made using the same models described in the previous simulation, and percent correct results were generated as in Jenkins (1983).

5.3. Simulation results and discussion

Figure 4 shows that humans show a steady increase in the discrimination of an expanding symmetrical strip from noise, as the width of symmetrical strip increases, with performance levelling off at around 0.22 deg. All of the models approximately show this behaviour but their predictions do not deteriorate as sharply as human data: humans are very poor when strip width falls to around 0.1 deg. This discrepancy is probably due to two characteristics of the alignment models' operation. Firstly it is noise-free with respect to the number and position of blobs. One might expect that smaller samples of tiny blobs would give worse estimates of alignment, either because blobs would fail to be registered, or because any noise on the alignment measurement would become large in relation to the width of the blobs. Secondly, there is no uncertainty of the location of the axis — all models know exactly which image column to use. Human subjects, even given that the axis did not shift between stimuli, probably have some error on locating the axis from trial-to-trial.

Figure 5 shows that as the width of a symmetrical strip increases (to a maximum value of only 20% of the total pattern width) subjects find it increasingly hard to discriminate patterns from pure symmetry. At the point where subjects can no longer discriminate, however, 80% of the pattern is symmetrical. Consequently, the correlators predict perfect discrimination for the widths tested. The alignment models, on the other hand, predict the steady decrease in discrimination performance. The fit of the O + A model is the closest.

The final simulation was of human discrimination of symmetrical textures containing strips of noise around the axis, from noise textures (Fig. 6). Our ability to perform this discrimination deteriorates with increasing width of the strip, and at a noise width of around 1.0 arcmin subjects are approaching chance performance. Again the correlators predict no change in performance because patterns contain so much symmetry. The alignment models' sensitivity to the *location* of the symmetry is again shown in the collapse of performance as noise width increases. Interestingly, however, the I + A model collapses too quickly; isotropic filtering mechanisms are insufficiently sensitive to the structure when the noise strip becomes wide. The

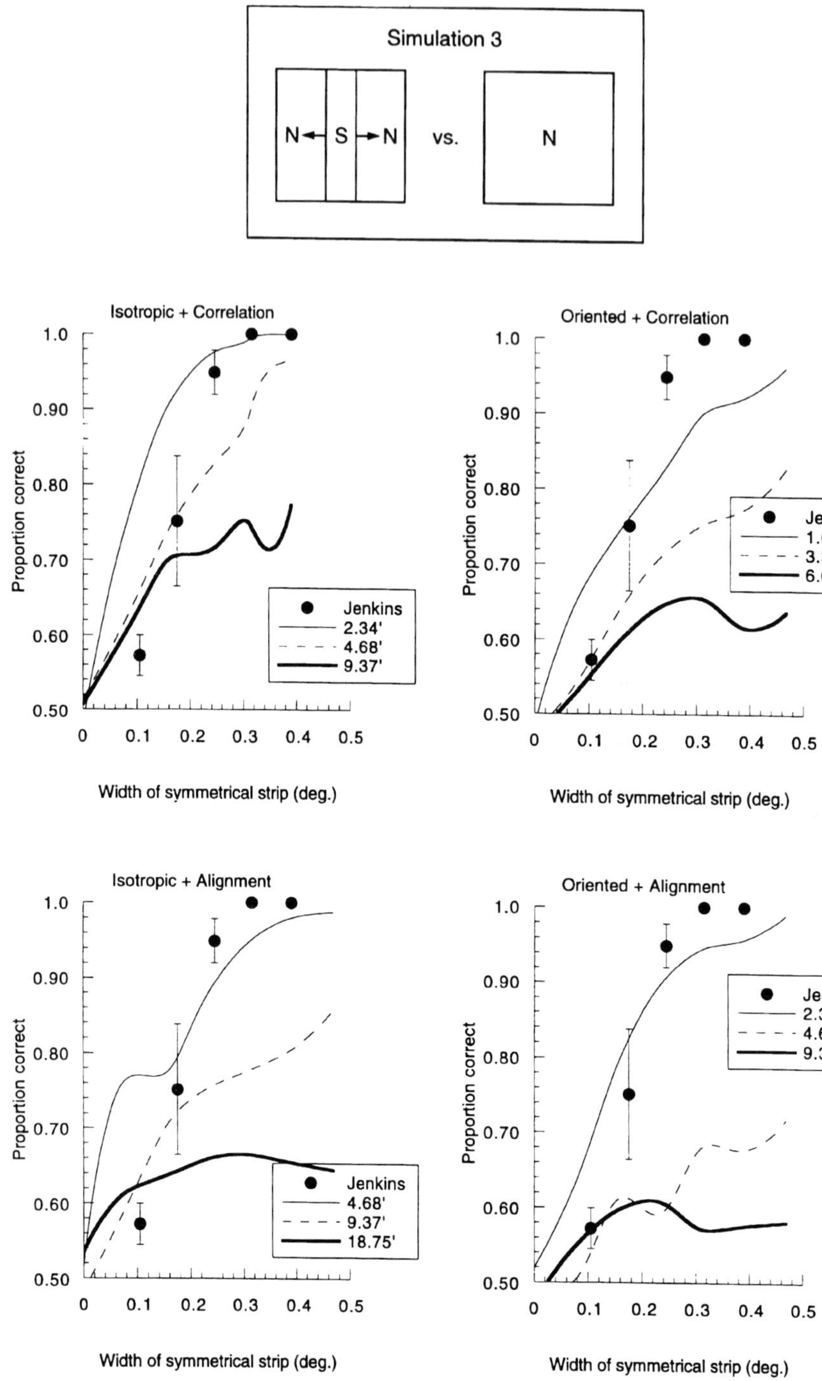

Figure 4. (Top part) Schematic representation of the task in Simulation 3. 'S' indicates symmetry, 'N' represents noise. (Bottom part) Solid symbols show human discrimination of such textures from pure noise. All models produce improved behaviour as a function of the width of the band.

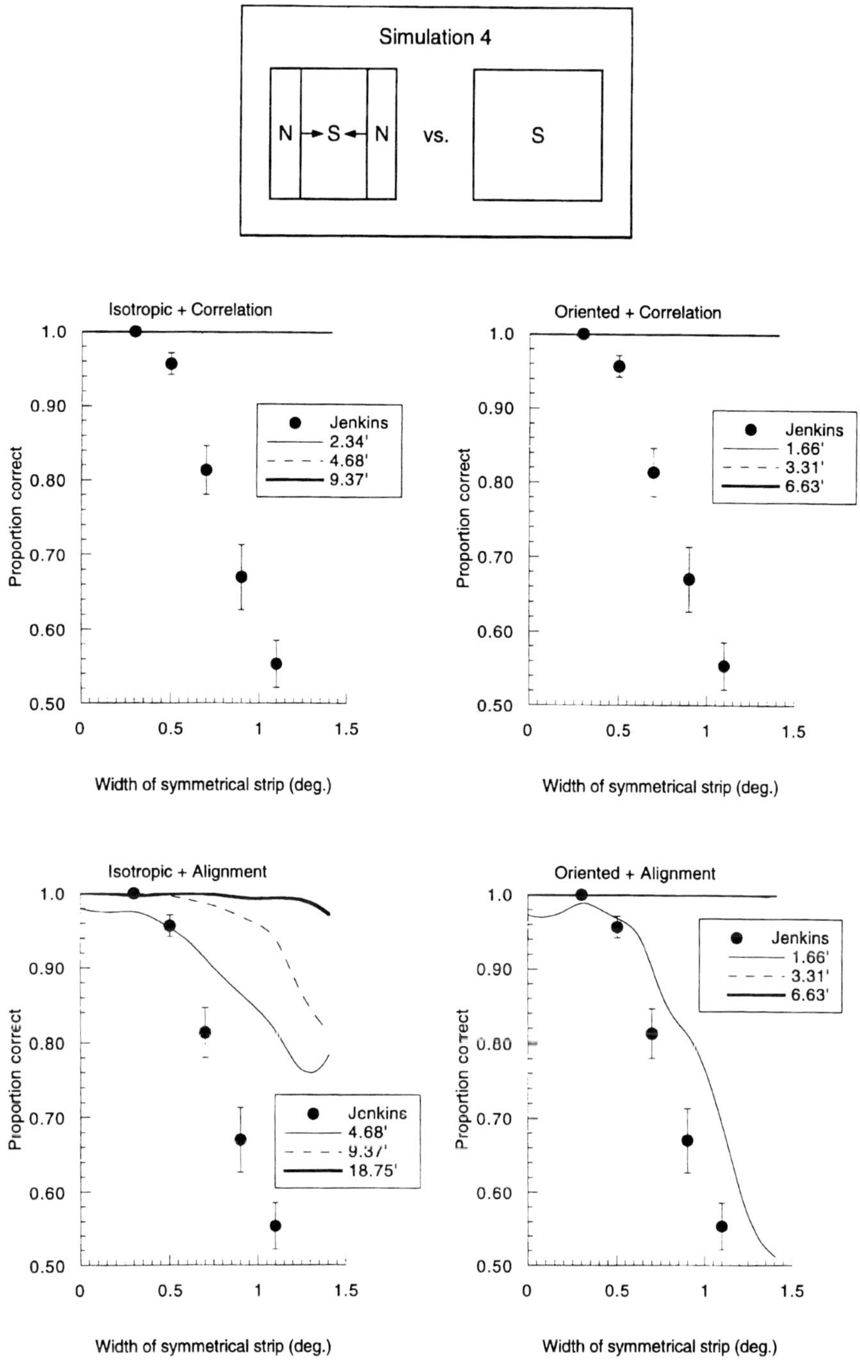

Figure 5. (Top part) Schematic representation of the task in Simulation 4. (Bottom part) Human performance for discrimination of an embedded band of symmetry from pure symmetry is represented by solid symbols. Only the performance of the alignment models show a dependence on the width of the symmetrical strip.

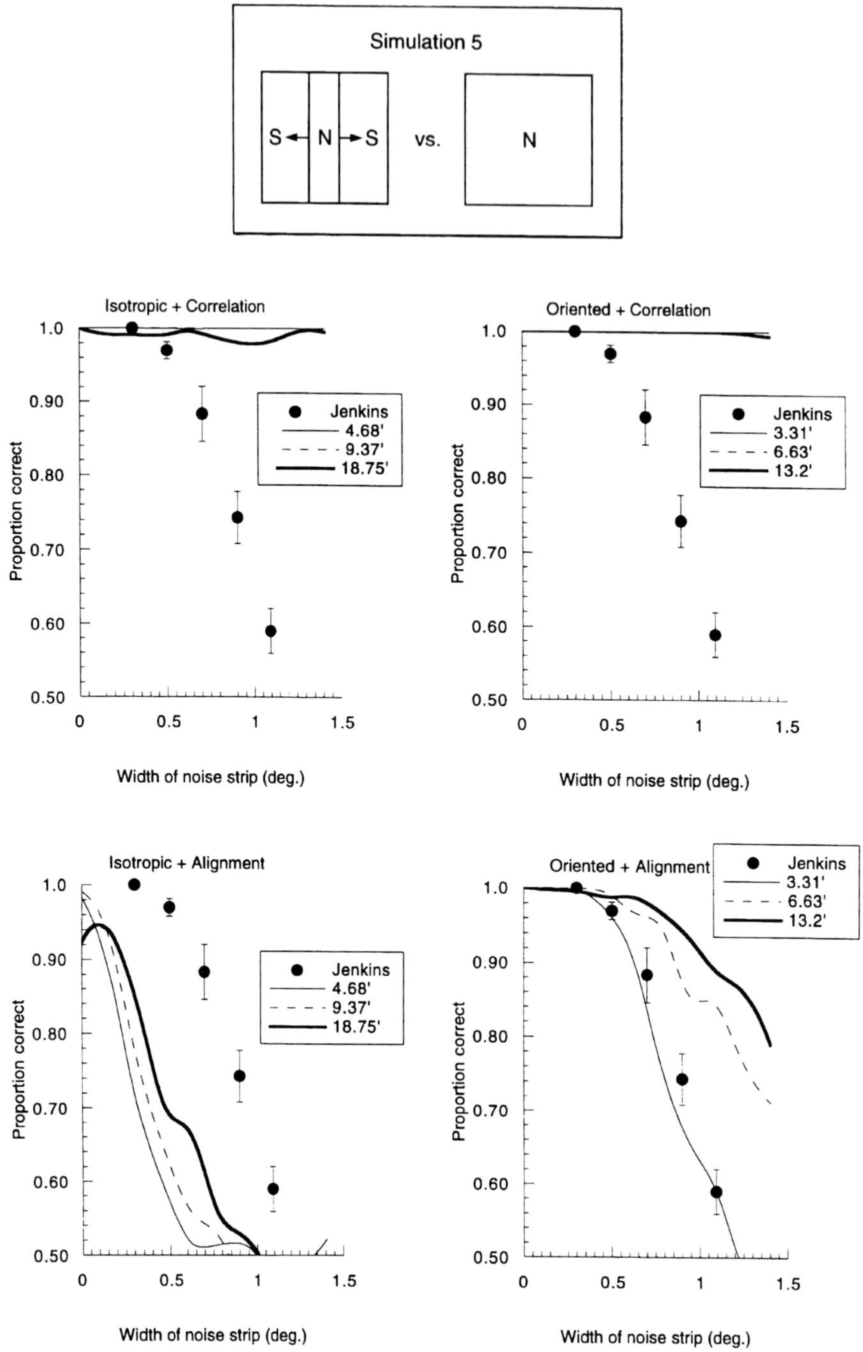

Figure 6. (Top part) Schematic representation of the task in Simulation 5. (Bottom part) Human discrimination of such textures from pure noise. Although all models are capable of detecting symmetry in these patterns, only the alignment models' performance breaks down, with increasing width of the noise strip, in accord with human data.

O + A model captures the best of both models. It can overperform, but certain scales (s.d. = 3.31 arcmin) produce patterns of discrimination which match human data. It is interesting, and somewhat counterintuitive, that an alignment model can detect correlation when the area around the axis of symmetry is pure noise. Note however that the filter size producing the best predictions are larger than those used in the previous two simulations. This also suggests an explanation for why structure at the periphery is more efficiently detected than structure embedded between axis and boundary (Barlow and Reeves, 1979). Small filters are responsible for detecting structure at the axis, large filters are responsible for structure at the boundaries. (We return to this point in the Conclusion section.)

To summarise: results from Simulation 4 indicated that correlators overestimate discrimination of symmetrical strips embedded in noise from pure symmetry. Alignment models account for human data well. Simulation 5 showed that only the O + A model could account for the deterioration shown by human subjects in discriminating symmetrical textures containing an expanding strip of noise around the axis from noise. The correlators again systematically overestimate, and the I + A model systematically under-estimates, human performance. Only the O + A model provides consistently good fits to the data across all simulations.

6. CONCLUSIONS

The results of the simulations presented in this paper lead to a number of tentative conclusions. First, we have shown the adequacy of the alignment measure for detecting symmetry. Second, we have shown that a model based on alignment but using an isotropic filter is inadequate, especially since it fails to reach the human levels of performance. Third, we have shown that the correlation measure of symmetry does not match human data closely with either form of filter, in several cases being much better in its performance than are the subjects. Fourth, we have shown that an alignment measure, after filtering with oriented filters, produces data that is in close agreement with subjects for all tasks considered.

For each simulation, results from a range of different spatial scales are reported. In every case the output of a single filter is used to reach a psychophysical decision and hence a psychometric function. It might be thought that combining information from the outputs of different filters would be a way of improving performance, for example through probability summation. In practice, this is not found to be the case. The reasoning is as follows. In reaching a psychophysical decision, the main causes of an incorrect response are 'spurious' near-symmetry responses in the noise targets. These are most awkward in filters that are not well matched, spatially, to the spatial structure of the target. Thus combining responses from different filters increases the probability of detecting the symmetry, but also increases the probability of detecting spurious symmetry. Probability summation therefore would not improve performance.

To conclude, we shall discuss why the alignment model with oriented filters succeeds when the others do not and then we shall go on to place the model in the context of a general-purpose visual process. Our claim is that symmetry detection, when

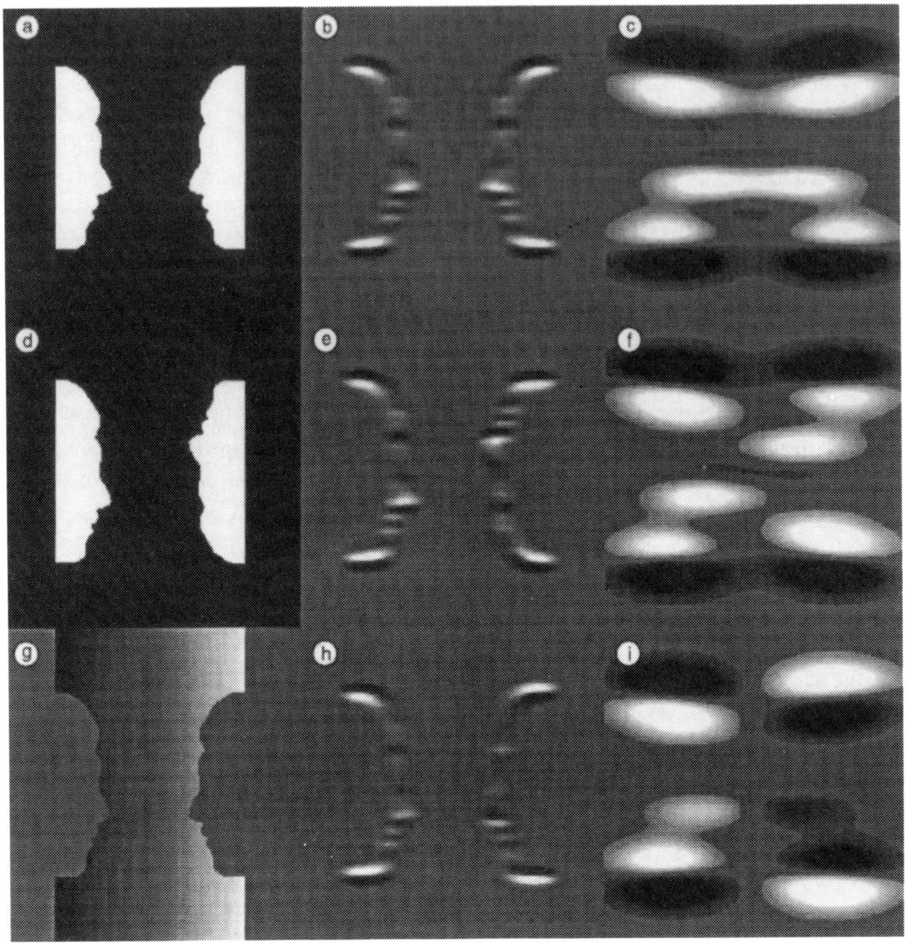

Figure 7. (First column) Variants on the Rubin vase: (a) standard vase, (d) figure with one inverted face, (g) figure with vase shaded by an intensity gradient. (Second column) Fine scale. (Third column) Coarse scale horizontally filtered versions. Note the clustering of blobs around the facial profiles in the second column, and the alignment of blobs around the centre of the symmetrical vase in (c).

conceived of as being performed by the alignment model, is a special case of a more general, object-detection mechanism.

First, we consider why the alignment model, with oriented filters, does match human performance. This model is based on the proposition that parallel and aligned correlations in images are worth detecting, a point we return to below. In practice, most randomly produced bilaterally symmetric patterns have structure of this type. They need not; for example, a checkerboard pattern will have bilateral symmetry that lacks continuous correlations. The critical data here, are the effects of which parts of the target are symmetric. Given the need for continuous correlations, it is not surprising that the alignment model detects symmetry about a mid-line efficiently, but does not detect symmetry when it does not extend over a sufficiently large area around the mid-line. A correlation model, on the other hand, has no specific requirement for the correlation to extend through the mid-line region. The alignment model only works with oriented filters. Isotropic filters do not produce extended regions of response.

For most of the simulations that we have run, the choice of spatial scale for the oriented filter is not critical for the alignment model. In general, human data match the performance of a fairly fine-scale filter. For some simulations (1, 2), the choice of filter spatial scale barely affects model performance. In Simulations 3 and 4, subjects match the performance of fine-scale filters. In Simulation 5, subjects match the performance of a slightly coarser scale filter. There is thus no one spatial scale at which human data can be uniformly modelled, although a parsimonious account would be that fine-scale filters ($s \approx 1.5$ arcmin) are used in all cases except Simulation 5. The cue in Simulation 5 is structure that is further from the mid-line of the pattern, and it is perhaps not surprising that a coarser scale serves best here. A similar point is discussed below and illustrated in Fig. 7.

Now, we consider whether the alignment model is plausible in the context of a visual system. The main components of the model are oriented filters and an alignment detection mechanism. Oriented filters are known to exist in visual cortex, but to our knowledge an alignment mechanism has not been sought, although the 'non-Cartesian' receptive fields reported by Gallant *et al.* (1993) may be suitable candidates. In this respect, it is important to note that a simple linear receptive-field type of weighting function will not suffice to detect alignments. Any mechanism that serves the function we propose must be broadly insensitive to the separation between the aligned stripes and to the sequence of their polarities. Since varying the separation and polarity of the aligned stripes would lead to variations in the spatial-frequency amplitude and phase contents of the pattern, such a requirement would need a receptive field that was broadly tuned and had complex-cell types of properties.

Our final claim is that the proposed alignment mechanism is a general-purpose strategy for detecting objects, or salient structures, in images and that symmetry is a special case of this. Figure 7 shows an example of the logic of our argument. The first column of the figure shows three variants of the Rubin vase figure. The first image shows a normal version; the second contains an inverted face profile; the third, a version where the 'vase profile' is shaded with a continuous intensity gradient.

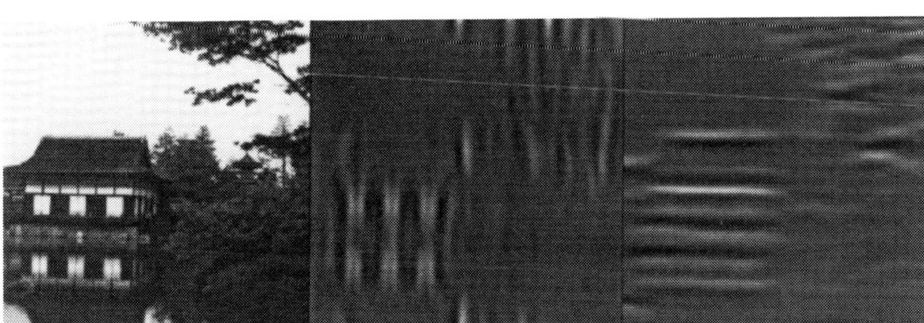

Figure 8. A scene containing natural and man-made objects, and versions filtered with horizontal and vertical filters.

The second and third columns show fine and coarse-scale filtered versions of each image. There are two types of alignment pattern revealed in the filtered images: one is related to the 'vase profile' (Fig. 7c) and one is related to the 'face profiles' (Fig. 7b, e, h). The vase case can be regarded as representing symmetry detection, but the second cannot. They both, however, can be regarded as alignment detection. In this sense, symmetry can be seen to be a special case of a more general-purpose structure detection.

Finally, Fig. 8 shows the results of filtering a real image with horizontal and vertical oriented filters. Note that the responses have aligned stripes along salient objects, man-made and natural. This point is elaborated in Watt (1994).

Acknowledgements

We thank Trish Carlin and Ian Paterson for many helpful comments on this work. This work was funded by the SERC (GR/H53181).

NOTES

1. Jenkins used *dynamic* dot displays. Each dot was sequentially generated (every 122 μs) and very briefly displayed (1.5 μs). Thus 16446 points per sec were generated and plotted although, at any one moment, there was only one dot pair on the screen. In the simulation described, static patterns were used and matched for *perceived* density: reported as 26 points per deg^2.

REFERENCES

Barlow, H. and Reeves, B. (1979). The versatility and absolute efficiency of detecting mirror symmetry in random dot displays. *Vision Res.* **19**, 783–793.

Bruce, V. and Morgan, M. (1975). Violations of symmetry and repetition in visual patterns. *Perception* **4**, 239–249.

Campbell, F. and Robson, J. (1968). Application of Fourier analysis to the visibility of gratings. *J. Physiol.* **197**, 551–566.

Daugman, J. (1985). Uncertainty relation for resolution in space, spatial frequency, and orientation optimised by two-dimensional cortical filters. *J. Opt. Soc. Am.* **2**, 1160–1169.

Gallant, J., Braun, J. and Van Essen, D. (1993). Selectivity for polar, hyperbolic, and Cartesian gratings in macaque visual cortex. *Science* **259**, 100–103.

Jenkins, B. (1982). Redundancy in the perception of bilateral symmetry in dot textures. *Percept. Psychophys.* **32**, 171–177.

Jenkins, B. (1983). Component processes in the perception of bilaterally symmetric dot patterns. *Percept. Psychophys.* **34**, 433–440.

Julesz, B. (1975). Experiments in the visual perception of texture. *Scient. Am.* **232**, 34–43.

Kanade, T. and Kender, J. (1987). Mapping image properties into shape constraints: Skewed symmetry, affine-transformable patterns, and the shape-from-texture paradigm. In: *Readings in Computer Vision: Issues, Problems, Principles, and Paradigms*. M. Fischler and O. Firschein (Eds). Morgan Kaufmann, California, pp. 257–267.

Koeppl, U. (1993). Local orientation versus local position as determinants of perceived symmetry. *Perception* **22** (Suppl.), 111.

Locher, P. and Wagemans, J. (1993). Effects of element type and spatial grouping on symmetry detection. *Perception* **22**, 565–587.

Marr, D. (1976). Early processing of visual information. *Proc. Roy. Soc. Lond.* **B 275**, 483–534.

Marr, D. (1982). *Vision*. Freeman, San Francisco, CA.

Phillips, G. and Wilson, H. (1983). Orientation bandwidths of spatial mechanisms measured by masking. *J. Opt. Soc. Am.* **A1**, 226–232.

Press, W., Teukolsky, S., Vetterling, W. and Flannery, B. (1992). *Numerical Recipes in C: The Art of Scientific Computing*. Cambridge University Press, Cambridge.

Sachs, M., Nachmias, J. and Robson, J. (1971). Spatial frequency channels in human vision. *J. Opt. Soc. Am.* **61**, 1176–1186.

Voorhees, H. and Poggio, T. (1987). Detecting textons and texture boundaries in natural images. In: *Proceedings of the First International Conference on Computer Vision*. pp. 250–258.

Watt, R. (1991). *Understanding Vision*. Academic Press, London.

Watt, R. (1994). A computational examination of image segmentation and the initial stages of human vision. *Perception* (In press).

Wilson, H. and Gelb, D. (1984). Modified line-element theory for spatial-frequency and width discrimination. *J. Opt. Soc. Am.* **A1**, 124–131.

Yeshurun, Y., Reisfeld, D. and Wolfson, H. (1992). Symmetry: A context free cue for foveated vision. In: *Neural Networks for Perception Vol. 1*. H. Wechsler (Ed.). Academic Press, London, pp. 477–491.

Zhang, L. (1991). *Symmetry perception in human vision*. PhD Thesis, University of Trieste.

Human Symmetry Perception, pp. 209–225
C. W. Tyler (Ed.)

Modelling symmetry detection with back-propagation networks

CYRIL LATIMER,[1,*] WENDY JOUNG[1] and CATHERINE STEVENS[2]

[1] *Department of Psychology, University of Sydney, New South Wates 2006, Australia*
[2] *Department of Psychology, University of Queensland, Queensland 4072, Australia*

Abstract—This paper reports experimental data and results of network simulations in a project on symmetry detection in small 6 × 6 binary patterns. Patterns were symmetrical about the vertical, horizontal, positive-oblique, or negative-oblique axis, and were viewed on a computer screen. Encouraged to react quickly and accurately, subjects indicated axis of symmetry by pressing one of four designated keys. Detection times and errors were recorded. Back-propagation networks were trained to categorize the patterns on the basis of axis of symmetry, and, by employing cascaded activation functions on their output units, it was possible to compare network performance with subjects' detection times. Best correspondence between simulated and human detection-time functions was observed after the networks had been given significantly more training on patterns symmetrical about the vertical and the horizontal axes. In comparison with no pre-training and pre-training with asymmetric patterns, pre-training networks with sets of single vertical, horizontal, positive-oblique or negative-oblique bars speeded subsequent learning of symmetrical patterns. Results are discussed within the context of theories suggesting that faster detection of symmetries about the vertical and horizontal axes may be due to significantly more early experience with stimuli oriented on these axes.

1. INTRODUCTION

Our ability to detect mirror symmetry continues to be of great interest to experimental psychologists (Julesz, 1971; Goldmeier, 1972; Corballis and Roldan, 1975; Palmer and Hemenway, 1978; Royer, 1981; Masame, 1983, 1984, 1985, 1986, 1987, 1988; Field and Golubitsky, 1992; Wenderoth, 1994). Various forms of symmetry are depicted in Fig. 1a, and the order of detection times usually observed for these is: multiple axes (fourfold) < vertical and horizontal < vertical < horizontal < double oblique < negative oblique < positive oblique < 90° centric < 180° centric < asymmetric (Royer, 1981). However, it has been noted that, although symmetry about the vertical axis is frequently said to be more salient, every possible ordering can be found in the

*To whom correspondence should be addressed.

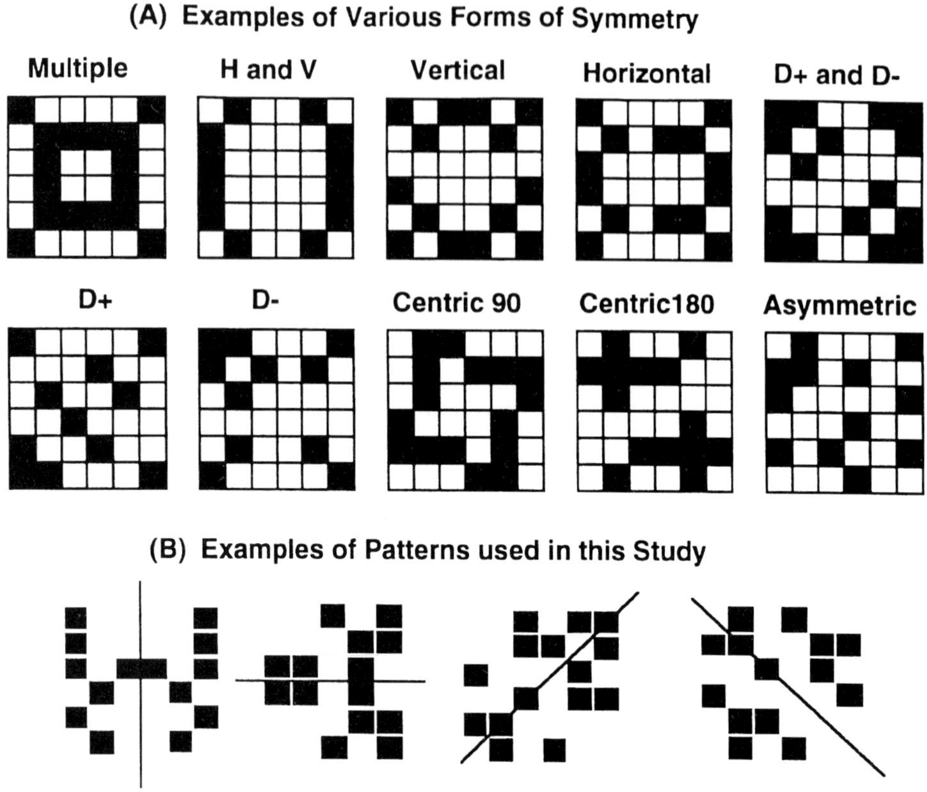

Figure 1. (A) Examples of the different forms of symmetry. (B) The 6 × 6 patterns used in the present study with axes of symmetry (not shown to subjects) superimposed.

literature, and this is probably due to differences in stimulus characteristics and other factors unique to a particular study (Locher and Wagemans, 1993, p. 569 and p. 575).

Mach (cited in Corballis and Roldan, 1975) suggested that saliency of symmetry about the vertical axis was a consequence of the bilateral symmetry of the brain, and that in order to detect horizontal symmetry, we may have to rotate a pattern mentally until its axis of symmetry is in line with vertical. Indeed, tilting of the head shifts the reaction time function in the direction of head tilt. Accordingly, it has been proposed that retinal information, mapped onto a phenomenal coordinate system, can be mentally rotated for test against a template for vertical symmetry 'embedded symmetrically in the brain' (Corballis and Roldan, 1975, p. 131). It has also been proposed (Julesz, 1971) that symmetry detection requires a point-by-point process of comparison based on neural anatomy organised symmetrically around the centre of the fovea. The notion seems to be that detection of symmetry may be mediated by processes that detect and compare mirror arrangements of pattern elements across particular axes. Others have argued for a two-stage model, the first stage of which is a 'crude but rapid analysis of symmetry in all orientations simultaneously.' Somehow a probable axis of symmetry is selected and a detailed comparison of halves about this axis for mirror identity is initiated (Palmer and Hemenway, 1978, p. 700).

Royer (1981) suggests that symmetry may be encoded integrally, and that the order of observed reaction times mentioned above is consistent with a serial, hierarchical comparison model. When decoding is required in detection tasks, subjects perform serial hierarchical decomposition of the symmetry code in the order specified above. Interest has also focused on the possibility that judgments of symmetry, rather than being based on detection of the axis itself, may depend upon the detection of equal spacing of figure components about the axis. The elements of a bilaterally symmetrical figure are equally spaced along lines orthogonal to the axis of symmetry, and it has been demonstrated that non orthogonal arrangements of figural components around an axis (skewed symmetry) can have regular effects on detection time and errors (Wagemans *et al.*, 1991, 1992, 1993; Wagemans, 1992, 1993).

With renewed interest in connectionism, attention has returned to the possibility that many of the reliably observed differences in reaction time to classes of visual patterns may be explicable in terms of differential experience and learning, and the promise of connectionism for psychology has always been that these learning processes can be embodied in networks of simple neuron-like elements that take inputs and produce outputs according to clearly specifiable activation functions and learning rules. Hebb (1949), for example, argued that the recognition of a shape as belonging to particular class, or the ability to attach labels to shapes were the results of learning and experience.

More recently, there has been an attempt to explain the Stroop Effect within a parallel distributed framework, where speed of response is seen to depend on strength of a processing pathway, and where strength increases with training (Cohen *et al.*, 1990). Simple back-propagation networks were given significantly more training at responding to word than to color inputs. Because of this unequal training, the networks developed stronger connections between input and output units responsible for word reading, and, as a result, responded more rapidly to word reading than to color naming. Stronger connections in the word processing pathways allow activation to increase more rapidly in output units dedicated to reading responses, and, if the *rate of increase in activation* of these units is considered across time, it can be used as a measure of 'reaction time' in a network. Activation builds up more rapidly in output units with stronger connections (see Cohen *et al.* (1990) for details).

A similar approach is taken in this project, and the question is asked: may not the faster detection times for multiple, vertical, and horizontal symmetry be the result of the predominance of stimuli oriented in these directions in the environment and the resultant development of a higher population of cells, neural structures or pathways tuned maximally to these orientations? The existence of visual anisotropies (meridional astigmatism) has been known since the nineteenth century (Howard, 1982); visual acuity is higher for vertical and horizontal stimuli than for obliques, but whether anisotropies are caused by innate or environmental factors is still a matter of debate. Certainly, at a neurophysiological level, it has been shown that, together with innate determinants, visual experience also plays an important role in determining the orientation selectivity and preferred orientation of cortical cells (see Howard (1982, pp. 114–119, 132–146) for a review of findings). For example, early exposure to horizontally and vertically striped environments can have profound effects on the development and response of cells in the visual cortex of kittens (Blakemore and Cooper, 1970; Blakemore and Mitchell, 1973).

Artificial neural networks provide an ideal platform for modelling the effects of learning and experience on mechanisms that may mediate symmetry detection. Networks can be subjected to varying amounts of training on different stimulus orientations, and the effects of training can be observed in the codes and representations that emerge in hidden unit space. For example, biased training with patterns symmetric about the vertical axis can not only produce stronger connections between input units and output units that signal vertical symmetry, but also can lead to development of more hidden units whose main function is to code vertically symmetric arrangements in the input array. In current theories of symmetry detection, very little in the way of clearly specified mechanisms have been reported. The present study seeks to fill this gap by establishing a basis for possible mechanistic explanation of symmetry detection.

2. SIMULATIONS OF SYMMETRY DETECTION

In mirror-symmetry detection by a network with six input units, two hidden units and one output unit, McClelland *et al.* (1986) found that solution weights from inputs to hiddens were arranged symmetrically. However, Minsky and Papert (1988, p. 253) have expressed concern about such small architectures on the grounds that solutions produced by them do not scale in a reasonable manner, because increases in the size of the retina are accompanied by exponential increases in the size of weights (see also Quinlan, 1991, p. 259). Additionally, such networks require inordinate numbers of training trials. On the other hand, using the Boltzmann learning algorithm, it has proved possible to find sets of weights that convert hidden units into higher-order feature detectors for the solution of symmetry problems. Again, weights on connections from input units to hidden units arranged themselves symmetrically about appropriate axes (Sejnowski *et al.*, 1986). After approximately 100 000 presentation of training patterns constructed within 10×10 binary matrices, their network achieved a correct classification rate of 90%.

To date, there has been one study comparing the performance of networks and human subjects in symmetry detection. Cohen *et al.* (1986) explored the possibility of simple incremental learning algorithms having the capacity to manifest stimulus generalizable symmetry detection and stimulus-specific repetition priming in the same network. Using the Boltzmann learning algorithm and binary patterns constructed within a 9×9 binary array, the authors argue that their behavioral and simulation data support the conclusion that repetition priming is not a separate mechanism and does not depend on an already structured system. Instead, priming may be an integral feature of learning processes. The current project seeks to model human symmetry detection processes in back-propagation networks, and by so doing, complements the earlier work of Cohen and co-workers.

3. DETECTION EXPERIMENT

We report one in a series of experiments on human symmetry detection. The aim of the experiment was generation of a detection-time function for 80 symmetrical patterns (20 vertical, 20 horizontal, 20 positive- and 20 negative-oblique) for comparison with detection-time functions obtained from neural-network simulations.

3.1. Method

3.1.1. Subjects. The 16 subjects were all volunteer first-year students enrolled in Psychology at the University of Sydney and none had participated in symmetry experiments before. All subjects had emmetropic or suitably corrected vision.

3.1.2. Stimuli. Patterns were constructed in a 6×6 matrix with the constraint that, across patterns, any cell had a probability of 0.4 of being filled. There were, on average, 14.4 filled elements per pattern. Figure 1b depicts examples of the patterns with axes of symmetry (not shown to subjects) superimposed.

3.1.3. Procedure. Subjects sat at a Macintosh SE computer and viewed the patterns center screen from a distance of 50 cm. Stimulus presentation and response recording were all programmed using MacLab software (Costin, 1988). Measuring 1 cm $\times$ 1 cm, the patterns subtended a visual angle of 1 deg, and, after a practice set of 20, were presented randomly within three blocks of 40 trials each. The first two blocks were designated training blocks and contained 80 trials with 20 patterns at each axis of symmetry. The third block, designated a test block, comprised 40 previously unseen patterns — 10 examples of each axis of symmetry. Very small patterns were used to avoid the necessity for eye movements during symmetry detection, and because of their small size, no fixation point was deemed necessary. Subjects were encouraged to respond to patterns as quickly and accurately as possible by pressing one of four keys representing the four possible axes of symmetry. Key designation was counterbalanced to control for any effects of handedness. To achieve a measure of simple, motor-reaction time, subjects also received one block of 40 trials in which they were required simply to press one of the four keys when a small black square appeared center screen. Throughout these control trials, subjects simply cycled through the four keys. This measure of motor-reaction time was subtracted from detection times on the experimental trials to provide a pure measure of detection time for the four axes of symmetry.

3.1.4. Results. Differences in errors and reaction times in the four experimental conditions were tested by planned contrasts between means. Detection time for vertical symmetry was faster than to the other symmetries in both training and test blocks, $F(1,45) = 52.534$, $p < 0.05$ and $F(1,45) = 29.321$, $p < 0.05$. Similarly, horizontal symmetry was detected faster than the oblique symmetries, $F(1,45) = 8.370$, $p < 0.05$ and $F(1,45) = 20.903$, $p < 0.05$. There were no differences in detection time between the positive- and negative-oblique patterns,

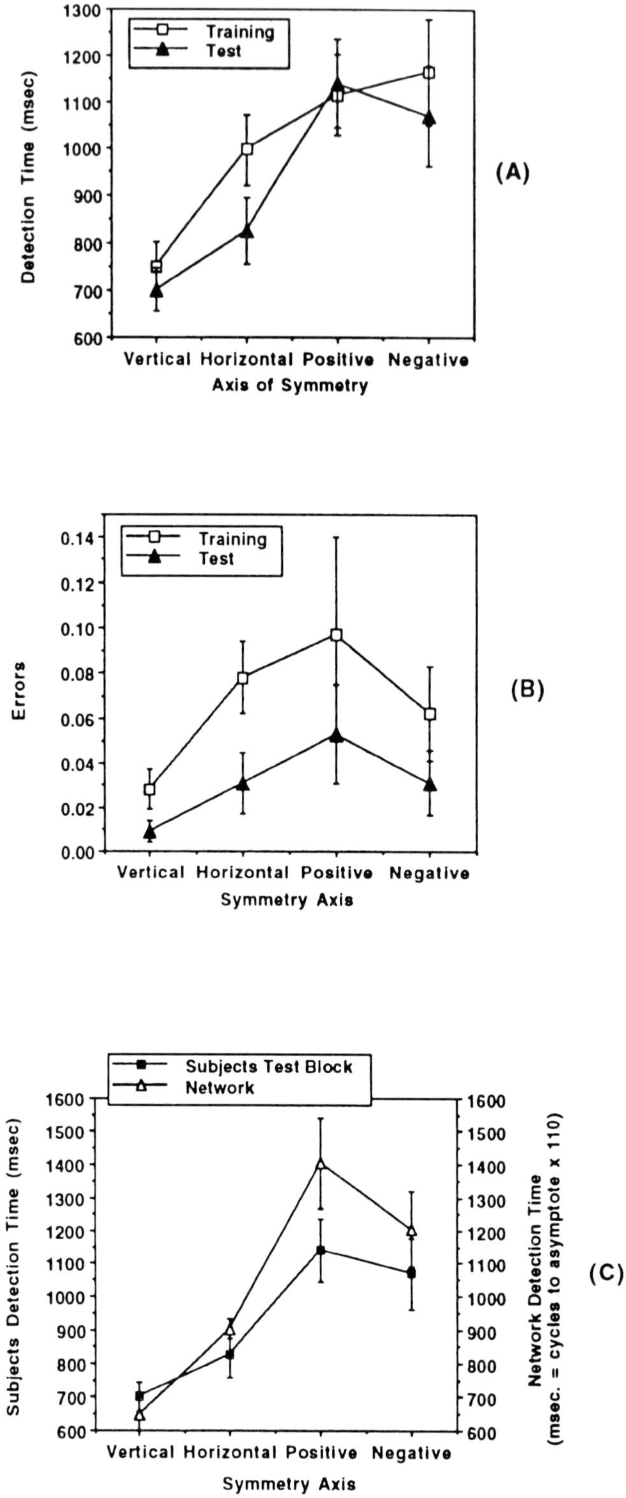

Training
Test
Detection Time (msec)
1300
1200
1100
1000
900
800
700
600
Vertical Horizontal Positive Negative
Axis of Symmetry
(A)
Training
Test
Errors
0.14
0.12
0.10
0.08
0.06
0.04
0.02
0.00
Vertical Horizontal Positive Negative
Symmetry Axis
(B)
Subjects Test Block
Network
Subjects Detection Time (msec)
1600
1500
1400
1300
1200
1100
1000
900
800
700
600
Network Detection Time
(msec. = cycles to asymptote x 110)
Vertical Horizontal Positive Negative
Symmetry Axis
(C)

$F(1, 45) = 0.174$, $p > 0.05$ and $F(1, 45) = 1.009$, $p > 0.05$. In the error data, there were fewer errors made on vertical symmetry trials than on other symmetries in the test blocks, $F(1, 45) = 4.388$, $p < 0.05$. Significance levels were maintained after Epsilon correction (Geisser and Greenhouse, 1958), and Figs 2a and 2b present graphically the detection-times and errors on training and test blocks for each axis of symmetry.

4. NETWORK SIMULATIONS

A back-propagation network with 36 input units, nine hidden units and four output units (one for each of the axes of symmetry) required 8000 presentations to categorize the 80 patterns with 100% accuracy. Other network architectures could have been used, but the back-propagation algorithm was chosen because it has been used extensively in network simulations of perception and cognition (Quinlan, 1991), and, with cascaded activation functions, it readily simulates reaction time (McClelland *et al.*, 1986). The use of nine hidden units was not arbitrary; through a process of elimination, it was determined that back-propagation networks with less than nine hidden units were unable to categorize the patterns. Networks were trained on a Macintosh IIci using the NeuralWorks Professional 4 network simulator.[1]

As may be expected with such a small training set, the network did not generalize very well, and classified correctly just over half of a set of 40 unseen test patterns. In future simulations, much larger samples of patterns are planned — at least more patterns than weights in the network. However, the main concern at this stage of the project was to replicate in the network the experimental conditions of the subjects. In support of this aim, a second network with 10 hidden units was given *biased training* on the previously described set of patterns, with 200 instances of the vertical, 100 instances of the horizontal, and 20 instances each of the positive- and negative-oblique patterns, i.e. a ratio of $10:5:1:1$. It was thought that this network's performance might simulate more closely the heterogeneous distribution of stimulus orientations in the environment of the subjects.

The exact distribution of orientations in the environment is not known, but estimates have been made from samples. Switkes *et al.* (1978) applied Fourier analysis to photographs of indoor and outdoor scenes from urban and rural environments. In the range dominated by low frequency components, and across all environments, there was a greater preponderance of horizontals and verticals than any other orientation. Within environments, there was no difference between vertical and horizontal components. In the high-frequency range, horizontal and vertical orientations again outnumbered all

Figure 2. (A) Mean detection time in training and test blocks for the four axes of symmetry. (B) Error proportions in training and test blocks for the four axes of symmetry. (C) Detection times of subjects in their test block of trials and of the biased network for the four axes of symmetry. Detection time for the network is based on the number of cycles required to drive an appropriate output unit to asymptote multiplied by a constant (110). Multiplication by the constant allows plotting of subject and network detection-time functions on the same axes.

other orientations, but verticals also outnumbered horizontals. Given the 10 : 5 : 1 : 1 biased training, the network learnt the vertical and horizontal patterns quickly, but required some 30 000 training trials overall to reduce error on the oblique patterns, and even then, was unable to classify correctly five positive- and one negative-oblique pattern.

5. EFFECTS OF TRAINING ON HIDDEN-UNIT SPACE

There is at least one important caveat to the following analysis of hidden-unit space. The networks under analysis were trained on relatively small sets of patterns and generalized to just 50% of an unseen set of patterns. They cannot therefore be regarded as having acquired the truly *general* characteristics of symmetry in the 6×6 binary matrices, not to speak of the general characteristics of symmetry itself. Nonetheless, the analysis reveals the sort of representations that may evolve in more highly dimensional simulations of symmetry detection.

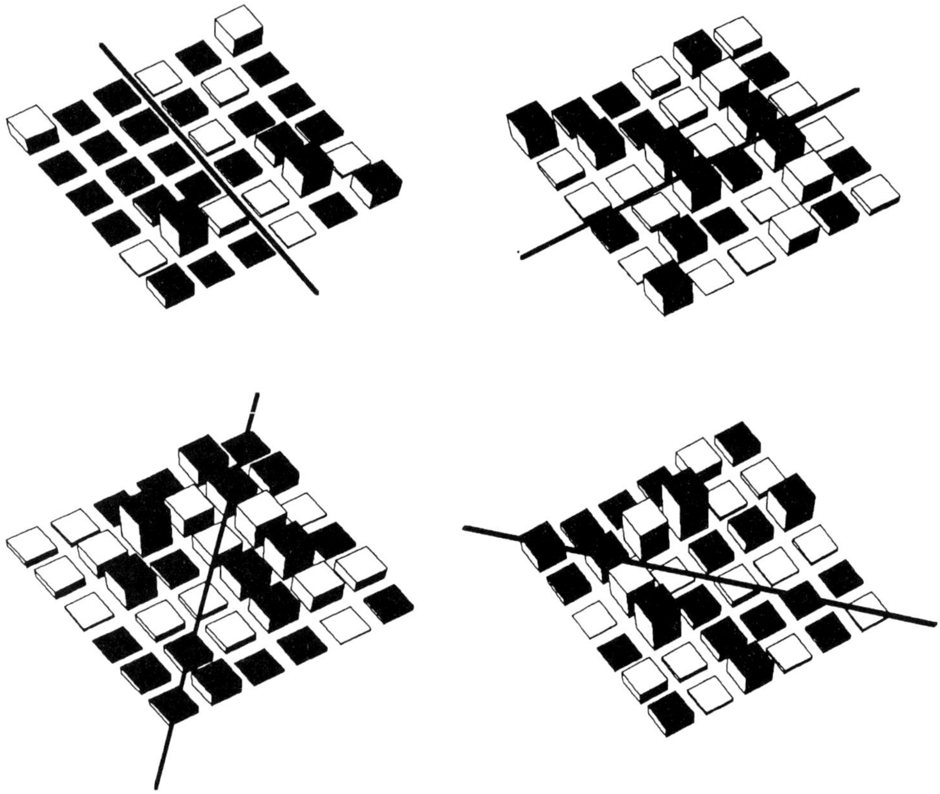

Figure 3. Three-dimensional Hinton diagrams of the weights on connections from the input array to four hidden units. Column height represents weight magnitude and black-topped weights represent negative, while white-topped columns signify positive weights. Axes of symmetry of the class of patterns for which each unit appears to be responsible have been superimposed; the view is from above the lower-left corner of the input array.

Methods for partitioning hidden-unit space concentrate mainly on the patterns of activation produced by inputs (Sejnowski and Rosenberg, 1988; Elman, 1990; Bloesch and Wiles, 1991). However, in the present simulations with relatively small numbers of hidden units, statistical techniques are inappropriate and attention was given to weights as well as hidden-unit activations. In low-dimensional, hidden-unit space, it is sufficient to trace major lines of influence from inputs through weights and activations to outputs. Clearly, in a distributed representation, it is difficult to single out hidden units with sole responsibility for signalling the presence of one axis of symmetry. Nonetheless, as with the symmetry-detecting networks reported earlier, the networks in this project developed symmetrical arrangements of weights on connections between input and hidden units. Figure 3 depicts three-dimensional Hinton diagrams of the weights to four of the hidden units in the unbiased network, and the axes of symmetry of the class of patterns for which they appear to be mainly responsible, have been superimposed. The diagrams are viewed from above the lower-left corner of the input array and patterns. Column height represents weight magnitude, and black-topped columns represent negative, while white-topped columns represent positive weights.

A consideration of the activations produced in the hidden units by patterns from each class of symmetry and the weights from each hidden unit to each output suggested that, in the unbiased network, three hidden units coded mainly vertical symmetry while the other symmetries were served mainly by two hidden units each. A different picture emerged from consideration of weights and activations in the biased network; more hidden units and a greater range of levels of activation of those units became associated with the prevailing vertical patterns. This finding is evidenced in Figs 4 and 5 which show box-and-whisker plots of the activation levels of hidden units for the four classes of patterns.

The horizontal markers in the boxes depict median activations. Again, consideration needs to be given to the size and sign of weights from hiddens to outputs as well as hidden unit activations, and when this is done, four of the ten hidden units appear to code verticals (Units 1, 3, 8, and 10). Hidden Unit 6 coded mainly horizontals, and it can be seen in Fig. 4 that this unit appears to be locked on for horizontals. Hidden Units 5 and 9 were mainly responsible for coding negative obliques, and Unit 7 appeared to code positive obliques. Hidden Unit 4 served both vertical and horizontal patterns, but the remaining Unit 2 can only be described as *ahorizontal*. This unit (largely inactive for horizontals) had an exceptionally high negative weight to the horizontal output unit and positive weights to the vertical and oblique output units. Overall, the biased training resulted in close to 50% of the hidden units being harnessed by the more numerous vertical patterns, although it should be noted that, in comparison with the oblique patterns, the relatively more numerous horizontal patterns did not achieve the same ascendancy.

Seizure of the hidden units by the prevalent vertical patterns is a possible connectionist analog of results reported in neurophysiological studies of the effects of restricted visual environments on the development of cells in the visual cortex of kittens (Blakemore and Cooper, 1970; Blakemore and Mitchell, 1973). For example, Blakemore and Cooper (1970) describe how two kittens, reared during the first five months of their lives in the dark except for a few hours each day spent in cylindrical

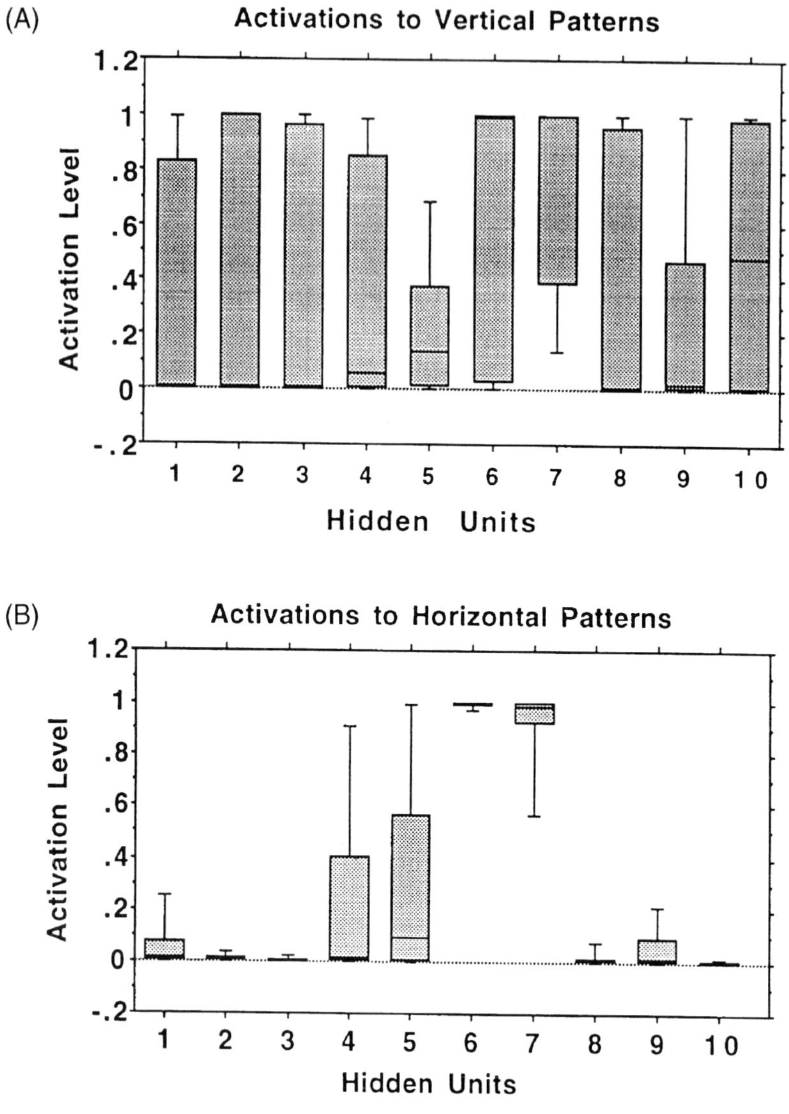

Figure 4. Ranges and levels of hidden-unit activation in the network given biased training. (A) Activation ranges to patterns symmetrical about the vertical axis. (B) Activation ranges to patterns symmetrical about the horizontal axis.

chambers with white and black bars at one orientation, had no cortical neurons which responded to the orientation perpendicular to the bars that they were reared in when young. Additionally, almost all of their cortical cells responded to orientations within 45 deg of the conditioned angle, and, there being no evidence of regions of silent cortex, the effect could not be explained by conditioning having produced degeneration of unused cells. The more extensive range of activation produced by vertical orientations may also be an analog of increased human visually evoked potentials (VEPs) to vertically oriented stimuli. Beh (1990) presented retinally vertical, horizontal and

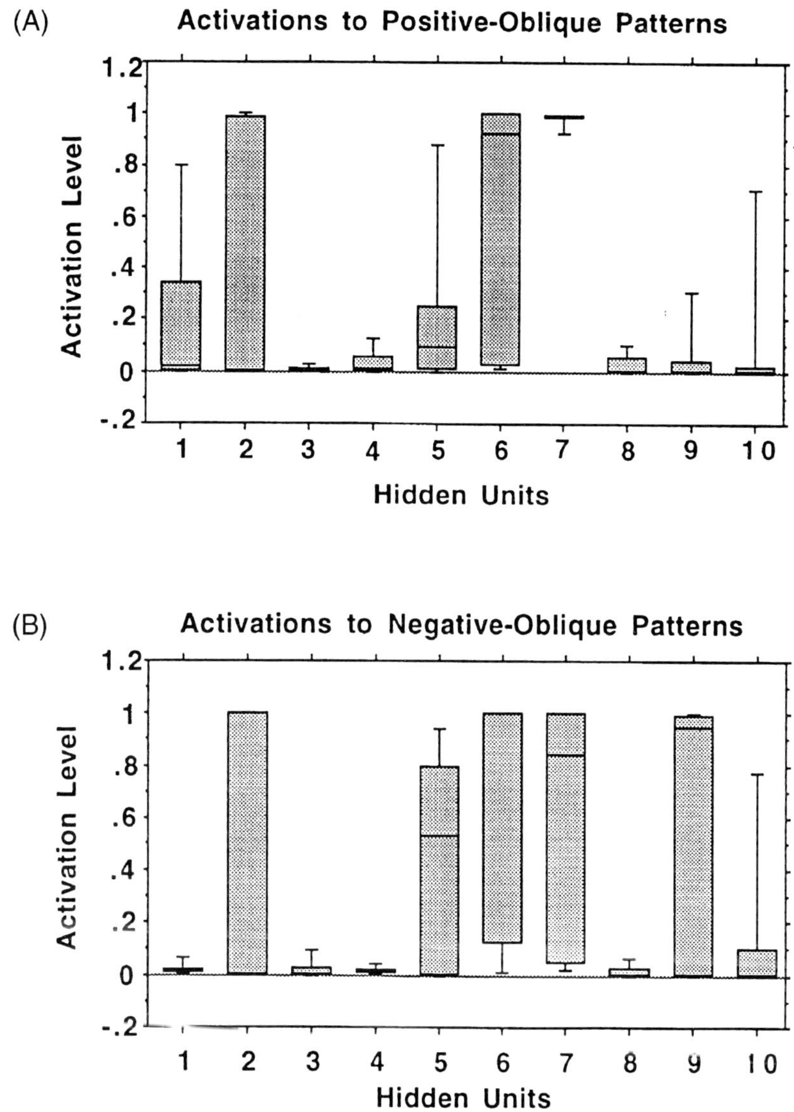

Figure 5. Ranges of hidden-unit activation in the network given biased training. (A) Activation ranges to patterns symmetrical about the positive-oblique axis. (B) Activation ranges to patterns symmetrical about the negative-oblique axis.

diagonal bar stimuli to subjects and found significantly higher VEP amplitudes to retinal verticals regardless of whether subjects adopted an upright or supine position. Additionally, Fiorentini *et al.* (1972) reported that, after five days, anisotropies in contrast sensitivity decreased in subjects who wore prisms that tilted the visual world by between 30 and 45 deg. Importantly, initial differences in the amplitude of cortical potentials evoked by vertical and oblique gratings were reduced by the period of exposure to visual tilt.

6. COMPARISON OF NETWORK AND HUMAN PERFORMANCE

There is no generally accepted measure of network response to compare with the detection-time functions produced by subjects — see Quinlan (1991, pp. 150–151) for discussion of this issue. In determining the response strength of output units in the present simulations, the sigmoid activation functions of these units in the trained networks were changed to cascaded activation functions (McClelland and Rummelhart, 1988). Each output unit takes on activation (a) based on inputs from the hidden units according to the following equation

$$a_{ir}(t) = k_r \sum_j W_{ij} a_{js}(t) + \left(1 - k_r\right) a_{ir}(t-1), \tag{1}$$

where t denotes time, r and s index the receiving output units and the sending hidden units, while i and j index units within layers. The rate constant k governs the rate at which the output unit activations reach the value that their summed input is driving them toward. In the current simulations, k was set to 0.05. Thus, when an input pattern is presented to the trained network, and its signal propagated forward through the hidden units to the output units, the output activations do not reach asymptote immediately but rather build up slowly at a rate determined by the input and the rate constant k. With k set at a small value, it is possible to observe, for each input pattern, the rate of increase of activation of the output units (measured by the number of cycles to reach asymptote) and to use this value as a measure of response strength.

The unequal training of the network resulted in units in the vertical and horizontal pathways having a greater number of trials in which to increase their connection strengths, and in which to be harnessed by the prevailing vertical and horizontal patterns. Stronger connections and more dedicated hidden units produce larger changes to the net inputs and activations of the vertical and horizontal output units, and thus allow activation in these units to build up more quickly than activation in the oblique output units. The more rapid the accumulation of activation in the correct output unit, the faster the correct output unit will reach asymptote, and, in this way, connection strength and domination of hidden units determine speed of processing.

On the assumption that the number of cycles required for a particular output unit to reach asymptote is a valid estimate of response strength in the network to a particular axis of symmetry, it was possible then to compute the mean and standard error (across input patterns) of the network's detection time for the four classes of symmetry. The detection times of the subjects in their test block of trials and the simulated detection times of the biased network are shown in Fig. 2c. It is necessary to multiply the network detection times by a constant (110) in order to plot the detection-time functions of the subjects and the network on the same axes. As may be seen in Fig. 2c, the biased training has produced levels of network response to the input patterns in the order of detection times observed for the human subjects.

7. PRE-TRAINING WITH BAR STIMULI

Given the possibility that differences in human reaction time to different classes of symmetry may be due to prolonged experience with carpentered environments and the heterogeneous distribution of stimulus orientations, networks were given pre-training with vertical, horizontal, positive- or negative-oblique bars clamped onto the 6×6 input array. For example, one hypothesis was that vertical pre-training would simulate pre-exposure to vertical orientations and sensitize and desensitize appropriate output units to vertical, horizontal, and oblique arrangements of elements in the input array. In consequence of this pre-training with vertical bars, a network given subsequent training on vertically symmetrical patterns that contain vertical elements it has already learnt, should display a faster learning rate than a network that has not been given vertical pre-training. However, if, as in studies reviewed earlier, vertical symmetry is conceived of as horizontally oriented elements arranged around a vertical axis, then the possibility exists that pre-training on vertical bars may speed subsequent learning of horizontally symmetric rather than vertically symmetric patterns. Similar arguments apply to the horizontal and opposing oblique orientations and axes of symmetry.

There were 18 bar stimuli for each of the four orientations. Six stimuli at each orientation consisted of the 6×6 input array with a single complete row, column or diagonal of filled squares. The remaining 12 stimuli for each orientation were single rows, columns or diagonals in the input array with either odd or even cells filled. Pre-training consisted of teaching the network to turn on the vertical-symmetry output unit and to turn off the three output units responsible for the other axes of symmetry when it received a vertical-bar pattern, and training continued until the RMS error reached zero. Likewise, other networks were pre-trained to turn on and off appropriate output units to horizontal and oblique-bar stimuli. Retaining their pre-trained weights, the networks were then trained as before to categorize the 80-pattern set in terms of axis of symmetry.

In comparison with a network given no pre-training, and a network given training with random or asymmetric patterns, the networks given pre-training with the vertical or the oblique sets of bars recorded faster initial learning of the appropriate symmetries. Pre-training with the horizontal bars produced faster learning of both horizontal and positive-oblique symmetries, suggesting the possibility that, in the small sample of patterns used, these symmetries may not be independent. However, the effects of pre-training require a more elaborate investigation and analysis than can be reported here. Further simulations were run in which networks were pre-trained to turn the *vertical* output unit on and the other three outputs off when they received *horizontal* bar stimuli. *This pre-training with horizontal bars also speeded learning of vertical symmetry*. Indeed, as Fig. 6 illustrates, in comparison with no pre-training or pre-training with random or asymmetric patterns, pre-training with either positive-oblique or negative-oblique bars also speeded subsequent learning of *vertical* symmetry. Figure 6 shows the RMS error functions for the vertical output unit only of networks learning the 80-pattern set after such pre-training. Four RMS error functions in Fig. 6 are from networks learning the 80-pattern set after

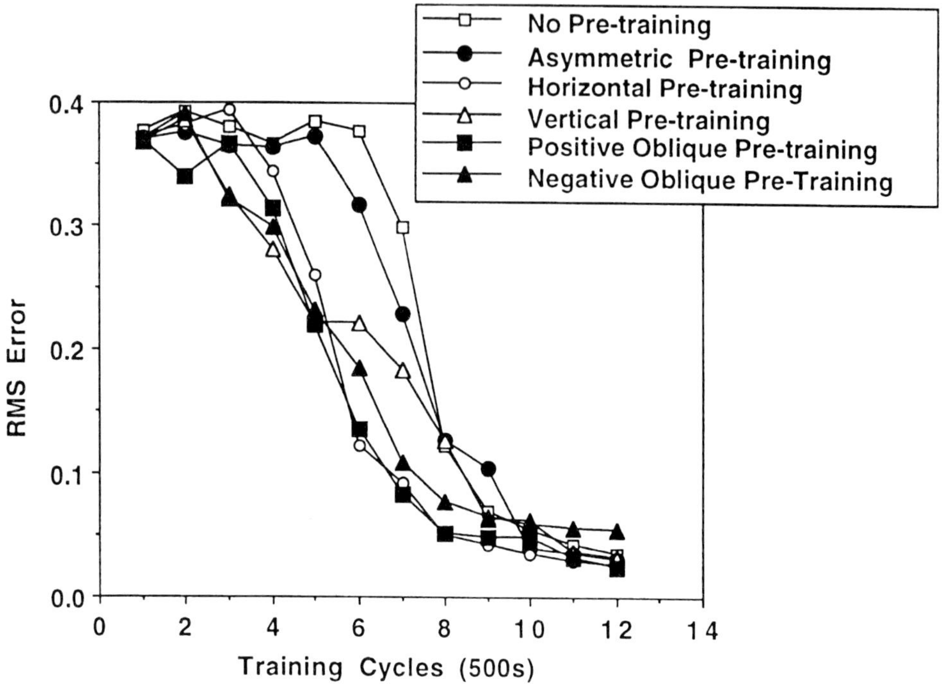

Figure 6. RMS error across training cycles on the *vertical* output unit of networks given no pre-training or pre-training on random asymmetric patterns or pre-training with vertical or horizontal or oblique bar stimuli prior to learning to categorize patterns from the four axes of symmetry.

being pre-trained to turn on their *vertical output unit* to variously oriented bar stimuli. Of the remaining two RMS error functions, one is from the vertical output unit of a control network given no pre-training, and the other is from the vertical output unit of a network given pre-training with random asymmetric patterns. These results were consistently obtained after starting the networks from a variety of different sets of random weights.

8. DISCUSSION AND CONCLUSIONS

With a sample of 16 subjects, a detection-time function was generated for a set of 80 patterns from four classes of mirror symmetry, and, with suitably biased training, it was shown how back-propagation networks, with cascaded activation functions on their output units, could produce response strengths in the order observed in the subjects' detection-time function. Descriptive analysis of hidden unit space revealed that biased training results in the appropriation of hidden units by the prevailing vertical stimuli — an effect paralleled in neurophysiological studies of the effects of biased early experience on the development of cells in the visual cortex of animals, and studies which demonstrate significantly higher amplitudes of VEPs to retinally vertical-bar stimuli. By comparison with no pre-training or pre-training with sets of random, asymmetric patterns, pre-training networks with bar stimuli produced initial faster learning of patterns whose axes of symmetry were consistent with the orientation of the bars used in pre-training. However, it was also demonstrated that pre-training with horizontal or oblique-bar stimuli also speeded subsequent learning of vertical symmetry. The general pattern of results lends support to theories which propose that the often observed order of response times in symmetry detection tasks may be the legacy of our prolonged experience with the heterogeneous distribution of stimulus orientations in the built environment. It should also be noted that, in contrast with some of the theories of symmetry detection reviewed earlier, the modelling approach adopted here provides a clear and testable specification of the learning mechanisms that may mediate our detection of symmetry.

8.1. Future directions

The simulations reported here were necessarily of small scale, but they do, nonetheless, uncover possible directions for further experimental investigations of symmetry detection.

8.1.1. Psychological implications of the analysis of hidden unit space. Consideration of the distribution of weights from inputs to hidden units, such as those depicted in Fig. 3, indicates the elements or features that networks weight heavily during their learning of each symmetry class. It may be that human subjects apply similar weightings during detection of the same patterns. This prediction could be tested in detection experiments by misplacement, blurring or deletion of elements highly weighted by networks, and, in comparison with similar manipulations of lowly-weighted elements, should produce longer detection times and more errors. A related conjecture is that, during symmetry detection, subjects' attention should be directed to the elements assigned high weight by networks. With patterns increased to a size requiring eye movements, it is possible to record attentional shifts during symmetry detection, and, with measures such as scan-path order, cumulative fixations and fixation time, test whether or not subjects attend to the predicted regions of the patterns (Latimer, 1988, 1990; Latimer and Stevens, 1993).

8.1.2. Pre-training networks with bar stimuli. Pre-training networks with bar stimuli, demonstrated that prior experience with vertical-, horizontal- or oblique-bar stimuli can speed subsequent learning of patterns symmetrical about appropriate axes. However, it was also found that, by comparison with pre-training on random asymmetric patterns, pre-training on either vertical, horizontal or oblique bars also speeded subsequent learning of vertical symmetry. From these preliminary results, it appears that pre-training with bar stimuli *in general* enhances learning of vertical symmetry, but additional experimental work on the effects of the same pre-training schedules on subsequent learning of horizontal and oblique symmetries is required.

8.1.3. Pre-training subjects with orientation judgments. Further in the area of human experimentation, and following from the results of network pre-training with bar stimuli, it may also be hypothesized that provision of extended practice at judging whether or not bar stimuli are exactly oriented at vertical, horizontal or oblique should produce faster and more accurate detection of the various symmetries.

8.1.4. Larger samples of patterns. One obvious difficulty with the simulations reported here has been that replication of human experimental conditions necessitated the use of a small sample of training patterns. Simulations employing larger samples of patterns and larger arrays of input units (20 × 20) should reveal more about the *general* properties of symmetry and symmetry detection and provide a broader base for experimental investigations with human subjects.

Acknowledgements

This research was supported by an Australian Research Council Postdoctoral Scholarship awarded to the third author. We would like to express gratitude for the most helpful and discerning comments provided by two anonymous referees.

NOTES

1. NeuralWorks Professional 4 is a product of NeuralWare Inc., Penn Center West, Building IV, Pittsburgh, PA 15276, USA.

REFERENCES

Beh, H. C. (1990). Visual ERPs and stimulus orientation. In: *Seventeenth Conference of the Australian Experimental Psychology Society*. Australian National University, Canberra.

Blakemore, C. and Cooper, G. F. (1970). Development of the brain depends on the visual environment. *Nature* **228**, 477–478.

Blakemore, C. and Mitchell, D. E. (1973). Environmental modification of the visual cortex and the neural basis of learning and memory. *Nature* **241**, 467–468.

Bloesch, A. and Wiles, J. (1991). Data reduction and display techniques for representations in hidden unit space. In: *First Indiana Conference on Dynamics in Cognition: Dynamic Representation in Cognition*. Indiana University, Bloomington, Indiana.

Cohen, J. D., Dunbar, K. and McClelland, J. L. (1990). On the control of automatic processes: A parallel distributed processing account of the Stroop effect. *Psychol. Rev.* **97**, 332–361.

Cohen, N. J., Abrams, I. T., Tabor, W. S. and Sejnowski, T. J. (1986). Skill learning and repetition priming in symmetry detection: Parallel studies of human subjects and connectionist models. In: *8th Annual Conference of the Cognitive Science Society*. Lawrence Erlbaum, pp. 23–44.

Corballis, M. C. and Roldan, C. E. (1975). Detection of symmetry as a function of angular orientation. *J. Exp. Psychol: Human Percept. Perform.* **1**, 221–230.

Costin, D. L. (1988). MacLab: A Macintosh system for psychology labs. *Behavior Res. Methods, Instruments and Computers* **20**, 197–200.

Elman, J. L. (1990). Finding structure in time. *Cognitive Sci.* **14**, 179–211.

Field, N. and Golubitsky, M. (1992). *Symmetry in Chaos*. Oxford University Press, New York.

Fiorentini, A., Ghez, C. and Maffei, L. (1972). Physiological correlates of adaptation to a rotated visual field. *J. Physiol.* **227**, 313–322.

Geisser, S. and Greenhouse, S. (1958). An extension of Box's results on the use of the F distribution in multivariate analysis. *Ann. Math. Stat.* **29**, 885–891.

Goldmeier, E. (1972). Similarity in visually perceived forms. *Psychol. Issues* **8**, Whole No. 29.

Hebb, D. O. (1949). *The Organisation of Behavior: A Neuropsychological Approach*. John Wiley and Sons, New York.

Howard, I. P. (1982). *Human Visual Orientation*. John Wiley and Sons, New York.

Julesz, B. (1971). *Foundations of Cyclopean Perception*. University of Chicago Press, Chicago.

Latimer, C. R. (1988). Eye-movement data: Cumulative fixation time and cluster analysis. *Behavior Res. Methods, Instruments and Computers* **20**, 437–470.

Latimer, C. R. (1990). Eye-movement indices of form perception: Some methods and preliminary results. In: *Eye to Mind: Information Acquisition in Perception, Search and Reading*. R. Groner, G. d'Ydewalle and R. Parham (Eds). North-Holland, Amsterdam.

Latimer, C. R. and Stevens, C. J. (1993). Eye movement measures and connectionist models of form perception. In: *Cognitive Approaches to Human Perception*. S. Ballesteros (Ed.). Lawrence Erlbaum Associates, Hillsdale, NJ.

Locher, P. J. and Wagemans, J. (1993). Effects of element type and spatial grouping on symmetry detection. *Perception* **22**, 565–587.

Masame, K. (1983). Detection of symmetry in complex patterns: Is symmetrical projection to the visual system necessary for the perception of symmetry? *Tohoku Psychologica Folia* **42**, 27–33.

Masame, K. (1984). Detection of symmetry in relatively simple patterns. *Tohoku Psychologica Folia* **43**, 42–48.

Masame, K. (1985). Perception of symmetry in patterns constructed from two kinds of elements. *Tohoku Psychologica Folia* **44**, 59–65.

Masame, K. (1986). Rating of symmetry as continuum. *Tohoku Psychologica Folia* **45**, 17–27.

Masame, K. (1987). Judgment of degree of symmetry in block patterns. *Tohoku Psychologica Folia* **46**, 43–50.

Masame, K. (1988). Sampling of patterns varying on degree of symmetry with subject-generating method. *Tohoku Psychologica Folia* **47**, 57–62.

McClelland, J. L. and Rummelhart, D. E. (1988). *Explorations in Parallel Distributed Processing: A Handbook of Models, Programs and Exercises*. MIT Press, Cambridge, MA.

McClelland, J. L., Rummelhart, D. E. and The PDP Research Group (1986). *Parallel Distributed Processing: Explorations in the Microstructure of Cognition*. MIT Press, Cambridge, MA.

Minsky, M. and Papert, S. (1988). *Perceptrons: An Introduction to Computational Geometry*. MIT Press, Cambridge, MA.

Palmer, S. E. and Hemenway, K. (1978). Orientation and symmetry: Effects of multiple, rotational, and near symmetries. *J. Exp. Psychol: Human Percept. Perform.* **4**, 691–702.

Quinlan, P. T. (1991). *Connectionism and Psychology: A Psychological Perceptive on New Connectionist Research*. University of Chicago Press, Chicago.

Royer, F. L. (1981). Detection of symmetry. *J. Exp. Psychol: Human Percept. Perform.* **7**, 1186–1210.

Sejnowski, T. J., Kienker, P. K. and Hinton, G. E. (1986). Learning symmetry groups with hidden units: Beyond the perceptron. *Physica* **22D**, 260–275.

Sejnowski, T. J. and Rosenberg, C. R. (1988). Learning and representation in connectionist models. In: *Perspectives in Memory Research*. M. Gazzaniga (Ed.). MIT Press, Cambridge, MA, pp. 135–178.

Switkes, E., Mayer, M. J. and Sloan, J. A. (1978). Spatial frequency analysis of the visual environment: Anisotropy and the carpentered environment hypothesis. *Vision Res.* **18**, 1393–1399.

Wagemans, J. (1992). Perceptual use of nonaccidental properties. *Can. J. Psychol.* **46**, 236–279.

Wagemans, J. (1993). Skewed symmetry: A nonaccidental property used to perceive visual forms. *J. Exp. Psychol: Human Percept. Perform.* **19**, 364–380.

Wagemans, J., Van Gool, L., Swinnen, V. and Van Horebeek, J. (1993). Higher-order structure in regularity detection. *Vision Res.* **33**, 1067–1088.

Wagemans, J., Van Gool, L. and d'Ydewalle, G. (1991). Detection of symmetry in tachistoscopically presented dot patterns: Effects of multiple axes and skewing. *Percept. Psychophys.* **50**, 413–427.

Wagemans, J., Van Gool, L. and d'Ydewalle, G. (1992). Orientational effects and component processes in symmetry detection. *Q. J. Exp. Psychol: Human Exp. Psychol.* **3**, 475–508.

Wenderoth, P. M. (1994). The salience of vertical symmetry. *Perception* **23**, 221–236.

Human Symmetry Perception, pp. 227–236
C. W. Tyler (Ed.)

A network model for generating differential symmetry axes of shapes via receptive fields

MATTHEW A. KURBAT
Department of Psychology, University of Michigan, 525 E. University St., Ann Arbor, MI 48109, USA

Abstract—Some symmetries (e.g. bilateral, rotational, translational) only describe quite specialized shapes, but differential symmetry axes (e.g. Blum, *J. Theoret. Biol.* **38**, 205–287, 1973; Brady and Asada, *Int. J. Robotics Res.* **3**, 36–61, 1984) describe more general shapes. Such axes are of interest in part because they form the 'backbone' of generalized cylinder and other shape representations used in shape recognition (e.g. Marr, *Vision*, W. H. Freeman and Co., NY, 1982; Biederman, *Psychol. Rev.* **94**, 115–147, 1987). However, despite the popularity of these representations as psychological models, algorithms from machine vision for computing them have strong limitations as psychological models. This paper presents two versions of a network model, one of which is more plausible as a psychological model because it derives symmetry axes from the activations of idealized visual receptive fields.

1. INTRODUCTION

Some objects are symmetrical in various ways — e.g. objects may be bilaterally, rotationally, or translationally symmetric. While these symmetries only apply to specialized classes of shapes, differential symmetry axes (e.g. Blum, 1973; Brady and Asada, 1984) describe the shapes of more arbitrary objects. Differential symmetry axes reduce two-dimensional silhouettes or outlines to something akin to a 'stick-figure' representation. The essential idea underlying the original type of symmetry axis, called a symmetric axis transform or SAT (Blum, 1973), is that points that make up a SAT of a shape are centers of all discs that lie inside the shape and are doubly tangent to the shape's boundary (SAT discs are also called *maximal discs* — a disc with the same center location but larger radius than a given SAT disc would exceed the bounds of the shape). A SAT point is symmetrical in that the two or more boundary points of a shape that touch a SAT disc are equidistant from the center of the disc (the SAT point).

Later investigators (e.g. Brady and Asada, 1984; Leyton, 1992) have created other kinds of axes that are variations on the original idea. Most depend in some way on the idea of doubly tangent discs, but what distinguishes these other ideas from the SAT is the point inside each disc that is a symmetry axis point. In the SAT, this point is the center of the disc. In contrast, in Brady's Smoothed Local Symmetry (SLS) representation (e.g. Brady and Asada, 1984) symmetry axis points are found by determining points at which each SAT disc is tangent to the boundary, connecting each pair of points for a given disc with a straight segment, and then using the midpoints of these straight segments as SLS points (for a mathematical and pictorial comparison of various kinds of symmetry axes, see Rosenfeld, 1986).

Symmetry axis representations are perhaps best known in the psychology of visual perception for their role in generalized cylinder shape representations (e.g. Marr, 1982; Biederman, 1987). A traditional cylinder from elementary geometry is a shape with a straight axis, and a circular cross-section that is orthogonal to the axis and that has a constant size along the length of the axis. A generalized cylinder (e.g. Marr, 1982) generalizes cylinders in three ways: (1) the axis may curve; (2) the cross-section may be non-circular; (3) the cross-section may change size along the axis. Symmetry axes form the 'backbone' of generalized cylinders because they typically *are* the axes of generalized cylinders (e.g. Biederman, 1987). It is important to distinguish the use of 'axis' in representing objects' components (the sense intended in this paper) from the use of 'axis' as reference frame — the former are shape representations, while the latter are used in coordinate systems with respect to which shape representations are coded (e.g. Marr, 1982).

Despite the popularity of these representations in psychological models of shape representation, algorithms from machine vision for computing symmetry axes lack features that are desirable in a psychological model of symmetry axis generation (to be fair, such models were probably not intended to be psychologically plausible). To begin, it is not clear how one would attempt to neurally instantiate these algorithms. For example, one of Heide's (1984) algorithms first approximates the boundary of the shape in terms of discrete points, and then for each such point compares the point to all other boundary points to find those pairs of points to which a disc inside the figure is tangent (Brady and Asada (1984) use a variation on this method). It is unclear how one would map this computation onto the operation of real or idealized neurons — what different types of functions would such units have to perform to instantiate this algorithm, and how would their interactions produce a symmetry axis representation? Similar arguments can be made about other algorithms. In general, most existing algorithms for symmetry axis computation were designed for machine vision applications, and not as psychological models.

A related point involves the serial nature of these past algorithms. Neurons operate at times measured in milliseconds or tens of milliseconds; given that object recognition can occur in less than one second (e.g. Biederman, 1987), this task must be done in no more than hundreds of serial steps, a constraint known as the *100-step program constraint* (e.g. Feldman, 1985). Further, in the case of object recognition many of these steps must be devoted to processes other than the computation of symmetry axes. The brain computes through massive parallelism (e.g. Zeki, 1993); even the fastest versions of past serial algorithms require thousands of steps (Heide, 1984) to compute the symmetry axes for a single object, making them unlikely models of human performance.

Given that past algorithms have these drawbacks as psychological models, it is not surprising that in presenting their psychological model of object recognition, Hummel and Biederman (1992, p. 487) claim that 'computing axes of symmetry is a difficult problem ... the solution of which we are admittedly assuming'. Their model, a connectionist instantiation of Biederman's (1987) theory, does not compute axes at all, but is given them as input. Given that past models of symmetry axis generation were developed only for machine vision uses, and that Hummel and Biederman found no ready way to adapt these models or create a new one for their purposes, it would be an advance to devise a model that is better motivated as a model of human visual symmetry axis generation.

The purpose of the current paper is to present such a model, one that seems more psychologically plausible than past models because it derives symmetry axes from the response properties of simulated mammalian visual neurons. The goal of this model is to produce output (in the form of axes – 'stick figure' representations) that could serve as input to a model of recognition.

2. THE MODEL

2.1. *Background for SAM* (*Symmetry Axis Model*)

The intuition behind the model is based on the notion of a receptive field (RF), i.e. an operator that takes a weighted contribution of points falling in a given region, a common example of which is the center-surround cell (e.g. Kuffler, 1953). These cells have an increased or decreased firing rate to light within a certain radial distance $r1$ from the center of the receptive field, and a decreased or increased (depending upon cell type) firing rate to light at some distance greater than $r1$ but less than $r2$. In other words, such a cell has a maximal firing rate in response to a light or dark disc of a certain size and location, and its firing rate decreases when the input deviates from this pattern (as the response properties of such cells are radially symmetric, they will be termed *radial receptive fields* or RRFs). The essential idea underlying SAM is that, as each RRF fires maximally to a disc of a certain size and location, the location of each center-surround cell that fires near maximum is approximately a SAT point for a shape (this notion makes sense for shapes of fairly homogeneous intensity against a contrasting background; generalization to other cases is treated in the Discussion). Various investigators (e.g. Rose, 1979; Marr, 1982) have shown how such cells might detect bar-like shapes when their receptive fields have the same sizes and are aligned in rows; the current proposal is an extension of that idea to more general shapes, with cell locations serving as SAT point locations. This proposal is developed first in a preliminary version that computes an approximate SAT via non-biological step-function 'RRFs', and second in a version that computes axes similar (in a fashion described below) to SATs via biological RRFs.

2.2. *NOBSAM* (*Non-Biological version of SAM*)

This version of the model is interesting in that it computes an approximate SAT via a network algorithm that permits parallel implementation, albeit using non-biological

RRFs. It takes as input silhouettes of shapes against a white background. The (x, y) plane on which input is represented is tessellated into a square grid (pixels). This grid is the first layer (L1) of the model, and the activation of a cell at location (x, y) is denoted as $a(1, x, y)$. A given L1 cell has discrete activation values — an L1 cell is active ($a(1, x, y) = 1$) if the (x, y) location of the cell is part of the silhouette (the pixel is black), and the cell is inactive ($a(1, x, y) = 0$) if the (x, y) location of the cell is not part of the silhouette (the pixel is white). RRFs of various radii are located at each (x, y) position in the visual field — these RRFs are the second layer (L2) of the model. In the other words, above each (x, y) location of an L1 cell, there is an RRF in L2 with each integer valued radius for integers between 3 and 20 (for the current implementation), inclusive. The previous radii apply to the centers of the RRFs; for the surround of a center of radius r, the annulus-shaped surround begins at distance r and ends at distance $2r$. Each RRF $F(x, y, r)$, where r is the radius of the center (not the surround), is step-functional in that activation is 1 if the center of the RRF is entirely within the figure (covers only black pixels) and the surround includes some background (i.e. white pixels), and is 0 otherwise. Such step-functional RRFs and activation values imply that points are exactly axis or non-axis points, as is true in standard axis algorithms, but untrue in the biological version of SAM presented below.

Activations of L1 and L2 are determined at $t = 1$ and $t = 2$, respectively. At $t = 3$, each active RRF inhibits slightly smaller neighboring RRFs. Specifically, a given active receptive field F1 $= (x, y, r+1)$ completely inhibits (sets activation $= 0$) the RRF F2 $= (x, y, r)$, and does the same for all of F2's horizontal, vertical, and diagonal neighbors at radius r. The intuition behind this method is this: inhibition of smaller neighboring RRFs makes sense when the center disc of the larger active (hence interior to the shape) RRF completely contains the center disc of the smaller disc, as the smaller disc is thus not a maximal disc. After $t = 3$ the centers of active RRFs are necessary and sufficient to approximately represent the SAT points of the shape. They are necessary in that if a slightly larger active RRF F2 that covered active RRF F1 existed, then F2 would have inhibited F1 — which is impossible because F1 is active at $t = 3$; thus RRFs active after $t = 3$ are maximal discs in the discrete approximation of the plane, and hence necessary. Proof of their (approximate) sufficiency follows from proof of the claim that inhibited RRFs are (approximately) unnecessary, so active RRFs are sufficient. This proof divides into two cases of inhibited RRFs. For RRF F1 $= (x, y, r)$ that is inhibited by a horizontal or vertical neighbor RRF F2 of radius $(r+1)$, F1 is clearly unnecessary because F2's center disc contains F1's, so F1's RRF cannot be a maximal disc. For RRF F1 $= (x, y, r)$ that is inhibited by a diagonal neighbor RRF F2 of radius $(r + 1)$, F1 is approximately but not exactly unnecessary. Specifically, F2's center has a radius one unit larger than those of the diagonal neighbors it inhibits, but the distance between the center $(x1, y1)$ of F2's center disc and the center $(x2, y2)$ of each diagonal neighbor's center disc is root two, so F2's center disc almost but not quite covers the center disc of each diagonal neighbor (for each diagonal neighbor of F2 with a radius one unit smaller than that of F2). In this case, a regular hexagonal grid of L1 cells would be more useful than the current square grid, as the distance between a cell and one of its neighbors in a hexagonal grid is the same as the distance between the cell and any other neighbor, thus allowing correct inhibition of (and only of) non-maximal discs. In any case, the resulting 'gaps' in the symmetry axis that result from the square grid used here may largely be overcome via a method similar to that developed by Ullman (1991). Essentially, this method uses an extra layer of eight oriented cells centered at each (x, y) location, cells corresponding to each of eight different equally spaced orientations (every 22.5 deg). If a given cell receives sufficient activation from active axis points beneath it, it feeds back to the RRFs below it, activating all of them and 'filling in' gaps. The axis produced by the resulting model for the shape of a hammer is presented, and contrasted with the more biologically-oriented version of the model, in the discussion of simulations below. (Details of this filling in process were omitted to allow further discussion of the biological version, which is intended as the psychological model; the non-biological version is included for comparison.)

2.3. *BIOSAM (Biological receptive field version of SAM)*

BIOSAM is similar to NOBSAM in several respects: (1) silhouettes of shapes against a contrasting background serve as input (generalization to non-silhouettes is treated in the Discussion section); (2) layer 1 of the model is again a 2D array of binary units (but activation values are opposite those in the non-biological version: $a(1, x, y) = 0$ for black pixels — those in the figure — and $a(1, x, y) = 1$ for white pixels — those in the background); (3) RRFs of various radii are located at each (x, y) position in the visual field and constitute the second layer (L2) of the model.

BIOSAM also differs from NOBSAM in several key respects. BIOSAM's RRFs are center-inhibitory surround-excitatory difference of Gaussians (DOG), a common model for center-surround cells (e.g. Rodieck, 1965; Rose, 1979). Thus input is weighted as a function of distance from the RRF's center according to the equation:

$$f(r) = -\frac{k1}{sq1}\exp\left\{-\left(\frac{r}{rc}\right)^2\right\} + \frac{k2}{sq2}\exp\left\{-\left(\frac{r}{rs}\right)^2\right\}.$$

Here r is the distance from a given point to the center of the RRF, $sq1$ is the square root of rc, $sq2$ is the square root of rs, rc is the parameter of the Gaussian center, rs is the parameter of the Gaussian surround, and $k1 = 1$ and $k2 = 1/3$ are parameters (chosen here by trial runs of BIOSAM, and of similar relative size to values used by Rose, 1979). Thus input to each RRF is (numerically) the DOG weighted average of the L1 values (i.e. analytically, the DOG is convolved with the input image and the result is integrated twice). There were 25 RRFs at each (x, y) location, one for each integer value of rc from 1 to 25 inclusive (25 being sufficiently large for stimuli used in the simulation), and $rs = 2rc$ (following Rodieck, 1965). The input-output function for each RRF was a threshold function. Output for each RRF was 0 for input below the threshold, and 1 for values at or above the threshold. The threshold value was chosen to be 4 for simulation results presented below, based on trial runs of the model. While a literal threshold is intended only as a first approximation and is not strictly realistic, a more realistic high-gain sigmoid has similar input-output properties and so would have likely given similar results. Use of the same threshold despite different RRF radii is possible because DOG receptive fields respond in a scale-invariant fashion (Koenderink and van Doorn, 1990), i.e. maximal response of DOG RRFs is the same regardless of scale.

The time course of activation is as follows: activation levels of the L1 cells are determined at $t = 1$, then these determine activation of L2 cells at $t = 2$, at which time the (x, y) locations of active L2 cells determine symmetry axis points. Thus in BIOSAM, unlike in NOBSAM, larger RRFs do not inhibit the smaller RRFs they cover. Instead, each (x, y) location is an axis point if and only if any of the RRFs centered at that location is active. Symmetry axes produced by BIOSAM will be referred to as 'bioaxes' in discussion below.

2.4. Simulations

Both versions of the model were tested on various stimuli; many of the results are presented in Fig. 1. The figure's top row shows a hammer at left and the results of the non-biological version at right and biological version at center. Results of both versions are juxtaposed to highlight differences between standard symmetry axes (exemplified by NOBSAM) and BIOSAM. Specifically, in standard versions a given point is either (exactly) an axis point, or it is not; in contrast, the use of a threshold allowing L2 cells near (not just at) peak activation to produce axis points means that points that are near, but are not exactly, SAT points are treated as axis points in the biological version of the model, making axes in that version slightly thicker than axes in the non-biological version.

The SAT has been criticized (e.g. Marr, 1982) for being too sensitive to small perturbations in the outline. However, this problem also occurs in other sorts of symmetry axes, and it arises only when there is a single representational scale, so it may be countered by providing multiple scales of representation (Brady and Asada, 1984). Row 2 of the figure shows a bar shape at far left, and just to its right is the axis resulting from a standard SAT algorithm. Moving to the right of the SAT in row 2, we have subsets of the original SAT points with the information at finer representational scales progressively removed as one moves increasingly rightward in the row. In other words, doubly-tangent circles whose centers are SAT points in the original SAT shown second from left in row 2 are all such circles with a radius greater than or equal to 2 pixels; to the right of this SAT is the subset of points with minimum radius of 3, to the right of this figure in turn is the subset of points with minimum radius of 4, and so forth, until finally at the far right of row 2 only points that are centers of doubly-tangent circles with minimum radius of 8 are left. Thus, in the original SAT, representational scale of a SAT point can be indicated by the radius of its maximal disc — the greater the radius, the coarser the scale. In BIOSAM, the scale of an axis point is indicated by rc for the RRF generating it. At the left of row 3 the same bar shape from row 2 is shown, to its right is its bioaxis, and to the right of the bioaxis we have subsets of bioaxis points of increasingly coarse scale, corresponding to the SAT points of increasingly coarse scale above them (note again that the bioaxis tends to be thicker than the SAT in

Figure 1. Shapes of various objects, and the results of symmetry axis simulations using those shapes (see text for explanation).

places). At left in row 4 we have the same bar shape in rows 2 and 3 with a small notch in it, resulting in a shape almost identical to Marr's (1982) Fig. 5-2 designed to show the sensitivity of the SAT to small perturbations in the boundary. To the right of the notched bar in row 4 we have the bioaxis, and then subsets of the bioaxis with increasingly coarse scale similar to the preceding two rows. Note that, once points at the finest scales are removed, the bioaxes for the bar in row 3 and those for the notched bar in row 4 are qualitatively the same, so their shapes would be virtually the same at coarser scales of representation; thus the bioaxis overcomes Marr's boundary perturbation problem if one uses information available at coarser scales (presumably, other axis types such as the SAT could overcome the problem in a similar fashion).

Moving to row 5, at left we have a shape that is less elongated than the previous examples, to its right we see first its bioaxis that includes finer scales, and then further right the bioaxis only at coarser scales. Note that finer scale version includes branches terminating near each of the convex curvature extrema, a feature common to standard symmetry axes (Leyton, 1992), but that these are lost at the coarser scales and only the main axis remains — thus major features of the shape are captured at coarser scales, and minor features at finer scales. At right in row 5, and in rows 6 and 7, we have five more examples of shapes and their corresponding bioaxes (the finest scale included for the flamingo in row 5 is $rc = 2$; the finest scale for the letters of the alphabet in rows 6 and 7, as well as for other bioaxis examples in previous rows, is $rc = 3$). In all the examples, bioaxes qualitatively capture elongations of their corresponding shapes in a fashion like that of previous types of axis. (An interesting exploration for future research would involve seeing how the model scales with number and distribution of cells employed.)

3. DISCUSSION

This paper provides two versions of a model for generating symmetry axes from the activations of different sorts of RRFs, axes which in the case of the latter version could then serve as input to a model of human pattern recognition. One natural question to ask about any sort of model of human performance is 'is it consistent with human data?' The primary grounds for positing axial shape representations in recognition are computational and phenomenological — it is possible to create general shape representations that are based on axes (e.g. generalized cylinders), and axial or 'stick-figure' renditions of many shapes appear qualitatively like the original shapes (e.g. Marr, 1982). In fact, empirical work that implicates axes in real-time object recognition has been provided only recently (in the form of priming experiments by Sanocki and Ling, 1994); but even this evidence implies only that axes play a role, and does not provide more specific constraints (there is virtually no empirical work on the role of differential symmetry axes in representing shapes for recognition, making this a fertile area for future research).

Other constraints that bear on models of axis generation may be used to compare the biological version of SAM to two possible alternatives. One such alternative would be a neural network implementation of the 'grassfire' or 'erosion' method of SAT generation (Blum, 1973). In this method, the shape is eroded from its boundary and the points where two erosion fronts meet are taken as axis points (the method of creating multiple scales in SAM simulations might be adapted for use in such a model). If it were possible to map such a scheme onto known neural hardware, would the resulting model be plausible as a psychological model? While the answer to this question depends in part on the details of the model and the neural mapping, two issues reduce its appeal relative to BIOSAM. First, bioaxes in BIOSAM are computed in a constant number of discrete time steps regardless of object size, consistent with the fact that recognition is scale-invariant (Biederman and Cooper, 1993). However, the computation time of axes produced by an erosion algorithm would be an increasing function of object size, violating the scale constraint. Second, given the speed at which recognition occurs, erosion might simply be too slow in some cases to be psychologically plausible (although this could depend on the particular implementation used).

A second alternative model is Burbeck and Pizer's (1994) recent proposal for a multiscale symmetry axis. In their model, an object's boundary is detected at multiple scales, and connections (hence axis points) are made only between those detectors whose scale is appropriate to the object's width at that location (their proposal is difficult to summarize briefly — for details see Burbeck and Pizer, 1994). This model is consistent with their claim that axis width ought to be proportional to object width (this is untrue of older axis types reviewed in the Introduction). This constraint is also true of bioaxes (see, e.g. the flamingo in Fig. 1; this property is implied by the scale invariance of DOG RRFs — Koenderink and van Doorn, 1990), and so does not distinguish the two. On the other hand, this constraint was obtained via experiments in which subjects were to judge whether particular points inside a figure were to the left or right of the middle (which relates to axes in that axis points are in the middles of figures). The accuracy of subjects' judgments was proportional to the widths of the shapes used, a result that Burbeck and Pizer interpreted as evidence that axes are proportional to width. However, it is not clear that results obtained in a slow, deliberative task such as the one described should constrain fast and relatively automatic real-time object recognition. Moving to other constraints, Burbeck and Pizer's model also satisfies the scale invariance constraint; however, they do not map their model onto neurophysiology, whereas SAM is grounded in this fashion.

On the other hand, their model operates on grey-scale images, whereas SAM does not yet do this. It is thus worth concluding by mentioning how SAM might be extended to operate on such images. In brief, Marr (1982) showed how rows of center-surround cells that detected edges could be combined to detect bars in grey-scale images, so a new version of SAM might be made to detect the bars in a similar fashion, with centers of bars being the axis points (neurally, this might correspond to elongated cortical receptive fields of the type discovered by Hubel and Wiesel, 1962). A working implementation of this extension would then allow SAM to perform general-purpose symmetry axis computation.

Acknowledgements

I would like to thank Patricia Reuter-Lorenz, Jun Zhang, and two anonymous reviewers for helpful comments.

REFERENCES

Biederman, I. (1987). Recognition-by-components: A theory of human image understanding. *Psychol. Rev.* **94**, 115–147.

Biederman, I. and Cooper, E. (1992). Size invariance in visual object priming. *J. Exp. Psychol: Human Percept. Perform.* **18**, 121–133.

Blum, H. (1973). Biological shape and visual science (Part 1). *J. Theoret. Biol.* **38**, 205–287.

Brady, M. and Asada, H. (1984). Smoothed local symmetries and their implementation. *Int. J. Robotics Res.* **3**, 36–61.

Burbeck, C. and Pizer, S. (1994). Object representation by Cores. Tech. Report TR94-160, Department of Computer Science, University of North Carolina at Chapel Hill..

Feldman, J. (1985). Connectionist models and their applications: Introduction. *Cognitive Sci.* **9**, 1–2.

Heide, S. (1984). A hierarchical representation of shape from smoothed local symmetries. Unpublished Master's thesis, Department of Mechanical Engineering, MIT.

Hubel, D. and Wiesel, T. (1962). Receptive fields, binocular interaction, and functional architecture in the cat's visual cortex. *J. Physiol.* **166**, 106–154.

Hummel, J. and Biederman, I. (1992). Dynamic binding in a neural network for shape recognition. *Psychol. Rev.* **99**, 480–517.

Koenderink, J. and van Doorn, A. (1990). Receptive field families. *Biol. Cybernet.* **63**, 291–297.

Kuffler, S. (1953). Discharge pattern and functional organization of mammalian retina. *J. Physiol.* **16**, 37–68.

Leyton, M. (1992). *Symmetry, Causality, Mind.* MIT Press, Cambridge.

Marr, D. (1982). *Vision.* W. H. Freeman and Co., New York.

Rodieck, R. (1965). Quantitative analysis of cat retinal ganglion cell response to visual stimuli. *Vision Res.* **5**, 583–601.

Rose, D. (1979). Mechanisms underlying the receptive field properties of neurons in cat visual cortex. *Vision Res.* **19**, 533–544.

Rosenfeld, A. (1986). Axial representations of shape. *Computer Vision, Graphics, and Image Processing* **33**, 156–173.

Sanocki, T. and Ling, X. (1994). Major axes and object recognition. *Invest. Ophthalmol. Vis. Sci.* **35** (Suppl.), 1625.

Ullman, S. (1991). The visual recognition of three-dimensional objects. In: *Attention and Performance XIV.* D. Meyer and S. Kornblum (Eds). MIT Press, Cambridge, pp. 79–98.

Zeki, S. (1993). *A Vision of the Brain.* Blackwell Scientific Publications, London.

Human Symmetry Perception, pp. 237–264
C. W. Tyler (Ed.)

On the generalization of symmetry relations in visual pattern classification*

INGO RENTSCHLER, ERHARDT BARTH, TERRY CAELLI,[1]
CHRISTOPH ZETZSCHE and MARTIN JÜTTNER

Institute of Medical Psychology, University of Munich, Goethestrasse 31, 80336 München, Germany
Fax: 89-5996615, E-mail: ingo@imp.med.uni-muenchen.de
[1]*Department of Computer Science, Curtin University of Technology, Perth, WA, Australia*

Abstract—Human observers were trained to criterion in classifying compound Gabor signals that were pairwise mirror-images of each other (left and right versions). They were then tested with 18 thresholded versions of the learning set. For the initial learning stage, the probability of confusing mirror-image patterns was significantly higher than that of confusing nonsymmetric patterns. This difference was greatly reduced for the test data, thus indicating a high degree of generalization of the concept of symmetry. The generalization to dark-only and light-only test signals, as well as to dark-*and*-light test signals, was found to be equally good. The behavioural data were analyzed in terms of various types of signal representation (pixelwise, Laplacian pyramid, curvature pyramid, ON/OFF, local maxima of Laplacian and curvature operators) and a minimum-distance rule. Results were suggestive of the existence of a structure-oriented 'blob-code'. Whether such a code could be used in conjunction with simple classifiers or should be transformed into a propositional scheme of representation operated upon by a rule-based classification process remained an open question.

1. INTRODUCTION

In a series of experiments we have investigated to what extent human observers classify signals composed of compound Gabor waveforms according to evident physical signal representations (Fourier descriptors) and conventional pattern classification strategies (Caelli *et al.*, 1987; Rentschler and Caelli, 1990; Shapley *et al.*, 1990; Rentschler *et al.*, 1994; Jüttner and Rentschler, 1996). Our results have demonstrated, that *internal* representations used by human observers for such Gaussian modulated grating patterns do not always fit with obvious *physical* feature states. This is particularly true when symmetry relations between patterns (Rentschler and Caelli, 1990; Shapley *et al.*, 1990) or extrafoveal viewing conditions (Jüttner and Rentschler, 1996) are involved.

*A preliminary version of this paper has been read at the *16th Conference on Visual Perception* (*ECVP*), held at the University of Edinburgh, UK, 25 August–29 August, 1993 (*Perception* **22** (Suppl.) 1993, 46).

To account for such deviations between human and formal classification performance, we have adopted the view that behavioural data reflect the operation of a (formal) classifier on some internal signal representation rather than on the physical input signals *per se* (Rentschler *et al.*, 1994). Clearly, this concept hinges on the assumptions made for formally relating the physical and internal signal representations. So far, we used linear filter operations for feature extraction (evenness and oddness pattern components), and we assumed, that the perceptual process of internal feature measurement gives rise to the occurrence of additive stochastic error signals. The corresponding random variables describe the additional degrees of freedom of bias and variance within the framework of a parametric Bayesian classifier. Estimates of internal feature states can then be obtained by using standard techniques of applied optimal estimation (Gauss, 1809; Gelb, 1974). Resulting class means and/or covariance matrices, the so-called *virtual prototypes* (Rentschler *et al.*, 1994), provide insight into the structure and dimensionality of internal signal representations.

Virtual prototype analysis treats physical feature states as a given and models human classification behaviour via the variation of pattern proximity in feature space. The proximity can be measured as Euclidean distance in the case of the (least squares) Minimum Distance Classifier (e.g. Duda and Hart, 1973; see also Caelli *et al.*, 1987), or be defined in terms of the scatter of samples around class means (Mahalanobis distance; e.g. Duda and Hart, 1973, Chapter 2; see also Jüttner and Rentschler, 1996). This approach is consistent with Shepard's proposal of a monotonic relationship between distance in some multidimensional vector space and stimulus *similarity*, i.e. the principle of multidimensional scaling (Shepard, 1962a, 1962b).

The nature of our previous approach, however, is limited in the sense that the feature representation of the signals is obtained from linear filtering, and that linear superposition of the signal statistics and of the stochastic components of the process of internal feature measurement is being employed. Thus we are to expect, that there exist problems in human visual pattern recognition which are too complex to be solved in this way.

One such problem is the perceptual confusion of left and right, which occurs with the two profiles of a head or pairs of shoes (see Mach, 1922; Gross and Bornstein, 1978). Such patterns or shapes are carried *into each other* by reflection in a line or plane. For two patterns $f_1(x, y)$ and $f_2(x, y)$, this type of *symmetry relation* can be expressed by the equality $f_1(x, y) = f_2(-x, y)$. It is not to be confused with the *bilateral symmetry* of a single pattern $f(x, y)$, which is carried *into itself* by reflection, i.e. $f(x, y) = f(-x, y)$.

In foveal vision, pairs of compound gratings with symmetry relations between them (briefly 'mirror-image gratings'; Rentschler and Treutwein, 1985) and mirror-image compound Gabor signals (Caelli *et al.*, 1986; Rentschler *et al.*, 1988) are more difficult to distinguish than nonsymmetric signals, whereas they are virtually indistinguishable both in extrafoveal vision (Rentschler and Treutwein, 1985; Bennett and Banks, 1987) and in texture perception (Caelli *et al.*, 1986; Rentschler *et al.*, 1988). Similar results have been obtained with pairs of mirror-symmetric patterns composed of line elements (Saarinen, 1987; 1988).

Furthermore, the problem of perceptually confounding mirror-image signals is not restricted to pattern *discrimination*, i.e. to the range of liminal pattern differences. We noticed this in an earlier study, where observers learned to *classify* compound Gabor signals, which were easily distinguished when being presented pairwise (Rentschler and Caelli, 1990; Shapley *et al.*, 1990). In an experimental situation, where each pattern in one pattern class (i.e. signal cluster in feature space) had its mirror-symmetric counterpart in another class, misclassification into nonsymmetric classes rarely occurred. Much more frequent were misclassifications into classes containing mirror-symmetric signals, thus indicating a greater perceptual similarity of mirror-image patterns than is predicted on the basis of distance in feature space.

Thus, it appears that there exist visual tasks that human observers solve by relying on internal representations more complex than those obtained from linear filtering. We examined the nature of such representations by training observers in classifying the samples of a learning set of signals with mutual symmetry relationships (the same one as used by Rentschler and Caelli, 1990, and by Shapley *et al.*, 1990), and then testing their ability to generalize to thresholded versions thereof. The choice of this experimental paradigm was motivated by the observation that in texture perception observers seem to encode compound Gabor signals in terms of the occurrence of specific blobs or parts of the patterns defined by size, shape, contrast parameters such as light and dark polarity, and (relative) position (see Caelli *et al.*, 1986). Related coding schemes have been considered by several authors (Koenderink, 1984; Watt and Morgan, 1985). If blob structures were indeed 'building blocks' of the visual quality of form, generalization of learned classification behaviour to blob-only transformed patterns should be relatively easy. What is not clear then is the question of whether symmetry relations existing among such learning patterns could be generalized as well. These two issues were under investigation here.

2. METHODS

2.1. Stimuli

Compound Gabor signals were generated on a 128×128 pixel format with a linear gray-level to luminance function. Intensity profiles were defined by

$$G(x, y) = L_0 + \exp\left\{ -\frac{1}{\alpha}(x^2 + y^2)\right\}\left[a\cos(2\pi f_0 x) + b\cos(2\pi 3 f_0 x - \phi)\right], \quad (1)$$

where L_0 determines the mean luminance, α the space constant of the Gaussian aperture, a the amplitude of the fundamental, b that of the third harmonic, and ϕ the phase angle of the latter. Thus the 2D images consisted of a fundamental cosine waveform and its third harmonic modulated by an isotropic Gaussian aperture that decayed to $1/e$ in 32 pixels.

The Gabor signals were displayed on a raster monitor linked to a digital image processing system (Barco TVM 3/3.2, P4 phosphor; Videograph, LSI 11/73). Space

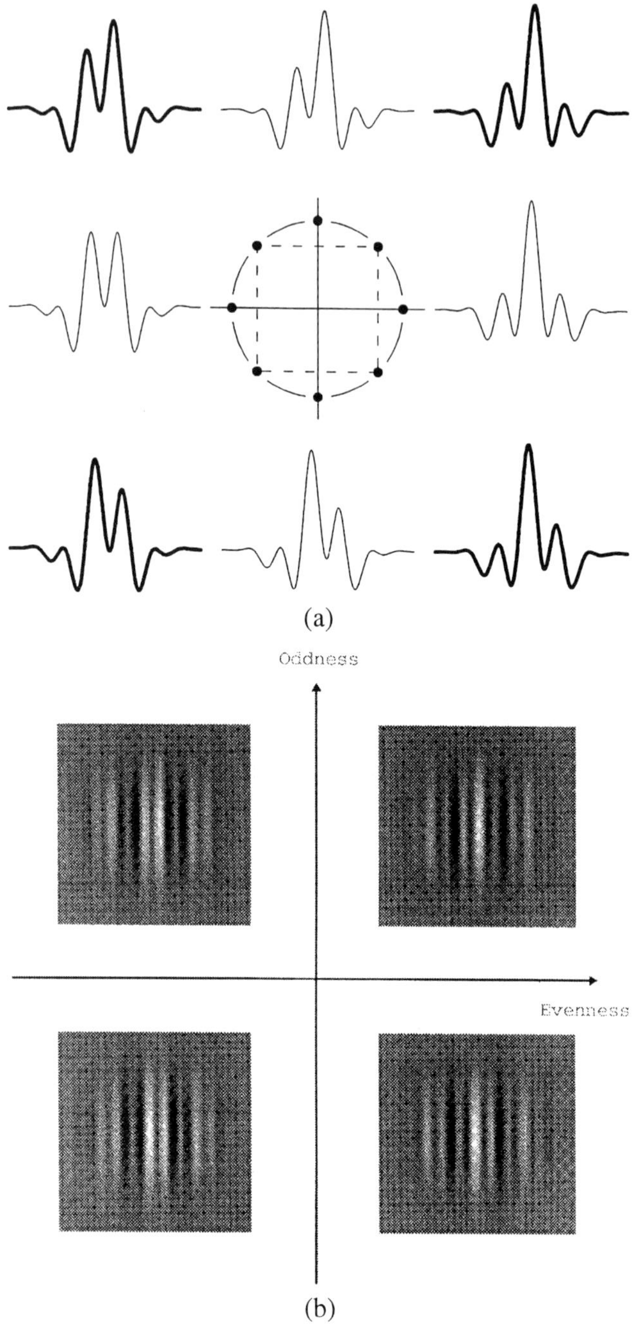

Figure 1. (a) Luminance profiles of compound Gabor signals with identical image power. Bilaterally symmetric profiles are on the abscissa, odd-symmetric profiles on the ordinate (thin lines). Profiles that lack bilateral symmetry but are pairwise mirror-symmetric are on the main diagonals (thick lines). (b) Two pairs of mirror-symmetric learning patterns located at same distance from the origin on the main diagonals of the evenness/oddness feature space.

average luminance (DC) was kept constant at 60 cd m^{-2}. The stimulus patterns subtended a visual angle of 1.6 deg at a viewing distance of 165 cm, and the fundamental spatial frequency was 2.5 cyc deg^{-1}.

2.2. Set of twelve learning patterns

The restriction of the signal variation to b and ϕ allowed the use of a two-dimensional feature space with the Cartesian coordinates $\xi = b\cos\phi$ ('evenness') and $\eta = b\sin\phi$ ('oddness'). A pattern vector (ξ_i, η_i) in that space uniquely determines a compound Gabor signal $G_i(x, y)$ as specified by Eqn (1) with parameters $b = \sqrt{\xi_i^2 + \eta_i^2}$ and $\phi = \operatorname{atan}(\eta_i/\xi_i)$ (Fig. 1). Given the definition of the function $y = \operatorname{atan} x$ in the interval $-\pi/2 < x < \pi/2$ and the evenness of the function $y = \cos x$, it follows that

- the reflection of pattern vector (ξ_i, η_i) in the ξ-axis of feature space yields the mirror-image signal $G_j(x, y)$ of $G_i(x, y)$, i.e. $G_i(x, y) = G_j(-x, y)$.
- the reflection of pattern vector (ξ_i, η_i) in the η-axis of feature space yields a pattern $G_k(x, y)$ that differs from $G_i(x, y)$, in a π-shift (contrast inversion) of the third harmonic.

These facts led to the construction of a set of 12 learning signals that were clustered in feature space in groups of three signals about the corners of a square (Fig. 2). This

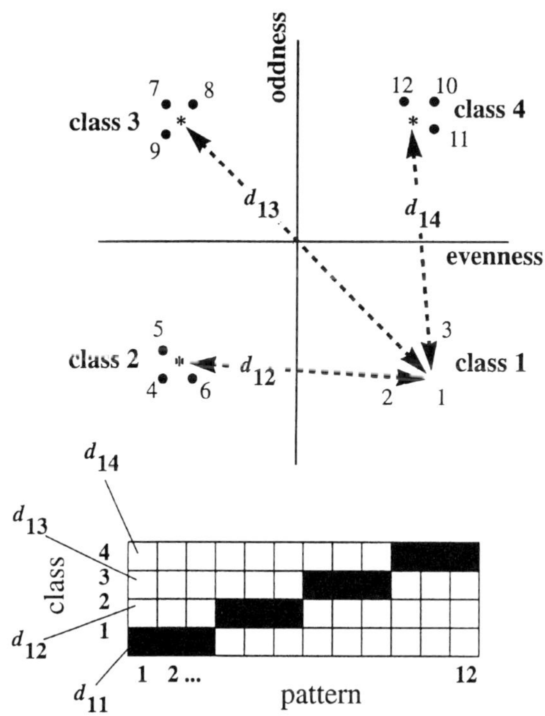

Figure 2. The learning set of twelve signals as defined in the evenness- and oddness-feature space (top). Symbolic representation of the distances between the 12 patterns of the learning set and the 4 mean pattern vectors (class prototypes; bottom).

square was centred on the origin, with its sides being parallel to the feature axes. For any given signal there existed a mirror-image signal (inflection of the oddness coordinate) and a signal with inverted contrast of the third harmonic (inflection of the evenness coordinate). These latter two signals had the *same* Euclidean distance from the former signal. This implied, that 8 of the 12 learning signals had the same lower image power, whereas four signals had the same slightly higher power.

2.3. *Eighteen sets of test signals*

Novel test signals were generated by means of gray-level modification of the learning set. The nonlinear image distortions consisted of dividing the range of 256 gray-levels into three regions, which were arranged symmetrically around DC. The width of the middle range was ±10, 20, or 40 gray-levels. Within one or two of the resulting regions, all gray-levels were set to 127 (i.e. DC), whereas they remained untouched elsewhere (Fig. 3). This resulted in $3 \times 6 = 18$ different sets of learning signals. Thus we obtained 18 *blob versions* for each of the 12 original patterns. The appearance of these blob-only patterns can be seen in Fig. 4, where the 18 blob versions corresponding to the first signal of the learning set are shown. The arrangement of the 6 lines corresponds to the six types of blob transformations shown in Fig. 3. Within one line of Fig. 4 the threshold range is ±10, 20, and 40 from left to right.

2.4. *Psychophysical procedure*

The procedure of supervised learning and classification consisted of a variable number of learning units and employed the *learning set* of 12 original compound Gabor signals. The latter was partitioned into four classes of three signals each. One learning unit contained three subsequent presentations, in random order, of the learning set with a 200 ms exposure duration for each pattern. Following each presentation, a number was displayed for 1 s with an ISI of 500 ms, specifying the class to which the pattern belonged. Each learning unit ended with a test of how well the subject was able to classify the 12 learning signals. Only one exposure per sample was used here. Viewing was always binocular.

The learning procedure continued until the subject had achieved an error-free (100% correct) classification in one learning unit. Following the initial learning stage, the subject was then tested with the first test set. Each pattern of this test set occurred 50 times in random sequence and was classified according to the previously learned classes. Thereafter subjects were re-trained to 100% correct with the learning set and were tested 50 times with each sample of the second test set. This procedure of alternating units of re-learning and testing was continued until the subjects had been tested with each of the 18 test sets.

To complete this program, a subject had to undergo an average number of 120 learning units, each consisting of $3 \times 12 = 36$ signal presentations with instruction plus 12 classification decisions, and a total 10 800 test decisions. The program was fulfilled during 20 to 30 experimental sessions, which were distributed over a period of about 3 months.

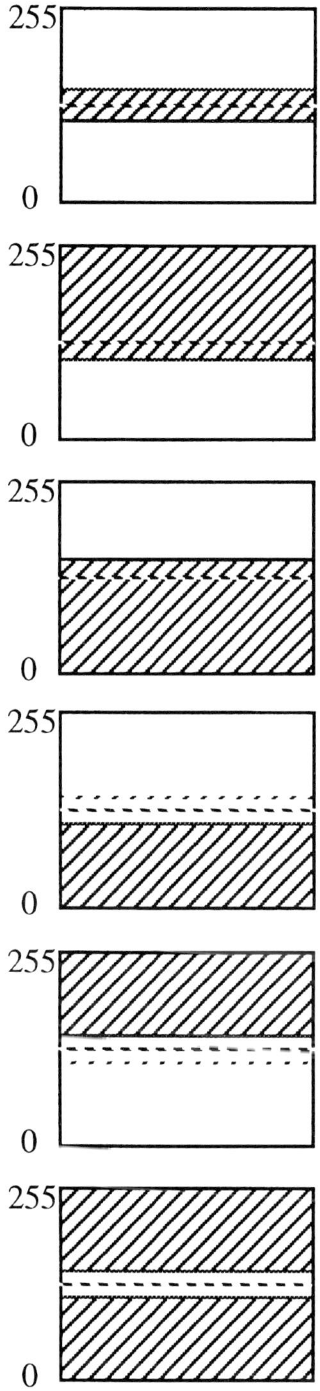

Figure 3. Definition of the 6 distortion types used for generating the test signals. The width of the gray-level interval around DC (i.e. level 127) was ±10, 20, or 40, leading to the 18 test signals shown in Fig. 4 (for details see text).

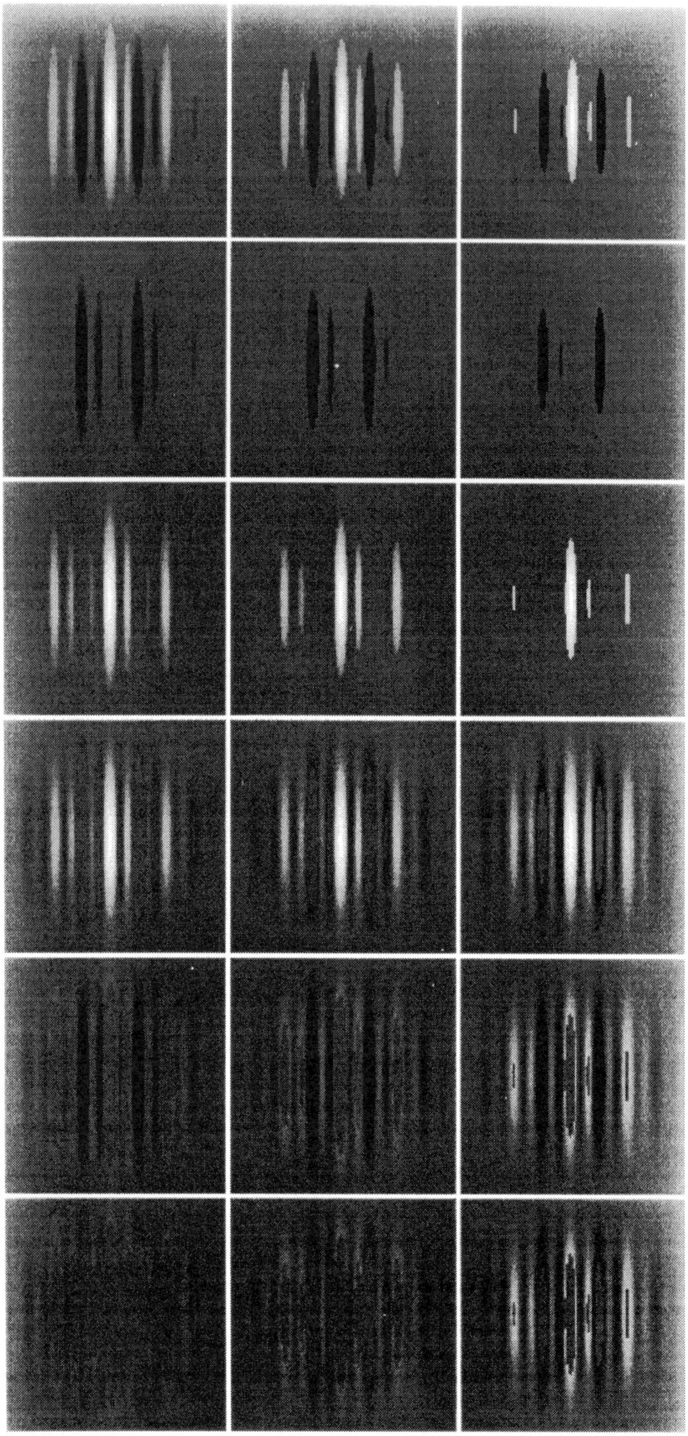

Figure 4. The 18 distorted versions of the first pattern of the learning set.

2.5. *Subjects*

Six paid subjects participated in the experiments, 3 females (G.A., P.E. and R.B.) and 3 males (J.N., H.S. and M.A.). Their age ranged between 20 and 35 years and they had normal or corrected to normal vision. Three subjects (J.N., H.S. and R.B.) had participated in the earlier experiments on supervised learning and classification of mirror-image signals (see Rentschler and Caelli, 1990, and by Shapley *et al.*, 1990), whereas the remaining three had never participated in psychophysical experiments before.

3. RESULTS

Generalization results for the 18 test sets are shown in Fig. 5 with the relative classification frequencies ('classification probabilities') being plotted against the numbers of the twelve test images. Only cumulative data over the group of six observers are shown, although the latter were not entirely consistent in their judgments. The reason for this restriction is our interest in more general aspects of generalization behaviour, against which details of individual characteristics shall be neglected. However, individual classification strategies are being considered in another study focusing on the possible applicability of recognition-by-parts strategies from machine intelligence (Jüttner *et al.*, 1996). The data on supervised learning of the original signal configuration (learning set) are not shown either. We simply state that subjects were to classify a given set of test signals once they had reached a criterion of 100% correct in classifying the learning set.

For judging the observed generalization performance of our six observers, it is important to note that the 18 types of blob transformations enacted on the learning signals are pairwise complementary with respect to the light and dark contrast polarities. The three blob transformations characterized in rows 1 of Fig. 4 *removed* narrow ranges of gray-levels around DC, whereas those of rows 6 exclusively *preserved* the same ranges. Thus it can be said that, in an informal sense, these transformations removed or preserved 'zero-crossings' (although a low-resolution Laplacian operator might have the same zero crossings when applied to the patterns in row 1). The three transformations characterized in rows 2 removed the *upper ranges* of gray-levels (light image components) plus some range around DC. Those of rows 3 removed the *lower ranges* of gray-levels (dark image components) plus the same range around DC. The three transformations of rows 5 differ from those of rows 2, in that they preserved the gray-level ranges around DC. The same difference exists between the transformations of rows 3 and 4 of Fig. 4. Briefly, both pairs of rows 2 and 5, as well as 3 and 4, correspond to light and dark image blob transformations, respectively, with the transformations 4 and 5 preserving the 'background' of gray levels around DC, i.e. the range of zero-crossings.

The overall conclusion from the psychophysical data shown in the same sort of array in Fig. 6 is, that the subjects did surprisingly well in classifying the test signals, although the transformations resulted from rather drastic image distortions. Scrutiny of the classification data, however, reveals some interesting characteristics of classification behaviour.

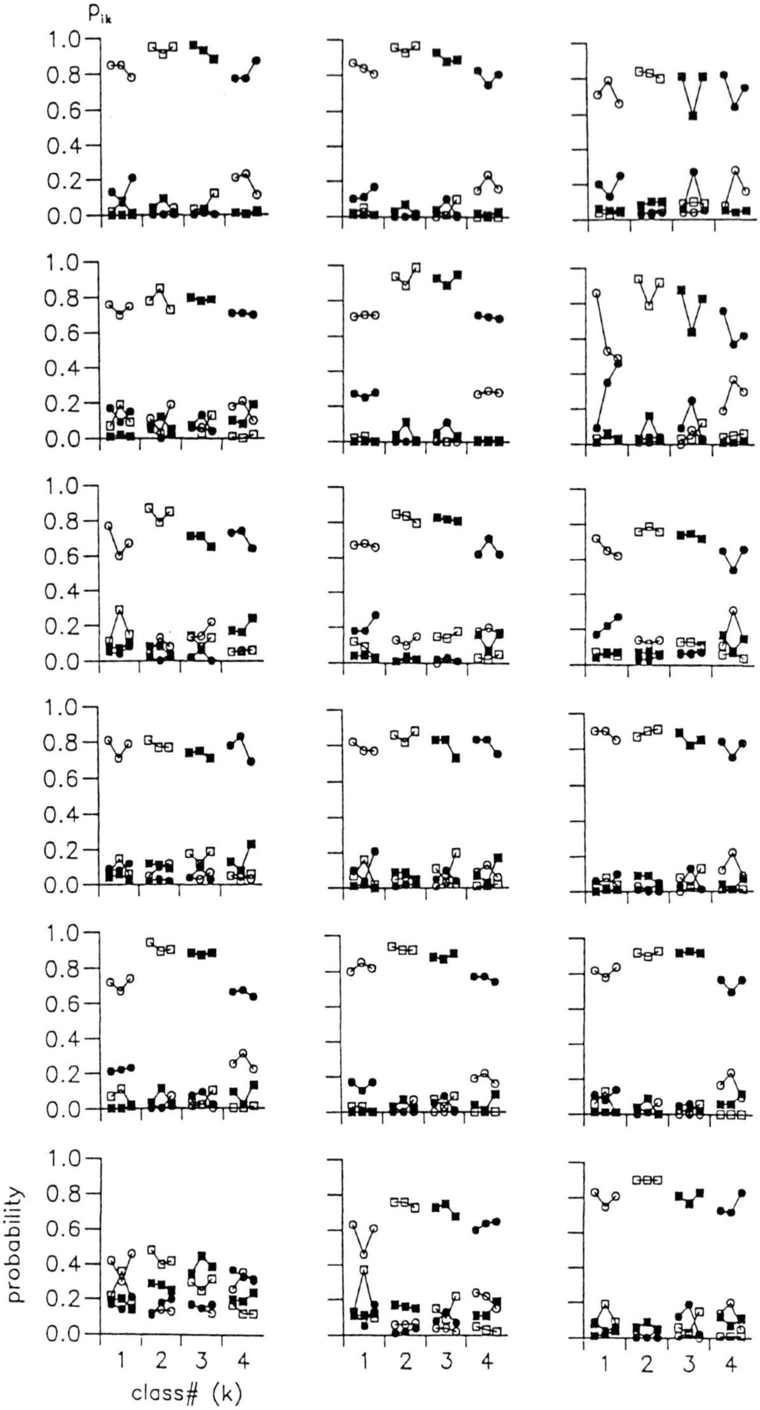

Figure 5. Psychophysically measured generalization performance (classification probabilities) for the 18 sets of test signals pooled over six observers.

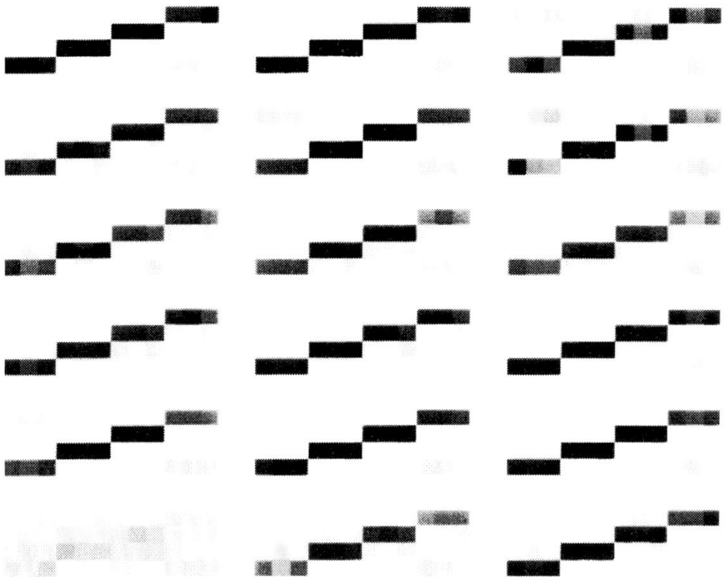

Figure 6. Generalization data as shown in Fig. 5. Classification probabilities are gray-level coded (dark shade indicates high probability) in order to allow a comparison with distance structures such as shown in Fig. 2.

3.1. Effect of symmetry relations

For analyzing the effect of symmetry relations on classification (generalization) behaviour it is useful to consider the two main types of classification errors made by the subjects. They consisted of either erroneously attributing a learning or test signal to the class containing the mirror-image of the latter (*symmetry error*) or of attributing it to the class that contains the pattern with the contrast inverted third harmonic (*non-symmetry error*; see Section 2.2).

For the classification data of the initial learning stage (Fig. 7), the probability of a symmetry error averaged across observers proved to be distinctly higher than that of a non-symmetry error ($t(46) = 3.48$, $p < 0.001$). This finding reflects the difficulties observers had in learning the concept of symmetry relations between patterns and pattern classes. Once they had learned the class concepts, they could be generalized to most test sets, though not to all of them. The analysis of the generalization data revealed that the rates of symmetry vs non-symmetry errors differed significantly ($\alpha = 0.05$) for only 4 out of the 18 test sets. The respective test sets were those characterized in Fig. 4 in the first row (left and centre) and in the second row (centre and right) from the top. For all other sets of test patterns we have no reason to reject the notion that both error types occurred equally often.

Since all the threshold operations employed preserved the symmetry relations between the patterns of the learning set, we conclude that the *concept of symmetry* could be partially but to a great extent be generalized within the present context of classification learning.

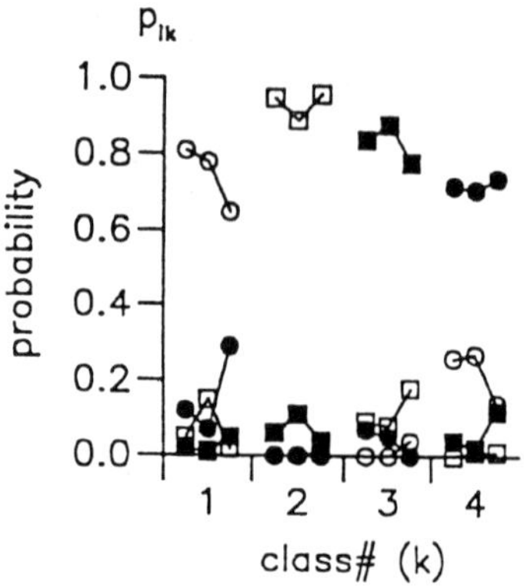

Figure 7. Classification of the learning set. Data cumulated over the six observers and all learning units that were needed to reach 100% correct prior to entering the generalization experiment.

3.2. Effects of 'blob coding'

A comparison of the respective classification performances corresponding to the 'no zero-crossings' and the 'zero-crossings only' conditions (rows 1 and 6 of Fig. 5) suggests that the processing of light and dark image components (blobs) is particularly important for the generalization of classification performance. However, it is also clear, that the latter decreased the more image gray-levels were lost from the range around DC (Fig. 5, row 1, rightmost graph, and row 6, leftmost graph).

Furthermore, the test patterns employed for obtaining the data shown in rows 2 or 3 of Fig. 5 contained dark or light blob configurations only (see rows 2 and 3 of Fig. 4). The resulting classification performance is good, with a somewhat better performance for test patterns with dark (row 2) blobs. The classification data shown in rows 4 and 5 are again equivalent with respect to contrast polarity but are generally somewhat better than those depicted in rows 2 and 3. The latter superiority obviously stems from the additional image information provided by preserving the gray-level ranges of zero-crossings in the test signals.

3.3. Conclusions

Summarizing, we note three surprising properties of the set of generalization data plotted in Figs 5 and 6. *First*, the ability of classifying mirror-image signals was acquired through learning and could be generalized for the majority of test sets. *Second*, generalization performance to blob-only versions of the learning signals was, with a few exceptions, excellent. *Third*, generalization performance was equally good for test patterns with bipolar blob structure (dark *and* light blobs), with dark blobs only, and with light blobs only.

As such, these results are not sufficient for drawing direct conclusions about the nature of the visual representations involved. However, the first result suggests that generalization is obtained from the same combination of pattern representation and classification rule that has been build up via learning of the original pattern configuration. The second result is clearly compatible with our hypothesis that compound Gabor signals are encoded in terms of their blob structure. The third result implies, that dark *or* light blob pattern components are sufficient for the classification of compound Gabor signals. Hence it is compatible with the hypothesis that the learning and test patterns are internally represented with 'labelled' contrast polarity of blobs. Yet it could also be that the patterns are contained in two different subsystems subserving the representation of the dark and light structures, respectively. This stronger view receives support from the introspective reports of our subjects and from individual classification data, which show that subjects tend to rely for generalization on only one polarity (cf. also Treisman and Gelade, 1980). Neither labelled nor separate ON/OFF representations can be explained within the theoretical framework of linear systems (e.g. Gaskill, 1978), as will be shown in the following section.

Taken together, we have reasons to assume that the classification of compound Gabor signals in a general context cannot be modelled in terms of the virtual prototype approach in its present linear form. We discuss in the following section ways of dealing with these difficulties.

4. COMPUTATIONAL ASPECTS OF PATTERN GENERALIZATION

4.1. Classification and generalization criteria

A straightforward assumption about pattern classification *and* generalization is that learning and test patterns can be represented in one and the same vector space. The basic but certainly not sufficient criterion for classification is then, that learning patterns are separable, that is, can be distinguished at all. If generalization is possible, it can further be inferred that the test patterns are located within the class boundaries established through learning (*Generalization Criterion* G) though the discriminant function may be arbitrarily complex (nonlinear).

In technical pattern recognition, the use of *linear* discriminant functions (sets of hyperplanes) is common, and forms the basis of minimum-distance-based classification schemes (e.g. Duda and Hart, 1973). This restriction (*Generalization Criterion* G1) delimits the range of suitable pattern configurations considerably, but does not imply an *intra-class clustering* of learning and test patterns. The distance between test and learning signals of a given class may be much larger than both the intra- and the inter-class distances of the learning patterns, provided the former is less than the between (inter-class) distances of the test signals of a given class and the learning signals of all other classes. The simplest model for the description of our data would thus consist of a single visual representation followed by a minimum distance classifier. Provided a suitable representation has been found, such a model would account both for the fact, that the test signals appear perceptually quite remote from (dissimilar to) the learning set, and for the result that they can be classified correctly.

In our view, however, an even stronger restriction (*Generalization Criterion* G2) appears reasonable as a basis of human generalization capabilities, namely that the learning and test patterns of each class are *conjointly* clustered. This implies, that the intra-class distances between learning and test patterns are comparable to the intra-class distances between the individual learning signals of the class, while the inter-class distances to all other patterns are substantially greater. Simply stated, this hypothesis would assume the existence of a visual representation wherein the test and the learning patterns are more similar to each other than they are to any pattern of the other classes. The separability of classes via linear discriminant functions is then obvious. However, it is also clear that the global perceptual dissimilarity between learning and test patterns must then be accounted for by assuming a further representation (or metric) that makes this property explicit.

To explore the extent, to which such restrictions apply, it seems reasonable to begin with the computation of *distances* between all the patterns involved and consider relationships between distances. Various distance measures are possible, but with respect to the following it is immaterial whether Euclidean distances, correlations, or angles (correlations normalized by vector length) between pattern vectors are considered.

We can then proceed to consider how different types of pattern representations (metric feature spaces) conform with the criteria established above. From a technical point of view, the consideration of Euclidean distances in signal space is a first step. The results of corresponding computations are shown in Fig. 8, left, where for each of the 18 sets of test patterns the Euclidean distances between the 12 test signals and the original pattern classes (mean pattern vectors or prototypes) are plotted. A gray-level code is used to visualize the distances, whereby increasing darkness indicates decreasing distance, i.e. proximity between test patterns and learned prototypes.

A comparison of behavioural classification data and such simulated distances is not trivial. In particular, a full specification of the classification scheme and of the noise sources within the system would be necessary in order to enable a direct comparison (see Rentschler *et al.*, 1994). To keep the computations manageable we have restricted ourselves to two kinds of first-order approximation. First, we computed distance matrices like the ones shown in Fig. 8 for all representations. The values in these matrices do not directly correspond to experimentally obtained percent-correct values (as illustrated in Fig. 6) but are related to the latter as follows: a strong diagonal structure in the distance matrices (as in some matrices in Fig. 8, left) indicates *clustering*, and thereby, *linear separability* (G2, and thereby G1, will be satisfied). This is compatible with the occurrence of strong diagonals in the percent-correct matrices (Fig. 6), indicating almost perfect classification performance of the subjects. It should be noted, however, that this relation is not one-to-one, i.e. perfect linear separability may well be possible *without* the emergence of such dominant diagonal structures. To take this into account, we have also computed the correct classifications obtained with a linear classifier being optimal with respect to the learning patterns. The results of these simulations are discussed in the following section.

Already the simple computation of Euclidean distances seems to be a satisfactory model, since the diagonal structures in Fig. 8, left, compare reasonably well with the behavioural data (Fig. 6), even with respect to the strong clustering criterion G2. Thus it is also possible to satisfy G1, i.e. to find hyperplanes in signal space, which delimit the four classes defined by the clustering of the 12 original patterns in such a way that all but two blob versions (indicated in Fig. 8, left, by means of a frame) can be classified perfectly.

The direct visual comparison, however, of test and prototype images via zero-shift cross-correlation is an unlikely assumption for the assessment of pattern similarity.

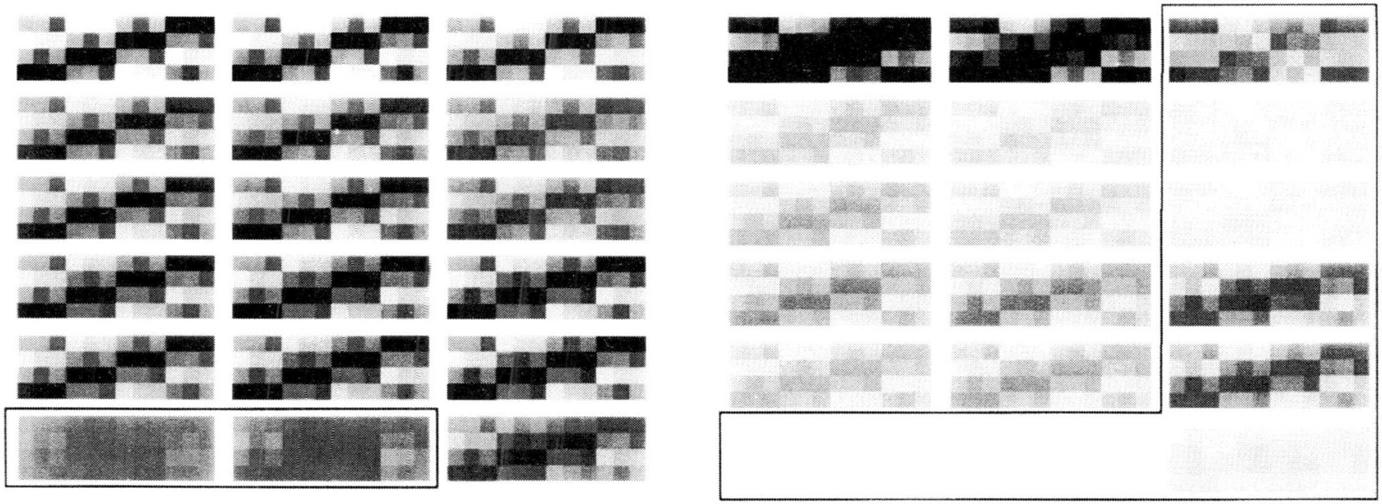

Figure 8. Euclidean distances between original classes (prototypes or mean vectors) and the 18 sets of test signals. The shown gray-values, normalized to 256 levels across all 18 blob versions, are proportional to the distances (the darker the closer), and the arrangement of the plotted distances is as follows. The rows of each block contain the 12 distances between the 12 distorted patterns and one pattern of the learning set (see Fig. 2). The resulting 12 by 4 matrices are computed for the 18 distortion types and displayed with the same arrangement as the corresponding patterns in Fig. 4. The results shown on the right are obtained by shifting the 12 signals relative to the prototypes of the 4 classes (mean vectors) such that the plotted values are distances which are minimal with respect to the shift. The minimum-distance rule would imply that each pattern has to be closer to the correct prototype than to all other prototypes (classification criterion 61; see text). The blob versions which do not yield 100% correct classification according to this criterion are marked.

The *full* cross-correlation information is needed to represent human performance in image matching (alignment; Caelli and Rentschler, 1986), and this suggests the use of *maximum* correlation (minimum distance with respect to shift-operations) as a behaviourally more realistic distance measure (Fig. 8, right). Switching to this measure changes the situation, in that the patterns shown in the third column and those shown in the sixth row of Fig. 4 are clearly no longer consistent with the proposed strong generalization criterion G2. Evaluation according to the minimum distance rule reveals that there is even a considerable violation of the weaker criterion G1. The greatest violation occurs for the dark-blob patterns in column 3, row 2, which are classified with 58% correct only. The experimental data (Fig. 6; the means are shown in the last column of Table 1) show no such steep decrease in performance (73% correct are obtained for this distortion type). We conclude that the simple cross-correlation model accounts for basic aspects, like the good generalization of the moderately thresholded dark-only and light-only blob patterns. It does not predict, however, behavioural performance for the extremely thresholded blob versions, which require the use of thin line elements for correct classification (right column of Fig. 4).

A further assumption, specific to the present context, would be that the visual system performs operations which are similar to the operations by which the test sets themselves were generated. Suppose that we chose one of the distortion types shown in Figs 3 and 4 as a model for the preprocessing of all the patterns. It can then be seen that a proper subset of the set of test patterns could be mapped on identical 'representations', such that the strong criterion G2 is satisfied. Our simulations show, however, that the results of such operations depend critically on the precise value of the gray-level thresholds used. The implementation of this type of algorithm in biological vision systems is not plausible.

4.2. *Visual representations*

We thus proceed to consider biologically more realistic visual representations. Based on standard assumptions about the function of early visual pathways, we have chosen two representative types of isotropic image representations since there exist no differences in orientation between the patterns employed here. The first one is a Laplacian pyramid (LAP) representation (Burt and Adelson, 1983), which corresponds to the simple centre-surround mechanisms of visual neurophysiology (see Fig. 12 in the Appendix). The second one is a 2D pyramid based on the clipped-eigenvalues (CEV) operator as introduced in our previous work (Zetzsche and Barth, 1990a, 1990b; Barth *et al.*, 1993a). It provides a model of the function of end-stopped neurons. A brief description of the latter formalism, which again uses different levels of resolution, is found in the Appendix (see also Fig. 12).

Based on these representations we added two further computational elements of visual processing. The first one is the ON/OFF separation as motivated in Section 4.3 which leads to the LON/LOFF, CON/COFF representations as described in Fig. 10 (the ON/OFF representations and later the MAX/MIN representations are used in combination with a switching circuit for choosing the appropriate representation with respect to the polarity of the input). The second one is related to the concepts of gain control and local phase (a ratio of even and odd measures) implemented here as the CEVG and LAPG operators (with the ON/OFF versions CONG/COFFG, LONG/LOFFG) defined in the Appendix. The 'G' refers to 'gradient' since the difference between the CEV and CEVG (and equally between LAP and LAPG) operators is that the former were divided by the gradient of image intensity. In the present context such an operation reduces the differences between blobs of different size and contrast.

Furthermore, we considered representations, where the local maxima of the Laplacian and of the CEV-operator are extracted (so-called MAXLAP and MAXCEV representations which will be described in more detail in Section 4.3). To obtain a distance measure for patterns in these representations, the positions of the maxima in scale space were correlated (after slight low-pass-filtering).

For all the representations listed above, we computed and plotted the distance patterns in the same way as shown in Fig. 8, right, i.e. under the assumption of maximum cross-correlation or minimum distance. (To reduce computational complexity, we computed the full transforms of the 128 by 128 pixel images but then extracted only the middle row of the transforms before computing the distances.) An overview of the results is given in Table 1 for the 18 test sets (in the arrangement of Fig. 4) and the representations specified. The numbers in the table denote the percentage of patterns of a given test set (of 12 patterns), which would be classified correctly according to the minimum-distance rule. In the last column the mean behavioural data are included, again with the caveat, that they cannot be compared directly with the simulation results (see Section 4.1).

The analysis of distance structures summarized in Table 1 failed to yield unequivocal results. In general, most representations can account for the basic effects, such as the good generalization of the dark-only and light-only blob patterns with low

Table 1.
Simulated classification results for different representations and mean behavioural performance (in % correct)

Pixelwise			LAP			CEV			LON/LOFF			CON/COFF		
100	100	66	100	100	58	100	100	83	100	100	83	100	100	91
100	100	58	100	100	50	100	83	58	83	83	16	91	83	50
100	100	66	100	100	66	91	91	83	83	83	83	91	91	66
100	100	83	100	100	83	100	100	83	100	100	100	91	91	100
100	100	83	100	100	83	100	100	66	83	83	100	91	91	83
0	0	0	0	0	0	25	16	16	33	0	25	25	25	33

LONG/ LOFFG			CONG/ COFFG			MAXLAP/ MINLAP			MAXCEV/ MINCEV			Behavioural performance		
100	100	100	100	100	100	100	100	50	100	91	41	87	87	75
66	66	8	83	75	33	83	91	58	50	50	25	75	82	73
100	83	66	100	100	58	75	41	58	66	50	50	72	74	69
100	100	100	100	100	100	83	91	100	83	75	66	76	81	85
66	50	66	83	83	66	66	75	100	66	66	75	78	84	85
41	16	8	33	25	25	8	8	66	33	25	58	39	66	81

and medium thresholds (Fig. 4, left and mid columns). Furthermore, some difficulties observed with the pixelwise representation and Euclidean distances, such as the pronounced drop off of correct classification for the extremely thresholded test patterns (see Section 2.3 and Fig. 4, right column) could be ameliorated by employing biologically more plausible representations. However, some representations, especially MAXCEV/MINCEV, seem to perform worse than the pixelwise representation according to the minimum distance rule used here for the simulations.

Still, many representations predict better classification for light-blobs only (row 3) than dark-blobs only (row 2) whereas the behavioural performance is rather better for dark blobs, and, in addition, it is higher for the test set in row 6, column 3, than for both the light-only and dark-only blobs. Further, many representations show a greater variability across test pattern types than was observed with the behavioural results. Whether the above arguments can be used to judge the usefulness of the representations, depends on the full statistical model used for the decision process. We conclude that the present predictions neither provide clear hints as to the nature of the visual representation(s) underlying generalization, nor answer the question as to the decision strategies employed by our observers (minimum-distance rule or others).

4.3. Towards more abstract codes

The previous results are less surprising, given the possibility that both our assumptions on the nature of visual representations and of decision rules (classifier types) underlying the generalization of form are too simplistic to accommodate all aspects of human performance. Yet it seems reasonable to assume that *the formidable power of biological pattern recognition originates predominantly from the brain's ability to construct and modify internal representations via learning of pattern structure while relatively simple decision rules are being maintained.* The fact, that expertise in sophisticated recognition tasks (e.g. face recognition in foreign ethnicities, radiological diagnosis, plant classification) depends on long learning periods but thereafter allows rapid decision making, seems to support this view. The above-mentioned 'rectification' of the internal representation of mirror-symmetric patterns through learning and the resulting ease of generalization to blob-only versions thereof may be another example of this. Furthermore, this notion follows classical theories of perceptual learning (Gibson, 1969), according to which the learning process involves the development of a structural pattern representation in terms of distinctive features and feature relations. Our further treatment of the generalization problem will thus emphasize the issue of pattern *structure* as revealed in different representations, while the formal nature of decision behaviour will not be further pursued.

Accordingly, we now reassess the representations under consideration. In particular, we consider basic neurophysiological and computational aspects, and will refer to the introspective reports of our subjects concerning their *recognition strategies.* Based on the earlier hypothesis, that the subjects' 'internal code' consisted of a relatively abstract description of the patterns, seemingly involving light and dark blob elements with explicit positional relationships ('blob algebra', see Caelli *et al.*, 1986, for an analysis in the context of texture perception), we suggest that the location of blob features in scale space (the MAXCEV representation) provides a suitable representational basis for characterizing generalization performance.

That the treelike configuration of 2D points in scale space could capture the essential image structure was first suggested by Zetzsche *et al.* (1993). Subsequent experimental and theoretical investigations of this hypothesis revealed that the maxima of curvature do indeed yield a consistent tree-structure in scale space (Barth *et al.*, 1993b, 1993c). The algorithm, which we propose for computing multi-scale blob features, bears similarities with coding schemes proposed earlier by several authors (Koenderink, 1984; Watt and Morgan, 1985). We propose that the stability of blob features across scale space to be the computational basis for determining the similarity of blob structures between learned classes and test patterns. Here the argument for the need of a stability analysis in scale space is essentially the same as that used by Rosenfeld and Thurston (1971) in the context of edge detection, and by Marr and Hildreth (1980) for evaluating the location of zero-crossings, namely, the positional uncertainty associated with the determination of such descriptors at only one level of resolution. The usefulness of 'parsing scale space' for stability analysis in the field of object recognition has been demonstrated by Bischof and Caelli (1988).

The determination of local maxima at multiple levels of resolution has the advantage that it can account for the more abstract notion of blob algebra in the sense that predicates such as the 'existence of a large blob in the centre and a smaller one at the right of it' can be formulated consistently, the size of a blob being encoded by its position in scale space. For these reasons, we computed the locations of the local maxima of the clipped-eigenvalues (CEV) operator for a sequence of different resolutions, i.e. 16 scales spanning two octaves (the same frequency range as in Fig. 12 with a denser sampling for better illustration). These locations of the blobs were plotted for a horizontal cut at half-height through the learning and test images as a function of scale.

The plots of the maxima across scale (Fig. 9, MAXCEV representation) reveal a relative invariance with respect to light-blob operations (Fig. 9, from left to right), thus providing a basis for the proposed requirement of *similarity of form* assumed to underlie generalization performance. This invariance is closely related to the strong generalization criterion G2, which requires a sufficient similarity between the representations of the corresponding training and test patterns. Yet the MAXCEV representation still allows the discrimination between patterns of different classes (Fig. 9, from top to bottom). The minima (MINCEV representation) display the same type of stability for dark-blob operations. This suggests, that the two representations can be employed for obtaining a pattern recognition scheme with the required degrees of invariance (generalization) and differentiation (discrimination between classes).

Similar results, but with a reduced invariance to blob operations, can be obtained by using the Laplacian for generating representations in scale space (MAXLAP representation). This approach has the general problem of insufficient ON/OFF-separation, which leads to 'false responses' in the MAXLAP representation of dark patterns and in the MINLAP representation of bright patterns (see Fig. 10). The use of the 2D operator is sufficient to avoid this problem but we do not know yet whether it is also necessary here. However, it seems that there exists a relationship between clean

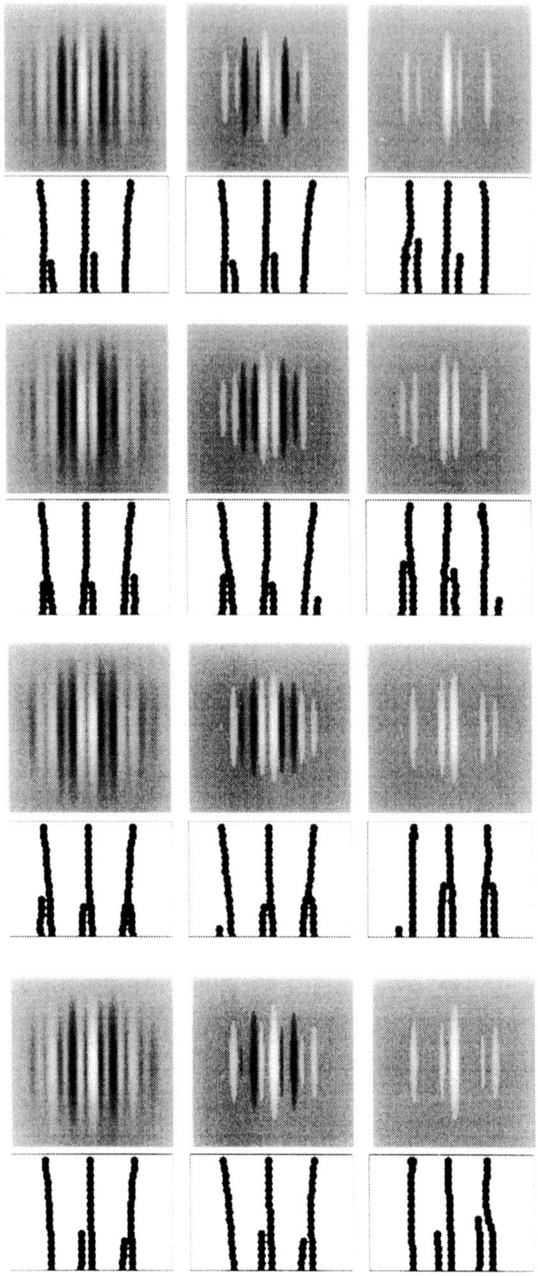

Figure 9. Locations of bright blobs plotted as a function of scale are shown below the corresponding images. The arrangement is as follows: from top to bottom the pattern number varies (the first patterns of each class are shown, i.e. patterns 1, 4, 7, and 10—see Fig. 2), and from left to right the distortion type (mid patterns of rows 1 and 3 in Fig. 4). The locations of the blobs were obtained in two steps. First, we computed the 2D-representations as shown in Fig. 12 on the right (the same range of scales was used but with a finer sampling of 16 scales). Second, the middle row of the images were extracted and those locations were selected for plotting, for which left and right neighbours were smaller (local maxima).

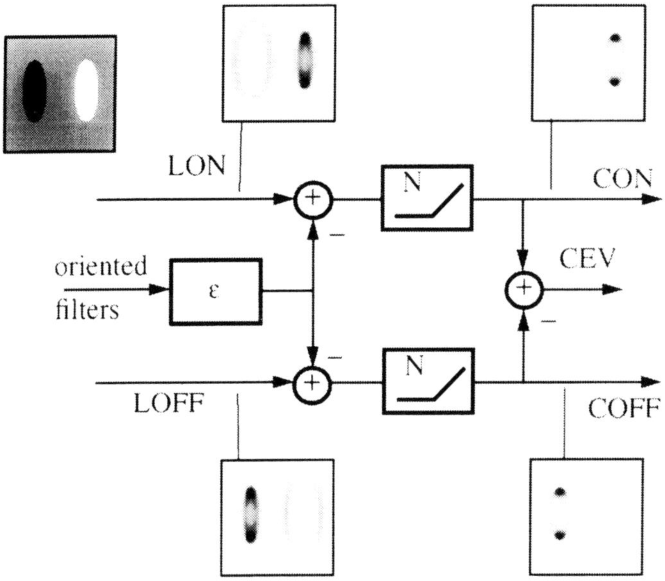

Figure 10. The Laplacian and the CEV operator (for details see Appendix). 'LON' and 'LOFF' denote the ON and OFF components as obtained by one-way rectification of the Laplacian. The eccentricity ε (defined in Eqn (5)) results from nonlinear inhibition between oriented filters. The resulting 'CON' and 'COFF' components of the CEV operator differ in two ways from the 'LON' and 'LOFF' components. First, they are 2D-operators, i.e. they equal zero for straight patterns, and respond only to curved features like corners and line ends. Even more important for the current context, they differ with respect to the ON/OFF separation as follows. A 'CON' operator responds only to light-on-dark features, whereas a 'LON' operator will respond also to dark-on-light features as illustrated for the image shown top left.

ON/OFF separation and the suppression of straight patterns, which 2D operators yield by definition. An alternative nonlinear method for the construction of separate ON and OFF representations would be a model of the MIRAGE-type as proposed by Watt and Morgan (1985), which starts with a set of spatial filters, the outputs of which are combined in such a way that negative and positive response components are carried separately. Whether such a model can also avoid the emergence of 'spurious' blobs remains to be investigated.

According to Fig. 9 the MAXCEV representation has the essential features proposed here as a basis of successful generalization: sufficient similarity within classes, as well as sufficient dissimilarity between classes. The principle illustrated in Fig. 9 has, of course, its limits. Figure 11 shows the MAXCEV and the MINCEV representations for all 12 patterns of a given blob type including the most critical blob operations for which the similarity of the representations decreases. Nevertheless, we think that there still is sufficient similarity within classes, as well as sufficient dissimilarity across classes to allow for a better agreement between data and predictions than obtained by the other representations in Table 1. However, the above mentioned features cannot be utilized by simple correlation and minimum distance rule (see again Table 1). One possible way of taking account of this difficulty is to develop a notion of similarity based on *explicit structural properties* of the patterns.

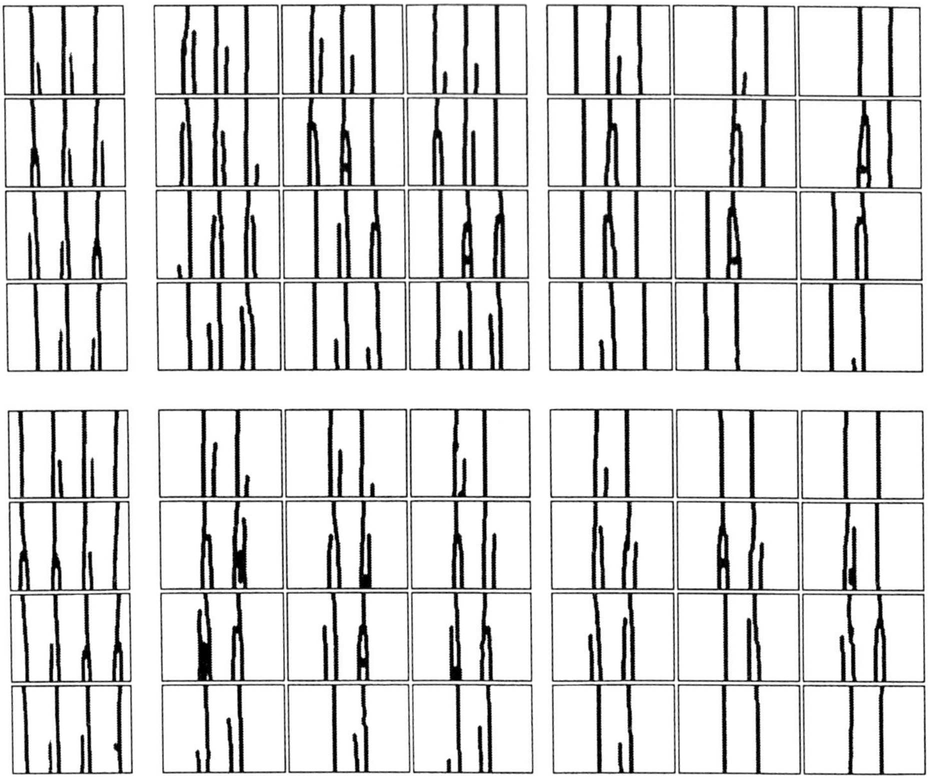

Figure 11. MAXCEV representations (upper half) and MINCEV representations (lower half) for complete sets of 12 test patterns (arranged as in Fig. 1) and for two blob types with medium and extreme thresholds (columns 2 and 3 in Fig. 4). The left column shows the 4 prototypes (class means of the original patterns). The upper 3 by 4 blocks show the MAXCEV representations of the light-only blob patterns with medium threshold (left block, corresponding to row 3 column 2 in Fig. 4) and extreme threshold (right block, corresponding to row 3 column 3 in Fig. 4). The lower two blocks show the MINCEV representations of the dark-only blob patterns, again with medium and extreme thresholds. The locations of maxima are here not illustrated by plotting circles (as in the previous two figures) but by setting pixels to one or zero, depending on whether they correspond to maxima or not, and then low-pass filtering the resulting binary images (the images shown here have been used to obtain the distance measures in Table 1). The first column of the upper left 3 by 4 block shows the same maxima as the right column in Fig. 9.

Accordingly, future research should investigate more abstract codes that can be derived from the MINCEV/MAXCEV representation. Such codes may, for example, explicitly represent both attributes of individual blobs (e.g. position or size) and relational attributes of pairs of blobs (e.g. distance or size ratio). It may then be possible that relatively simple classifier structures acting on such representations account for the experimental data.

Alternatively, such structure-oriented codes may be used for a propositional scheme of representation operated upon by a rule-based classification process. Indeed, Jüttner *et al.* (1996) demonstrated that such an approach could predict basic aspects of both the individual and group performance in learning and generalizing the patterns in the present experiment. Their study, however, is based on the decomposition of patterns into parts defined in the image domain, thus avoiding assumptions about stimulus encoding in terms of known filter properties of the visual system. Hence the present study and that by Jüttner *et al.* (1996) are complementary in the sense that their emphasis is more on aspects of early vision and cognition, respectively. Which aspects of these two concepts have to be combined to provide a more realistic scheme of visual classification and generalization is a question for future research.

5. DISCUSSION

We examined the extent to which human observers are able to generalize classification performance learned with compound Gabor signals that are pairwise mirror-symmetric to each other. The test signals used were gray-scale thresholded versions of the learning signals. Subjects were trained to criterion in classifying the learning signals, and then tested with the learning sets one at a time. We found that generalization was possible to a remarkably large extent. This implied that the ability to recognize symmetry relations was generalized as well. Without further learning, our subjects showed also excellent classification capabilities for bipolar blob-only versions of the learning signals (i.e. patterns containing both light-*and*-dark blob information), as well as for test patterns with light blobs *or* dark blobs only.

These psychophysical findings suggest that the learning signals are internally represented in a way which is similar to the representation of the distorted test signals. Briefly, the signals seem to be internally represented in terms of a 'blob code', as has been conjectured by Caelli *et al.* (1986) in the context of texture discrimination. Furthermore, the observed equivalence for generalization performance of dark-only and light-only blob codes, taken together with the introspective reports of our subjects, suggests that labelled or even separate representations exist for the two types of contrast polarity.

Assuming separate encoding processes for dark and light pattern components is consistent with neurophysiological results concerning the existence of two functionally separate neural subsystems. Indeed, a 'darkness' (OFF) system and a 'lightness' (ON) system have been discussed by neurophysiologists in an attempt to find correlates for the perception of darkness and brightness (for reviews see Fiorentini *et al.*, 1990; Schiller, 1992). These channels have been related previously to the visual tasks of detecting and discriminating light increments and decrements. Our results suggest that ON- and OFF-type subsystems may be functionally separate even at the level of spatial visual representations underlying pattern recognition, i.e. cognitive functions.

As a first step in our computational analysis we investigated whether our data can be described within the standard framework for technical pattern recognition. Since distances in feature space are the essential values required for the application of minimum-distance classifier schemes, we started with the evaluation of distance measures with respect to the image pixel configurations and with respect to different visual representations (Laplacian and CEV operators including ON/OFF separation and gradient normalization). The simulation results showed that all representations can basically capture the generalization of the light-only and dark-only blob patterns but fail to predict human performance for the extreme blob operations. We concluded that both the assumptions about representation and classifier are insufficient for adequately characterizing human strategies of pattern classification and generalization.

This situation led us to reconsider the structural properties of possible representations with respect to basic neurophysiological and computational aspects as well as the introspective reports of our subjects. We found that, mainly due to the consistent separation of ON and OFF signals, the MINCEV/MAXCEV representation best matches the notion of an ideal blob-code which accommodates the proposed requirement of *pattern similarity* assumed to underlie the generalization of form. Thus it is important to note that the suggested representation is introduced not in an ad-hoc fashion for this special context but in accordance with the view of several authors regarding general aspects of human pattern recognition (Marr and Hildreth, 1980; Watt and Morgan, 1985; Bischof and Caelli, 1988; Zetzsche *et al.*, 1993).

However, the ability to establish *pattern similarities* must be complemented by *pattern discrimination*, since the subjects can distinguish between the learning and the test patterns. This complementarity is already explicit in Pavlovian conditioning. On the one hand, a subject will emit a conditioned response to the conditioned stimulus but, to some extent, to similar stimuli as well. On the other hand, a subject can also be trained to discriminate stimuli by pairing the conditioned and the unconditioned stimuli, while similar stimuli are presented unpaired (see Brown and Herrnstein, 1975, Chapter 3). Similarly, Marr and Nishihara (1978, p. 272) list 'stability and sensitivity' among the criteria for judging the effectiveness of shape recognition systems: 'To be useful for recognition, the degree of similarity between two shapes must be reflected in their descriptions but at the same time even subtle differences must be expressible. These opposing conditions can be satisfied only if it is possible to decouple stable information that captures more general and less varying properties of shape, from information that is sensitive to the finer distinctions between shapes.' A possible neurophysiological basis of this functional complementarity has recently be found by Fujita *et al.* (1992) in the columnar structure of the infero-temporal cortex (IT) of the macaque. In this context, we suggest that the structure of maxima across scale captures the 'stable information' whereas the full representation allows for 'finer distinctions' between learning and test patterns.

Further, we have argued that the locations of maxima in scale space are an adequate basis for more abstract codes, which would involve the extraction of more explicit structural properties of such representations. Whether the resulting representations can then already be handled by relatively simple classifier schemes, or whether they require a more complicated subsequent elaboration of a propositional scheme operated upon by a rule-based classification process, as suggested by Jüttner *et al.* (1996), remains a topic for future research.

Acknowledgement

The study has been supported by a grant from the Deutsche Forschungsgemeinschaft (DFG-Re 337/7) to I.R. and C.Z.

APPENDIX

Laplacian and curvature representations of blob-patterns

Such representations are shown in Fig. 12 (Laplacian pyramid on the left and 2D-pyramid on the right).

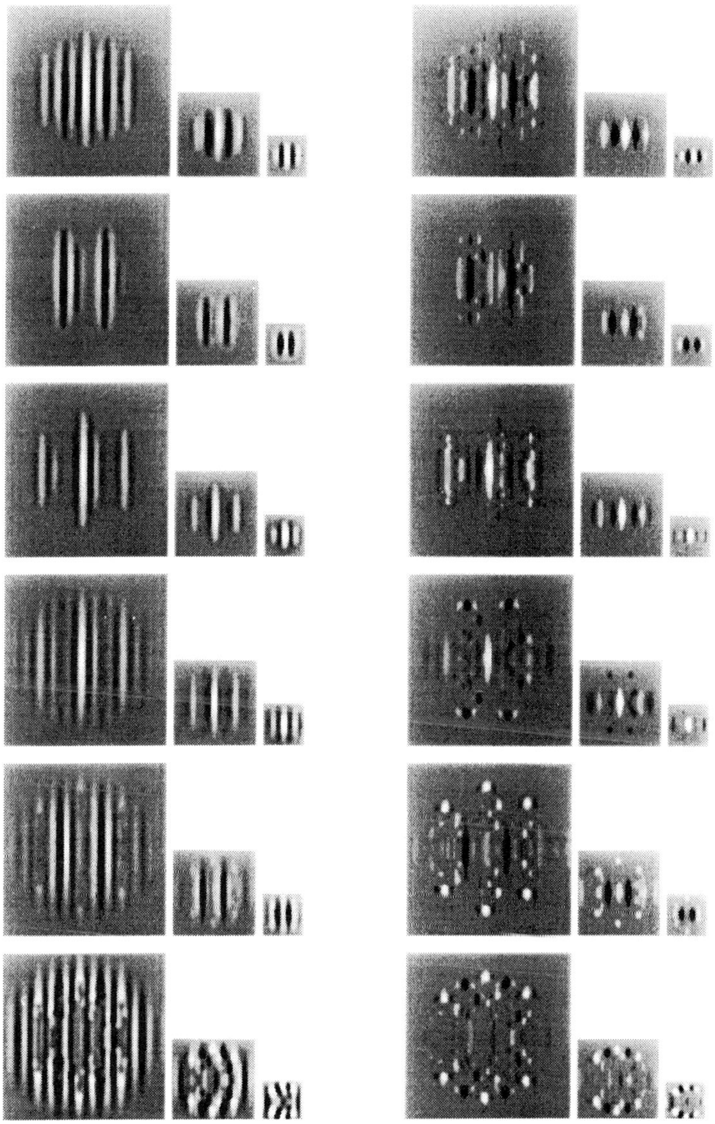

Figure 12. Laplacian-pyramid (left) and 2D-pyramid (clipped-eigenvalues operator as defined by Eqn (4); right) representations on three different scales for 6 distorted versions of the first original pattern (these distortions are shown in the mid column of Fig. 4). The representations include a threshold of 5% of the maximum and are the basis for the computation of the predicted classification results shown in Table 1 and the local maxima shown in Fig. 9.

The clipped-eigenvalues operator (CEV)

The dot-responsive cell model as introduced by Zetzsche and Barth (1990a, 1990b) is derived from an interpretation of the image intensity function as a surface to which predicates from differential geometry can be applied. In order to extent these purely geometrical predicates towards the variability found in visual neurons we have developed a nonlinear filter framework for the modeling of a broad class of 2D selective operators (Zetzsche and Barth 1990a, 1990b). We start from the determinant of the Hessian, which can be written as

$$G = \left[\frac{1}{4}(l_{xx} + l_{yy})^2\right] - \left[\frac{1}{4}(l_{xx} - l_{yy})^2 - l_{xy}^2\right] = (\nabla^2 l)^2 - \varepsilon^2, \qquad (2)$$

where $l(x, y)$ denotes image intensity at position (x, y), $\nabla^2 l$ is the Laplacian of l, and ε is the 'eccentricity'. The eigenvalues of the Hessian are then given as

$$d_{1,2} = \nabla^2 l \pm \varepsilon. \qquad (3)$$

From these we can derive the clipped-eigenvalues operator (CEV), which is defined as

$$\mathrm{CEV} = d_2^+ - d_1^-, \qquad (4)$$

where the clipping operation $d^\pm$ takes the positive and negative values respectively (one-way rectification). Note that CEV equals the Laplacian for isotropic patches where $\varepsilon = 0$. Rewriting the eccentricity ε in terms of second-order derivates (even symmetric receptive fields) yields

$$\varepsilon = \frac{1}{4}(l_{xx} - l_{yy})^2 + \frac{1}{4}(l_{uu} - l_{vv})^2, \qquad (5)$$

where the coordinates (u, v) are rotated relative to (x, y) by 45 deg. Thus, the eccentricity can be seen as resulting from four oriented filters l_{xx}, l_{yy}, l_{uu}, l_{vv} with vertical, horizontal, and oblique orientations—see Fig. 10. The operator CEV has been described and generalized in Zetzsche and Barth (1990b) and Barth *et al.* (1993a), and the relationships to other curvature estimates have been discussed. The multi-scale version of CEV is obtained by blurring the intensity $l(x, y)$ with a Gaussian kernel. Equivalently, one could use blurred convolution kernels for computing the derivatives.

The gain controlled operators (CEVG and LAPG)

They are defined as

$$\mathrm{CEVG} = \frac{\mathrm{CEV}}{l_x^2 + l_y^2 + c}$$

and

$$\mathrm{LAPG} = \frac{\nabla l}{l_x^2 + l_y^2 + c}, \qquad (6)$$

with the value of the constant c chosen as 10% of the maximum of $l_x^2 + l_y^2$.

REFERENCES

Barth, E., Caelli, T. and Zetzsche, C. (1993a). Image encoding, labelling and reconstruction from differential geometry. *CVGIP: Graphical Models and Image Processing* **55**, 428–446.

Barth, E., Ferraro, M. and Zetzsche, C. (1993b). On the structure of 2D-curvature scale space. *Perception* **22** (Suppl.), 21.

Barth, E., Zetzsche, C., Ferraro, M. and Rentschler, I. (1993c). Fractal properties from 2D-curvature on multiple scales. In: *Geometric Methods in Computer Vision II*. B. Vemuri (Ed.). Proceedings SPIE **2031**, pp. 87–99.

Bennett, P. J. and Banks, M. S. (1987). Sensitivity loss in odd-symmetric mechanisms and phase anomalies in peripheral vision. *Nature* **326**, 873–876.

Bennett, P. J. and Banks, M. S. (1991). The effects of contrast, spatial scale, and orientation on foveal and peripheral phase discrimination. *Vision Res.* **31**, 1759–1786.

Bischof, W. F. and Caelli, T. (1988). Parsing scale-space and spatial stability analysis. *CVGIP* **42**, 192–205.

Brown, R. and Herrnstein, R. J. (1975). *Psychology*. Methuen, London.

Burt, P. J. and Adelson, E. H. (1983). The Laplacian pyramid as a compact image code. *IEEE Trans.* **COM-31**, 532–540.

Caelli, T. M. and Rentschler, I. (1986). Cross-correlation model for pattern acuity. *J. Opt. Soc. Am.* **A3**, 1948–1956.

Caelli, T. M., Hübner, M. and Rentschler, I. (1986). On the discrimination of micropatterns and textures. *Human Neurobiol.* **5**, 129–136.

Caelli, T., Rentschler, I. and Scheidler, W. (1987). Visual pattern recognition in humans. I. Evidence for the existence of adaptive filters. *Biol. Cybernet.* **57**, 233–240.

Duda, R. O. and Hart, P. E. (1973). *Pattern Classification and Scene Analysis*. Wiley, New York.

Fiorentini, A., Baumgartner, G., Magnussen, S., Schiller, P. H. and Thomas, J. P. (1990). The perception of brightness and darkness. Relations to neuronal receptive fields. In: *Visual Perception. The Neurophysiological Foundations*, Ch. 7. L. Spillmann and J. S. Werner (Eds). Academic Press, San Diego.

Fujita, I., Tanaka, K. and Cheng, M. I. K. (1992). Colums for visual features of objects in monkey inferotemporal cortex. *Nature* **360**, 343–346.

Gaskill, J. D. (1978). *Linear Systems, Fourier Transforms and Optics*. Wiley, New York.

Gauss, C. F. (1809). *Theoria Motus*. English translation (1963): *Theory of the Motion of the Heavenly Bodies about the Sun in the Conic Sections*. Dover, New York.

Gelb, A. (1974). *Applied Optimal Estimation*. MIT Press, Cambridge, MA.

Gibson, E. (1969). *Principles of Perceptual Learning and Development*. Appleton, New York.

Gross, C. G. and Bornstein, M. H. (1978). Left and right in science and art. *Leonardo* **11**, 29–38.

Jüttner, M. and Rentschler, I. (1996) Reduced perceptual dimensionality in extrafoveal vision. *Vision Res.* **36**, 1007–1022.

Jüttner, M., Caelli, T. and Rentschler, I. (1996). Recognition-by-parts: A computational approach to human learning and generalization of shapes. *Biol. Cybernet.* (In press).

Koenderink, J. J. (1984). The structure of images. *Biol. Cybernet.* **50**, 363–370.

Mach, E. (1922). *Die Analyse der Empfindungen. 9.* Aufl. G. Fischer, Jena.

Marr, D. and Nishihara, H. K. (1978). Representation and recognition of the spatial organisation of three-dimensional shapes. *Proc. R. Soc. London* **B200**, 269–294.

Marr, D. and Hildreth, E. (1980). Theory of edge detection. *Proc. R. Soc. London* **B207**, 187–217.

Rentschler, I. and Treutwein, B. (1985). Loss of spatial phase relationships in extrafoveal vision. *Nature* **313**, 308–310.

Rentschler, I., Hübner, M. and Caelli, T. (1988). On the discrimination of compound Gabor signals and textures. *Vision Res.* **28**, 279–291.

Rentschler, I. and Caelli, T. (1990). Visual representations in the brain: Inference from psychophysical research. In: *Synergetics of Cognition*. J. Haken and M. Stadler (Eds). Springer, Berlin, pp. 233–248.

Rentschler, I., Jüttner, M. and Caelli, T. (1994). Probabilistic analysis of human supervised learning and classification. *Vision Res.* **34**, 669–687.

Rosenfeld, A. and Thurston, M. (1971). Edge and curve detection for visual scene analysis. *IEEE Trans. Comput.* **C20**, 562–569.

Saarinen, J. (1987). Perception of positional relationships between line segments in eccentric vision. *Perception* **16**, 583–591.

Saarinen, J. (1988). The effect of exposure duration on the analysis of spatial structure in eccentric vision. *Spatial Vision* **3**, 1–7.

Schiller, P. H. (1992). The ON and OFF channels of the visual system. *Trends Neurosci.* **15**, 86–92.

Shapley, R., Caelli, T., Grossberg, S., Morgan, M. and Rentschler, I. (1990). Computational theories of visual perception. In: *Visual Perception: The Neurophysiological Foundations*. L. Spillmann and J. S. Werner (Eds). Academic Press, San Diego, pp. 231–272.

Shepard, R. N. (1962a). The analysis of proximities: multidimensional scaling with a unknown distance function. i. *Psychometrika* **27**, 125–140.

Shepard, R. N. (1962b). The analysis of proximities: multidimensional scaling with a unknown distance function. ii. *Psychometrika* **27**, 219–246.

Treisman, A. and Gelade, G. (1980). A feature integration theory of attention. *Cognit. Psychol.* **12**, 97–136.

Watt, R. J. and Morgan, M. J. (1985). A theory of the primitive spatial code in human vision. *Vision Res.* **25**, 1661–1674.

Zetzsche, C. and Barth, E. (1990a). Fundamental limits of linear filters in the visual processing of two-dimensional signals. *Vision Res.* **30**, 1111–1117.

Zetzsche, C. and Barth, E. (1990b). Image surface predicates and the neural encoding of two-dimensional signal variation. In: *Human Vision and Electronic Imaging: Models, Methods, and Applications*. B. Rogowitz (Ed.). Proceedings SPIE **1249**, pp. 160–177.

Zetzsche, C., Barth, E. and Wegmann, B. (1993). The importance of intrinsically two-dimensional image features in biological vision and picture coding. In: *Digital Images and Human Vision*. A. B. Watson (Ed.). MIT Press, Cambridge, MA, pp. 107–138.

Human Symmetry Perception, pp. 265–287
C. W. Tyler (Ed.)

A model for global symmetry detection in dense images*

FRANÇOYS LABONTÉ,[1] YERUCHAM SHAPIRA,[1] PAUL COHEN[1]
and JOCELYN FAUBERT[1,2]

[1] *Perception and Robotics Research Group, Department of Electrical and Computer Engineering, Pavillon André-Aisenstadt, Ecole Polytechnique de Montréal, PO Box 6079, Station 'Centre-Ville', Montréal, Québec, Canada, H3C 3A7*

[2] *Ecole d'Optométrie, Université de Montréal, PO Box 6128, Station 'Centre-Ville', Montréal, Québec, Canada, H3C 3J7*

Abstract—In this paper, a model is proposed for bilateral symmetry detection in images consisting of dense arrangements of local features. The model is elaborated on the basis of a psychophysical experiment showing that grouping precedes and facilitates symmetry detection. The proposed computational model consists of three stages: a grouping stage, a symmetry-detection stage, and a symmetry-subsumption stage. Reliance upon a preliminary grouping stage enables a significant reduction of the computational load for detecting symmetry. An implementation of the model is described, and results are presented, showing a good agreement of the model performance with human symmetry perception.

1. INTRODUCTION

In this paper, we propose a model for global bilateral symmetry detection in dense images. Dense arrangements of local features, such as dots or oriented segments, are encountered in various situations such as images of textured scenes, brightness gradient fields, stereo disparity fields, motion flow fields, and Glass patterns.

Different classes of symmetry may be present in an image: bilateral (also known as mirror symmetry), rotational, or repetitive. Multiple symmetries, i.e. identities under more than one transformation, may be also present. Unlike its mathematical equivalent, visual symmetry may be characterized by a variable degree of exactness. It may be approximate, incomplete (missing elements), and apply only to restricted regions of the image.

Symmetry relations can be local or global. Local symmetry refers to small neighborhoods of the image, and can usually be detected through a limited number of comparisons between nearby features. Global symmetry concerns regions of larger

*Part of the work in this paper was presented at the IEEE 4th International Conference on Computer Vision, Berlin 1993.

extents (although usually not affecting the entire image) and requires an examination of far-away image elements, followed by a comparison stage that involves the establishment of symmetry relations between compatible elements.

Since introduced by Mach (1906/1959), symmetry detection in images has drawn considerable attention in terms of both experimental studies and computational models. This interest probably comes from the fact that symmetry could be helpful in the accomplishment of a large number of visual tasks.

For instance, symmetry might be used for shape representation, either by providing a means for efficient encoding of pictorial information (Attneave, 1954), or by facilitating the establishment of object-centered representations involving perceptual reference frames (Palmer, 1983). Several representations based on local symmetry have been proposed in computer vision, such as the Symmetrical Axis Transform (Blum and Nagel, 1978), the Smoothed Local Symmetry Transform (Brady, 1983), and the Local Rotational Symmetry Transform (Fleck, 1986). Recently, Subirana-Vilanova (1990) proposed a method to extract curved axes of symmetry from an image, for the efficient description of object shapes.

Symmetry might also facilitate the inference of the 3D structure of symmetrical objects from their image projections. Kanade and Kender (1983) suggest a method that uses the amount of symmetry skewness of projected object contours. Nalwa (1989) uses local bilateral symmetry in line drawings, and its invariance under changes of viewpoint under orthographic projection, to infer the presence of surfaces of revolution. Ulupinar and Nevatia (1988) analyze the constraints of two kinds of symmetry, parallel and bilateral, under orthographic projections, to characterize zero-Gaussian curvature surfaces.

Symmetry has been also used for several other tasks, such as detection of interest points in an image (Reisfeld *et al.*, 1990), and efficient image encoding (Kumar *et al.*, 1983). It was also found to be useful in other cognitive tasks, such as evaluation of pattern interest, complexity and pleasingness (Day, 1968), numerosity judgments in patterns (Howe and Jung, 1987), and memory encoding (Attneave, 1955). Recently, Leyton (1992) proposed a general theory of human perception according to which symmetry plays a central role in the cognition of shape. Through a series of principles relating such notions as symmetry, curvature and morphogenesis, he showed that symmetry constitutes a fundamental element in shape-history recovery, and thus in shape memorization.

Symmetry has been also considered by many to be a fundamental grouping property of perceptual organization, as suggested by the Gestalt psychologists. Elements sharing symmetry relations tend to aggregate and be perceived as figure rather than ground. However, the domination of symmetry by other grouping properties, such as convexity and good continuation, demonstrates the weakness of the organizational power of this property. Recent studies (Jenkins, 1983; Pomerantz and Kubovy, 1986; Pashler, 1990) call into question the role of symmetry as a strong grouping property. They suggest rather that symmetry detection could operate among features obtained from other grouping properties, thus implying a precedence of grouping over symmetry perception.

In this paper, we propose to elucidate this question by studying the strategies used by the visual system to detect global bilateral symmetry in dense images. As Section 2.1 reveals, two separate mechanisms may exist for symmetry detection, one for simple shapes and another one for dense patterns (Julesz, 1971); here we concentrate on the latter. The problem of detecting global symmetry in dense images has not received much attention, and some interesting issues have not been studied; in particular the strategy followed by the visual system to integrate non-central, non-contiguous symmetry information in dense images. In Section 2.2, we present a psychophysical experiment that we conducted to investigate this issue. Results of our experiment, in agreement with recent studies (Jenkins, 1983; Pomerantz and Kubovy, 1986; Pashler, 1990), show that grouping facilitates and actually precedes symmetry detection.

We also propose a computational model for global symmetry detection in dense images. This model, elaborated on the basis of the psychophysical experiment reported in Section 2.2, is presented in detail in Section 3. The proposed model consists of three stages: a grouping stage, in which neighboring compatible elements are aggregated, a comparison stage, in which clusters formed in the grouping stage are systematically compared to detect local axes of symmetry, and a subsumption stage, in which local axes are merged into global ones. The implementation details of the model and results are reported in Section 4, showing that the model performance agrees in principle with human symmetry perception. The paper ends with a discussion of further issues that might be investigated.

2. HUMAN SYMMETRY PERCEPTION IN DENSE IMAGES

2.1. Previous work

Some aspects of human symmetry perception in the case of complex high-frequency patterns have been studied by Julesz (1966, 1971; see also Julesz and Chang, 1979), and in the case of moderately complex random-dot patterns (100 dots) by Barlow and Reeves (1979). Barlow and Reeves' results indicate that approximate symmetry can still convey the visual impression of symmetry, provided the deviation from exact symmetry is not too large. They suggest that the mechanism of symmetry detection is not a highly accurate one, and considers dots positioned at ± 6 arcmin of visual angle from the exact symmetrical position as contributing to the symmetry effect. The authors also studied the effects of introducing violations (i.e. suppression) of symmetry at different locations in the pattern, and of changing the location of the axis of symmetry. They found that violations are more easily detected near the axis of symmetry, and that symmetry perception becomes less sensitive when the location of the axis is not central. Jenkins (1982) also discusses the importance of the central region near the axis of symmetry. According to his conclusions, for completely symmetrical images, only the symmetry information located within a strip of 1.1 deg of visual angle centered on the axis of symmetry is exploited by the human visual system. Outside of this strip, symmetry information does not contribute to symmetry detection when the symmetrical portions are contiguous. All the preceding results were obtained in experimental conditions involving brief exposure times (of the order of 100 ms). Other findings reported by Julesz (1971) indicate that bilateral symmetry is detected more quickly than rotational symmetry, and that vertical symmetry is detected faster than horizontal symmetry. Also, multiple symmetry (with respect to more than one axis at the same time) is detected faster than simple symmetry (one axis only). Zucker (1986) suggests the presence of two separate mechanisms in low-level vision, one for contour detection, the other for texture analysis, thus implying that symmetry detection in textures would be accomplished by a different process than the one used in contour images. Julesz (1971) also suggests the existence of two different mechanisms for symmetry perception.

Given that global symmetry relations often involve non-adjacent symmetrical regions, it would be of particular interest to determine how non-central, non-contiguous symmetry information is detected by the visual system. In such cases, the preceding results about the importance of the region near the axis of symmetry probably do not apply, since special grouping configurations along the axis of symmetry are not created with non-contiguous symmetry regions. Another important issue is the processing time required for symmetry detection. It might be that, for certain conditions, symmetry detection requires longer periods of time (of the order of 1 s). Even if the aforementioned experiments show that, for particular conditions, symmetry can be detected with brief exposure times, detection performance might be different for more typical vision situations, involving longer processing times.

2.2. Experiment with human symmetry detection

Bilateral symmetry can be defined as the reflection of a region about a given axis. In symmetry-detection studies with dense patterns (e.g. random dots), symmetrical images generally are created by reflecting one half of the image about a central vertical axis. Patterns generated according to this method usually contain salient grouping configurations along the central axis. An example of a figure containing such configurations can be seen in Fig. 1. Results of an experiment on symmetry detection using this kind of image indicate that subjects probably were using the presence of central coherent features as a clue to discriminate between symmetrical and asymmetrical patterns (Jenkins, 1983). This observation suggests that a preliminary cluster formation (grouping) stage might precede, or at least facilitate, symmetry detection.

In natural environments, symmetrical regions are not always contiguous, and therefore do not always contain central grouping configurations to facilitate symmetry detection. On the other hand, non-contiguous regions of symmetry often can be segregated from the background, and thus contain some form of grouping. It is reasonable to suppose that texture-related grouping could be used by the visual system to facilitate the symmetry-detection process for non-contiguous regions. In this section, we report the results of a psychophysical experiment that we conducted in order to investigate the effect of grouping on symmetry detection. The goal of the study was to establish the strategy followed by the visual system in order to elicit non-central, non-contiguous symmetry information from dense images, and to discover how the system deals with the computational load imposed by global symmetry detection.

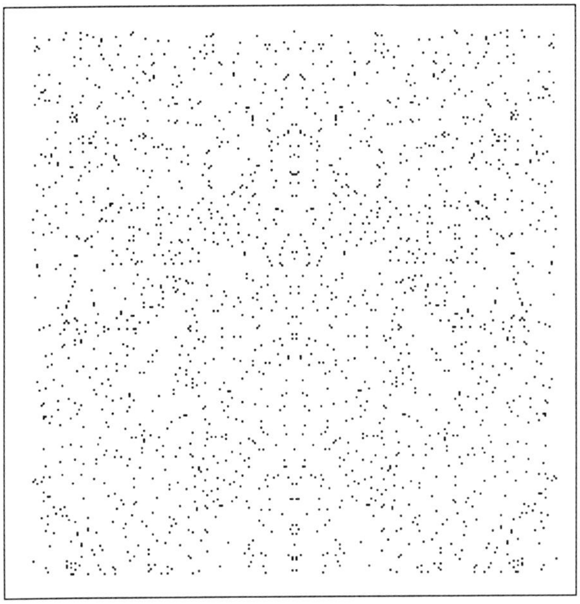

Figure 1. Dense random-dot image with grouping configurations along the central vertical axis of symmetry.

The experiment compared symmetry-detection performance when a significant amount of grouping was possible in the regions of symmetry by means of texture segregation (grouping condition), and when it was not (no-grouping condition), thus reducing the grouping possibilities. A reasonable amount of time (of the order of 1 s) was given to the subject in order to obtain results applicable to typical vision situations, as opposed to more constrained situations with brief exposure times (of the order of 100 ms). The stimuli consisted of dense images of oriented segments for which grouping was performed on the basis of the similarity of orientation of neighboring segments. Texture segregation, and thus grouping, was produced when segments inside a symmetry region had a different orientation from the background segments. Therefore, for the no-grouping condition, background segments and segments inside regions of symmetry had the same orientation, while for the grouping condition they had different orientations. The segment orientations in corresponding regions were always symmetrical and the subjects had to verify whether or not the position of the segments inside the regions were symmetrical.

The detection task was not simply a case of orientation discrimination of the texture segregation components; segment positions inside the regions of symmetry had to be compared. Detection performance was measured at three different distances from the central axis of symmetry. In the first case, the regions of symmetry were adjacent. In this situation, in addition to the controlled grouping/no-grouping factor, grouping configurations were created along the central axis of symmetry for the two conditions. Therefore, a significant difference in detection performance between the two conditions was not expected. In the second case, the regions of symmetry were positioned in the middle of each half image. In this situation, a significant difference in detection performance was expected since only the controlled grouping/no-grouping factor was involved. For the third case, the regions of symmetry were located at the outer extremities of the image. In this situation, additional grouping effects might have been caused by segregation at the texture/non-texture borders of the image.

2.2.1. Method

Stimuli. Computer-generated images were presented, on the center of a VGA monitor (640 by 480 pixels) measuring 42 cm by 31 cm, to subjects at a viewing distance of 57 cm from a fixation point placed in the middle of the screen. The mean luminance of the screen was 50 cd m^{-2}. An 80386 IBM computer running at 33 MHz with an 80387 math coprocessor was used to generate the images in real time. A chin-rest facilitated stabilization of head position and an I-SCAN system with infrared camera was used to monitor eye movements. Subjects' responses were recorded with the mouse of the computer: the left button corresponded to a *yes* response and the right button to a *no* response.

Images were made of 1000 white oriented segments on a black background. Image size was 26.4 by 26.4 deg of visual angle and the two regions inside each image were 6.6 by 13.2 deg. The three values for the distance between the regions were 0, 6.6, and 13.2 deg. Each stimulus could represent one of the two following types of images with one of the two types of grouping. Figure 2 shows examples of the stimuli that were used.

- *Symmetrical image*: the position and orientation of segments inside the regions were symmetrical about the central vertical axis of symmetry, and background segments were randomly located outside the regions.
- *Asymmetrical image*: the orientation but not position of segments inside the regions were symmetrical about the axis, and background segments were randomly located outside the regions.
- *Grouping condition*: the background segments were vertical and segments inside the regions were oriented at 45 or 135 deg, depending on which region of symmetry they belonged to.
- *Non-grouping condition*: the background segments and segments inside a region were both vertical.

Control. To control for possible orientation effects, another grouping condition, where background segments were either oriented at 45 or 135 deg, and segments inside the symmetrical region were vertical, was also used for the distance of 6.6 deg. Figure 3 shows an example of a stimulus that was used to control for orientation effects. It is not possible to perform the same type of control for the non-grouping condition since inclined segments (45 or 135 deg) cannot be used at the same time in the background and in the regions of symmetry without creating texture segregation.

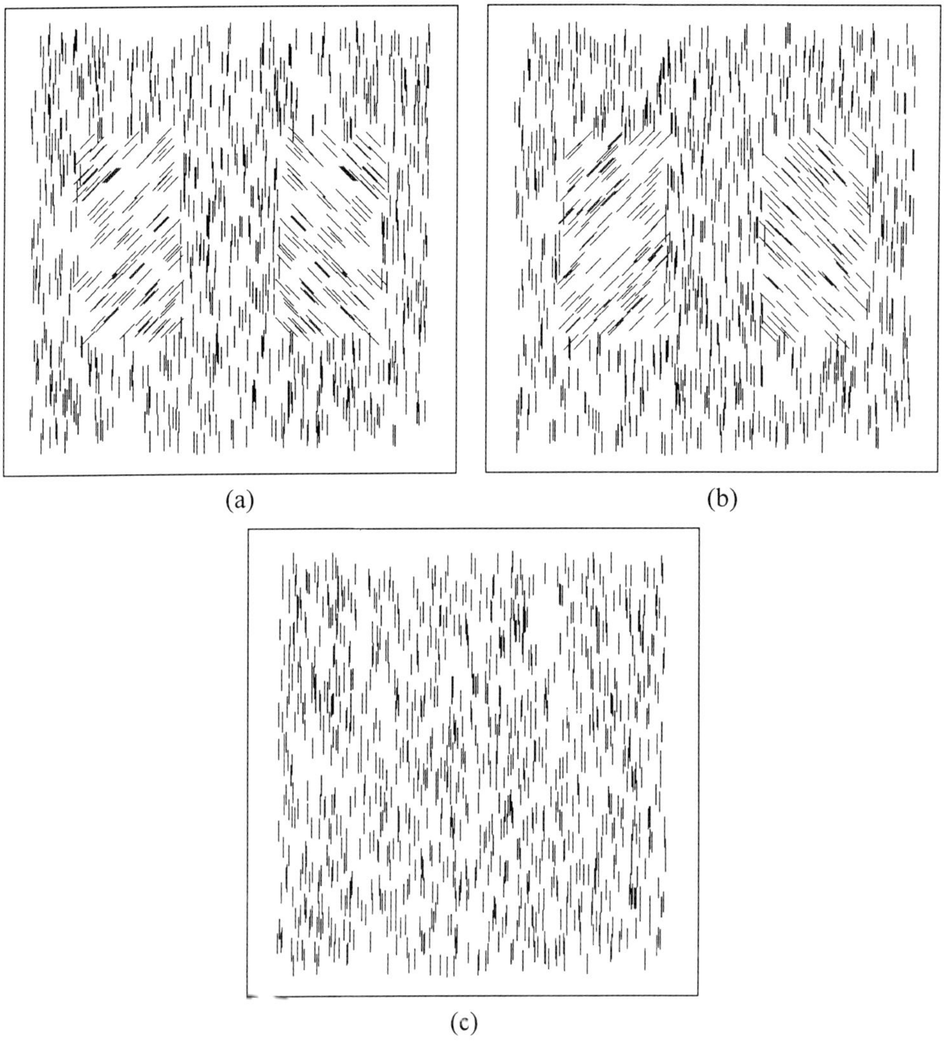

Figure 2. Examples of stimuli used: (a) symmetrical with grouping, (b) asymmetrical with grouping, and (c) symmetrical without grouping. The stimuli are shown with a distance of 6.6 deg of visual angle between the regions of symmetry.

Procedure. Five right-handed subjects with normal or corrected-to-normal vision were tested. They all agreed to perform the experiment without any remuneration. Subjects were students or staff members with the Perception and Robotics Laboratory at Ecole Polytechnique de Montréal or with Ecole d'Optométrie at Université de Montréal.

The experiment consisted in testing each subject with 7 independent blocks of 50 images: 3 for the grouping condition, with distances between the regions of 0, 6.6, and 13.2 deg respectively, 3 for the no-grouping condition, with the same distances as in the grouping condition, and 1 to control for orientation effects, with a distance of 6.6 deg (Fig. 3). In each block, the probabilities of presentation for symmetrical

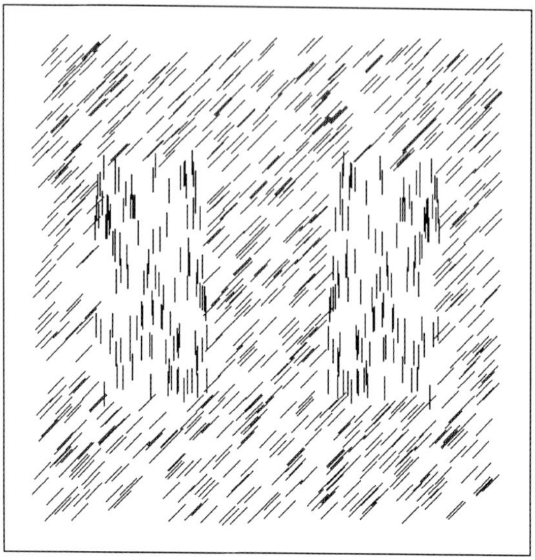

Figure 3. Example of stimuli used to control for orientation effects: symmetrical with grouping at a distance of 6.6 deg of visual angle.

and asymmetrical images were equal. The images were randomly generated but, for each block, all subjects were shown the same set of images in the same order. However, the order of presentation of blocks varied between subjects. This technique of presentation, for which the distance between the regions of symmetry is constant in each block, permits the elimination of position uncertainty since the subjects knew the exact position of the target regions at all times.

Subjects were asked to detect the presence or absence of symmetry in the images that were presented. They were told that the accuracy of their responses was important and that reaction times were not measured. They had to look at the fixation point in the middle of the screen and were told not to move their eyes. The presentation time of each image was 1.5 s. A new image was not shown until a response was recorded. The experimenter monitored the eye movements of the subjects and told them whenever they were moving their eyes. Before each block, example images were shown. During this presentation, the experimenter told the subjects whether or not the images were symmetrical. Subjects then performed practice trials during which they had to assess if the 10 presented images were symmetrical or not. Finally, for each block, subjects were tested with 50 images. It took, on average, 1 h and 10 min to perform the experiment (10 min per block).

Data analysis. The individual data at each distance, in terms of symmetry-detection rate as a function of false-positive response, are shown in Fig. 4. Figure 5 presents the averaged symmetry-detection rate of the 5 subjects as a function of grouping condition and distance. For the grouping condition, performance as a function of the distance between the regions of symmetry slightly decreases with increasing distance, while for the no-grouping condition, performance is V-shaped: high at a distance of 0 deg,

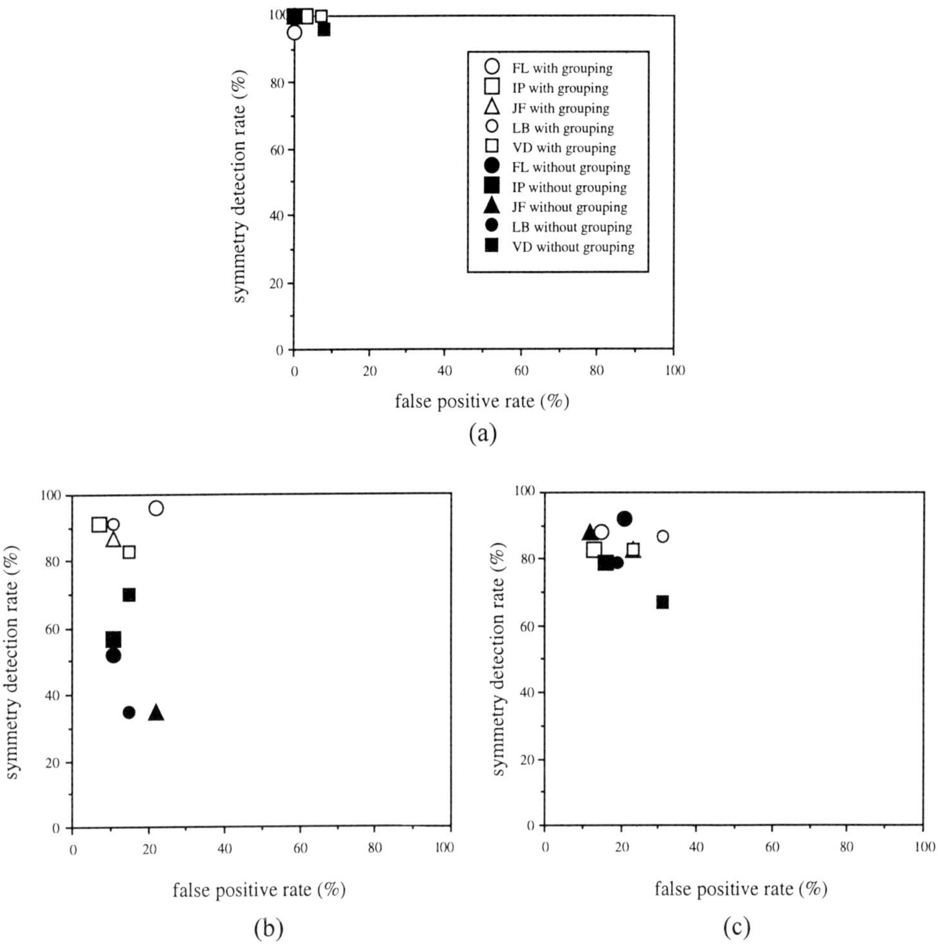

Figure 4. Comparison of symmetry detection rate between the grouping and the non-grouping conditions as a function of false positive detection rate at distances of 0, 6.6, and 13.2 deg of visual angle (a, b and c respectively).

significantly lower at 6.6 deg, and high again at 13.2 deg, consistent with the shape but not the range of Barlow and Reeves, (1979) data. Results of a two-factor ANOVA indicate that the effect of grouping is significant ($F(1, 24) = 26.35$, $p < 0.001$), the effect of distance is significant ($F(2, 24) = 36.40$, $p < 0.001$) and the interaction grouping-distance is significant ($F(2, 24) = 20.37$, $p < 0.001$). Results of a post-hoc Tukey test also show that, even if there is a monotonic decrease of symmetry detectability as a function of the distance in the grouping condition, this effect is not significant.

At a distance of 0 deg, performance is almost perfect for all subjects, regardless of the condition. At a distance of 6.6 deg, performance for the grouping condition is much better than for the non-grouping condition. This difference in detection cannot be attributed to a criterion shift because the false-positive rate remained fairly constant.

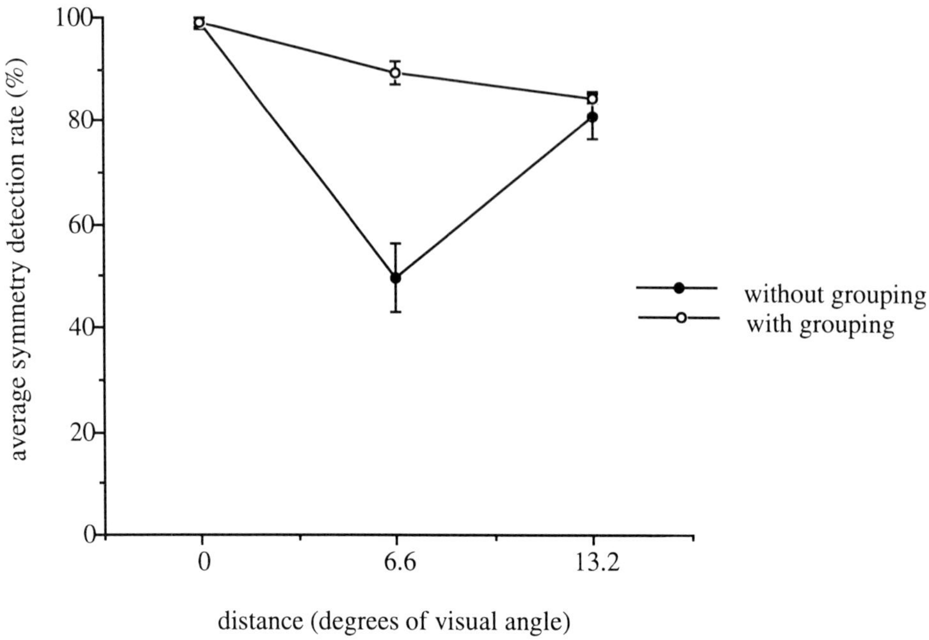

Figure 5. Averaged symmetry detection rate for the 5 subjects as a function of grouping conditions and of distance between regions of symmetry. Error bars represent ±1 SE.

At a distance of 13.2 deg, no major difference between the two conditions can be observed. Post-hoc comparisons between the conditions by a Tukey test indicate significant differences between the no-grouping condition at a distance of 6.6 deg and the five remaining conditions. A paired t-test comparison between the two grouping conditions at a distance of 6.6 deg (the control for orientation effects) shows no significant effect ($t(4) = 0.22$); thus orientation is not a factor.

2.2.2. Discussion. Symmetry-detection performance in dense images was measured as a function of the distance and of grouping of oriented segments. As expected, a significant advantage in detection performance for the grouping condition was observed at a distance of 6.6 deg, since only the controlled grouping factor was involved in the detection task at this distance. At a distance of 0 deg, the results show no significant difference in detection performance between the grouping and the non-grouping condition. As discussed at the beginning of Section 2.2, it has been suggested that central grouping configurations help symmetry detection. We suggest that, at a distance of 0 deg, regardless of the controlled grouping conditions, additional grouping configurations were already present along the central axis of symmetry. Thus, the facilitation effect of central grouping configurations was equally present for both conditions. At a distance of 13.2 deg, no significant difference in detection performance was observed. Barlow and Reeves (1979) mention that, in addition to the vivid impression created by the paired dots along the central axis of symmetry, the outline or the shape of the pattern is also important for symmetry detection. However, this explanation does not apply with the kind of stimuli that we used since the global shape of the image did not vary. We propose rather that grouping effects were present at the outer borders of the image. These groupings, due to segregation at the texture/non-texture border, helped the detection of symmetry at the edges.

In summary, we propose that at distances of 0 and 13.2 deg, groupings were already present, either along the central axis of symmetry or along the image borders, regardless of the grouping condition. Therefore the introduction of another grouping factor was superfluous. On the other hand, at a distance of 6.6 deg, no texture-related grouping effect based on orientation was available. Therefore, the introduction of a grouping factor improved symmetry-detection performance. We propose, based on our results, that grouping plays a significant role for symmetry detection not only near the central axis of symmetry, but also for non-adjacent regions of symmetry. Alternative explanations cannot account for the obtained performance for the following reasons. First, grouping was not an important factor for the localization of the regions of symmetry since, for each block of stimuli images, regions of symmetry were at the same position. Second, orientation effects induced by groupings did not play a significant role in the detection task since subjects had also to verify the segments' positions in the regions of symmetry. Third, low spatial-frequency components, which are known to significantly affect symmetry perception, were not involved since the spatial-frequency content of the stimuli was similar for both conditions. Fourth, small groupings created by the proximity of neighboring segments did not significantly affect detection performance since they were present equally in both conditions. On the other hand, texture segregation, which was facilitated in the grouping condition, improved symmetry-detection performance.

Our findings have important theoretical implications for the elaboration of a model for symmetry detection. They imply that, in addition to the fact that symmetry is a fundamental grouping property, as suggested by Gestalt psychologists, grouping could be used to detect symmetry. If Julesz (1971) is correct that there are two different processes for symmetry detection it could be that, in the case of dense images, general-purpose grouping mechanisms already used for other visual tasks perform the first stage of the symmetry detection process. Symmetry detection would then be a comparison of higher-order features obtained from grouping mechanisms. This hypothesis agrees with recent studies. (1) Owing to the weak organizational power of symmetry, other grouping factors could precede it. These factors would dominate symmetry, and symmetry would be used for grouping only when it is not in competition with them. Symmetry could be useful not at the level of grouping formation, but at a higher level where it would be detected among groupings created on the basis of other factors (Pomerantz and Kubovy, 1986). (2) Grouping mechanisms already used for other visual functions could perform the most important part of the symmetry detection task (Pashler, 1990). (3) Symmetry detection in random-dot images could be accomplished in three stages, including a grouping stage (Jenkins, 1983). (4) Symmetry detection could be accomplished by general grouping mechanisms, and therefore, no special symmetry-detection mechanism needs to be postulated (Wagemans *et al.*, 1991).

In the next section, we propose a computational model for symmetry detection in dense images. Our model agrees in principle with the results presented in this section since symmètry detection is preceded by a grouping stage which permits a significant reduction in the number of comparisons to be performed. Before presenting our model, the possible strategies for symmetry detection in images are reviewed.

3. COMPUTATIONAL MODELS

3.1. The possible strategies for symmetry detection

A number of computational models for symmetry detection have been proposed, either to model the human visual system or to provide efficient means of detecting symmetry in images. Most of the proposed models for the human visual system are not detailed enough to elaborate algorithms that could be simulated and tested on a computer. For instance, Palmer and Hemenway (1978) suggest that symmetry is detected in two stages. The first stage consists of a fast and global analysis to select a potential axis of symmetry by considering all the orientations of the axis at the same time. The second stage consists of an explicit comparison of the two halves of the stimuli to detect symmetry. From a computational point of view, simulation of the first stage of this model is not a trivial task.

Two principal strategies are used by most of the proposed methods for symmetry detection in images. The first one necessitates a systematic comparison of all the potentially symmetrical elements in the image (usually the individual pixels). Bigün (1988) illustrates this approach by presenting a convolution operator which works on the brightness values, and detects various situations of local symmetry in neighborhoods of circular, linear, hyperbolic and parabolic shapes. Reisfeld *et al.* (1990) propose, in a similar strategy, a method for the detection of various interest points in an image. In an attempt to model the human visual system, Royer (1981) proposes that symmetry relations present in a stimulus are represented integrally in a code consisting of classes and subclasses of symmetry. Two channels sensitive to vertical and horizontal orientations are used. By computing many transformations in parallel on the output of these channels, the type of symmetry present in an image can be identified. Palmer (1983) also proposes a model for the human visual system. The visual field is covered by a multitude of local spatial analyzers working in parallel. Invariance under Euclidean similarity transformations applied to the output of these analyzers indicates symmetry relations.

The second strategy is accomplished in two stages. First, more elaborate (and less dense) features than brightness values are extracted. Second, these features are systematically compared, in the search for symmetry relations. Wilson (1991), for example, presents a system to detect local symmetry among contours extracted from an image by comparing key-points on each contour (local extrema of curvature, middle points of straight sections and points of change in curvature sign). Jenkins (1983) proposes, as a plausible model for the human visual system, a three-stage model for symmetry detection in random-dot images. First, orientation uniformity of pairs of points is detected. Second, salient pairs are fused to form a new representation that corresponds to virtual lines joining the two points of each pair. Third, from this new representation, symmetry is detected. The orientation of the axis of symmetry is determined by joining the middle points of the virtual lines. Wagemans *et al.* (1991) suggest that, in addition to the first-order regularities consisting of orientation uniformity and middle point collinearity, second-order relations between pairs of symmetrical elements are used to detect symmetry. These second-order regularities correspond to geometric regularities (symmetric trapezoid and parallelogram connectivities) between virtual lines joining pairs of points. To justify their proposal, they note that the same first-order relations are present for bilateral symmetry and skewed symmetry. Since skewed symmetry is much more difficult to detect than bilateral symmetry, other factors (second-order regularities) must affect symmetry detection.

In the case of global symmetry detection in dense images, the first strategy is inadequate because of the excessive amount of computation required. Since global symmetry involves aggregates of local elements sharing compatible symmetry relations, it seems more appropriate to use, in accordance with the second strategy, a procedure that does not look immediately for instances of symmetry among local elements, but tries first to elicit aggregates of elements that constitute plausible candidates for global symmetry relations. Symmetry is then tested among these aggregates only. The advantages of such a procedure are clear: by making grouping a prerequisite to symmetry detection, a substantial reduction in computational cost is achieved and more global levels of symmetry relations may be achieved.

3.2. The proposed model

In view of the experimental results reported in Section 2.2 illustrating the precedence and the facilitation of grouping over symmetry detection, we propose a strategy for bilateral symmetry detection in dense images that consists of three successive stages:

1. *a grouping stage* in which clusters are formed among local elements presenting a sufficient level of mutual affinity;
2. *a symmetry-detection stage* in which pairs of symmetrical clusters are discovered and their axes of symmetry determined;
3. *a symmetry-subsumption stage* in which an attempt is made to detect even more global symmetries by comparing the various axes of symmetry previously found.

The strategy is computationally efficient and suitable for parallel implementation because only local computations are performed in the first stage, and the subsequent stages necessitate only a small number of comparisons (with respect to a systematic comparison of every element with all other elements) of higher-level features. The first stage involves local computations to evaluate the affinity between each local element and its various neighbors (relying on such properties as proximity, collinearity, and similarity of orientation), and the formation of clusters of elements by relaxation labeling. The second stage involves systematic comparisons between pairs of clusters in order to discover symmetrical pairs and compute the positions and orientations of the corresponding axes of symmetry. Clusters are compared on the basis of their compatibilities in orientation and dimensions. At this stage, the number of comparisons should be limited due to the small number of clusters. The third stage involves a fusion of the axes of symmetry previously determined. Each axis being characterized by its location and orientation, a Hough-transform type of procedure (Duda and Hart, 1973) is used to subsume the pairs of symmetrical clusters into more global symmetrical structures. It is important to observe that the plausibility of our model for the human visual system relies on experimental results for the first stage only. We do not claim that the human visual system necessarily performs the same computations as the ones we propose in our model. However, our model is plausible at a computational level: a first stage implying massive parallel computations performed by a general-purpose mechanism is followed by more specific subsequent stages that perform computations on a smaller number of higher-level features.

Before describing our model in more detail, it is worth mentioning that our model is in agreement with a recent research trend questioning the role of symmetry as a strong property of perceptual organization and suggesting rather that symmetry is detected among more abstract features obtained from general purpose grouping mechanisms (see Sections 2.2 and 3.1). Our model fits into this scheme, since the cluster-detection process can be seen as a general-purpose grouping mechanism useful for other visual tasks, and the cluster comparison process as a tool more specific to symmetry detection, which takes advantage of the small number of higher-level features that were created. It is also important to note that the validity of the model does not depend on the density of the features in the image. In fact, the term *dense images* is used to refer to texture-like images consisting of a large number of similar local features.

4. IMPLEMENTATION AND RESULTS

To illustrate the relevance and the performance of the above model we present an implementation for the particular case of dense images consisting of oriented line segments. Such images are common as intermediate representations in several fundamental vision mechanisms such as shape from shading, shape from texture, shape from stereo or shape from motion. It is furthermore believed that, in such cases, global symmetry may prove to be a powerful tool of scene interpretation.

4.1. The grouping stage

We use here a relaxation-labeling procedure in order to coalesce clusters of mutually compatible oriented segments. First proposed by Rosenfeld *et al.* (1976), relaxation labeling is a parallel, local, cooperative and iterative process that assigns labels to objects, together with weights indicating the levels of confidence of each association between an object and a label.

Let $A = \{a_1, a_2, \ldots, a_n\}$ denote the set of oriented segments to be labeled and $\Lambda_i = \{\lambda^0, \lambda^1\}$ the set of possible labels for segment a_i, where $\lambda_i = \lambda^0$ indicates that a_i belongs to a cluster and $\lambda_i = \lambda^1$ that it does not. Let Λ_{ij} designate the set of compatible pairs of labels for each pair of segments (a_i, a_j), $i \neq j$. The four possible pairs represent the three following situations: a_i and a_j both belong to a cluster

($\lambda_i = \lambda_j = \lambda^0$), one of a_i or a_j belongs to a cluster but not the other ($\lambda_i \neq \lambda_j$), and neither a_i nor a_j belong to a cluster ($\lambda_i = \lambda_j = \lambda^1$).

Let $r_{ij}(\lambda_i, \lambda_j)$ designate the compatibility function between λ_i and λ_j assigned to the neighboring segments a_i and a_j. The neighbors of segment a_i are all the Voronoi neighbors a_j that fall within an elliptic region centered on a_i and having the same orientation as a_i. The use of an elliptic restricting region is meant to favor Voronoi neighbors that are positioned longitudinally with respect to the current segment. A possible expression for the compatibility function might be:

$$r_{ij}(\lambda_i, \lambda_j) = \begin{cases} \dfrac{1}{1 + e^{K(|\Delta\theta| - W)}} & \lambda_i = \lambda_j = \lambda^0, \\ 1 - \dfrac{1}{1 + e^{K(|\Delta\theta| - W)}} & \text{otherwise,} \end{cases} \tag{1}$$

where $\Delta\theta$ designates the difference in orientation between segments a_i and a_j, K is a positive constant, and W represents the maximal orientation difference producing a mutual affinity.

According to the above expression, segments with a small orientation difference mutually encourage each other to belong to the same cluster, while those with a large orientation difference mutually discourage each other. Consequently, in a situation such as the one of Fig. 6, the two adjacent clusters would be merged into one single cluster. Such an undesirable effect may be avoided by introducing border elements into the labeling process, in order to separate adjacent clusters.

Let $B = \{b_{1,2}, b_{1,3}, \ldots, b_{i,j}, \ldots, b_{n,n-1}\}$ be the border elements defining cluster boundaries. A border element b_{ij} is defined between each pair of segments (a_i, a_j), $i \neq j$. The role of border elements is not to delineate exactly each cluster; it is

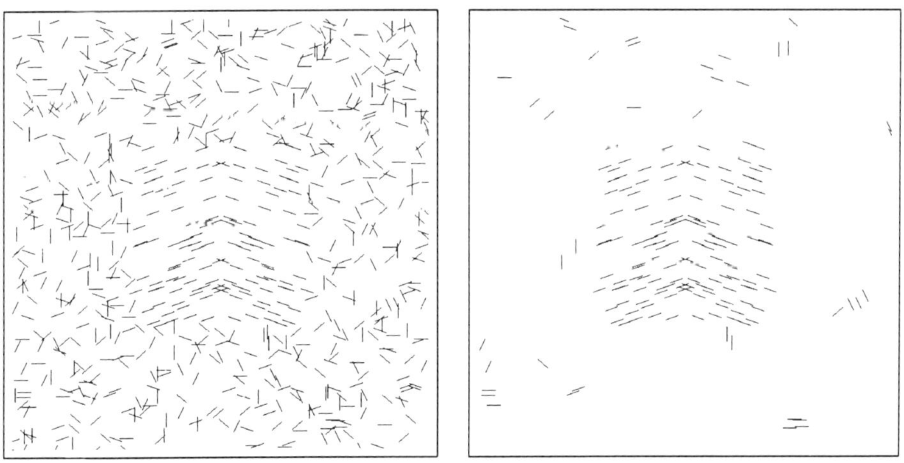

(a) (b)

Figure 6. (a) Original image and (b) result of the grouping process (without border elements).

rather to create a separation between segments belonging to different clusters. Let $\Omega_{ij} = \{\omega^0, \omega^1\}$ be the set of possible labels for each border element b_{ij}, ($\omega_{ij} = \omega^0$ and $\omega_{ij} = \omega^1$ indicate, respectively, the absence and the presence of a border element between segments a_i and a_j). The product $\Lambda_{ij} \times \Omega_{ij}$ represents the ensemble of compatible label triplets for each pair of segments (a_i, a_j), $i \neq j$. Five of the eight possible triplets represent the four following situations (the three other triplets represent incoherent situations): (1) a_i and a_j belong to the same cluster and there is no border element between them ($\omega_{ij} = \omega^0, \lambda_i = \lambda_j = \lambda^0$); (2) a_i and a_j belong to different clusters and there is a border element between them ($\omega_{ij} = \omega^1, \lambda_i = \lambda_j = \lambda^0$); (3) a_i or a_j belongs to a cluster while the other one does not and there is a border element between them ($\omega_{ij} = \omega^1, \lambda_i \neq \lambda_j$); (4) neither a_i nor a_j belong to a cluster and there is no border element between them ($\omega_{ij} = \omega^0, \lambda_i = \lambda_j = \lambda^1$).

Let $s_{ij}(\omega_{ij}, \lambda_i, \lambda_j)$ represent the new compatibility function characterizing the associations of ω_{ij} to b_{ij}, λ_i to a_i and λ_j to a_j. We propose the following expression for $s_{ij}(\omega_{ij}, \lambda_i, \lambda_j)$:

$$s_{ij} = \begin{cases} \dfrac{1}{1+e^{K(|\Delta\theta|-W)}} & \omega_{ij} = \omega^0, \ \lambda_i = \lambda_j = \lambda^0 \\ 1 - \dfrac{1}{1+e^{K(|\Delta\theta|-W)}} & \omega_{ij} = \omega^1, \ \lambda_i = \lambda_j = \lambda^0 \\ 0 & \omega_{ij} = \omega^0 \text{ with } \lambda_i \neq \lambda_j \\ & \omega_{ij} = \omega^1, \ \lambda_i = \lambda_j = \lambda^1 \\ 1 & \omega_{ij} = \omega^1 \text{ with } \lambda_i \neq \lambda_j \\ & \omega_{ij} = \omega^0, \ \lambda_i = \lambda_j = \lambda^1. \end{cases} \tag{2}$$

Neighboring elements with a small orientation difference mutually encourage each other to belong to the same cluster and discourage the appearance of a border element between them ($\omega_{ij} = \omega^0$, $\lambda_i = \lambda_j = \lambda^0$). Neighboring elements with a large orientation difference, which are each encouraged by their other neighbors to belong to a cluster, favor the appearance of a border element between them ($\omega_{ij} = \omega^1$ and $\lambda_i = \lambda_j = \lambda^0$). Whenever a_i is encouraged by its neighbors to belong to a cluster, and a_j is discouraged by its neighbors from belonging to a cluster (or the opposite), they favor the appearance of a border element between them ($\lambda_i \neq \lambda_j$). When neighboring elements are both discouraged by their neighbors from belonging to a cluster, they discourage the appearance of a border element between them ($\lambda_i = \lambda_j = \lambda^1$).

Let $p_i^k(\lambda)$ be the probability of associating λ with a_i, and $u_{ij}^k(\omega)$ be the probability of associating ω with b_{ij} at the kth iteration. These probabilities are changed in a parallel and iterative fashion according to the following expressions:

$$p_i^{k+1}(\lambda) = \frac{p_i^k(\lambda) q_i^k(\lambda)}{\sum_\lambda p_i^k(\lambda) q_i^k(\lambda)}, \quad \text{where } q_i^k(\lambda) = \sum_j \sum_{\lambda'} r_{ij}(\lambda, \lambda') p_j^k(\lambda') u_{ij}^k(\omega^0); \tag{3}$$

$$u_{ij}^{k+1}(\omega) = \frac{u_{ij}^k(\omega) t_{ij}^k(\omega)}{\sum_\omega u_{ij}^k(\omega) t_{ij}^k(\omega)}, \quad \text{where } t_{ij}^k(\omega) = \sum_\lambda \sum_{\lambda'} p_i(\lambda) p_j(\lambda') s_{ij}(\omega_{ij}, \lambda, \lambda'). \tag{4}$$

The expression for $q_i^k(\lambda)$ includes the term $u_{ij}^k(\omega^0)$, which represents the probability of absence of a border element. If this term was not present, two neighboring segments a_i and a_j with a small orientation difference would mutually encourage each other to belong to the same cluster. However, two neighboring elements a_i and a_j with a large orientation difference, which are each encouraged by their other neighbors to belong to different clusters, would mutually discourage each other from belonging to a cluster. This discouraging effect is undesirable and it would be preferable that each segment be encouraged to belong to its own cluster. Introducing the term $u_{ij}^k(\omega^0)$ has precisely the desired effect: when a_i and a_j have a small orientation difference, $u_{ij}^k(\omega^0)$ is high and its presence does not modify the probability values. However, when a_i and a_j have a large orientation difference, $u_{ij}^k(\omega^0)$ is low and attenuates the undesirable discouraging effect.

Figure 7 illustrates the results of the grouping process by the relaxation-labeling method, in the three following cases: two non-touching isolated clusters, two adjacent clusters, and two clusters with a smooth variation of orientation. The results are shown after 10 iterations, starting from a uniform label probability distribution ($p^0(\lambda^0) = p^0(\lambda^1) = 1/2$). In this experiment, K and W were set to 0.05 deg^{-1} and 20 deg respectively, and the label probability threshold was set to 0.9. As one can observe, the clusters are properly separated for the three cases. It is important to note that the role of border elements is not to delineate exactly the shape of the clusters but rather to create a separation between them. Even if some isolated lines also show up, they will be eliminated when clusters are identified.

The result of relaxation labeling is an ensemble of segments and an ensemble of border elements, from which clusters must be identified. Elements a_i and a_j belong to the same cluster when they are neighbors, when $p_i(\lambda_i = \lambda^0)$ and $p_j(\lambda_j = \lambda^0)$ are above the label probability threshold, and $u_{ij}(\omega^1)$ is under this threshold. All pairs of segments verifying these three conditions belong to the same cluster. Clusters that are too small (less than 5 segments) are eliminated.

4.2. *The symmetry-detection stage*

To detect instances of symmetry, the clusters resulting from the grouping stage have to be compared. Computational efficiency imposes the requirement to base the comparisons on a representation of each cluster in terms of a limited number of global parameters. In the case of clusters consisting of similarly oriented line segments, these parameters can be the size of each cluster, as well as the mean orientation and the mean position of its segments. However, clusters with smooth variations of orientations, such as the ones in Fig. 7f, cannot be represented by the mean orientation of their segments. Consequently, one must subdivide such clusters into smaller components with reduced orientation variations, while retaining their common identity for further restoration. Clusters are then systematically compared on the basis of location, orientation and size, in order to detect eventual symmetrical pairs.

4.2.1. Cluster subdivision. The eventual subdivision of clusters into subcomponents of small orientation variations is performed by means of a relaxation-labeling

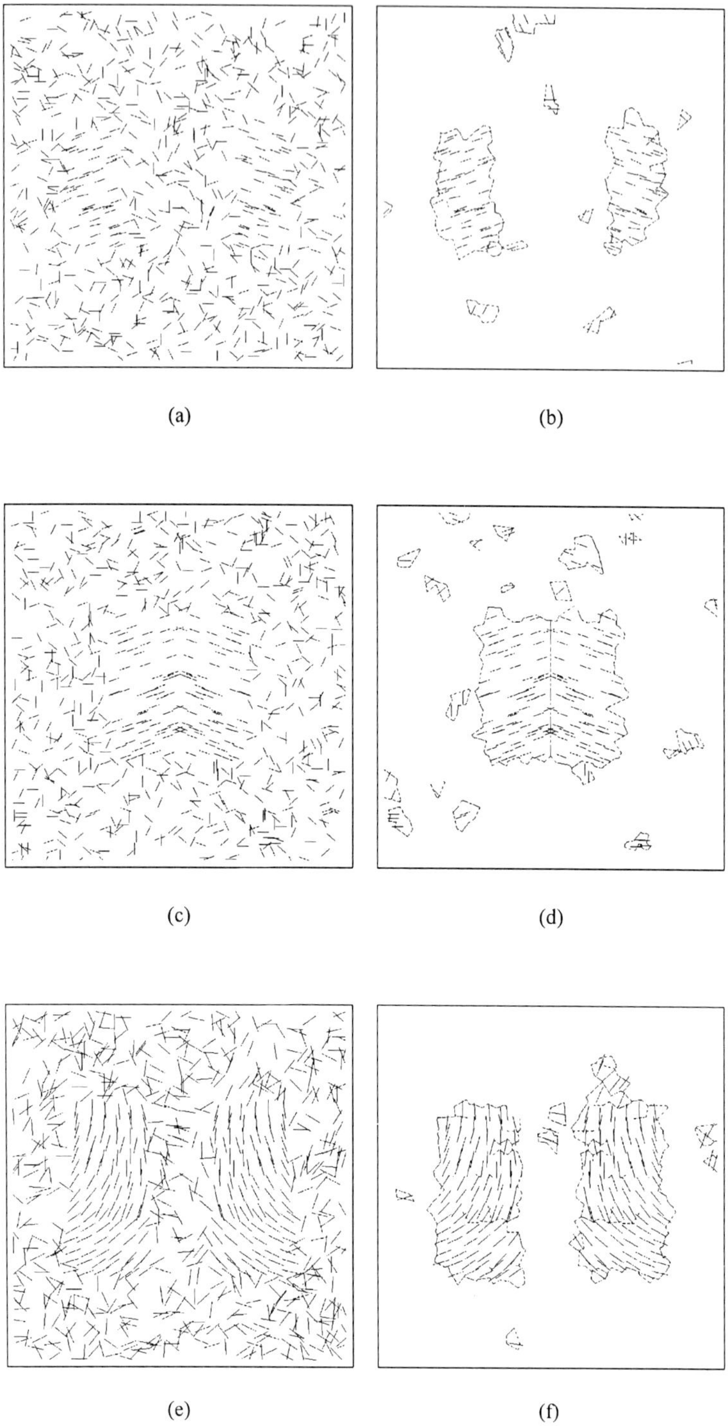

Figure 7. Original images (a, c, e), and the corresponding results of the grouping process (with border elements) (b, d, f).

method similar to the one presented in Section 4.1. Let $A = \{a_1, a_2, \ldots, a_n\}$ now designate only the line segments belonging to the cluster under consideration, and $\Lambda_i = \{\lambda^0, \ldots, \lambda^k\}$ be a set of labels representing k successive orientation intervals of width $\alpha = \pi/k$ between 0 and π (we choose here $k = 12$). At the end of the relaxation procedure, each label will eventually characterize a component whose mean orientation falls within the range of orientations it represents. The compatibility function between neighboring segments a_i and a_j is defined as:

$$r_{ij}(\lambda_i, \lambda_j) = \begin{cases} \dfrac{1}{1 + e^{K(|\Delta\theta| - W)}} & \lambda_i = \lambda_j, \\ 1 - \dfrac{1}{1 + e^{K(|\Delta\theta| - W)}} & \text{otherwise.} \end{cases} \tag{5}$$

The parameters are the same as in Section 4.1. For each line segment, the initial probability distribution is a normal distribution centered on the label corresponding to the segment orientation, with a small standard deviation. Figure 8 illustrates the results of the cluster-subdivision process applied to Fig. 7f, after 20 iterations. As desired, the original clusters with smooth variations of orientations have been subdivided into components with smaller orientation variations. Figures 7b and 7d are not modified by the cluster-subdivision process.

4.2.2. Cluster comparison. Each cluster is represented by its size as well as the mean position and the mean orientation of its segments. The size information is made of two measurements, namely the largest distances between two line segments in the mean orientation direction and in the direction normal to the mean orientation.

Clusters are systematically compared by searching for pairs of clusters of comparable sizes, whose orientations are identical under a reflection transformation around the medial axis of the line joining the cluster centers. Since the strategy is meant to detect approximate symmetry, exactness is not imposed for the preceding relations. Orientations and sizes need to correspond within certain tolerances. A reasonable tolerance on the orientations is the parameter W used in the compatibility function, since it represents the maximal amount of orientation difference characterizing the affinity between line segments. The tolerance used for size comparisons is twice the length of the line segments.

4.3. The symmetry-subsumption stage

Having compared all clusters or subcomponents, the last stage consists of trying to elicit global symmetry relations involving several pairs of clusters or components. In order to do so, one must simply compare the axes of symmetry established during the symmetry-detection stage, grouping collinear axes and the corresponding cluster pairs. A Hough-transform technique is used for that purpose (Duda and Hart, 1973), each axis being parameterized according to its distance r to a fixed point and its orientation Θ. The resolution of the Hough transform accumulator is chosen in order to allow sufficient tolerances on r and Θ (in our experiments 1/30 of the image size and 0.1 rad, respectively).

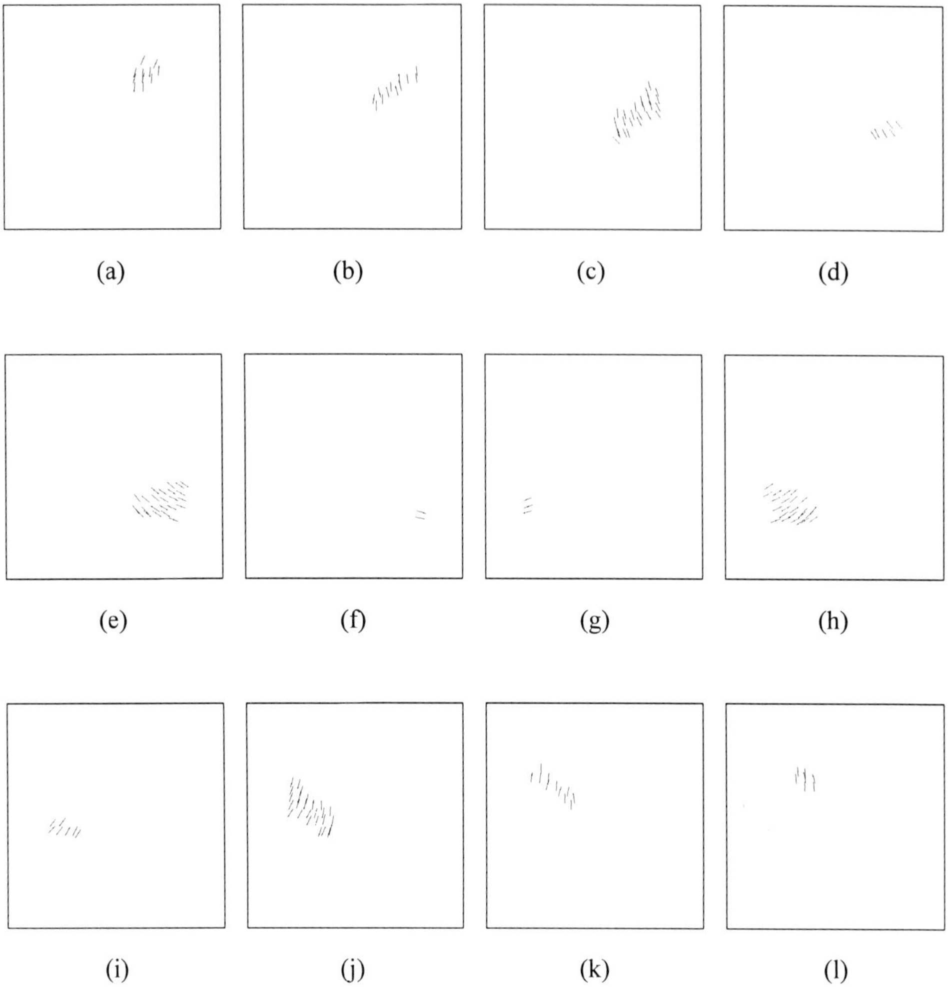

Figure 8. Results of the cluster subdivision process applied to the right cluster (a–f) and to the left cluster (g–l) of Fig. 7f.

Figure 9 illustrates the final results of the method. As one can observe, symmetry was detected in the case of Figs 7a, 7c, and 7e. Figure 9d illustrates an image containing two symmetrical regions, in which grouping cannot be achieved. Symmetry was not detected in this image, a fact which is in agreement with human symmetry perception.

5. CONCLUSION

Detecting symmetry in images is interesting because of the many possible roles that this property could play. One controversial issue concerns the role of symmetry as a grouping property. Gestalt psychologists suggest that symmetry is a fundamental

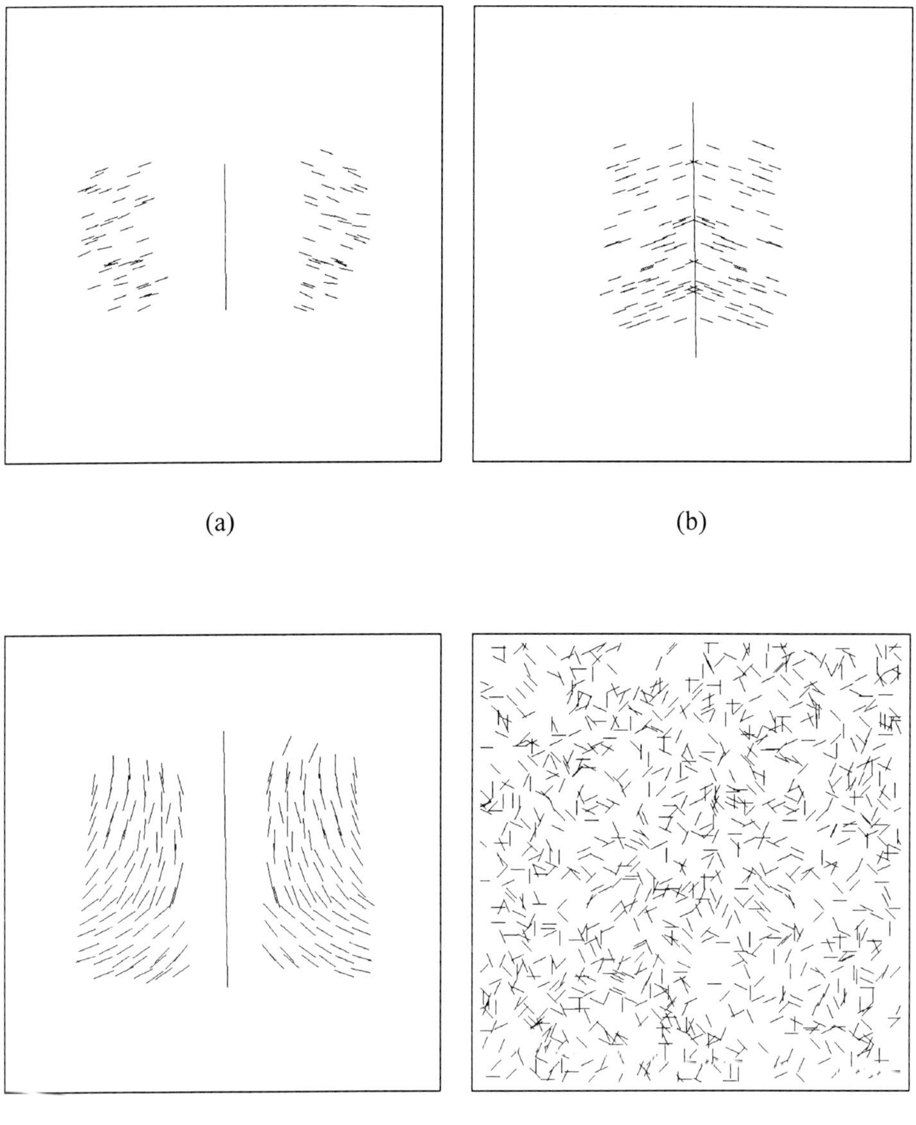

(a) (b) (c) (d)

Figure 9. Final results of the method applied to images 7a, 7c and 7e respectively (a, b, c); (d) an image containing symmetrical regions in which grouping cannot be achieved and for which symmetry was not detected.

grouping property of perceptual organization, while recent studies suggest rather that symmetry might be detected among higher level features, created on the basis of other grouping properties.

We have reported here a psychophysical experiment conducted in order to investigate this question. Results of the experiment show that the task of detecting symmetry is significantly easier when the elements in the stimuli can be grouped together according to properties such as similarity of orientation and proximity, and thus support the hypothesis that grouping precedes symmetry detection.

As in the case of other visual tasks (e.g. motion perception, shape perception, object recognition), more than a single mechanism may underly the ability of the human visual system to detect symmetry. Julesz (1971) has suggested two mechanisms for symmetry detection, one underlying detection in dense stimuli and the other suited for larger-scale stimuli.

We have proposed a three-stage computational model for the detection of symmetry in field-type dense images. In the first stage, relaxation labeling is used to group primitive elements of similar properties into a small number of clusters. This grouping stage may also be useful for other visual tasks. Pairs of clusters are then compared in the second stage of the model in order to establish local symmetry axes. These symmetry axes are then aggregated in the third stage (using Hough transform) to form global symmetry relations.

The model is in good agreement with human symmetry perception in dense images, in the sense that it deals not only with exact symmetry, but also with approximate symmetry. The reliance upon a preliminary grouping stage, consisting of local, cooperative computations, enables a significant reduction of the computational load entailed for the process as a whole (compared to a more direct approach to the problem).

The model has been presented and implemented in the context of texture images, providing satisfactory results. Other possible domains of relevance are gradient fields, optical flow fields or $2\frac{1}{2}$D representations. Indeed, interesting questions relate to the use of symmetry properties in the segmentation and interpretation of motion fields, or in the extraction of lighting conditions and shape information from needle diagrams produced by shape from shading algorithms. In the context of human symmetry perception, further studies are also required to investigate the relationship between grouping and symmetry detection, and the effect grouping may have on the complexity of the processes involved.

Acknowledgement

This work was supported by the Canadian Program of Networks of Centers of Excellence (IRIS Network, project A-4). JF is supported by NSERC operating grant OGP0121333 and FCAR grant 93-NC-0903.

REFERENCES

Attneave, F. (1954). Some informational aspects of visual perception. *Psychol. Rev.* **61**, 183–193.

Attneave, F. (1955). Symmetry, information and memory for patterns. *Am. J. Psychol.* **68**, 209–222.

Barlow, H. B. and Reeves, B. C. (1979). The versatility and absolute efficiency of detecting mirror symmetry in random dot displays. *Vision Res.* **19**, 783–793.

Bigün, J. (1988). Recognition of local symmetries in gray value images by harmonic functions. In: *Proc. 9th Int. Conf. on Pattern Recognit.* pp. 345–347.

Blum, H. and Nagel, R. N. (1978). Shape description using weighted symmetric axis features. *Pattern Recognit.* **10**, 167–180.

Brady, M. (1983). Criteria for representation of shape. In: *Human and Machine Vision*. J. Beck, B. Hope and A. Rosenfeld (Eds). Academic Press, New York.

Day, H. (1968). The importance of symmetry and complexity in the evaluation of complexity, interest, and pleasingness. *Psychonomic Sci.* **10**, 339–340.
Duda, R. O. and Hart, P. E. (1973). *Pattern Classification and Scene Analysis*. Wiley, New York.
Fleck, M. M. (1986). Local rotational symmetries. In: *Proc. Conf. Vision and Pattern Recognit.* pp. 332–337.
Howe, E. and Jung, K. (1987). Judgement of numerosity: Effects of symmetry and goodness in dot pattern arrays. *Acta Psychologica* **64**, 3–11.
Jenkins, B. (1982). Redundancy in the perception of bilateral symmetry in dot textures. *Percept. Psychophys.* **32**, 171–177.
Jenkins, B. (1983). Component processes in the perception of bilaterally symmetric dot textures. *Percept. Psychophys.* **34**, 433–440.
Julesz, B. (1966). Binocular disappearance of monocular symmetry. *Science* **153**, 657–658.
Julesz, B. (1971). *Foundations of Cyclopean Perception*. University of Chicago Press, Chicago.
Julesz, B. and Chang, J. J. (1979). Symmetry perception and spatial-frequency channels. *Perception* **8**, 711–718.
Kanade, T. and Kender, J. R. (1983). Mapping image properties into shape constraints: Skewed symmetry, affine-transformable patterns, and the shape-from-texture paradigm. In: *Human and Machine Vision*. J. Beck, B. Hope and A. Rosenfeld (Eds). Academic Press, New York.
Kumar, N., Mukherjee, A. K. and Chakraborti, N. B. (1983). On using symmetry properties for selecting transform components for image coding. *IEEE Trans. Acoustics, Speech, and Signal Processing* **31**, 749–752.
Leyton, M. (1992). *Symmetry, Causality, Mind*. MIT Press, Cambridge, Massachusetts.
Mach, E. (1906/1959). *The Analysis of Sensations and the Relation of the Physical to the Psychical*. Dover, New York. (Original publication in 1906).
Nalwa, V. S. (1989). Line-drawing interpretation: Bilateral symmetry. *IEEE Trans. Pattern Anal. and Machine Intell.* **11**, 104–108.
Palmer, S. E. (1983). The psychology of perceptual organization: A transformational approach. In: *Human and Machine Vision*. J. Beck, B. Hope and A. Rosenfeld (Eds). Academic Press, New York.
Palmer, S. E. and Hemenway, K. (1978). Orientation and symmetry: Effects of multiple, rotational, and near symmetries. *J. Exp. Psychol: Human Percept. Perform.* **4**, 691–702.
Pashler, H. (1990). Coordinate frame for symmetry detection and object recognition. *J. Exp. Psychol: Human Percept. Perform.* **16**, 150–163.
Pomerantz, J. R. and Kubovy, M. (1986). Theoretical approaches to perceptual organization. In: *Handbook of Perception and Human Performance*. K. R. Boff, L. Kaufman and J. P. Thomas (Eds). Wiley, New York.
Reisfeld, D., Wolfson, H. and Yeshurun, Y. (1990). Detection of interest points using symmetry. In: *Proc. 3rd Int. Conf. Computer Vision*. pp. 62–65.
Rosenfeld, A., Hummel, R. and Zucker, S. (1976). Scene labeling by relaxation operations. *IEEE Trans. Systems, Man, and Cybernet.* **6**, 420–433.
Royer, F. L. (1981). Detection of symmetry. *J. Exp. Psychol: Human Percept. Perform.* **7**, 1186–1210.
Subirana-Vilanova, J. B. (1990). Curved inertia frames and the skeleton sketch: Finding salient frames of reference. In: *Proc. 3rd Int. Conf. Computer Vision*. pp. 702–708.
Ulupinar, F. and Nevatia, R. (1988). Using symmetry for analysis of shape from contour. In: *Proc. 2nd Int. Conf. Computer Vision*. pp. 414–426.
Wagemans, J., Van Gool, L. and d'Ydewalle, G. (1991). Detection of symmetry in tachistoscopically presented dot patterns: Effects of multiple axes and skewing. *Percept. Psychophys.* **50**, 413–427.
Wilson, S. (1991). *Perceptual Organization and Symmetry in Visual Object Recognition*. Masters Thesis, University of British Columbia, Canada.
Zucker, S. W. (1986). Early processes for orientation selection and grouping. In: *From Pixels to Predicates*. A. P. Pentland (Ed.). Ablex Publishing Corporation, Norwood, New Jersey.

Human Symmetry Perception, pp. 289–301
C. W. Tyler (Ed.)

Continuous symmetry: a model for human figural perception

HAGIT ZABRODSKY[1,*,†] and DANIEL ALGOM[2]

[1]*Department of Psychology, Stanford University, Stanford, CA 94305, USA*
[2]*Department of Psychology, Bar-Ilan University, Ramat Gan, Israel 52900*

Abstract—Symmetry is usually viewed as a discrete feature: an object is either symmetric or non-symmetric. In this presentation, symmetry is treated as a continuous feature and a continuous measure of symmetry (the *Symmetry Distance*) is defined. This measure can be easily evaluated for any shape or pattern in any dimension. A preliminary study presented here shows that the Symmetry Distance is commensurate with human perceptual experience. Good correlation is found between the continuous symmetry values and the perceived goodness of figures.

1. INTRODUCTION

William Blake's classic poem sings of the 'fearful symmetry' of his celebrated 'tyger'. Why is the symmetry fearful? A possible reason is that although we perceive the biological symmetry, it is not perfect. The tiger still looks like a tiger when you view it in a mirror — just as does a human face — but the left-hand side of the tiger or your face is not precisely the same as the reflection of the right-hand side. Symmetries abound in the biological world and in our visual world; yet most of these symmetries are not perfect. In the visual world, loss of symmetry is further enhanced: Even perfectly symmetric objects lose their exact symmetry when projected onto the image plane or the retina due to occlusion, perspective transformations, or digitization. Clearly, the popular binary conception of symmetry (i.e. an object is either symmetric or is not symmetric) and exact mathematical definitions thereof (Weyl, 1952; Miller, 1972) are inadequate to describe and quantify the symmetries found in the natural world. Needed is a continuous measure of symmetry, one that would be able to quantify the 'amount' of symmetry entailed in Blake's 'tyger'. The development of

*This study was performed while HZ was at the Department of Computer Science, Bar-Ilan University, Ramat Gan, Israel.

†To whom correspondence should be addressed.

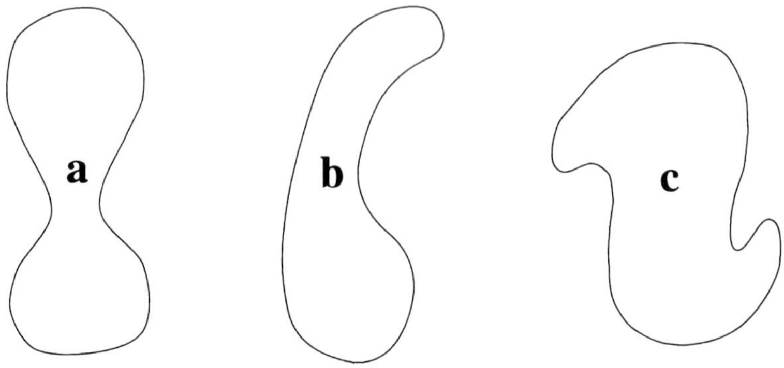

Figure 1. The continuous symmetry measure can compare the 'amount' of symmetry of different shapes and can compare the 'amount' of different symmetries of a single shape. Thus the notion that **a** is 'more' mirror symmetric than **b** can be quantified. Similarly, the notion that **c** is 'more' rotationally symmetric than mirror symmetric, can be quantified.

such a measure, and its evaluation against actual perception, form the theme of this article.

In this paper, a 'Symmetry Distance', capable of measuring and quantifying all types of continuous symmetries of objects is introduced. Our definition of the Symmetry Distance produces a versatile and simple tool that can supply a set of measures for any object reflecting the amount of different types of symmetry (rotational, reflectional, etc.) possessed by the object. The generality of this symmetry measure allows one to compare the symmetry distance of several objects relative to a single symmetry type and to compare the symmetry distance of a single object relative to various symmetry types. Thus the intuitive notion that the shape of Fig. 1a is 'more' mirror-symmetric than the shape of Fig. 1b, can be quantified. Similarly, the intuitive notion that the shape of Fig. 1c is 'more' rotationally symmetric (of order two) than mirror symmetric can be quantified.

This concept of continuous symmetry is commensurate with perceptual behavior. To demonstrate this we tested visual evaluation of figural goodness as mediated by symmetry. In Section 2, we define the Symmetry Distance and in Section 3 we briefly describe a method for evaluating the Symmetry Distance. In Section 4 we discuss continuous symmetry in terms of human perception and present some preliminary studies that show a good correlation with human perception of figural goodness.

2. A CONTINUOUS SYMMETRY MEASURE — DEFINITION

We define the *Symmetry Distance* (SD) as the minimum effort required to transform a given object into a symmetric object. This effort is measured by the mean of the square distances taken to move each point of the object from its location in the original object to its location in the symmetric object. Note that no a priori symmetric reference shape is assumed. In effect, we measure the distance from a given object to the set of all symmetric objects.

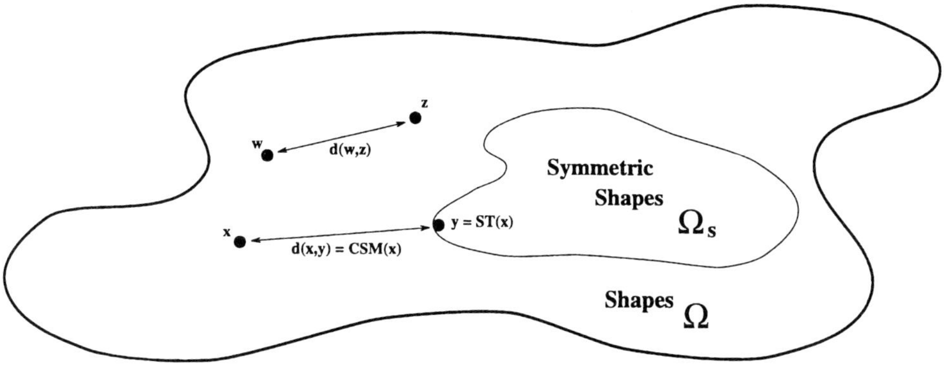

Figure 2. The space Ω of all shapes of a given dimension, where each shape P is represented by a sequence of n points. A metric d is defined on this space which serves as a distance function between every two shapes in Ω.

Denote by Ω the space of all shapes of a given dimension, where each shape P is represented by a sequence of n points $\{P_i\}_{i=0}^{n-1}$ (Fig. 2). We define a metric d on this space as follows:

$$d : \Omega \times \Omega \rightarrow R,$$

$$d(P, Q) = d(\{P_i\}, \{Q_i\}) = \frac{1}{n}\sum_{i=0}^{n-1} \|P_i - Q_i\|^2.$$

This metric defines a distance function between every two shapes in Ω.

We define the *Symmetry Transform* of a shape P, with respect to a given symmetry type G, as the shape $\hat{P}$ which is G-symmetric and closest to P in terms of the metric d.

The *Symmetry Distance* (SD) of a shape P with respect to a given symmetry type G, is now defined as the distance between P and its Symmetry Transform $\hat{P}$:

$$\mathrm{SD} = d(P, \hat{P}).$$

The SD of a shape $P = \{P_i\}_{i=0}^{n-1}$ is evaluated by finding the symmetry transform $\hat{P} = \{\hat{P}_i\}_{i=0}^{n-1}$ of P (Fig. 3d) and computing:

$$\mathrm{SD} = \frac{1}{n}\sum_{i=0}^{n-1} \|P_i - \hat{P}_i\|^2.$$

This definition of the Symmetry Distance implicitly implies invariance to rotation and translation. Normalization of the original shape prior to the transformation allows insensitivity to size (Fig. 3). We normalize by scaling the shape so that the maximum distance between points on the contour and the centroid is a given constant (in this paper all examples are normalized to 100). The normalization presents an upper bound on the mean squared distance moved by points of the shape. Thus the SD value is limited in range, where SD = 0 for perfectly symmetric shapes.

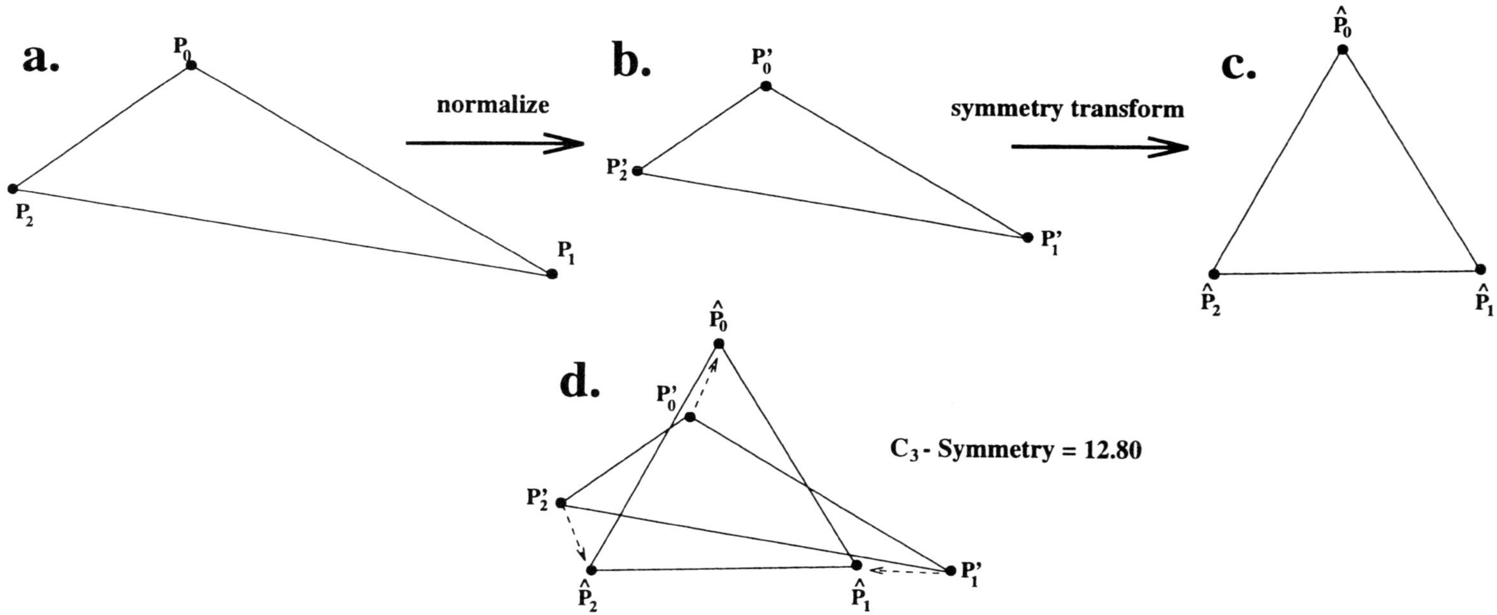

Figure 3. Calculating the Symmetry Distance of a shape: (a) Original shape $\{P_0, P_1, P_2\}$. (b) Normalized shape $\{P_0', P_1', P_2'\}$, such that maximum distance to the center of mass is constant (100). (c) Applying the symmetry transform to obtain a symmetric shape $\{\hat{P}_0, \hat{P}_1, \hat{P}_2\}$. (d) $\mathrm{SD} = (\| P_0' - \hat{P}_0 \|^2 + \| P_1' - \hat{P}_1 \|^2 + \| P_2' - \hat{P}_2 \|^2)/3$.

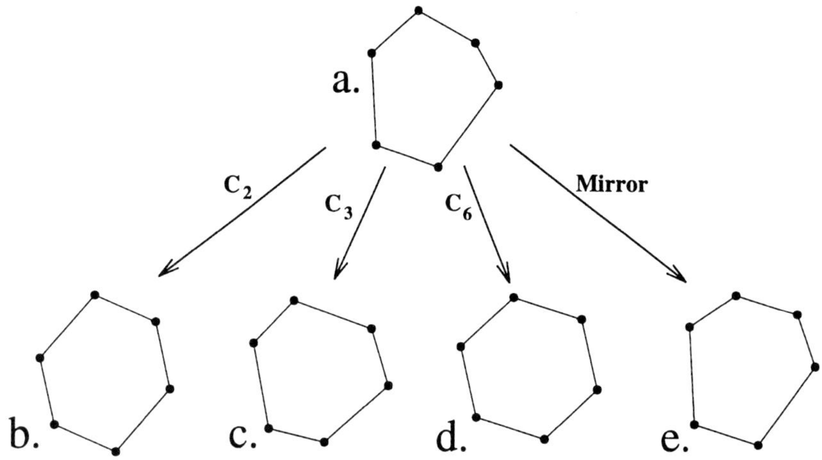

Figure 4. Symmetry Transforms and Symmetry Distances of a 2D polygon. (a) The 2D polygon. (b) Symmetry Transform of (a) with respect to C_2-symmetry (SD = 1.87). (c) Symmetry Transform of (a) with respect to C_3-symmetry (SD = 1.64). (d) Symmetry Transform of (a) with respect to C_6-symmetry (SD = 2.53). (e) Symmetry Transform of (a) with respect to Mirror-symmetry (SD = 0.66).

The general definition of the Symmetry Distance enables evaluation of a given shape for different types of symmetries (mirror-symmetries, rotational symmetries, etc.). Moreover, this generalization allows comparisons between the different symmetry types, and allows expressions such as 'a shape is more mirror-symmetric than rotationally-symmetric of order two'. An additional feature of the Symmetry Distance is that we obtain the symmetric shape which is 'closest' to the given one, the symmetry transform, enabling visual evaluation of SD.

An example of a 2D polygon and its symmetry transforms with respect to various symmetry types and the corresponding SD values are shown in Fig. 4. Note that shape 4e is the most similar to the original shape 4a and, indeed, its SD value is the smallest. The Symmetry Distance and symmetry transform have recently been applied to image processing and chemical applications (Zabrodsky *et al.*, 1993; Zabrodsky and Avnir, 1993).

3. EVALUATING THE SYMMETRY TRANSFORM

A simple geometric algorithm has been derived for evaluating the Symmetry Transform (and accordingly, the Symmetry Distance) of a shape represented by a sequence of points, with respect to any type of symmetry. For simplicity, an outline of the algorithm is given here for the case of rotational symmetry in 2D. For more details and extensions to other types of symmetry in higher dimensions, see Zabrodsky (1993).

Consider the simple case where a shape P is represented by n points and its Symmetry Transform with respect to rotational symmetry of order n (C_n-symmetry) is to be found. The following algorithm finds the C_n-symmetric configuration of points which is closest to P in the mean square sense (i.e. in terms of the metric d defined in Section 2):

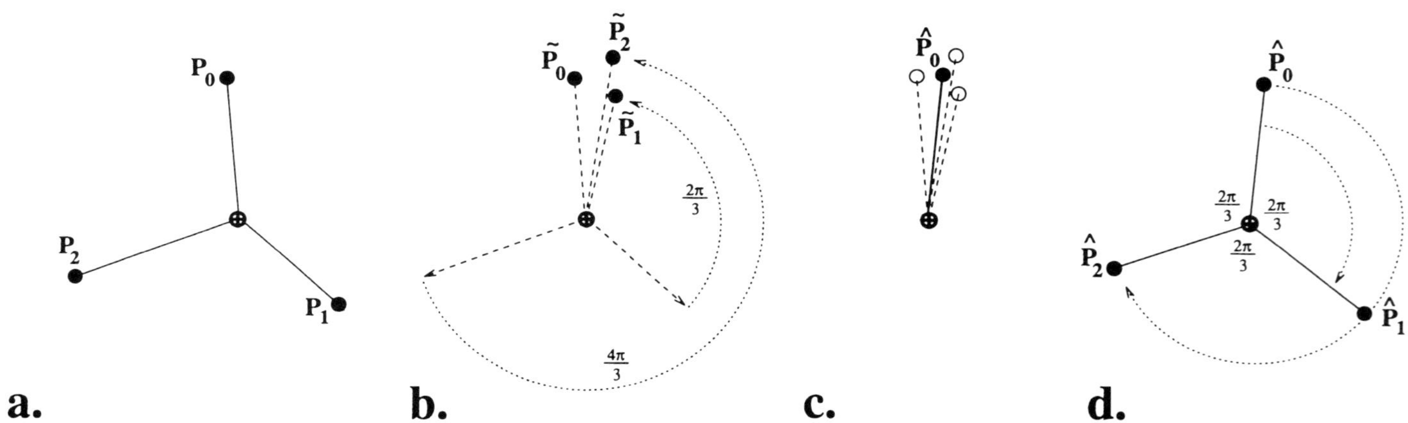

Figure 5. The Symmetry Transform of 3 points with respect to rotational symmetry of order 3. (a) Original 3 points $\{P_i\}_{i=0}^{2}$. (b) Fold $\{P_i\}_{i=0}^{2}$ into $\{\tilde{P}_i\}_{i=0}^{2}$. (c) Average $\{\tilde{P}_i\}_{i=0}^{2}$ obtaining $\hat{P}_0 = \frac{1}{3}\sum_{i=0}^{2}\tilde{P}_i$. (d) Unfold the average point obtaining $\{\hat{P}_i\}_{i=0}^{2}$. The centroid is marked by $\oplus$.

Algorithm for finding the C_n-symmetry transform:

1. *Fold* the points $\{P_i\}_{i=0}^{n-1}$ by rotating each point P_i counterclockwise about the centroid by $2\pi i/n$ radians obtaining the points $\{\tilde{P}_i\}_{i=0}^{n-1}$ (Fig. 5b).
2. *Average* the points $\{\tilde{P}_i\}_{i=0}^{n-1}$ obtaining point $\hat{P}_0$ (Fig. 5c).
3. *Unfold* the points by duplicating $\hat{P}_0$ and rotating clockwise about the centroid by $2\pi i/n$ radians obtaining the C_n-symmetric points $\{\hat{P}_i\}_{i=0}^{n-1}$ (Fig. 5d).

The set of points $\{\hat{P}_i\}_{i=0}^{n-1}$ is the symmetry transform of the points $\{P_i\}_{i=0}^{n-1}$; i.e. they are the C_n-symmetric configuration of points closest to $\{P_i\}_{i=0}^{n-1}$ in terms of the average distance squared. Proof of the correctness of this algorithm can be found in Zabrodsky (1993).

The common case, however, is that shapes have more points than the order of the symmetry. For symmetry of order n, the folding method can be extended to shapes having a number of points which is a multiple of n. A 2D shape P having qn points is represented as q sets $\{S_r\}_{r=0}^{q-1}$ of n interlaced points $S_r = \{P_{iq+r}\}_{i=0}^{n-1}$. The C_n-symmetry transform of P is obtained by applying the above algorithm to each set of n points separately, where the folding is performed about the centroid of all the points (Fig. 6).

The algorithm described above is general and extends to all symmetry groups in any dimension (see Zabrodsky, 1993). Specifically, for mirror-symmetry in 2D, if the axis of reflection is not specified, the optimal axis can be determined analytically.

As presented here, the input to the algorithm is a set of points; however, given a contour, an image or a 3D volume, the measure of symmetry can be evaluated by selecting points to represent these objects. The selection of the points must be such that the division into sets, as described in the above algorithm, can be performed. This requires every point in the collection of representation points to be matched, under symmetry, to other points in the collection. This is analogous to the correspondence

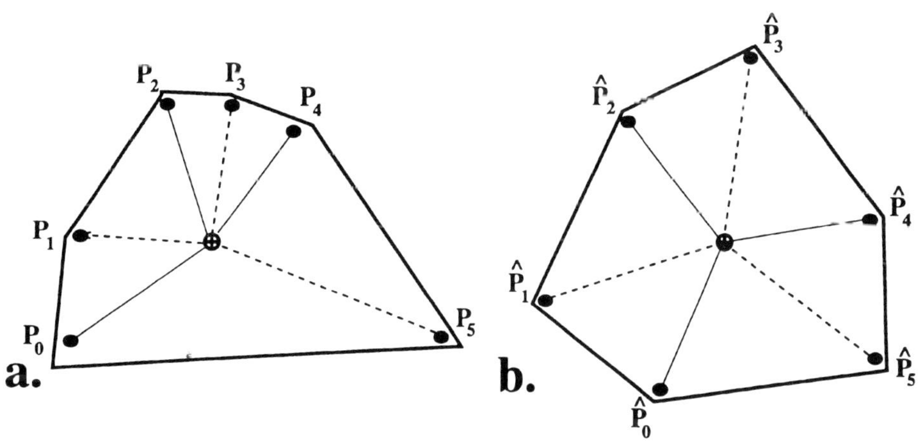

Figure 6. Geometric description of the C_3-symmetry transform for 6 points. The centroid of the points is marked by $\oplus$. (a) The original points shown as two sets of 3 points: $S_0 = \{P_0, P_2, P_4\}$ and $S_1 = \{P_1, P_3, P_5\}$. (b) The obtained C_3-symmetric configuration.

problem in stereo matching and motion determination. We deal with the selection of points by considering an object in two possible ways for the symmetry procedure with the underlying assumptions that (a) all matches of points are in the final collection of representation points, i.e. no points are missing nor are points redundant, and (b) the order of the points (on a contour for example) are preserved under the symmetry transform.

The two ways of considering an object for symmetry are:

1. The shape is a contour with no 'special' points, i.e. points at corners or high curvature points are not considered salient. Thus under the symmetry transform these points are not preserved (corners may round off, protrusions may disappear etc.). In these cases a contour sampling procedure is required. There are several ways to select a sequence of points to represent continuous 2D shapes; one such method is sampling at *equal distances*; however, contour length is not always meaningful, as in noisy or occluded shapes. In such cases, we sample points on a smoothed version of the contour and then project the sampled points back onto the original contour. The level of smoothing can vary and, for a high level of smoothing, the resulting shape becomes almost circular about the centroid in which case the sampling is reduced to sampling the original shape at *equal angles* about the centroid of the shape. (For details and examples of the various sampling procedures see Zabrodsky, 1993.) For any of the approaches mentioned above, it can be shown empirically that, as the density of sampled points increases, the symmetry distance obtained asymptotes to some value.
2. Salient points are considered and should be matched to salient counterparts. In this case the special points (vertices of polygons, corner points, etc.) are all chosen to represent the object. However, because of the above-mentioned assumptions, no spurious points (or unmatched points) are assumed to exist. Future work would extend the topological stage of the symmetry procedure to allow points to be discarded or added into the collection of representation points, in analogy to the correspondence problem, where unmatched feature points would be allowed (see Zabrodsky *et al.*, 1992 where this is partially done by deleting the extraneous points of occluding boundaries).

4. CONTINUOUS SYMMETRY AND HUMAN PERCEPTION

The notion of symmetry was introduced to psychological theorizing and research by Mach and later by Gestalt psychologists in the first half of this century. It has since been associated with the Gestalt constructs of 'good Gestalt' or 'figural goodness', referring to perception of the simplest, most stable figure possible. Given the same number and kind of physical components, some stimuli appear to be unique, simple, regular, or better organized than other stimuli. Those impressions are captured by the quality of 'goodness', the end product of inherent organizational principles that govern the perception of form. In that scheme, symmetry contributes to figural goodness. The contingency between the subjective experience of goodness and the objective feature of symmetry could not be pursued further, however, because the Gestalt psychologists have made no attempt towards an objective, quantitative analysis of figural goodness.

In a series of seminal papers in the 1950s, Attneave, Hochberg and their associates (e.g. Hochberg and McAlister, 1953; Attneave, 1954, 1955; Attneave and Arnoult, 1957) applied concepts of information theory by way of a more quantitative analysis of figural goodness. Good forms contain portions that are repetitive and predictable from other portions. Cast in informational terms, good forms are redundant, containing little information. A symmetric figure is redundant because parts of it can be predicted from other parts. Of all the possible forms of redundancy, symmetry is the easiest to quantify (cf. Attneave, 1959). Significantly, as Palmer (1991) has pointed out, the informational definition of 'goodness' fully conveys the original Gestalt notions of simplicity, order, and singularity.

Nevertheless, for all its originality and heuristic value, the approach espoused by theorists like Attneave, Hochberg or Berlyne (e.g. Berlyne, 1957, 1966) is limited in a crucial aspect. These theorists applied the informational analysis to components of single stimuli. However, as Palmer notes, such piecewise analysis does not fit well with the Gestalt style of explanation, emphasizing global properties and analyzable wholes. More seriously, in mathematical information theory itself, redundancy applies to sets of stimuli, and is not characteristic of unique stimuli (Garner, 1962). As Garner (1974) observed, 'information is a function not of what the stimulus is, but rather of what it might have been, of its alternatives' (p. 194). This is the major contribution of information theory to psychology: Presented with a stimulus, our perception is influenced by other stimuli that could have been presented. Here, too, lies the challenge to the theorists; namely, specifying the set of those other stimuli or identifying the alternatives that did not appear in any given trial. Garner's theory of *Rotation and Reflection* (R & R) *Subsets* was suggested to accomplish that goal in a truly Gestalt spirit. Along with Palmer, we believe that Garner's work has laid a solid foundation for future studies of pattern goodness and symmetry, including the present endeavor.

Garner suggested that good patterns are those which have few alternatives. Good patterns produce less variance than do 'bad' patterns when spatial transformations are applied to the figures. Quantitatively, the perceived goodness of a pattern is inversely related to the size of a subset of equivalent patterns that are obtained by applying the transformations. In Garner's view, the observer associates the pattern in question with the subset of its transformational variants. Good patterns come from small inferred subsets and poor patterns come from large ones (cf. Dember and Warm, 1979).

To test this theory, Garner and Clement (1963) prepared dot patterns, similar to those illustrated in Fig. 7 by placing 5 dots in an imaginary 3×3 matrix. They applied four rotations (angles of 0, 90, 180 and 270 deg) and four reflections (horizontal, vertical and two diagonals) to define the R & R *set* for each pattern. The subset of distinguishably different patterns within each set defines the R & R *subset* (cf. Palmer, 1991). Garner and Clement had one group of subjects rate the patterns for 'goodness', and another arrange the patterns into sets on the basis of perceived similarity. The sets did not need to have equal numbers of patterns; indeed, the total number of patterns placed in a set was one parameter of interest. Garner and Clement found that the

Dot Patterns			
R & R Subset size	1	4	8
Symmetry Distance Measures	v = 0.00 h = 0.00 hv = 0.00 vh = 0.00 C_2 = 0.00 C_4 = 0.00	v = 0.00 h = 12.24 hv = 6.12 vh = 6.12 C_2 = 12.24 C_4 = 12.24	v = 6.10 h = 9.76 hv = 4.88 vh = 4.88 C_2 = 3.66 C_4 = 6.71

Figure 7. Analysis of figural goodness for dot patterns in terms of R & R subset size, and in terms of Symmetry Distance measures developed here. v: vertical mirror symmetry, h: horizontal mirror symmetry, hv: diagonal top-left to bottom-right mirror symmetry, vh: diagonal top-right to bottom-left mirror symmetry, C_2: rotational symmetry of order two (120 deg), C_4: rotational symmetry of order four (90 deg).

higher the rating for goodness, the smaller is the similarity set. In many additional experiments (summarized in Garner, 1974), Garner and his associates confirmed that goodness of a figure was inversely related to the size of its R & R subset.

Garner did not couch his theory in terms of symmetry because he felt that subsets of equivalent patterns served better for perceptual explanation than 'the more restrictive concept of symmetry' (Garner, 1974, p. 17). In point of fact, however, Garner's R & R sets conform fully to the mathematical definition of symmetry including rotational and mirror symmetries (see also Palmer, 1991). Indeed, Palmer developed an alternative theory of figural goodness focusing on the transformations over which the patterns remain invariant. A figure is symmetrical with respect to a given transformation, if that transformation leaves the figure unchanged. One can list all such transformations, thus specifying the resulting *symmetry subgroups*. Garner's R & R theory only defines the *number* of the transformations, whereas Palmer's symmetry subgroup theory also refers to the *identity* of those transformations. The latter, as research has shown, does matter to perception (e.g. Royer, 1981; Rock, 1983). For instance, it has been repeatedly demonstrated that vertical symmetry influences perception to a greater extent than does horizontal symmetry (Rock and Leaman, 1963; Chipman, 1977; Palmer and Hemenway, 1978).

Our theory of continuous symmetry follows in the footsteps of Garner's and Palmer's contributions. Palmer's construct of symmetry subgroups is well taken, and our analysis provides a rich variety of symmetries, both rotational and reflectional (see again Fig. 7). At the same time, we also share Garner's concern with the limited use of traditional *binary* notion of symmetry. Although Palmer's analysis relates to different types of symmetry, for any given transformation, a figure is still either symmetric or asymmetric. In contradistinction, our theory employs a continuous measure of symmetry, and it applies to any figure or pattern in two or three dimensions. Indeed, generality of application is the hallmark of our approach.

To serve a useful role, of course, one must show that the present definition of symmetry is commensurate with perceptual experience. The following study provided a preliminary observation using Symmetry Distance for specification of the stimuli. We created 4 sets of 9 'random shapes', of which 9 stimuli (comprising a single set) are shown in Fig. 8. Within a set, each shape had different values of Symmetry Distance, with respect to mirror-symmetry and to rotational symmetry of order 2. The shapes were constructed by randomly choosing one of four possible radius lengths for each of the twelve points forming the shape, and spacing the points at $2\pi/12$ deg angles. Twenty observers provided ratings of goodness for all 36 shapes on a 20-point scale (with 20 and 1 standing for the best and the worst figures, respectively). Following Garner, we left the definition of goodness to the subject's discretion. Despite the completely arbitrary fashion of creating the figures and the small number of subjects, the results are encouraging. As the scatter plots in Fig. 9 show, the continuous measures of mirror symmetry and rotational symmetry covaried with the judgement of goodness. The correlation coefficients were −0.689 and −0.570, respectively, for reflectional and rotational symmetries. The multiple correlation of these two measures with experienced goodness equaled 0.90. In other words, the continuous measures

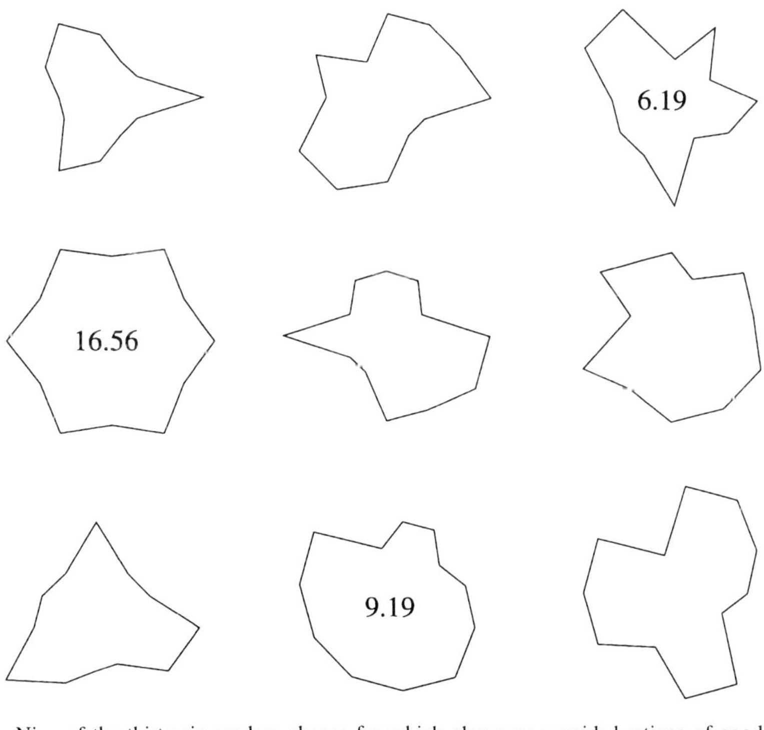

Figure 8. Nine of the thirty-six random shapes for which observers provided ratings of goodness. The shapes were constructed by randomly choosing one of four possible radus lengths for each of the twelve points of the shape and spacing the points at $2\pi/12$ deg angles. The highest, lowest and middle goodness ratings, averaged over all observers, are shown.

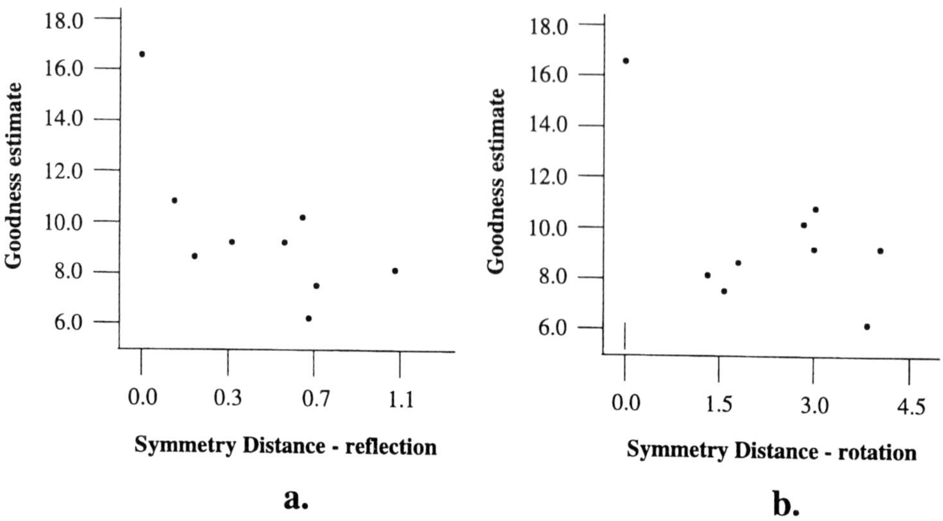

Figure 9. The dependence of perception of goodness on (a) reflectional and (b) rotational measures of Symmetry Distance. The plot displays the result for the set of shapes shown in Fig. 8. The judgements of goodness for each shape are averaged over all observers.

of symmetry accounted for over 80% of the variance of the goodness judgements, although much of that proportion is attributable to one highly symmetric figure.

Again, despite the moderating effect of the small number of subjects, these results are impressive. Equipped with the present measures, the Garnerian notion of inferred equivalence sets readily generalizes to many forms, not just to simple patterns created for the laboratory. So does Palmer's analysis in terms of symmetry subgroups. The results provide powerful support for a general perceptual principle: Figures are processed in terms of their uniqueness measured in units of symmetry distance.

REFERENCES

Attneave, F. (1954). Some informational aspects of visual perception. *Psychol. Rev.* **61**, 183–193.

Attneave, F. (1955). Symmetry information and memory for patterns. *Am. J. Psychol.* **68**, 209–222.

Attneave, F. (1959). *Applications of Information Theory to Psychology*. Holt, Rinehart, and Winston, New York.

Attneave, F. and Arnoult, M. D. (1957). The quantitative study of shape and pattern perception. *Psychol. Bull.* **53**, 452–471.

Berlyne, D. E. (1957). Uncertainty and conflict: A point of contact between information-theory and behavioural-theory concepts. *Psychol. Rev.* **64**, 329–339.

Berlyne, D. E. (1966). Curiosity and exploration. *Science* **153**, 25–33.

Chipman, S. (1977). Complexity and structure in visual patterns. *J. Exp. Psychol: General* **106**, 269–301.

Dember, W. N. and Warm, J. S. (1979). *Psychology of Perception*. Holt, Rinehart, and Winston, New York.

Garner, W. R. (1962). *Uncertainty and Structure as Psychological Concepts*. Wiley, New York.

Garner, W. R. (1974). *The Processing of Information and Structure*. Erlbaum, Potomac, MD.

Garner, W. R. and Clement, D. E. (1963). Goodness of pattern and pattern uncertainty. *J. Verbal Learning and Verbal Behavior* **2**, 446–452.

Hochberg, J. and McAlister, E. (1953). A quantitative approach to figural 'goodness'. *J. Exp. Psychol.* **46**, 361–364.

Miller, W. (1972). *Symmetry Groups and their Applications*. Academic Press, London.

Palmer, S. E. (1991). Goodness, Gestalt, groups, and Garner: Local symmetry subgroups as a theory of figural goodness. In: *The Perception of Structure*. G. R. Lockhead and J. R. Pomerantz (Eds). American Psychological Association, Washington, DC.

Palmer, S. E. and Hemenway, K. (1978). Orientation and symmetry: Effects of multiple rotational and near symmetries. *J. Exp. Psychol: Human Percept. Perform.* **4**, 691–702.

Rock, I. (1983). *The Logic of Perception*. MIT Press, Cambridge, MA.

Rock, I and Leaman, R. (1963). An experimental analysis of visual symmetry. *Acta Psychologica* **21**, 171–183.

Royer, F. (1981). Detection of symmetry. *J. Exp. Psychol: Human Percept. Perform.* **7**, 1186–1210.

Weyl, H. (1952). *Symmetry*. Princeton University Press.

Zabrodsky, H. (1993). *Computational Aspects of Pattern Characterization — Continuous Symmetry*. PhD Thesis, Dept of Computer Science, Hebrew University, Jerusalem, Israel.

Zabrodsky, H. and Avnir, D. (1993). Measuring symmetry in structural chemistry. In: *Advanced Molecular Structure Research*, Vol. 1. I. Hargittai (Ed.). (In press).

Zabrodsky, H., Peleg, S. and Avnir, D. (1993). Completion of occluded shapes using symmetry. In: *IEEE Conference on Computer Vision and Pattern Recognition*. New York, pp. 678–679.

Human Symmetry Perception, pp. 303–318
C. W. Tyler (Ed.)

Quantification of local symmetry: application to texture discrimination

YORAM BONNEH, DANIEL REISFELD and YEHEZKEL YESHURUN*
Department of Computer Science, Tel Aviv University, 69978 Tel Aviv, Israel

Abstract—Symmetry is one of the most prominent cues in visual perception as well as in computer vision. We have recently presented a Generalized Symmetry Transform that receives as input an edge map, and outputs a symmetry map, where every point marks the intensity and orientation of the local generalized symmetry. In the context of computer vision, this map emphasizes points of high symmetry, which, in turn, are used to detect regions of interest for active vision systems. Many psychophysical experiments in texture discrimination use images that consist of various micro-patterns. Since the Generalized Symmetry Transform captures local spatial relations between image edges, it has been used here to predict human performance in discrimination tasks. Applying the transform to micro-patterns in some well-studied quantitative experiments of human texture discrimination, it is shown that symmetry, as characterized by the present computational scheme, can account for most of them.

1. INTRODUCTION

Symmetry is among the most prominent spatial relations perceived by humans. Natural and artificial objects often give rise to the human sensation of symmetry, and this sense of symmetry is so strong that the Gestalt school considered symmetry as a fundamental principle of perception. Looking around us, we get the immediate impression that practically every interesting visual area consists of some generalized form of symmetry.

We have recently presented a generalized symmetry transform and demonstrated its application to detection of interest points in natural images (Reisfeld *et al.*, in press), in face recognition (Edelman *et al.*, 1992) and in normalization (Reisfeld and Yeshurun, 1992) tasks. Using this measure, we have suggested a computational model that takes as an input the intensity gradient at each image point, and generates activity maps of the generalized symmetry in different scales. Areas of different texture are thus characterized by different activity on one or more of these maps. The basis of our transform is the quantification of local spatial relations between image edges in a way

*To whom correspondence should be addressed.

that captures the amount of symmetry support at each point. For example, parallel centered intensity gradients strongly support a symmetry point between them while collinear gradients do not. This quantification can be interpreted as an estimation of a local 'Gestalt Glue'. It computes points of interest, where attention might then be directed, e.g. eyes of a person in a portrait or heads in a group of people (Reisfeld *et al.*, 1990). When applied more locally to images, it detects sharp corners as interest points by assigning high 'symmetry' value. Applied to texture micro-patterns, it quantifies every micro-pattern, and thus can be used as the basis for discrimination. In this paper, we present the results of applying our measure to some well-studied quantitative results in human psychophysics, and show that our simple model yields a good fit to human performance.

2. GENERALIZED SYMMETRY

In the usual mathematical notion, an object is regarded as symmetric if it is invariant to the application of certain transformations, called symmetry operations. A typical symmetry operation is the well-known reflectional (mirror) symmetry. In order to use these symmetry operations, it is necessary to know the shape of an object before we can estimate whether it is symmetric or not. However, we wish to quantify symmetry without any prior knowledge of objects, especially if the symmetry measure is used to detect regions of interest.

Our symmetry transform does not require the knowledge of the object's shape. It performs local operations on the edges of the image. Moreover, it assigns a continuous symmetry measure to each point in the image, rather than a binary symmetry label.

We first define a symmetry measure for each point. Let $p_k = (x_k, y_k)$ be any point ($k = 1, \ldots, K$), and denote by $\nabla p_k = (\partial p_k/\partial x, \partial p_k/\partial y)$ the gradient of the intensity at point p_k. We assume that a vector $v_k = (r_k, \theta_k)$ is associated with each p_k such that $r_k = \log(1 + \|\nabla p_k\|)$ and $\theta_k = \arctan((\partial p_k/\partial y)/(\partial p_k/\partial x))$. For each two points p_i and p_j, we denote by l the line passing through them, and by α_{ij} the angle counterclockwise between l and the horizon. We define the set $\Gamma(p)$, a distance weight function $D_\sigma(i, j)$, and a phase weight function $P(i, j)$ as

$$\Gamma(p) = \left\{(i, j)\Big|\frac{p_i + p_j}{2} = p\right\},$$
$$D_\sigma(i, j) = G_\sigma\left(\|p_i - p_j\|\right),$$
$$P(i, j) = \left(1 - \cos\left(\theta_i + \theta_j - 2\alpha_{ij}\right)\right)\left(1 - \cos\left(\theta_i - \theta_j\right)\right),$$

where $G_\sigma(t) = (1/\sqrt{2\pi\sigma})\exp-(t^2/(2\sigma^2))$ is the Gaussian, σ of D_σ corresponds to scale or a locality channel.

We define the contribution of the points p_i and p_j as

$$C(i, j) = D_\sigma(i, j)P(i, j)r_ir_j.$$

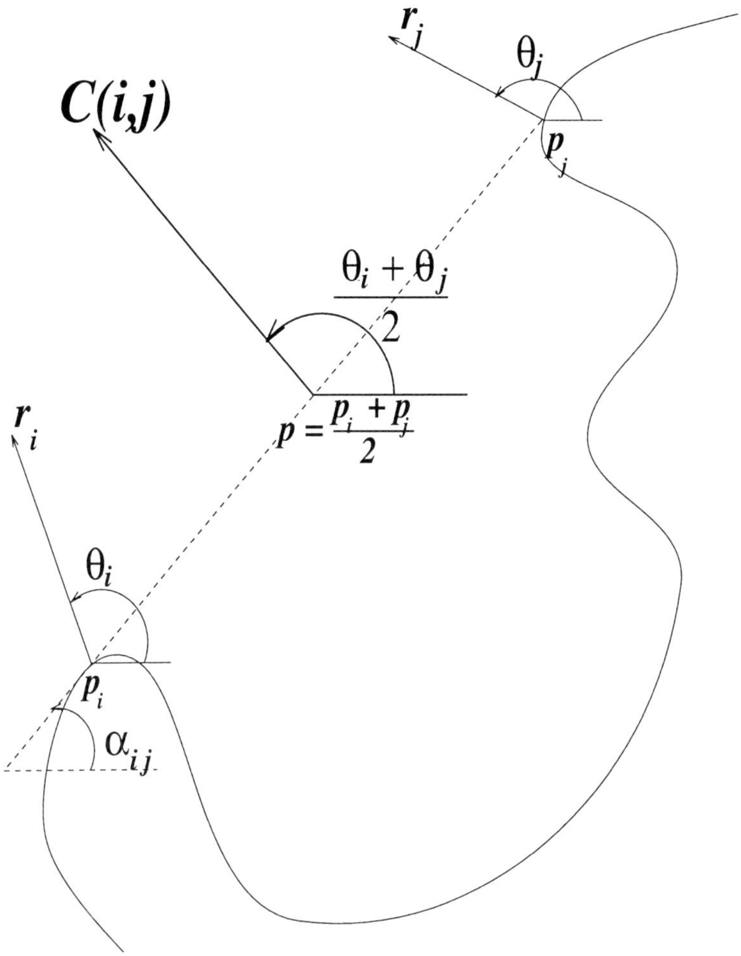

Figure 1. The contribution to symmetry of the gradients at p_i and p_j.

This measure easily can be normalized, and reflects the fact that each of its components modulates the other ones. The *symmetry magnitude* or *isotropic symmetry* $M_\sigma(p)$ of each point p is defined as

$$M_\sigma(p) = \sum_{(i,j)\in\Gamma(p)} C(i, j),$$

which averages the symmetry value over all orientations. We define the direction of the contribution of p_i and p_j as

$$\varphi(i, j) = \frac{\theta_i + \theta_j}{2}.$$

The *symmetry direction* is defined as $\phi(p) = \varphi(i, j)$ such that $C(i, j)$ is maximal for $(i, j) \in \Gamma(p)$. Thus, the symmetry of the point p is defined as

$$S_\sigma(p) = \big(M_\sigma(p), \phi(p)\big).$$

The demand that the symmetry transform be local is reflected by the Gaussian distance weight function, $D_\sigma(i, j)$. Different values for σ imply different scales, thus enabling convenient implementation of multi-resolution schemes. Note that the Gaussian defined above has circular isoclines, i.e. it has no preferred orientation. However, one can also define Gaussians with elliptic isoclines. This is useful when the transform is applied as a feature detector of elliptic regions such as eyes in human faces (Reisfeld and Yeshurun, 1992). In the experimental results presented in this paper we have used only circular Gaussians.

The phase weight function, $P(i, j)$ is composed of two terms. The first term, $1-\cos(\theta_i+\theta_j-2\alpha_{ij})$, achieves maximum symmetry when $(\theta_i-\alpha_{ij})+(\theta_j-\alpha_{ij}) = \pi$, i.e. when the gradients at p_i and p_j are oriented in the same direction towards each other. This is consistent with the intuitive notion of mirror symmetry. This expression decreases continuously as the situation deviates from the ideal one. Notice that the same measure is achieved for various object reflectance and lighting situations, which are discussed in (Reisfeld *et al.*, 1994).

The second term of $P(i, j)$, $1 - \cos(\theta_i - \theta_j)$, is introduced since the first term attains its maximum whenever $(\theta_i - \alpha_{ij}) + (\theta_j - \alpha_{ij}) = \pi$. This includes the case $\theta_i - \alpha_{ij} = \theta_j - \alpha_{ij} = \pi/2$, which occurs on a straight edge, which we do not regard as interesting. The current expression compensates for this situation.

The term $r_i r_j$ is high when there is a strong correlation between two large gradients. We use gradients rather than intensities since we are mainly interested in edges that relate to object borders. For instance, a uniform intensity wall is highly symmetric but probably not very interesting. In natural scenes we prefer to use the logarithm of magnitude instead of the magnitude itself, since it reduces the differences between high gradients, and therefore the correlation measure is less sensitive to very strong edges.

Sometimes it is necessary to detect points that are highly symmetric in multiple distinct orientations rather than in a principle one. We define such a symmetry as *radial symmetry* $RS(p)$ and its value can be evaluated using the formula:

$$RS_\sigma(p) = \sum_{(i,j)\in\Gamma(p)} C(i, j)\sin^2\big(\varphi(i, j) - \phi(p)\big).$$

This expression emphasizes contribution in the directions which are perpendicular to the main symmetry direction, and attains its maximum in a point that is surrounded by edges. Notice that, due to the continuous nature of the operator, the radial symmetry is not sensitive to gaps in the contour that surrounds the point p, and does not require this contour to be uninterrupted.

The above definition is highly sensitive to the direction of contrast, which is significant in evaluating symmetry of objects. Sometimes however, it is unimportant as

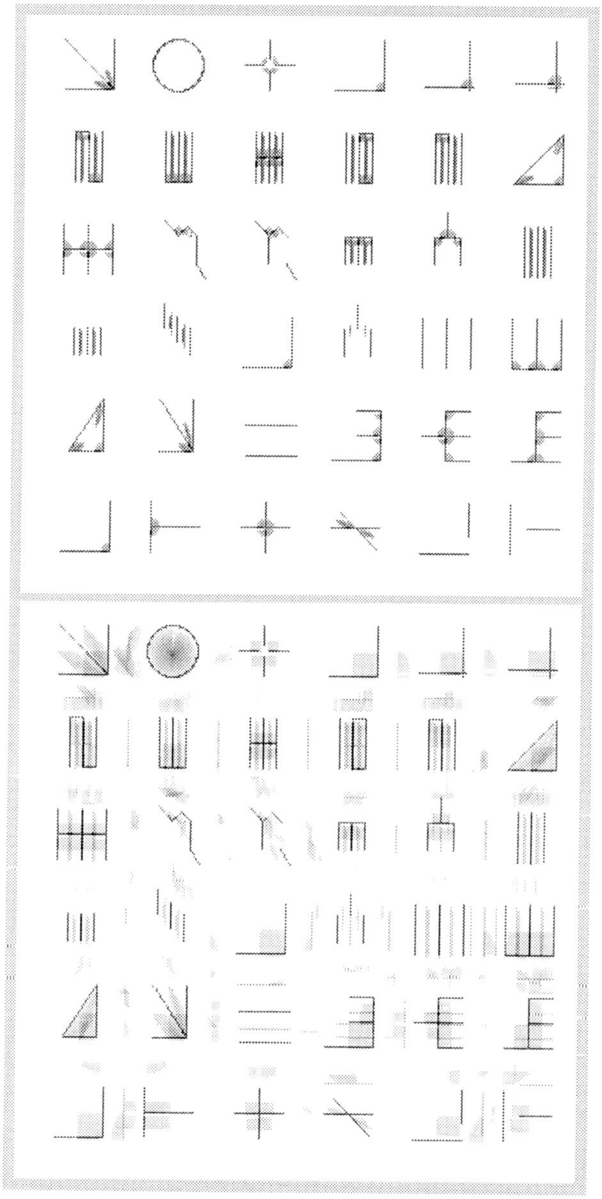

Figure 2. Some of the patterns used in the psychophysical experiments superimposed on their symmetry response. Right: the local channel ($\sigma = 2$); left: a wide channel ($\sigma = 4$). The logarithm of the response of the radial symmetry is shown as gray-level intensity.

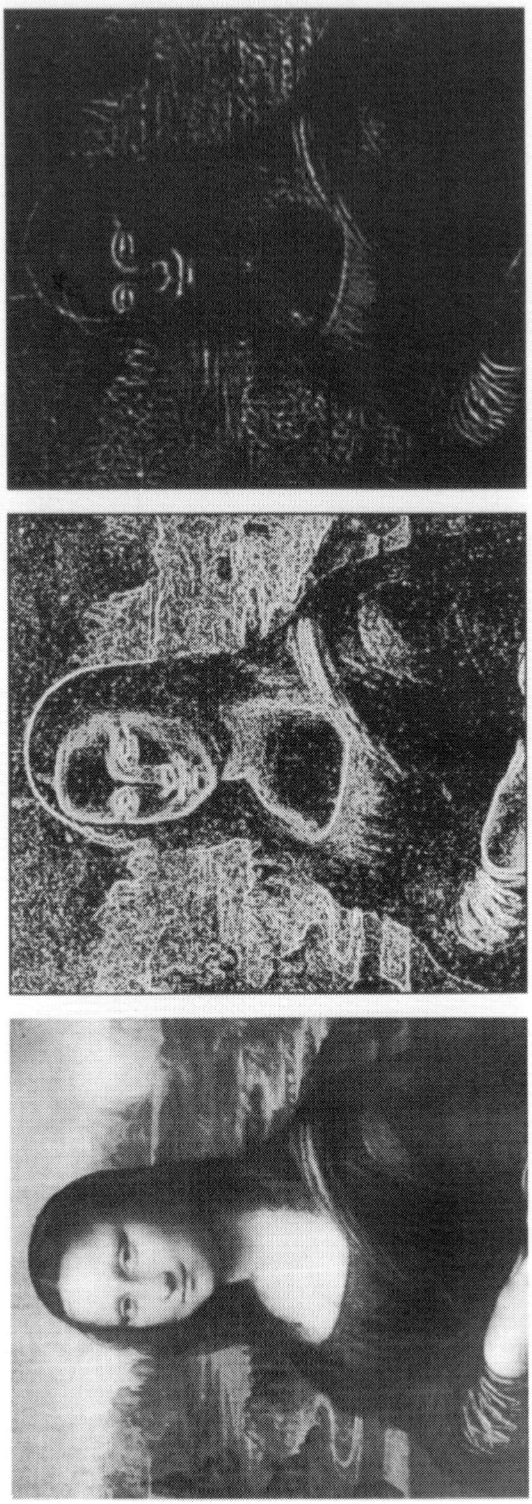

Figure 3. Detection of points of interest using the generalized symmetry transform. (Left to right: original image; its edges; isotropic symmetry.)

in the case of texture micro-patterns or line drawings. A line, for example, should be treated as a single edge and not as a thin object. We define a variant of the symmetry transform by modifying the phase weight function so that edge orientation is used and direction of contrast ignored. Let $u_k = (r_k, \theta_k)$ be the edge at p_k such that r_k is the edge magnitude and $\theta_k \in [0, \pi]$ is the edge orientation. The texture variant of the phase weight function is given by

$$P'(i, j) = G_{\sigma_0}(\theta_i - \alpha_{ij} - \pi/2)G_{\sigma_0}(\theta_j - \alpha_{ij} - \pi/2),$$

where G_σ is the Gaussian function as in the previous definition and the phase tuning σ_0 is fixed. When the edge orientations are collinear there is no support for symmetry. The strongest support is when the two edge orientations are both orthogonal to the virtual line connecting the points, i.e. parallel centered. This is captured by a product of two Gaussian tuning functions, which makes the phase function monotonic between these two extreme cases. The effect of this definition is similar to the one defined earlier when ignoring the direction of contrast. We use it in all the applications of the symmetry transform to textures.

In Fig. 3 we demonstrate the symmetry map produced by the transform on a (almost) natural image. As we show in Reisfeld *et al.* (in press), the symmetry transform generalizes most of the existing methods for detection of regions of interest in computer vision (e.g. high curvature, density of edges, junctions). The peaks of intensity in the symmetry map are used to index regions of interest, and in Fig. 3 the highest peaks of the intensity map are the facial features. In Fig. 2 we present the symmetry map (texture variant) of various texture micro-patterns, to demonstrate the motivation behind our approach, and to show that different micro-patterns are differently quantified by the transform. In the following we define in detail the specific measure used for discrimination.

3. MODELS OF TEXTURE DISCRIMINATION AND SEGMENTATION

Humans are able to discriminate between surfaces sharing the same average color and brightness but differing in small-scale luminance variations generally called texture. This ability is vital to segmentation since natural objects are often heterogeneous and their boundary cannot always be found by simple edge detectors. The classical work of Julesz claimed that texture discrimination could be explained in terms of global second-order statistic differences between points in the image (Julesz, 1975). Later work attributed the discrimination to first-order differences in features such as orientation, size and brightness of local texture elements (Beck, 1966, 1983; Julesz, 1981, 1986). The Texton theory (Julesz, 1981) specified these elements as textons, that are elongated blobs with specific color, orientation and size, line ends (terminators) and line crossings. Some theories (Marr, 1982; Beck *et al.*, 1983) claim that a hierarchical grouping process links the basic features according to Gestalt rules (e.g. proximity, similarity and good continuation) and texture boundaries are extracted from differences in these higher order elements. All these theories were based on a qualitative

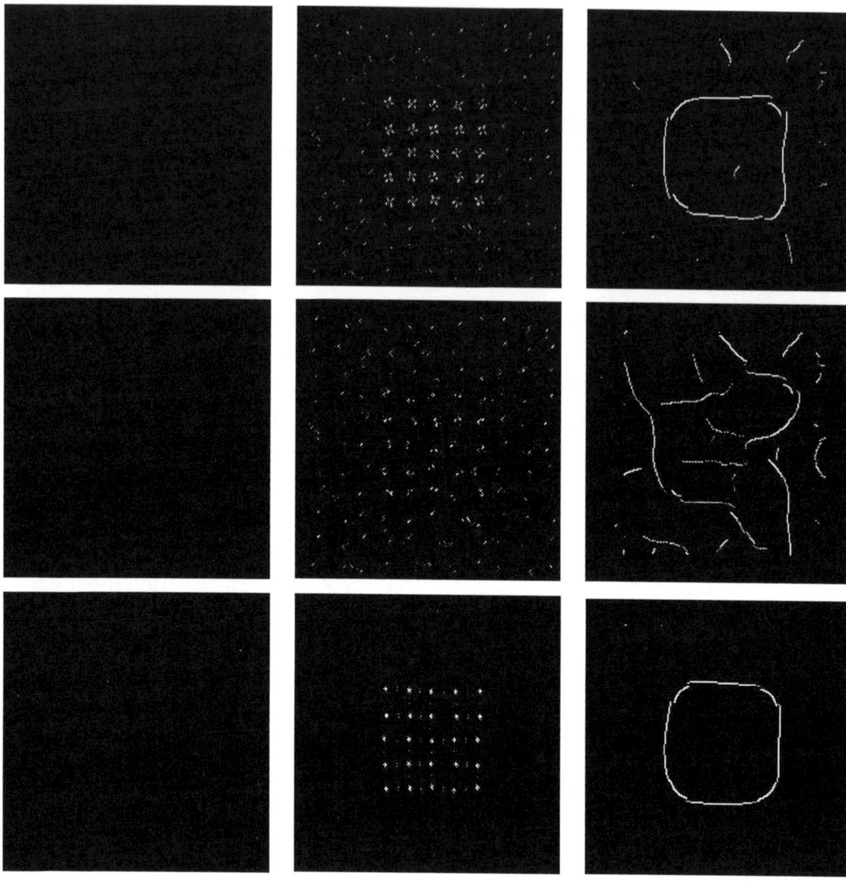

Figure 4. Artificial texture discrimination by the model. Left column: the textures; middle: local symmetry maps; right: computed boundaries.

distinction between effortless *preattentive* texture discrimination and the more time consuming *attentive* discrimination.

Recent psychophysical experiments have questioned the preattentive-attentive dichotomy and found graded discriminability in textures composed from randomly rotated patterns (similar to Fig. 4) (Gurnsey and Browse, 1987). Similar results were obtained in detection tasks (Krose, 1987). They provide us with a database of discriminability measures for a set of randomly rotated artificial pattern pairs. These observations raised the need for a computational model that discriminates between textures in a continuous manner and can be quantitatively compared with human performance. Such models (Fogel and Sagi, 1989; Malik and Perona, 1990) are based on linear filters followed by a non-linear stage that produces activity maps where texture boundaries becomes activity gradient. Discrimination is then achieved by combining the gradients in the different maps followed by a decision stage that corresponds to subject's decision as to whether a boundary exists or not. The effect of figure-ground asymmetry in discriminability is attributed to this decision stage (Rubenstein and Sagi, 1990).

Other recent studies of computer vision (Porat and Zeevi, 1989; Jain and Farrokhnia, 1991; Bovik, 1991; Dunn *et al.*, 1994) and human texture discrimination are concerned with segmentation based on orientation and spatial-frequency filters rather than with a direct quantification of micro-pattern's spatial configuration. Nothdurft (1991) demonstrated texture segmentation that arises from orientation differences rather than from the orientation features themselves. Landy and Bergen (1991) studied orientation and scale difference and developed a full-scale quantitative model based on linear filters in different orientations scales and energy summation. Their model matches orientation discrimination data but was not tested on micro-pattern textures. Hallett (1992) developed mesh-derived textures and studied the effect of orientation and spatial disorder. His results show that orientation and position jitter of textural element are often exchangeable, a fact that contradicts the common belief that de-emphasizes the role of element position (Julesz, 1984).

4. DISCRIMINATION OF TEXTURE MICRO-PATTERNS BY GENERALIZED SYMMETRY

Almost all psychophysical studies of texture discrimination use various artificial textures. The most popular paradigm uses split fields of randomly rotated micro-patterns. Figure 4 demonstrates the model's operation on such artificial textures. The top and middle rows demonstrate textures composed of X-L and L-T patterns. These textures have been widely studied (Julesz, 1984; Gurnsey and Browse, 1987; Krose, 1987; Bergen and Adelson, 1988; Fogel and Sagi, 1989; Malik and Perona, 1990) since the X-L texture is easily discriminated while the L-T requires more time and attention. This was measured (among other pattern pairs) by Gurnsey and Browse (1987) and in detection tasks that yield similar results by Krose (1987). The current explanations include the texton theory (Julesz, 1981) that attributes the X-L discrimination to a crossing texton, the size-tuning principle (Bergen and Adelson, 1988) that attributes discrimination to difference in size, and models that define filter-based mechanisms producing the desired results (Fogel and Sagi, 1989; Malik and Perona, 1990).

These well known examples are shown in Fig. 4, along with their corresponding symmetry maps, in order to clarify the intuition behind our approach. The response for the X pattern consists of four components, the T of two and the L consists only of a single component. If discriminability is attributed to difference in the total amount of symmetry, then X-L should be discriminated much more easily than the L-T. In order to make explicit use of the differences in the symmetry map, we used a simple low-frequency edge-detector that is applied to the symmetry map, to extract the texture boundaries. We emphasize that this stage was carried out as a demonstration only, since we do not present a complete texture segmentation algorithm, but rather a micro-pattern discrimination measure.

The boundary of the X-L texture is clearly visible (top right) while the boundary of the L-T texture is not clear, although some fragments of it can be noticed.

The bottom row of Fig. 4 demonstrates the discrimination of similar textures composed of X-O patterns studied by Krose (1987). The computation is identical to that used for the other textures except for a wider symmetry channel ($\sigma = 4$ instead of

$\sigma = 2$ used for the upper two rows). The model succeeds in computing explicit boundaries for these highly discriminable textures as displayed in the bottom right section. It is important to note that the discrimination of these artificial textures is a simple task and can be done in many different ways. The key point is not to segment these textures, but to match human performance in texture discrimination. Thus, we are looking for a measure that incorporates the response of all the symmetry channels, and can be compared to human performance.

4.1. The discriminability measure

In order to compare our model to human performance, we define a discriminability measure between micro-pattern pairs. The symmetry measure for a micro-pattern m, which is a small binary image, is the log of the sum of a symmetry measure SM (which can be the isotropic symmetry, the radial symmetry, etc.) of all of its points:

$$R_\sigma(m) = \log\left(\sum_{p \in m} SM_\sigma(p)\right).$$

The *discriminability measure* between two micro-patterns $m1$ and $m2$ in a channel determined by σ is defined as

$$DM_\sigma(m, n) = \|R_\sigma(m) - R_\sigma(n)\|.$$

The (total) discriminability is

$$DM(m, n) = \max_\sigma DM_\sigma(m, n),$$

which is the maximal response over all channels. This is a natural extension to models of brightness perception (e.g. Land and McCann, 1971), since the luminance and the symmetry edges might be computed in a similar way. In the following, we have actually used radial symmetry (RS_σ) as the specific symmetry measure, and two channels.

4.2. Results

We applied this computation to the patterns studied by Gurnsey and Browse (1987) and Krose (1987) using 30×30 pixel patterns, 2 channels with $\sigma = 7$ and $\sigma = 40$ (in pixels) and $\sigma_0 = 0.125$ for the phase function (see definition of the transform). Results appear in Fig. 5. The black bars represent human discrimination ability, where higher values represent easier discrimination. The gray bars represent the model's discriminability values after linear normalization. The pair numbers follow the data from Gurnsey and Browse (1987) and can be used for reference. The computed correlation for the bottom table which corresponds to the experiments of Gurnsey and Browse (1987) is $\rho = 0.93$. We also tested our model with data from Krose (1987) which appear in the top charts. The resulting correlation is $\rho = 0.86$ and $\rho = 0.98$ for the two experiments. These values are better than those reported previously (Krose, 1987; Fogel and Sagi, 1989).

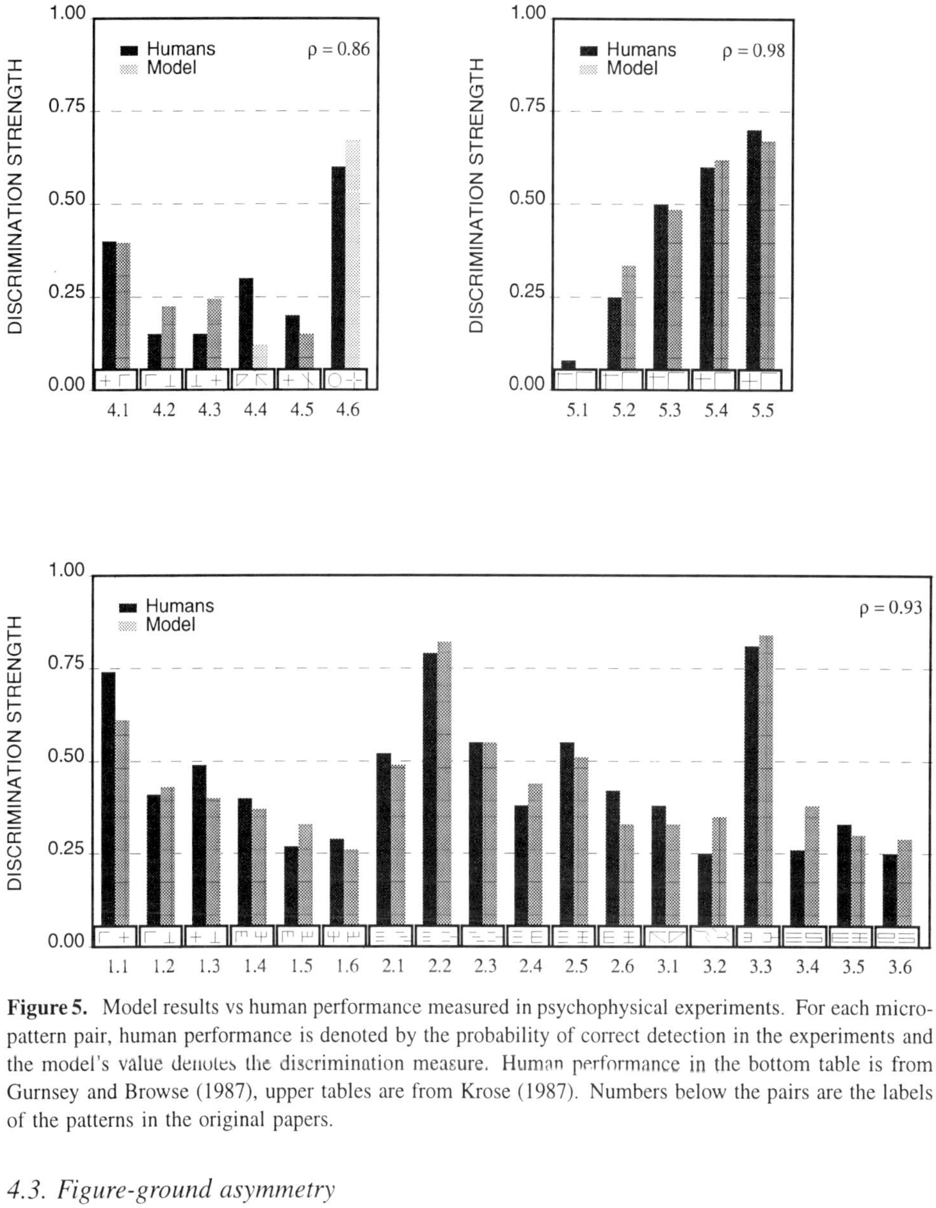

Figure 5. Model results vs human performance measured in psychophysical experiments. For each micro-pattern pair, human performance is denoted by the probability of correct detection in the experiments and the model's value denotes the discrimination measure. Human performance in the bottom table is from Gurnsey and Browse (1987), upper tables are from Krose (1987). Numbers below the pairs are the labels of the patterns in the original papers.

4.3. Figure-ground asymmetry

An asymmetry of figure-ground in texture discrimination was reported by Gurnsey and Browse (1987) and by Williams and Julesz (1992) and was modeled by Rubenstein and Sagi (1990). Asymmetry was also found in similar detection tasks (Triesman and Gormican, 1988). For example, it is easier to detect a square of L patterns embedded in surrounding X patterns than vice versa. In testing our model, we have used average-discriminability data reported by others (Fogel and Sagi, 1989; Malik and Perona, 1990). A more careful investigation of the model's computed symmetry values shows that patterns of higher symmetry are the preferred background. This is

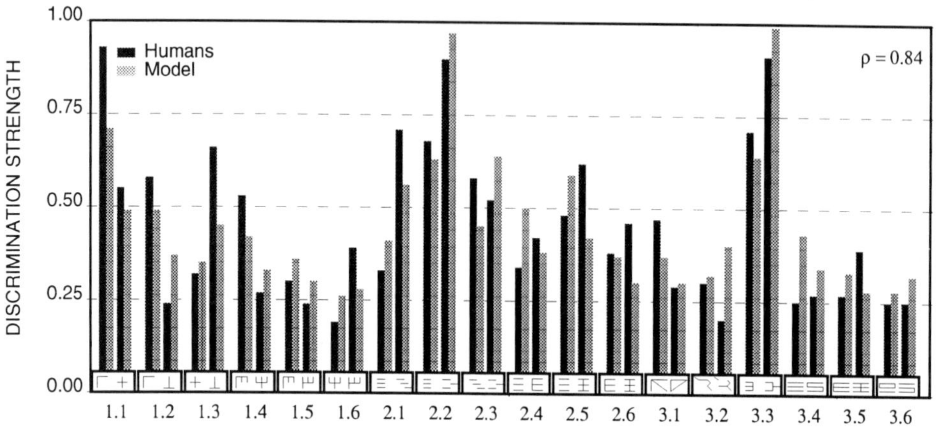

Figure 6. Model results vs human asymmetric performance measured in the psychophysical experiments of Gurnsey and Browse (1987). For each micro-pattern pair, there are two pairs of bars where the left/right pair corresponds to the left/right pattern being the target, respectively. Human performance is denoted by the probability of correct detection in the experiments and the model's value denotes the discrimination measure.

formulated as a modification to the discriminability measure defined earlier:

$$DM_\sigma(g, t) = |SM_\sigma(g) - SM_\sigma(t)| + k * \big(SM_\sigma(g) - SM_\sigma(t)\big),$$

where g, t are the ground and target patterns, respectively, and $k < 1$ is a weight constant. The total discriminability is taken to be the maximal response over all channels as in the symmetric model.

We applied the modified model to the data from Gurnsey and Browse (1987) using $k = 0.3$ and all other parameters as in the symmetric case. Results appear in Fig. 6. The graph is similar to the previous symmetric case except that for each pattern pair there are two pairs of bars. The left/right pair correspond to the left/right pattern being the target. Again, the black bars represent human discrimination ability and the gray bars represent the model's discriminability values. The computed correlation is $\rho = 0.84$, which is slightly smaller than reported previously (Rubenstein and Sagi, 1990). The value of k is not that critical and similar correlations were obtained for k between 0.1 and 0.4.

5. DISCUSSION

We have described a method for the quantification of local generalized symmetry in images that operates on edge images and produces a symmetry map. We have demonstrated its use in the detection of regions of interest, a task which is mainly relevant to computer vision, and presented a simple model, based on generalized symmetry, that shows good correlation with psychophysical data in texture-discrimination tasks.

Since the factors involved in the computation are few, we tested the significance of each in the final result. First we consider locality. Since the symmetry transform is local, as expressed in its distance-weight function, one may suspect that locality *per se* is sufficient and that discrimination may be based merely on evaluating the size of the micro-patterns and comparing it for both textures (Bergen and Adelson, 1988). Bergen and Adelson (1988) demonstrated this principle using the rectified Laplacian of a Gaussian. We have tested a variant of the symmetry model, which stresses this locality idea, by neglecting the phase component. This decreases the correlation to the psychophysical data of Gurnsey and Browse from 0.93 to 0.73 that the full model achieves. This degradation is mainly due to micro-patterns with identical second-order statistics like pair 3.1 in Fig. 5. It is clear, however, that the main factor in the discrimination of the textures we have tested is indeed size (as already noted by Gurnsey and Browse). Our model captures size with its local Gaussian weighting function and its non-linear operation (the conjunction of pairs of edge points). There might be alternative ways to capture it.

Next we consider radial symmetry and the need for actual symmetry maps. Using isotropic symmetry instead, correlation decreases to 0.85 for the Gurnsey and Browse data and 0.82, 0.98 for the data from Krose. Again, there is a specific problem with the patterns with identical second-order statistics, so the total correlation is somewhat misleading. There is, however, an advantage in using the isotropic symmetry. In this case the model is reduced to a simpler quantification of local spatial relations.

The figure-ground asymmetry effect is addressed in our work by modifying the symmetry measure to reflect the ratio between the background and foreground symmetry measures. The relative success of this formulation can be interpreted as another effect of size tuning: it is easier to detect large patterns among a background of small ones than the other way around. However, further inspection of figure-ground asymmetry evidence (Treisman and Gormican, 1988; Williams and Julesz, 1992) does not fully support this idea. Instead, it seems that the general principle behind all cases is that the more symmetric pattern is a 'better' background. For example, a circle is a more symmetrical pattern than ellipse, so it is easier to detect ellipses among circles as background than vice versa. A full quantified explanation of the perceptual asymmetries in discrimination and detection tasks is left for future work.

Next we would like to point out that the set of patterns used in texture psychophysics is somewhat biased. Most of it was constructed in order to demonstrate a specific principle or theory. Therefore, it is possible that these patterns, which are traditionally used as test data, are not sufficient for the purpose of assessing new algorithms and theories. We have, therefore, constructed a data base of 10^4 micro-patterns by generating patterns composed of three line segments whose position is selected at random (see Fig. 7). This might be the first step towards constructing a large database of artificial textures that could be used for evaluating computational models for texture discrimination.

We hypothesize that patterns that differ in their generalized symmetry are easily discriminated by humans and vice versa, and suggest that textures from this database be used to verify it. In order to exclude other factors, like differences in the second-order statistics, patterns with similar second-order statistics were selected by an automatic exhaustive search from this database of 10^4 randomly generated patterns. This was

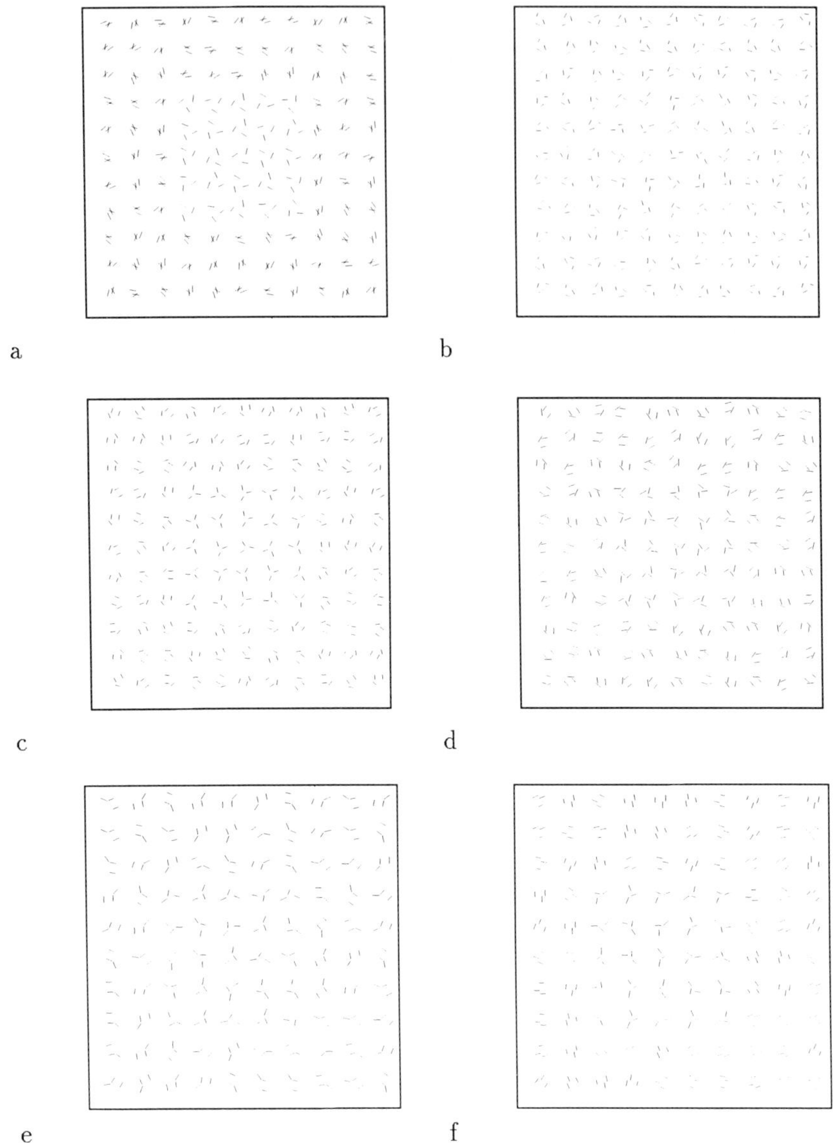

Figure 7. Randomly generated 3-line pattern textures: (a) strong discriminability predicted, probably due to the locality Gaussian; (b) very weak discriminability predicted; (c)–(f) textures with similar second-order statistics, predicted to have intermediate discriminability.

done by an explicit computation and comparison of the dot-distance distribution (neglecting the orientation since the patterns are randomly rotated). Figure 7 demonstrates some of these textures tested by the model. Panel (a) consists of textures predicted to have high discriminability. This may be attributed to the different size of the patterns captured by the locality Gaussian. Panel (b) consists of a texture predicted to be almost indiscriminable (having tiny discriminability measure). Panels (c), (d), (e), (f) consist of textures with similar second-order statistics predicted to have some discriminability. We suggest that a systematic and quantitative analysis of randomly generated patterns might prove useful for further evaluation of models and algorithms for texture discrimination.

Our model should be compared with previous models (Krose, 1987; Bergen and Adelson, 1988; Fogel and Sagi, 1989; Rubenstein and Sagi, 1990; Malik and Perona, 1990; Landy and Bergen, 1991; Buf and Heitger, 1993) in terms of fitting the psychophysical data, computational efficiency and biological plausibility. The model fits the average figure-ground discriminability measured by Gurnsey and Browse (1987) better than any other model (not all provide a quantitative match). Computationally, our model is somewhat different from the filter-based models mentioned previously (Fogel and Sagi, 1989; Malik and Perona, 1990; Landy and Bergen, 1991). These models are defined in the level of neural mechanism inspired by cells in the visual cortex. Their discrimination is based on the responses of filters whose size is not smaller than a micro-pattern, such that the main computation is done by the filters. When larger patterns are used, larger receptive fields are assumed, ignoring the computation carried out by the cells with the smaller receptive fields. Indeed, our approach is based on a simple and intuitive quantification of generalized symmetry, and does not explicitly involve cortical models. However, if we examine the biological plausibility of our model, it turns out that it fits perfectly a view of the primary visual cortex as a network that analyzes local spatial configurations of edges (Zucker *et al.*, 1989). According to this model, V1 can be viewed in a level that is higher than a set of filters; namely, it is a network where every cell responds to a given range of edge orientation and size, but there are lateral connections between cells that give rise to the notion of curvature, for example. The same network that estimates curvature can quantify every image area according to our algorithm, thus enabling a direct segregation of texture. Another alternative is the computation of symmetry maps as defined by the model. Each symmetry cell in such a map integrates pairs of V1 simple cells in the proper orientations using Sigma-Pi connections (sum of the product of each pair). There is currently no evidence for such maps although some similarity can be found in V2 cells that respond to illusory contours, probably by integrating orientation selective cells in V1 (Heitger *et al.*, 1992).

While offering a good fit for psychophysical data, symmetry quantification should be viewed as part of a larger process. The generalized symmetry model may be a part of an hierarchical texture-processing system in which the lowest level might compute simple texton differences (orientation, size, color, contrast) in low resolution; the next level quantifies local spatial relations based on local generalized symmetry; and the highest level may perform grouping processes as suggested by Beck (1983).

REFERENCES

Beck, J. (1966). Effect of orientation and of shape similarity on grouping. *Percept. Psychophys.* **1**, 300–302.

Beck, J. (1983). Textural segmentation, second-order statistics, and textural elements. *Biol. Cybernet.* **48**, 125–130.

Beck, J., Prazdny, K. and Rosenfeld, A. (1983). A theory of textural segmentation. In: *Human and Machine Vision*. J. Beck, B. Hope and A. Rosenfeld (Eds). Academic Press Inc., New York, pp. 1–38.

Bergen, J. R. and Adelson, E. H. (1988). Early vision and texture perception. *Nature* **333**, 363–364.

Bovik, A. C. (1991). Analysis of multichannel narrow band filters for image texture segmentation. *IEEE Trans. Signal. Proc.* **39**, 2025–2043.

Buf, J. D. and Heitger, F. (1993). A neural network for detecting symmetry orders. In: *Proc. 7th Int. Conf. on Image Analysis*, Bari, Italy.

Dunn, D., Higgins, W. E. and Wakely, J. (1994). Texture segmentation using 2D Gabor elementary functions. *IEEE Trans. Pattern Anal. Machine Intell.* **PAMI-16** (2), 130–149.

Edelman, S., Reisfeld, D. and Yeshurun, Y. (1992, May). Learning to recognize faces from examples. In: *Second European Conference on Computer Vision*, S. Margherita, Ligure, Italy.

Fogel, I. and Sagi, D. (1989). Gabor filters as texture discriminator. *Biol. Cybernet.* **61**, 103–113.

Gurnsey, R. and Browse, A. (1987). Micropattern properties and presentation conditions influencing visual texture discrimination. *Percept. Psychophys.* **41** (3), 239–252.

Hallet, P. E. (1992). Segregation of mesh-derived textures evaluated by resistance to added disorder. *Vision Res.* **32** (10), 1899–1911.

Heitger, F., Rosenthaler, L., Heydt, R. V. D., Peterhans, E. and Kubler, O. (1992). Simulation of neural contour mechanisms: from simple to end-stopped cells. *Vision Res.* **32** (5), 963–981.

Jain, A. K. and Farrokhnia, F. (1991). Unsupervised texture segmentation using Gabor filters. *Pattern Recog.* **23**, 1167–1185.

Julesz, B. (1975). Experiments in the visual perception of texture. *Scient. Amer.* **232**, 34–43.

Julesz, B. (1981). Textons, the elements of texture perception and their interactions. *Nature* **290**, 91–97.

Julesz, B. (1984). Toward an axiomatic theory of preattentive vision. In: *Dynamic Aspects of Neocortical Function*. G. M. Edelman and W. E. Gall (Eds). Neurosciences Research Foundation, pp. 585–612.

Julesz, B. (1986). Texton gradients: The texton theory revisited. *Biol. Cybernet.* **54**, 245–251.

Krose, B. J. (1987). Local structure analyzers as determinants of preattentive texture discrimination. *Biol. Cybernet.* **55**, 289–298.

Land, E. and McCann, J. J. (1971). Lightness and retinex theory. *J. Opt. Soc. Am.* **61** (1), 1–11.

Landy, M. S. and Bergen, J. R. (1991). Texture segregation and orientation gradient. *Vision Res.* **31**, 679–691.

Malik, J. and Perona, P. (1990). Preattentive texture discrimination with early vision mechanisms. *J. Opt. Soc. Am.* **7** (5), 923–932.

Marr, D. (1982). *Vision*. Freeman.

Nothdurft, H. (1991). Texture segmentation and pop-out from orientation contrast. *Vision Res.* **31** (6), 1073–1078.

Porat, M. and Zeevi, Y. (1989). Localized texture processing in vision: analysis and synthesis in the Gaborian space. *IEEE Trans. Biomedical Eng.* **36**, 115–129.

Reisfeld, D., Wolfson, H. and Yeshurun, Y. (1990, December). Detection of interest point using symmetry. In: *Third International Conference on Computer Vision*. Osaka, Japan, pp. 62–65.

Reisfeld, D., Wolfson, H. and Yeshurun, Y. (In press). Context free attentional operators: the generalized symmetry transform. *Int. J. of Computer Vision*.

Reisfeld, D. and Yeshurun, Y. (1992, September). Robust detection of facial features by generalized symmetry. In: *Proceedings of the 11th International Conference on Pattern Recognition*. The Hague, Netherlands.

Rubenstein, B. S. and Sagi, D. (1990). Spatial variability as a limiting factor in texture discrimination tasks: Implications for performance asymmetries. *J. Opt. Soc. Am.* **7**, 1632–1643.

Treisman, A. and Gormican, S. (1988). Feature analysis in early vision: evidence from search asymmetries. *Psycholog. Rev.* **95**, 15–48.

Williams, D. and Julesz, B. (1992). Filters versus textons in human and machine texture discrimination. In: *Neural Networks for Perception*, Vol. 1. H. Wechsler (Ed.). Academic Press Inc., pp. 145–175.

Zucker, S. W., Dobbins, A. and Iverson, L. (1989). Two stages of curve detection suggest two styles of visual computation. *Neural Comput.* **1**, 68–81.

Human Symmetry Perception, pp. 319–329
C. W. Tyler (Ed.)

A continuum of non-Gaussian self-similar image ensembles with white power spectra

JULIAN S. JOSEPH[1,*] and JONATHAN D. VICTOR[2]

[1] *Vision Sciences Laboratory, Department of Psychology, Harvard University, Cambridge, MA 02138, USA*

[2] *Department of Neurology and Neuroscience, Cornell University Medical College, 1300 York Avenue, New York, NY 10021, USA*

Abstract—We introduce a scaling procedure that acts on visual textures to produce new textures having the same resolution, display size, and mean contrast power. We derive the form of the scaling fixed-point textures (image ensembles) resulting from repeated application of this scale transformation to the 'even' texture with an arbitrary amount of sporadic decorrelation. The result is a continuum of scaling fixed-point or 'self-similar' image ensembles, ranging from a strongly non-Gaussian white texture with higher-order spatial correlations at one extreme to Gaussian white noise at the other. The simple construction of a continuum of self-similar ensembles possessing phase correlations provides a tool for investigating human perception of structure in the absence of useful length scales. The fixed-point textures have luminance histogram differences, a direct result of their higher-order spatial properties. This suggests that scaling might afford an extension of our understanding of IID (independent identically distributed) texture discrimination (Chubb *et al.*, *J. Opt. Soc. Am. A* **11**, 2350–2374, 1994) to more general texture discrimination tasks in which spatial correlations play a role.

1. INTRODUCTION

To recognize objects, the visual system needs to abstract signals into a form that is left invariant by certain transformations, such as translations in the plane, rotations, and changes in light sources. Changes in the scale of an object or surface come about naturally by variations in viewing distance from the surface in question. It is these scale changes that are the topic of this paper.

The human ability to recognize objects and distinguish surface properties is robust with respect to changes in scale over a wide range. The approximate scale-invariance of spatial-frequency discrimination Weber fractions attests to this ability (Campbell *et al.*, 1970; Hirsch and Hylton, 1982).

*To whom correspondence should be addressed.

Any symmetry operation has objects that transform in a simple way under its transformation. Image ensembles with translation-invariant statistics are well known. Examples include ensembles of sinusoidal gratings with some distribution of amplitudes and randomly chosen phases, which have been proven to be a powerful tool in the investigation of spatial vision. Discrimination between translation-invariant image ensembles cannot be facilitated by the use of localization mechanisms, because there are no position cues to help with the task. Similarly, discrimination between self-similar or 'scale-invariant' image ensembles receives no benefit from detection mechanisms that are sensitive to absolute spatial scale. It should be noted that the ensemble of natural scenes appears to have the self-similarity property (Carlson, 1978; Burton and Moorhead, 1987; Field, 1987), a fact that has been used in theories of retinal encoding (Atick and Redlich, 1990, 1992) and of the scaling properties of striate cell populations (Li and Atick, 1994). More generally, however, self-similar ensembles are useful tools for investigating the visual perception of the texture of a surface in the absence of cues as to its scale.

We are then led to ask: What images or image ensembles are left unchanged by variations in the viewing distance? One way to find an image ensemble that is invariant under some transformation is to apply the transformation over and over and see if a limit is approached. Consider the independent identically distributed (IID) textures of Chubb and Landy (1990, 1991) and Chubb *et al.* (1994). After stepping back a long way, all the IID ensembles with the same mean and variance approach the same Gaussian white-noise ensemble, by the Central Limit Theorem. Any ensemble of gratings with a finite correlation length (non-zero bandwidth) will approach the same limit, regardless of the original frequencies of the gratings. We also know that any Gaussian ensemble whose power spectrum depends on the spatial frequency as $|\vec{k}|^{-\alpha}$ is self-similar (Rensink, 1986). Discriminations between such Gaussian power-law ensembles have been performed in one dimension (Rensink, 1986) and in two dimensions (Knill *et al.*, 1990).

While these Gaussian self-similar ensembles may tell us a great deal about scale-independent processing of visual signals, they cannot tell us about processing of spatial phases. This is because, in a Gaussian ensemble, all the Fourier components are statistically independent of one another and have randomly distributed phases. It has been observed that, in the majority of photographs taken of real objects and scenes, independent variation of the Fourier phases by any appreciable amount completely destroys the visual percept. By contrast to this situation, human perception of the image is relatively robust with respect to variation of the Fourier amplitudes, as long as the phases are kept constant (Oppenheim and Lim, 1981; Piotrowski and Campbell, 1982; Shapley *et al.*, 1990). In non-Gaussian image ensembles, such as the type we construct here, the Fourier components are not independent of each other and the relative phases of different Fourier components are not randomly distributed. This means that, in this type of image ensemble, some of the phase structure that is apparently so important to the identification of naturally occurring scenes, faces, objects, and so on is represented by the non-Gaussian nature of the ensemble statistics. These image ensembles would be uniquely suited for studying the role of spatial phase or higher-order correlations in a context devoid of length scales. We will construct a continuum of self-similar image ensembles that differ from each other by their degree of non-Gaussian character.

What image ensembles do we know that are likely to go to non-Gaussian fixed-points upon repeated application of a scale transformation? A good candidate would be an ensemble characterized by higher-order spatial correlations among different points in the visual field, such that the correlations do not decay as we consider points that are more widely separated. A family of such candidates is provided by the so-called 'even' texture with varying degrees of 'sporadic decorrelation' (Victor and Conte, 1989, 1991). The even texture was originally presented by Julesz and co-workers (Julesz *et al.*, 1978; Caelli *et al.*, 1978) in a counter-example to the now-famous Julesz conjecture (Julesz, 1962), which posits that two textures easily segregate and are rapidly discriminable if and only if they possess different power spectra. The even/random discrimination can be done rapidly although the two ensembles share the same power spectrum. Spatial nonlinearities appear to play an important role in the performance of such an 'iso-dipole' discrimination (Victor and Conte, 1989, 1991). It has recently been shown that such nonlinearities are present in primary visual cortex at a broad range of length scales (Purpura *et al.*, 1994). As may be readily apparent at this stage, we will be using the term 'texture' to mean 'image ensemble', rather than a single textured image. As Yellott (1993) and Victor (1994) have emphasized, the problem of discriminating between images based on their autocorrelation functions is very different from the problem of discriminating between image ensembles based on their ensemble correlation functions. We focus on the latter problem here, believing it to be biologically relevant under circumstances in which the organism must determine 'What kind of stuff am I looking at?'

In addition to revealing a family of phase-correlated self-similar ensembles, scaling textures with higher-order correlations is also useful because of its effect on the histogram of the luminance at a single point in the visual field. Chubb, Econopouly, and Landy have shown that the class of IID textures with equal variance and mean luminance is perceptually one-dimensional, so that only one detector at each point in the visual field is needed to discriminate between them (Chubb and Landy, 1990, 1991; Chubb *et al.*, 1994) After these authors, we will refer to this as the CEL detector. We would like to use this as a point of departure in understanding more about the perception of spatial correlations.

One way to make a connection between the IID textures and textures possessing spatial correlations is to apply a scaling transformation. Consider the result of 'stepping back' from a texture and restoring the resolution (check size), the display size, and the standard deviation of the luminance in the ensemble to their original values. In other words, suppose we step back from an underlying texture and display image samples that are digitized to have the same check size as the original texture. This generally leads to a change in the ensemble mean contrast power ([standard-deviation/mean]^2), but this can be restored to its original level by an overall amplification of the deviations from the mean. The display size of the image samples that are presented can also be the same as in the original ensemble. When this 'stepping back' transformation is performed, the luminance histogram of the ensemble generally changes. We know that all the equi-variant IID textures eventually become Gaussian (by the

Central Limit Theorem), and therefore indiscriminable. However, for textures with spatial correlations, there are other possible fates. Suppose the higher-order correlations of the texture have infinite extent, as is the case for the even texture with any amount of sporadic decorrelation. Stepping back from the texture results in a new texture that also has an infinite length scale for the higher-order correlations. As we will see, stepping back to infinity while maintaining resolution, display size, and mean contrast power can lead to a non-Gaussian ensemble that is a fixed-point of the scaling transformation. This means that histogram differences that appear when two ensembles are scaled may persist even after an infinite degree of scaling, and differences of higher-order spatial correlations can as well. These permanent histogram differences can be traced to the different spatial correlation properties of the two original ensembles; histogram differences are inherited from spatial correlation differences through scaling.

2. THE SCALING TRANSFORMATION

The precise statement of the scale transformation we are using is as follows. Suppose that we want to scale a translation-invariant texture (the 'old' ensemble) by a factor of 2, and that all the images in the ensemble are composed of checks or pixels within which the luminance is uniform. Let us say the images drawn from the ensemble in the course of an experiment are $M \times M$ checks in size. Scaling the old ensemble by a factor of 2 produces the 'new' ensemble, and we will define the scaling transformation in terms of how one would generate sample images from the new ensemble. Let us state at the outset that we consider the mean luminance of the ensemble to be subtracted out beforehand, so that it is not changed by the transformation. We would:

1) Generate an image from the old ensemble that is $2M \times 2M$ checks in size.
2) Shrink it by a factor of 2 in each linear dimension.
3) Average the luminance in each 2×2 block of checks throughout the image. This returns the image to its original resolution.
4) Multiply the luminance in each check by a common amplification factor, chosen so that the ensemble mean contrast power ($[\text{standard-deviation/mean}]^2$) is the same as before the transformation.

Step (2) is what we intuitively think of as stepping back from a texture, but for many vision experiments it is desirable to keep the display size constant so that eccentricity differences are not an experimental factor. That is the purpose of step (1); it insures that we will end up with the same display size as in the old ensemble after we shrink the image. Step (3) is performed in order to keep the spatial resolution of the texture held fixed. Once we perform this local average, however, the contrast naturally drops from its original value. We would like the mean contrast power to be unchanged by the scaling procedure; there are several reasons for this. First, it is the simplest statistical property of the texture, aside from the mean luminance (which is unaffected). We know that in the case of IID textures, a simplification in the description of human performance occurs when the mean contrast power is held constant within the texture class being studied (Chubb *et al.*, 1994). Second, work by Moulden *et al.* (1990) has shown that the standard deviation of the luminance in random-dot images is the best predictor of their adapting power (as opposed to the Michelson contrast, for example). Finally, human performance in a variety of visual tasks has already been investigated as a function of the contrast; we would like to learn something more about the perception of luminance correlations across space. These considerations lead us to step (4); we will simply increase the contrast (standard deviation/mean) back up to its value before the scaling. The scaling transformation we have defined is a version of the blocking procedure introduced in statistical mechanics by Kadanoff (1966; see also Ma, 1976).

3. A CONTINUUM OF SCALING FIXED-POINT IMAGE ENSEMBLES

We apply this transformation to the 'even' (or 'standard') texture with an arbitrary amount of sporadic decorrelation. This is a class of textures that has been in use for some time (Victor and Conte, 1989, 1991). An $M \times M$ sample from the fully correlated even texture is easily generated. We pick either $+1$ or -1 with equal probability for each $x = 0, \ldots, M-1$ and for each $y = 0, \ldots, M-1$. Call these values $f(x)$ and $g(y)$. The image is then $I(x, y) = L_0[1 + c\, f(x)\, g(y)]$, where c is the contrast and L_0 the mean luminance. This image ensemble has a white power spectrum, i.e. the correlation between two different points is 0. However, it has higher-order spatial correlations. The correlation among four points forming the corners of a rectangle is non-zero. These higher-order correlations remain no matter how large the size of the rectangle; one can say that the higher-order correlations have an infinite extent, or are characterized by an infinite length scale. Sporadic decorrelation is introduced by reversing the contrast of each check independently with probability ε. Samples from the even texture with varying degrees of sporadic decorrelation are shown in Fig. 1. The histogram of the luminance in the individual checks is same for all of these textures; there is equal probability for the two possibilities $L_0(1 + c)$ and $L_0(1 - c)$. Sporadic decorrelation reduces the strength of the higher-order spatial correlations, but leaves the length scale of the higher-order correlations infinite. It also leaves the power spectrum white. At $\varepsilon = 0.5$, the higher-order correlations have been reduced to nothing, and we are left with the random texture. Since all the ensembles in this class have white power spectra, it is worth noting that under the scale transformation as we have defined it the power spectrum remains white. The entire power spectrum is held constant 'for free' if we start with a white ensemble. This is because for white ensembles the length scale of the power spectrum is effectively 0 (or as close to 0 as possible — one check length, the resolution limit).

We can apply the scale transformation repeatedly to the even texture with some amount of sporadic decorrelation, obtaining a succession of white image ensembles with long-range higher-order correlations. After k iterations, the resulting image ensemble is characterized by a probability distribution for the $M \times M$ image given

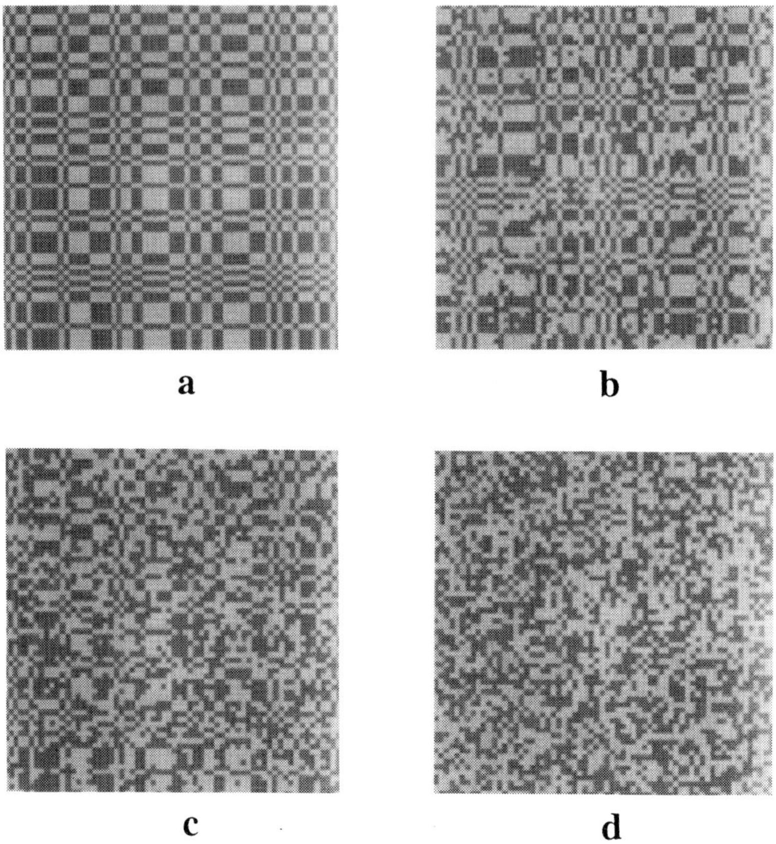

Figure 1. Examples of the even texture with varying degrees of sporadic decorrelation ε. The display is 64 × 64 checks in size, with a contrast of 0.2 and mean luminance of 75 cd m^{-2}. (a) $\varepsilon = 0$ (fully correlated even texture), (b) $\varepsilon = 0.075$ (partially decorrelated), (c) $\varepsilon = 0.15$, and (d) $\varepsilon = 0.5$ (random texture — completely decorrelated).

by

$$P_{\text{blocked}}[I] = \sum_{\{f\}} \sum_{\{g\}} \sum_{\{\eta\}} P_{\text{bin}}[f]\, P_{\text{bin}}[g]\, P_{\text{decorr}}[\eta \mid f, g]$$
$$\times \prod_{x,y=0}^{M-1} \delta\left(I(x, y) - \frac{1}{N} \sum_{i=xN}^{xN+N-1} \sum_{j=yN}^{yN+N-1} \left[f(i)\, g(j) + \eta(i, j) \right] \right), \tag{1}$$

where $N = 2^k$. We have momentarily dropped the mean luminance L_0 and the contrast c for the sake of notational simplicity; these will be restored shortly. Here the sums are over all possible values of all the $f(i)$, $g(j)$, and $\eta(i, j)$ for $i, j = 0, \ldots, MN-1$. P_{bin} is the independent one-half probability for each binary possibility ± 1. We have treated the sporadic decorrelation as an additive noise, with probability distribution $P_{\text{decorr}}[\eta \mid f, g]$. This depends on f and g, but that turns out not to be a problem.

In the limit of infinitely many iterations, we invoke the Central Limit Theorem, arriving at the scaling fixed-point (self-similar) image ensemble described by

$$\begin{aligned}\lim_{N\to\infty} P_{\text{blocked}}[I] &= P_{\text{fixed-pt}}[I] \\ &= \int D\bar{f}\, D\bar{g}\, D\bar{\nu}\, G_{1\text{D}}[\bar{f}]\, G_{1\text{D}}[\bar{g}]\, G_{2\text{D}}[\bar{\nu}] \\ &\times \prod_{x,y=0}^{M-1} \delta\Big(I(x,y) - \bar{f}(x)\bar{g}(y)(1-2\varepsilon) - \bar{\nu}(x,y)\sqrt{4\varepsilon(1-\varepsilon)}\Big).\end{aligned} \tag{2}$$

Here $G_{1\text{D}}$ and $G_{2\text{D}}$ are one- and two-dimensional Gaussian white-noise processes, respectively, with zero mean and unit variance. For convenience, we use the functional integral notation $\int D\bar{f}$ to represent integration over all possible functions $\bar{f}(x)$ on $x = 0, \ldots, M-1$; similarly for $\bar{g}(y)$ and $\bar{\nu}(x, y)$. The variable ε is the decorrelation probability characterizing the original sporadically decorrelated even texture.

This scaling fixed-point texture is easily constructed. A sample image from the fixed-point ensemble is obtained by the following procedure.

1) From a Gaussian distribution with zero mean and unit variance, independently choose values for $\bar{f}(x)$, $\bar{g}(y)$, and $\bar{\nu}(x, y)$, where $x, y = 0, \ldots, M-1$.
2) The sample image is

$$I(x,y) = \Big[\bar{f}(x)\,\bar{g}(y)(1-2\varepsilon) + \bar{\nu}(x,y)\sqrt{4\varepsilon(1-\varepsilon)}\,\Big]cL_0 + L_0. \tag{3}$$

The ensemble-mean contrast power of this ensemble is c^2, as it must be, because this was the contrast power of the original unscaled texture. For any ε between 0 and 0.5, the ensemble thus constructed is a fixed-point of the scaling transformation, possessing a white power spectrum and higher-order correlations of infinite range. The continuum consists of a graded transition between two extremes. At one end, $\varepsilon = 0$, is a strongly non-Gaussian texture with salient higher-order structure. At the other, $\varepsilon = 0.5$, lies Gaussian white noise. Figure 2 shows samples from fixed-point ensembles with varying degrees of structure along this continuum.

It turns out that the $\varepsilon = 0$ 'product' texture has previously been generated by Victor and Conte (1993) for different reasons; they were studying the effects of various kinds of phase-jitter on the VEP. This texture is the result of phase-shifting the Fourier components of the even texture by angles $\beta(\vec{k}) = \beta_1(k_1) + \beta_2(k_2)$, where $\vec{k}$ is the spatial-frequency vector with components k_1 and k_2, and $\beta_1(k_1)$ and $\beta_2(k_2)$ are randomly chosen phases. This restricted type of phase-jitter preserves much of the phase-correlation structure, leaving us with a strongly non-Gaussian texture. We have performed 50 ms discrimination between this non-Gaussian texture and the Gaussian white-noise texture, and found the performance to be close to 100%. Among other things, this means there is a just-noticeable difference (JND) somewhere along the scaling fixed-point continuum for discrimination from unstructured Gaussian white

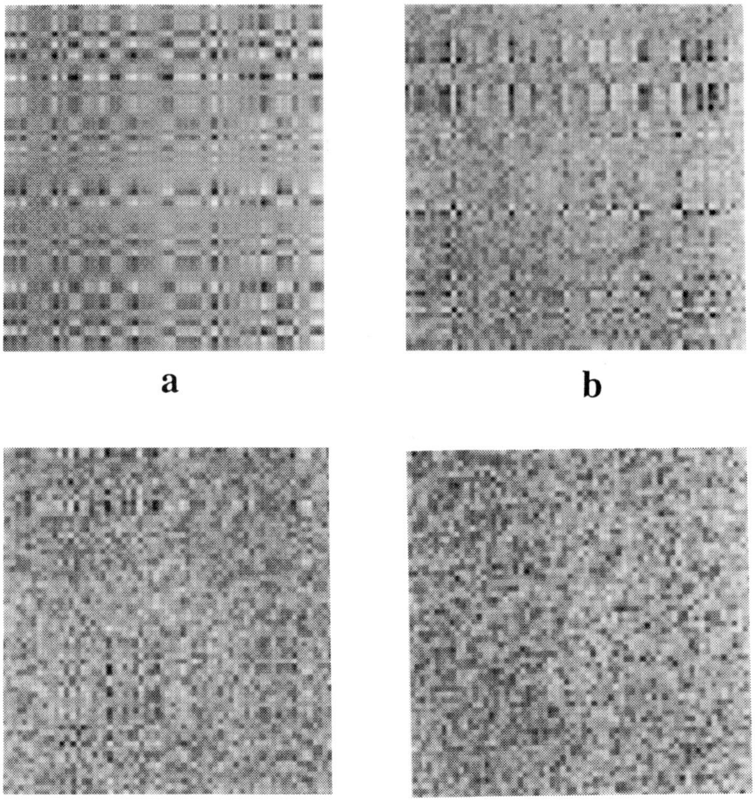

a b c d

Figure 2. Sample images from the scaling fixed-point ensembles approached by the ensembles illustrated in Fig. 1 (sporadically decorrelated even textures). The mean luminance, mean contrast power, display size, and check size are the same as in Fig. 1. (a) $\varepsilon = 0$ (strongly non-Gaussian texture), (b) $\varepsilon = 0.075$, (c) $\varepsilon = 0.15$, and (d) $\varepsilon = 0.5$ (Gaussian white noise).

noise. This raises a number of questions. What is the relationship between the JNDs within the continuum of unscaled textures and the JNDs for the corresponding scaling fixed-point ensembles? Does scaling or stepping back diminish performance when we hold the contrast, resolution and display size constant, or is performance essentially robust? How do the discrimination thresholds compare before and after scaling? What, if any, is the contribution of the CEL detector to the discrimination between these fixed-point ensembles and to the difference in the discrimination performance before and after scaling?

These questions suggest a series of experiments in which near-threshold performance between sporadically decorrelated even textures is compared with the performance in discriminating between their corresponding scaling fixed-point ensembles. Previously identified mechanisms may contribute to any performance changes we might observe when the texture discriminands undergo infinite scaling; these may include mecha-

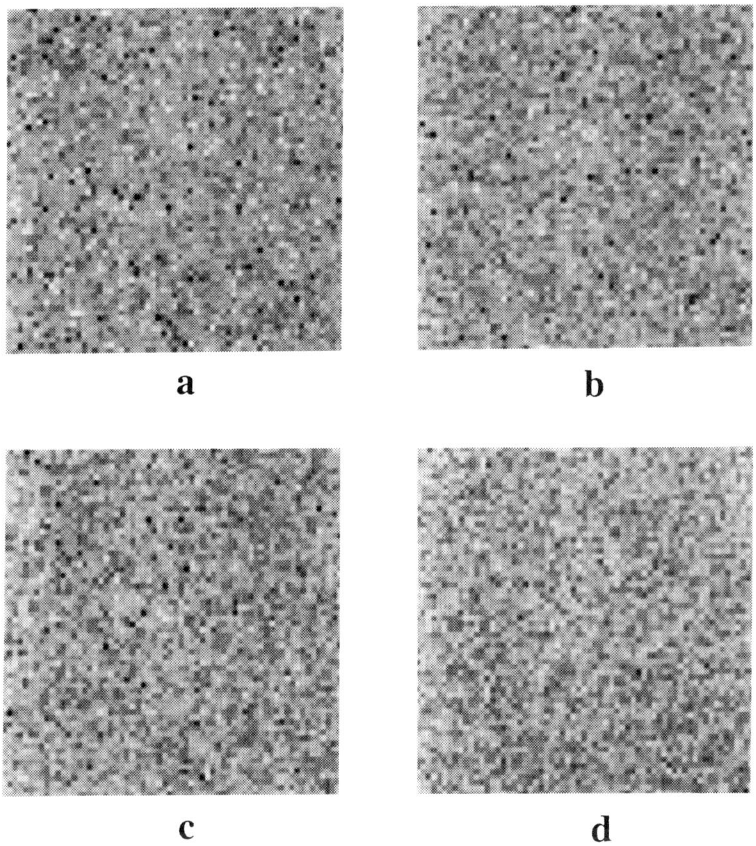

Figure 3. Examples of the IID (independent identically distributed) textures having the same luminance histograms as the textures shown in Fig. 2. (a) $\varepsilon = 0$, (b) $\varepsilon = 0.075$, (c) $\varepsilon = 0.15$, and (d) $\varepsilon = 0.5$ (Gaussian white noise).

nisms sensitive to local contrast, 'blobs', or other properties that can change during the scaling. The contribution of the CEL detector in particular can be ascertained by performing discrimination between pairs of IID textures, in which the luminance histograms of the IID textures are chosen to coincide with those of the spatially structured scaling fixed-points. This is easily done: at each point (x, y) in the image, we generate independent Gaussian random variables $f(x, y)$, $g(x, y)$, and $\nu(x, y)$ with unit variance, and compute a sample image by

$$I(x, y) = \left[f(x, y)\, g(x, y)(1 - 2\varepsilon) + \nu(x, y)\sqrt{4\varepsilon(1 - \varepsilon)} \right] cL_0 + L_0. \qquad (4)$$

Figure 3 shows photos of IID samples with the same luminance histograms as the scaling fixed-points of Fig. 2. The histogram is non-Gaussian with a positive kurtosis.

The peak at the central gray level becomes lower and less sharp as the decorrelation parameter is increased, eventually bringing the histogram to a Gaussian shape.

While the original unscaled textures had precisely the same luminance histogram as each other, the corresponding scaling fixed-points can have markedly different histograms. It is the inter-ensemble differences in spatial correlation properties that are responsible for these histogram differences which appear upon scaling and persist even after infinite scaling. This means that, even if the CEL detector is completely useless for discriminating some texture pair at one viewing distance, it can become useful upon stepping back from the textures, and this usefulness can persist indefinitely as we step back further.

4. SUMMARY

We have presented a combination of theoretical tools and experimental approaches to shed light on the problem of visual texture perception, particularly the perception of phase structure in a scale-invariant context. We defined a scale transformation that keeps the resolution, display size, and mean contrast power of an image ensemble constant under variations in viewing distance. Applying this stepping back transformation to even textures with sporadic decorrelation, we arrived at a continuum of non-Gaussian self-similar image ensembles that share the same power spectrum (white) but differ in the degree of their phase correlations. We have observed that the textures at the extremes of this continuum are easily discriminable. This raises the question of how much performance loss (if any) is suffered by stepping back to infinity while the contrast is maintained. A detector sensitive to the luminance histogram at a single point in space can make a contribution to the discrimination between these textures; this contribution can be measured by an IID discrimination task with the same luminance histograms as the scaling fixed-point textures.

Acknowledgements

J.S.J. thanks K. P. Purpura for helpful discussions, M. S. Landy for constructive comments on an earlier version of the manuscript, the National Eye Institute for fellowship 1-F32-EY06531-01, and K. Nakayama for his hospitality during the writing of this paper. J.D.V. was funded by National Eye Institute grant EY7977.

REFERENCES

Atick, J. J. and Redlich, A. N. (1990). Towards a theory of early visual processing. *Neural Comput.* **2**, 308–320.

Atick, J. J. and Redlich, A. N. (1992). What does the retina know about natural scenes? *Neural Comput.* **4**, 196–210.

Burton, G. J. and Moorhead, I. R. (1987). Colour and spatial structure in natural scenes. *Applied Optics* **26**, 157–170.

Caelli, T. M., Julesz, B. and Gilbert, E. N. (1978). On perceptual analyzers underlying visual texture discrimination: part II. *Biol. Cybernet.* **29**, 201–214.

Campbell, F. W., Nachmias, J. and Jukes, J. (1970). Spatial-frequency discrimination in human vision. *J. Opt. Soc. Am.* **60**, 555–559.

Carlson, C. R. (1978). Thresholds for perceived image sharpness. *Photographic Science and Engineering* **22**, 69–71.

Chubb, C. and Landy, M. S. (1990). Orthogonal distribution analysis: A systematic approach to the study of texture perception. *Invest. Ophthalmol. Vis. Sci.* **31** (Suppl.), 561.

Chubb, C. and Landy, M. S. (1991). Orthogonal distribution analysis: A new approach to the study of texture perception. In: *Computational Models of Visual Processing*. M. S. Landy and J. A. Movshon (Eds). MIT Press, Cambridge, MA, pp. 291–301.

Chubb, C., Econopouly, J. and Landy, M. S. (1994). Histogram contrast analysis and the visual segregation of IID textures. *J. Opt. Soc. Am. A* **11**, 2350–2374.

Field, D. J. (1987). Relations between the statistics of natural images and the response properties of cortical cells. *J. Opt. Soc. Am. A* **4**, 2379–2394.

Hirsch, J. and Hylton, R. (1982). Limits of spatial-frequency discrimination as evidence of neural interpolation. *J. Opt. Soc. Am.* **72**, 1367–1374.

Julesz, B. (1962). Visual pattern discrimination. *I.R.E. Trans. Information Theory* **IT-8**, 84–92.

Julesz, B., Gilbert, E. N. and Victor, J. D. (1978). Visual discrimination of textures with identical third-order statistics. *Biol. Cybernet.* **31**, 137–140.

Kadanoff, L. P. (1966). Scaling laws for Ising models near T_c. *Physics (N.Y.)* **2**, 263–272.

Knill, D. C., Field, D. and Kersten, D. (1990). Human discrimination of fractal images. *J. Opt. Soc. Am. A* **7**, 1113–1123.

Li, Z. and Atick, J. J. (1994). Toward a theory of the striate cortex. *Neural Comput.* **6**, 125–144.

Ma, S.-K. (1976). *Modern Theory of Critical Phenomena*. Benjamin/Cummings Publishing Company, Reading, MA.

Moulden, B., Kingdom, F. and Gatley, L. F. (1990). The standard deviation of luminance as a metric for contrast in random-dot images. *Perception* **19**, 79–101.

Oppenheim, A. V. and Lim, J. S. (1981). The importance of phase in signals. *Proc. IEEE* **69**, 529–541.

Piotrowski, L. N. and Campbell, F. W. (1982). A demonstration of the visual importance and flexibility of spatial-frequency amplitude and phase. *Perception* **11**, 337–346.

Purpura, K. P., Victor, J. D. and Katz, E. (1994). Striate cortex extracts higher-order spatial correlations from visual textures. *Proc. Natl. Acad. Sci. USA* **91**, 8482–8486.

Rensink, R. A. (1986). On the visual discrimination of self-similar random textures. Dept. of Comp. Sci. Tech. Rep. 86-16, Univ. of British Columbia, Vancouver, B.C., Canada.

Shapley, R., Caelli, T., Grossberg, S., Morgan, M. J. and Rentschler, I. (1990). Computational theories of visual perception. In: *Visual Perception: The Neurophysiological Foundations*. L. Spillman and J. B. Werner (Eds). Academic Press, New York, pp. 417–448.

Victor, J. D. (1994). Images, statistics, and textures: implications of triple correlation uniqueness for texture statistics and the Julesz conjecture: comment. *J. Opt. Soc. Am. A* **11**, 1680–1684.

Victor, J. D. and Conte, M. M. (1989). Cortical interactions in texture processing: Scale and dynamics. *Vis. Neurosci.* **2**, 297–313.

Victor, J. D. and Conte, M. M. (1991). Spatial organization of nonlinear interactions in form perception. *Vision Res.* **31**, 1457–1488.

Victor, J. D. and Conte, M. M. (1993). Lack of global power spectral contributions to isodipole discrimination, as assessed by manipulations of spatial phase. *Invest. Ophthalmol. Vis. Sci.* **34** (Suppl.), 1238.

Yellott, J. I. (1993). Implications of triple correlation uniqueness for texture statistics and the Julesz conjecture. *J. Opt. Soc. Am. A* **10**, 777–793.

Human Symmetry Perception, pp. 331–347
C. W. Tyler (Ed.)

Symmetry as a depth cue

LEONID L. KONTSEVICH
Smith-Kettlewell Eye Research Institute, 2232 Webster St., San Francisco, CA 94115, USA

Abstract—Properties of 3D perception from line drawings are discussed. From them, with dimension counting, it is shown that there exist 3D objects, any depiction of which would be interpreted by the visual system incorrectly. Two examples of such undepictable objects are proposed. Analysis of the examples indicated that 3D interpretation of the line drawings is driven by a tendency to make the interpretation as symmetrical as possible. The algorithmic basis for symmetry search and making use of the additional images provided by symmetry is discussed: it follows from the result that the correspondence between the object points imposed by any symmetry necessarily obeys the rigidity constraint. A model framework for view combination within the same image and between different images in the visual system is proposed; it provides a consistent explanation of a number of 3D perceptual phenomena. This model, in particular, suggests a way to unify such 3D information sources as mesh plots, shading and texture within a single mechanism.

1. INTRODUCTION

That line drawings on flat surfaces can evoke three-dimensional (3D) percepts has been known as long as graphic art has existed. Until now, however, our understanding of the principles underlying this kind of depth perception has remained incomplete despite a variety of efforts in this direction (Binford, 1981; Kanade, 1981; Barrow and Tenenbaum, 1981; Lowe, 1987; Malik, 1987; Pentland and Kuo, 1990; Sinha and Adelson, 1993). The problem with the 3D interpretation of line drawings is there is no readily identifiable module that processes depth cues such as kinetic depth or binocular stereopsis; there is little consensus as to what depth cues are encoded in line drawings.

This paper has two goals. The first is to analyze the basic properties of 3D interpretation from line drawings. Although these properties look trivial, they lead to the conclusion that there are 3D shapes for which a line drawing from any angle creates a distorted impression of the original object. Two examples of such shapes will be presented.

The observation that depth is seen best for drawings of symmetrical objects leads to the conjecture that symmetry is the predominant cue that drives 3D interpretations from drawings. From this conjecture, I analyze how spatial symmetry can be detected from a single 2D drawing, what additional information about the object it provides, and how this information can be used.

The second goal of the paper is to generalize the approach that will be developed for symmetry. I show that, in order for symmetry to be a depth cue, there must exist a mechanism that establishes within a single image partial correspondences that obey the rigidity constraint. The conjecture that this mechanism may work with different images generates a consistent explanation of a number of phenomena related to 3D shape perception.

2. CONVENTIONS

First we define the terms used in this study.

Objects. The objects are sets of isolated points (nodes) in space, some of which are connected by edges. These objects include polyhedra.

Projection. We assume the projection method known as *weak perspective*, which combines orthographic projection and arbitrary scaling. This projection provides a good approximation for the central projection employed by the eye's optics when the distance between object and observer is at least 2–3 times larger than the object size. In weak perspective projection, scaling is added to orthography to simulate change of the object's angular size with a distance.

Shape similarity. We say that the objects have *identical* shapes if they can be aligned with a combination of translation, rotation and scaling. Scaling is included in this definition to make the shape tolerant to the scaling that occurs in weak perspective projection.

3. OBSERVATIONS RELATED TO SEEING DEPTH FROM DRAWINGS

Before discussing depth perception from line drawings, I present a number of empirical observations to provide the common ground on which the subsequent speculations are based. Most of the following observations are well known and appear in the literature.

Observation 1. The Necker cube (Necker, 1832) provides a compelling demonstration of the existence of a depth percept evoked by the line drawings. It allows two possible 3D interpretations of which alternate to produce a vivid sensation of depth right after the flip, see Fig. 1(a).

Observation 2. The strength of the depth percept depends on the presentation conditions. Binocular viewing certainly reduces the depth sensation from line drawings because stereopsis tells the observer that the object is flat. The texture of the paper and the observer's knowledge that he is looking at a flat sheet of paper also flattens the percept. Evidently, the best way to present line drawings is to present them monocularly, glowing in the dark (Gregory, 1969, p. 139).

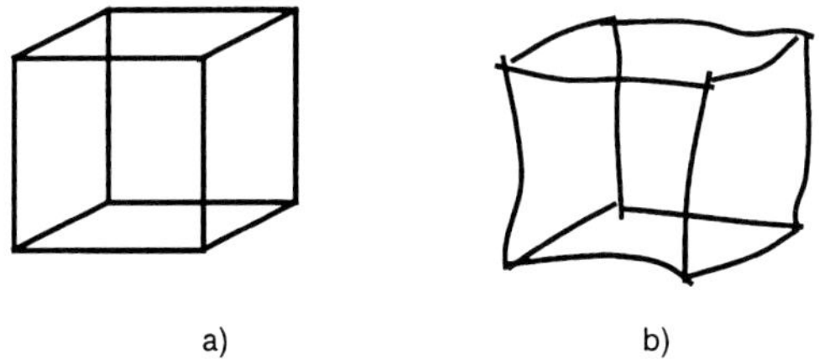

Figure 1. a) The Necker cube. This image has two 3D interpretations that involuntarily flip providing compelling sense of depth right after the shape invertion. b) The strength of 3D percept is does not depend substantially of the quality of drawing.

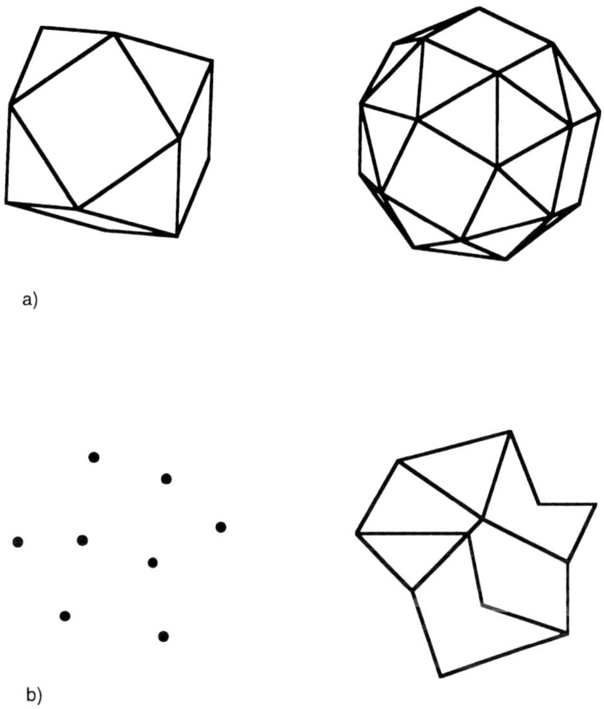

Figure 2. a) Examples of objects where the observers easily see depth although most of them never saw these objects before. b) Some examples that do not produce an obvious depth percept.

Observation 3. Some drawings produce strong depth percepts (Fig. 2(a)); others hardly can be associated with any 3D shape (Fig. 2(b)).

Observation 4. The strength of the depth percept does not depend substantially on the accuracy of the drawing. The depth seen in Fig. 1(b) is as good as for the carefully drawn Necker cube in Fig. 1(a).

Observation 5. The strength of the depth percept is not conditioned solely by familiarity with the object. Most of the readers have never seen the particular polyhedra shown in Fig. 2, however they can easily comprehend their shape.

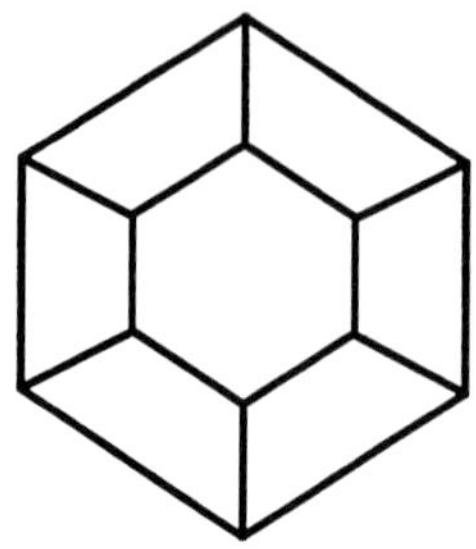

Figure 3. The 3D interpretation of this drawing is a truncated pyramid. Its height is unstable.

Observation 6. Typically, the observer can distort at will the shape of any depth percept. However, if a drawing is viewed with no conscious effort to impose some predetermined 3D structure, the perceived shape may fluctuate, as in Fig. 3, or gravitate to one or a few discrete states. This paper is concerned with the properties of such stable 3D percepts.

Observation 7. The stable percept assumes a generic view. In this case, small distortions of the image cause small distortions in the 3D percept (compare with the togetherness and separateness demonstration by A. Ames described in Ittelson, 1952, pp. 38–39).

Observation 8. Some drawings have a number of stable 3D percepts among which the observer can switch at will. Sometimes, as in the case of the Necker cube, such switches happen involuntarily. The switch between stable percepts is discrete in contrast with the shape variations observed in unstable percepts.

Observation 9. The number of stable 3D percepts for a given drawing is finite.

Observation 10. The shape of the percept does not change with translation, rotation and scaling of the image as long as the latter remain frontoparallel.

4. THE EXISTENCE OF UNDEPICTABLE OBJECTS

We provide a dimension-counting argument to show that there exist undepictable objects; this result is the opposite of the Penrose and Penrose (1958) drawings, which are depicted but do not exist. Let us compare the dimensions of the 3D object with the 2D projection spaces. Each node on the 2D projection is determined by two parameters: X and Y coordinates. Thus, the projection of N nodes is determined by $2N$ independent parameters. (For time being I do not take into account the need to specify edges.)

According to Observation 10, translation, rotation and scaling of the projection do not change the shape of the 3D percept. These transformations specify a four-dimensional class of projections that evoke percepts with the same shapes. Therefore, the family of different 3D interpretations that can be evoked by N-point projections can be specified by $2N - 4$ independent parameters. (Also, for the time being I do not take into account that projections can have more than one interpretation.)

Consider the family of different 3D shapes comprised of N points in space. Each point is specified by 3 parameters; thus, the N-point arrangements can be determined by $3N$ parameters. Because we define the 3D shapes as similar if they match after translation, rotation and scaling to align one shape with another, the class of similar shapes is specified by 7 independent parameters. Thus, the arrangements of N points in 3D space produces $(3N - 7)$-parameter family of different shapes.

A stable percept of a 2D projection (as specified in Observation 6) defines a mapping of the image space into the object space. If a projection does not have any stable 3D percepts, the mapping of this projection is undefined. According to Observation 7 the mapping is smooth (gomeomorphic) in the vicinity of the of the projection with stable 3D percept. Therefore, the result of this mapping is a $(2N - 4)$-dimensional surface (manifold) in $(3N - 7)$-dimensional space of possible shapes.

The presence of edges and the possibility of several stable percepts (Observation 9) easily can be added to our account. In this case any particular arrangement of points on the projection corresponds to a finite number of edge arrangements each of which, correspondingly, may correspond to a finite number of shapes. Thus, any node arrangement corresponds to a finite number of shapes and in the vicinity of each shape the mapping is smooth. Therefore, in general, 3D interpretations constitute a $(2N-4)$-dimensional surface with a finite number of layers in $(3N - 7)$-dimensional space of shapes.

The manifold of the interpretations cannot fill the space of shapes when it has smaller dimensionality than the space, i.e. when $2N - 4 < 3N - 7$. This inequality holds for all $N \geqslant 3$. Therefore, we arrive at the following conclusion:

There are objects that are undepictable: any stable 3D interpretation of any view of such objects is different from the true shape of the object.

This counterintuitive result puts limits on graphic art: it is inherently unable to represent all possible objects of real world.

Because the inequality holds already for N as small as four, there must be very simple undepictable objects. Two examples of such objects are presented below.

5. EXAMPLES OF UNDEPICTABLE OBJECTS

5.1. Tetrahedron

Consider a tetrahedron ABCD, one edge of which (AB) is much shorter than other edges as shown in Fig. 4(a). Three examples of such tetrahedrons with different orientations of the shortest edge are shown in Fig. 4(a), (b), (c) (the visible face BCD is the same for all three to underscore the effect of the edge orientation). The 3D percepts of these examples share a common feature: all of them are symmetrical relative to the plane passing the points C, D, and the midpoint of the segment AB. The

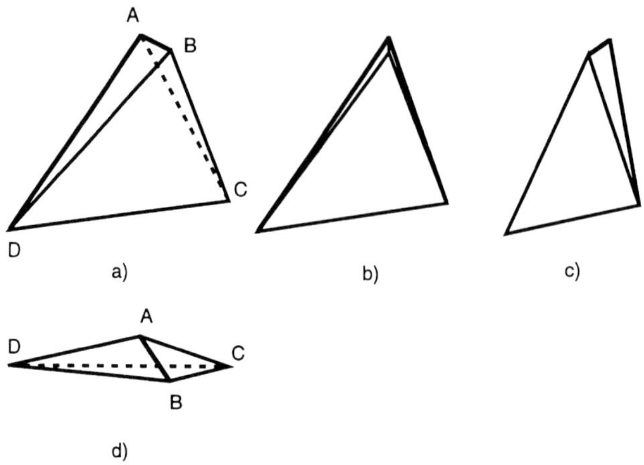

Figure 4. An undepictable pyramid. a)–c) The 3D percepts of these three lateral views of the same tetrahedron share the same feature: the shortest edge AB is perceived as perpendicular to edge CD. d) Top view of the tetrahedron. This view illustrates that the edges AB and CD may be oblique, which is never perceived from the lateral views.

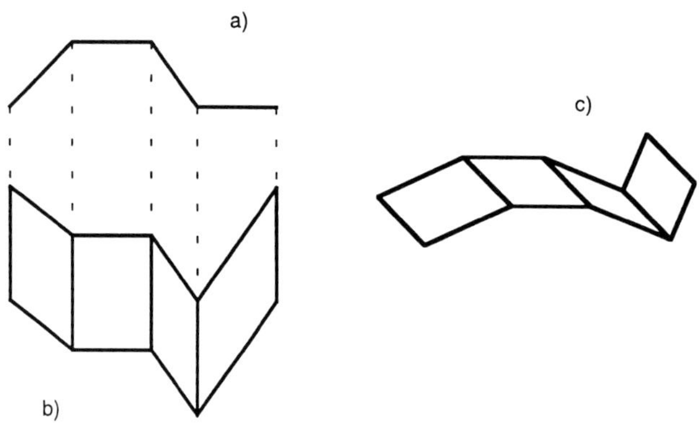

Figure 5. a), b) Side and top view of the object. Its edges are parallelograms. c) A general view of this object. The sides of this object on any projection are perceived as rectangles.

symmetrical edges are seen as equal on the examples; i.e. AC = BC and AD = BD where they are visible. In the percept, the shortest edge AB is perpendicular to edge CD.

Consider a real tetrahedron whose shortest edge AB is oblique to edge CD. Figure 4(d) depicts a top view of the object shown in Fig. 4(a). In almost all projections of this tetrahedron, the edge AB will be the shortest and, therefore, it will be interpreted in 3D space as perpendicular to the edge CD, which is incorrect. Therefore, the constructed tetrahedron is misperceived from most of its views.

The present example leaves a possibility that the object could be perceived correctly from a view where edge AB is not the shortest, although the number of such views relative the overall number of views is negligibly small. In the next example a slightly more complex object does not leave such a possibility.

5.2. *Folded band*

Consider a 3D object that consists of a number of parallel segments of equal length that are sequentially connected into a band. Suppose that the adjacent segments are shifted relative each other along their direction; physically the faces of this band are parallelograms but not rectangles. Frontal and top views of this object are shown in Fig. 5(a), (b). Any view of this object is interpreted as a straight band folded along the lines perpendicular to the band axis. The faces of this band tend to be perceived as rectangles. A generic view of the same object is shown in Fig. 5(c).

6. SYMMETRY

The attentive reader may have noticed a similarity between the 3D percepts evoked by the easily-interpreted drawings presented so far: they are all highly symmetrical. If the drawing evokes several stable depth percepts, they are all symmetrical. Incidentally, the undepictable objects described above also exploit symmetry: both of the objects are highly asymmetrical but any projection of those allows a symmetrical 3D interpretation. Thus, symmetry appears to play an important role in the choice of an interpretation from the continuum.

The only kind of symmetry considered thus far is the bilateral kind. One can ask what happens with other kinds of symmetries. To answer this question we first need to define such symmetries. We define the class of *symmetry transformations* as the class of affine transformations preserving angles. These transformations also can be described as superposition of bilateral reflection and the shape-preserving transformations, which are translation, rotation and scaling. A *symmetry operation* on an object is any symmetry transformation that transforms every point of the object into every other point of it (Hilbert and Cohn-Vossen, 1952, p. 87). We call the object *symmetric* if it allows at least one symmetry operation on itself. Thus, to put it in simpler terms, a symmetric object is one that coincides with itself after particular translation, rotation, scaling and reflection.

Figure 6 presents a number of the objects that are symmetrical in accord with the definition given. Taking into account the fact that Fig. 3 shows an unstable percept, we arrive at the following observation.

Observation 11. The presence of any symmetry operation that rotates the projection axis tends to generate a stable 3D percept.

Now we are prepared to ask two main questions about symmetry.

Question 1. How to establish that a 2D drawing allows a symmetrical interpretation?

Question 2. How to use this knowledge for depth reconstruction?

Preliminary versions of the answers can be found in Kontsevich (1992, 1993), Poggio and Vetter (1992), and Vetter and Poggio (1994). Here we reiterate these answers in more detailed form.

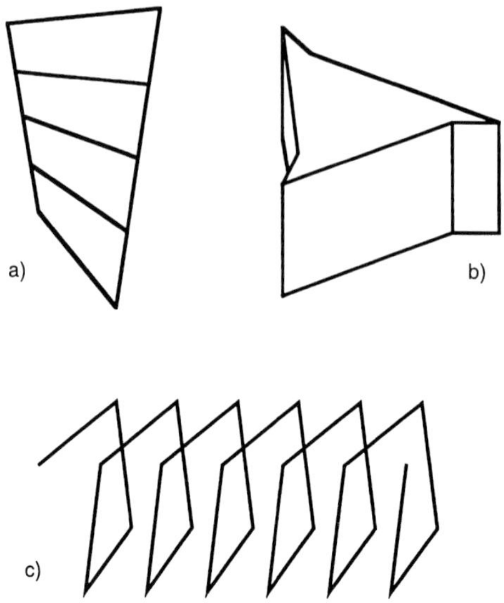

Figure 6. All symmetries tend to generate a stable 3D percept. a) Axial symmetry (generates the percept of a twisted surface). b) Rotational symmetry. c) Spiral symmetry, the superposition of rotation and translation.

7. SYMMETRY DETECTION

Consider a symmetrical object. Appropriate symmetry operations on this object transform every node of the object into every other node of the same object. Thus, the symmetry operation defines a *permutation* of the nodes in the object's 2D projection.

Symmetry transformations, as defined above, are a subclass of affine transforms. It is known from analytic geometry that affine transforms of metric space can be of two kinds: those that invert the orientation of the basis (negative signature) and those that do not (positive signature). Therefore, symmetry transforms also can be classified by their signatures.

Symmetry operations of positive signature can be decomposed into translation, rotation, and scaling. An object with such symmetry can be transformed into itself with rigid motion and scaling.

Symmetries of negative signature require the presence of an additional bilateral symmetry about arbitrary plane. Let this plane be the frontoparallel plane. Then the remaining translation, rotation and scaling components of the negative-signature symmetry operation transform the object into a copy of itself that is symmetrical with the original object relative the frontoparallel plane. This copy has the identical 2D projections with the original object.

Thus, the present analysis of symmetries of positive and negative signatures implies that:

The correspondence between nodes in the object projection imposed by any symmetry of the object must yield the rigidity constraint under the weak perspective projection assumption.

In other words, symmetry provides new views of the object that can be obtained from a single view of the object.

There are several methods to check the rigidity constraint for two weak perspective projections (Kontsevich *et al.*, 1987; Bennett *et al.*, 1989; Lee and Huang, 1990; Poelman and Kanade, 1992). The simplest one (Kontsevich, 1992, 1993) requires estimation of the rank of a system of homogeneous linear equations.

A check of whether a given projection allows symmetrical 3D interpretation can be implemented in a straightforward manner: to check the rigidity constraint for all permutations of the projection nodes — the edges impose an additional constraint because the permutations should be limited only to those which transfer edges into edges. If at least one permutation complies with this constraint, the projection has a symmetric interpretation.

When the presence of symmetrical 3D interpretation is established, this information may be used for shape reconstruction.

8. 3D RECONSTRUCTION OF A SYMMETRICAL OBJECT

The presence of symmetries for an object established from the analysis of its projection automatically provides new views of this object, one additional view for each symmetry operation found. Thus, we arrive at the domain of 3D reconstruction of depth from multiple views.

It has been proven mathematically that a minimum of three weak perspective views are required to reconstruct 3D shape of a solid object (for the more restrictive orthographic case, see Ullman, 1979; for weak perspective see Kontsevich *et al.*, 1987). Thus, if the object allows two or more symmetry operations, it contains at least two additional views and its shape can be reconstructed uniquely (up to a bilateral symmetry relative the frontoparallel plane) from a single projection. This explanation of depth perception from a single view works for almost all the examples presented in this paper (Figs 1, 2, 6).

There are, however, some exclusions that need to be explained. The 3D percepts of the undepictable objects in Figs 4 and 5 have only one bilateral symmetry operation. For them, only one additional view is available and all possible 3D interpretations comprise a one-parameter family (Bennett *et al.*, 1989). To arrive at some stable 3D percept, the visual system must choose a particular member from the family. There are two strategies for making such choice:

1) object-independent: to assume a value of the angle between the projection axes of the views;
2) object-dependent: to choose a member of the family that has some distinctive property; for example, to make the size of the object in the frontoparallel plane and its depth commensurable; for a similar strategy in kinetic depth perception, see Caudek and Proffitt (1993).

To answer which strategy the visual system uses to resolve the underconstrained symmetry cases, additional experimental studies are required.

9. GENERALIZATION OF THE APPROACH: PARTIAL CORRESPONDENCES

The explanation for the 3D perception of 2D images in the form presented would not provide a satisfactory solution for the vast majority of images people deal with in real life. Often the symmetry can be established only for parts of objects; also the symmetry information about rear side can be lost if the object is opaque. These two reasons illustrate that the additional views are not necessarily permutations of all object nodes and the corresponding parts may be different point sets. Thus, it should be concluded, a symmetry-based explanation for 3D perception from a single view assumes existence of the process that looks for *partial correspondences* within the image.

Assuming that such process exists (without any claims regarding its implementation), this process is reminiscent of the well-established processes that establish interocular (partial) correspondence for stereopsis and (partial) correspondence between the images coming in motion sequences for kinetic depth. Based on this similarity and on parsimony, I propose the following Partial Correspondence Conjecture:

There is a single mechanism in the visual system that establishes partial correspondences conforming to the rigidity constraint; this mechanism operates both within a single image and between different images.

The function of this hypothetical mechanism is to look for the correspondences of any preselected (perhaps this selection is provided by spatial attention) part of the current input image with other unspecified parts of this image and of other images stored within the visual system.

The essence of this mechanism is that it can establish long-range correspondences: this feature is crucial for symmetry detection (Wagemans *et al.*, 1993). For short ranges, as in the cases of kinetic depth and stereopsis, this mechanism may use local processing based on velocity and disparity detectors. However, the long-range ability to match is always present and, when the local processing fails (as in the case of kinetic depth evoked by long-range motion), the long-range processing takes off. Local processing is very sensitive to tiny disparities; the long-range component conversely is tolerant to position mismatch, as indicated by Observation 4.

The Partial Correspondence Conjecture states that the partial correspondence process is non-specific to the particular inputs; it matches all the images available. The mechanisms that use the correspondences also should not be input-specific. Therefore, the Partial Correspondence Conjecture can be continued:

There is a single mechanism in the visual system that combines multiple views and builds a 3D percept.

The possibility of a single processing stream for correspondence and 3D reconstruction has numerous unexplored consequences. First I consider possible origins of the additional views and then focus on interaction of the additional views in 3D reconstruction.

10. THE ORIGINS OF ADDITIONAL VIEWS

Additional views can come from the limited number of sources. Traditionally, each source is associated with a particular depth modality.

1. Stereopsis: the additional view comes from the second (non-dominant) eye.
2. Kinetic depth effect: the additional view comes from short-term memory.
3. Familiarity: the additional view comes from long-term memory.
4. Symmetry: the additional view comes from the same image and the same object.
5. Unnamed mechanism: the additional view may come from the same image but from a different object. For example, if the scene contains several copies of the same object in different orientations to the observer (such as, say, several chairs in a room), the structure can be reconstructed from the simultaneous comparison of different views of potentially the same object. This depth modality, to my knowledge, is not documented in the literature.

11. COMBINATION OF THE ADDITIONAL VIEWS

According to the second part of the Partial Correspondence Conjecture, the additional views to be combined may be of different origin. Perhaps the interaction between stereopsis and the two-frame kinetic depth effect elegantly demonstrated in Johnston *et al.* (1994) is the best illustration of this point. Both stereopsis and two-frame kinetic depth have only two images to compare. By disambiguating one-parameter families of possible solutions, the visual system in most conditions reconstructs a distorted shape of the object. For example, the shape is perceived flattened by stereopsis for small vergence angles; the kinetic depth effect also produces a flattened percept at small rotation angles. However, if motion and binocular cues are combined, they evoke a veridical 3D percept that has more depth in its shape than stereopsis and kinetic depth effect produce alone. This 'promotional' interaction is distinct from typical interaction of different cues where the resultant 3D percept is a weighted mean of the percepts evoked by each cue alone.

Another interesting interaction occurs between stereopsis and symmetry. Stereopsis provides one additional view and symmetry supplies the others. The interaction of these views was ingeniously demonstrated by Ames (Luneburg, 1950) who designed a model of a cube flattened along the viewing direction. This distorted cube, viewed monocularly, had the identical projection to a non-distorted cube and, because of the bias towards the symmetry, it was interpreted by the observers as a non-distorted cube. The same distorted cube model viewed binocularly was also perceived as a non-distorted cube as if stereopsis did not provide contradictory information. This incorrect disambiguation of depth was possible because the symmetric interpretation belonged to the one-parameter family of the solutions provided by stereopsis. The visual system thus arrived at 3D percept that was consistent with all additional views.

A phenomenon related to the Ames' experiment is micropsia or the toy-town illusion. If the photographs for a stereopair of a street are taken with a large base

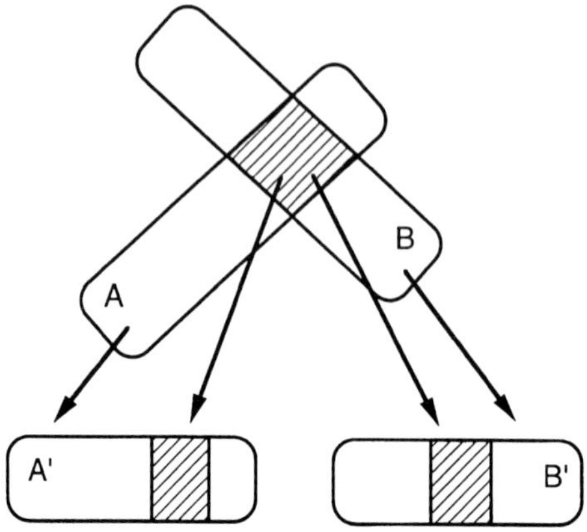

Figure 7. This diagram illustrates that if two intersecting parts A and B have one additional image each (A' and B', respectively), their intersection (the hatched area) has two additional images. If the intersection is sufficiently large, its depth is defined uniquely and the parameter values can be used to reconstruct 3D structure for the rest of A and B.

(relative to the interocular distance), then this stereopair creates an impression of an unrealistically small scene situated close to the observer's face. The shapes of the objects are perceived undistorted. Such a percept can arise only if the visual system somehow gets access to the angle between the projection axes for two stereopair projections. My account of this effect is based on the symmetry and/or familiarity of the objects observed. Real scenes usually contain symmetric and/or known object shapes like buildings, people, cars, etc. The additional views provided by these objects disambiguate the ambiguous vergence estimate.

This symmetry/familiarity based account for micropsia leads to an interesting prediction that can be tested in an experiment. If the stereogram depicts unfamiliar and non-symmetric objects, their shapes will be perceived distorted.

Our explanation of micropsia implies that the parts of a scene may disambiguate each other; in particular, a part of a scene may disambiguate the whole scene. This phenomenon is illustrated in Fig. 7. Suppose that intersecting parts A and B in the main view have the additional views A' and B' correspondingly. In this case the intersection of A and B effectively has two additional views; if it contains non-coplanar set of points, its 3D structure is defined uniquely. The unique solution for the intersection provides the parameter values for the families produced by pairwise comparison of A and A', and B and B', thus disambiguating depth for all points of A and B.

Similar disambiguation also takes place in degenerated cases where the intersection is insufficient to produce a unique 3D structure; for example, if the intersection consists of 2 or 3 points (as shown in Fig. 8). Such intersections constrain the free parameters of the families which must come up with the consistent solutions. The method for 'tailoring' of one parameter families in degenerate cases needs to be developed.

The last issue to be discussed in this section is what happens when the additional views are inconsistent with each other. In analyzing symmetry we considered two

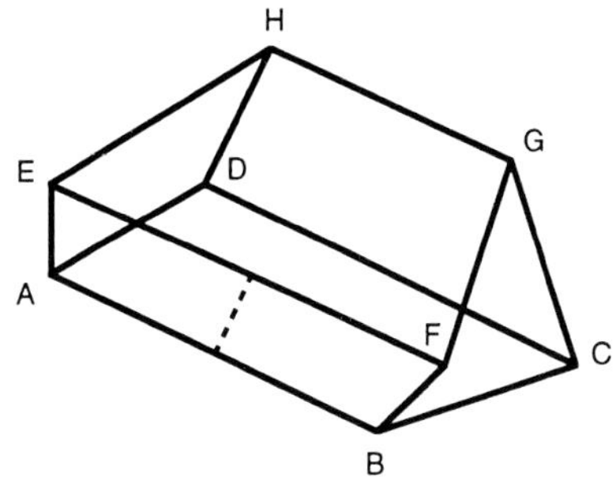

Figure 8. An example of 'tailoring' of two parts each of which has only one additional image and, therefore, allows a one-parameter family of 3D interpretations. One part consists of two trapezoids ABCD and CDHG and has one (medial) symmetry plane. The second part is a twisted band ABFE which has a symmetry axis shown by the dashed line. There is a unique 3D interpretation that makes both parts consistent in the common edge AB.

options: to select one additional view while dismissing all others and use heuristics to disambiguate the resultant one-parameter family, or to look for some compromise between all one-dimensional families defined by all additional views. Recent studies (Young *et al.*, 1993; Johnston *et al.*, 1994; Parker *et al.*, 1995; Landy *et al.*, 1995) suggest that the mechanism reconstructing depth from multiple views generally employs the second option. These studies demonstrate that depth cues that evoke highly inconsistent 3D percepts produce an intermediate 3D percept when presented together. The shape of this percept can be modified by varying the strength of these cues, for example by adding noise to one of them. The depth of the 3D percept can be predicted by the Bayesian rule, which imposes an important constraint on possible implementations of the structure-from-multiple-views mechanism.

12. MESH PLOTS

The Partial Correspondence Conjecture was based on the analysis and observation of relatively simple objects. If the presented logic is correct, it should be applicable also to complex objects such as mesh plots, which are usually produced by the intersection lines of some surface with the families of other surfaces (see Fig. 9(a), (b)). Mesh plots produce vivid depth percepts and for this reason they are widely used for surface representation. From where does their depth percept originate?

The Partial Correspondence Conjecture provides a possible answer. The correspondence mechanism finds correspondences for any particular part of the mesh plot with nearby parts of it that will usually have a similar 3D shape and grid arrangement. The small differences between these parts provide additional views that disambiguate the shape of the part being analyzed. After all parts of the mesh plot are analyzed this way, the local shapes of all the parts are reconstructed. These local shapes may, however, be inconsistent with any global solution: to combine them some regularization may be necessary.

The described approach is applicable to uni-dimensional mash plots and to the case where the section lines are smooth (vs discrete nodes and straight edges considered

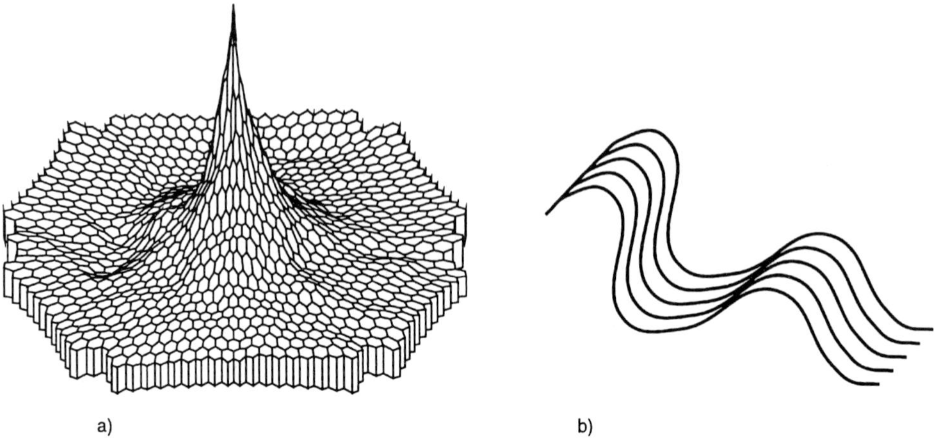

Figure 9. a) This hexagonal mesh plot depicts the retinal activity of normal eye (provided by Dr. E. Sutter, Smith-Kettlewell Eye Research Inst., San Francisco). b) The uni-dimensional mesh plot also evokes a strong 3D percept. Notice that the percept has a bilateral symmetry (the symmetry plane coincides with the medial section plane) which constrains the ridges and troughs to be orthogonal to the symmetry plane.

in this study). For example, the wavy surface shown in Fig. 9(b) is represented by smooth intersection lines. These lines are identical up to translation and obey the constraint imposed by bilateral symmetry. For this reason the ridges and troughs in the 3D percept are perpendicular to the section planes.

The proposed explanation is largely a guess to outline the principle on which mesh-plot interpretation could be based. If this guess proves valid, the partial correspondences approach can be expanded further.

13. STRUCTURE FROM SHADING AND STRUCTURE FROM TEXTURE

The phenomena of structure-from-shading and structure-from-texture at first glance look completely separate from the search of partial correspondences. Some hint that this is not true comes from the fact that texture cues interact with stereopsis in the same way as the depth modalities based on partial correspondences (compare Johnston *et al.*, 1993 and Johnston *et al.*, 1994). In the computer vision literature, the idea that a correspondence process is involved in shape-from-texture perception was suggested by Malik and Rosenholtz (1993). They related shape-from-texture with motion perception; I propose here another account reducing shading and texture-based shape perception to the 3D interpretation of mesh plots. For simplicity, only the case of shading will be considered; texture can be treated similarly.

A shaded image (Fig. 10(a)) can be converted into a set of equiluminant contour lines drawn for certain discrete levels of luminance (Fig. 10(b)). These lines constitute a unidimensional mesh plot that evokes a 3D percept. The core of the proposal is that shape perception of shaded surface is at deeper level of visual processing is shape perception of the mesh plot evoked by the half-tone image. This shape is certainly not a precise solution that would be provided by shape-from-shading algorithms; however, it is likely to resemble the object's true shape.

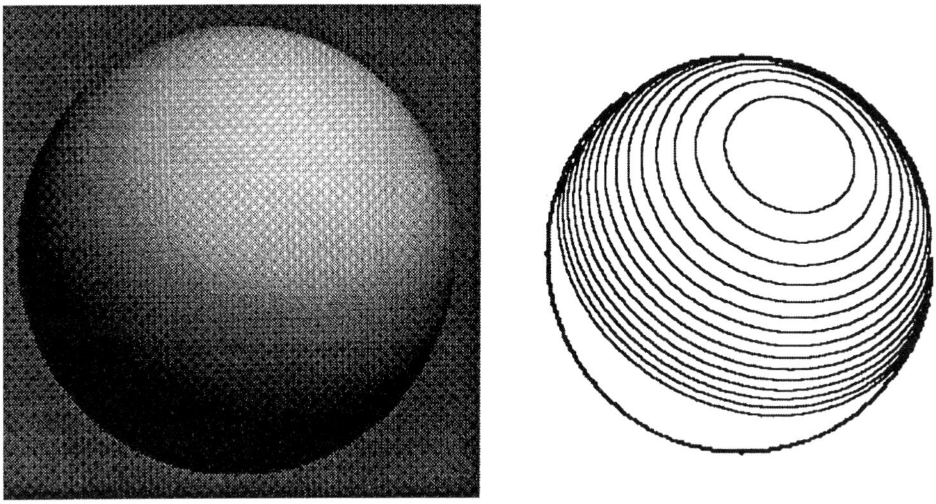

a) b)

Figure 10. a) A halftone image of the sphere. b) The meshplot produced by the isoluminant lines for the halftone image. Notice that the shape of 3D percept somewhat depends on orientation of section surfaces.

The shape of the mesh plot generally depends on the particular choice of the section surfaces and the family of equiluminant lines derived from the shaded image depends on the direction of the light source. Therefore, according to this account, the result of the shape-from-shading reconstruction should depend on the direction to the light. Such a dependence was recently shown by Koenderink *et al.* (1995), according to whom the shape of the 3D percept evoked by shading is elongated towards the light source. The real test of the proposed hypothesis would be to compare the distortions of the 3D percepts evoked by the shaded image and the mesh plot derived from it: my account predicts that they should be identical.

14. CONCLUSIONS

In this study, started with the analysis of 3D perception of drawings, I arrive at the conclusion that symmetry plays an important role in 3D vision. As a byproduct, I demonstrate that there are objects that never can be interpreted veridically from their projections. The analysis of symmetry drew us to the idea of the existence of a partial correspondence mechanism that operates throughout the whole visual system and followed by a depth-from-multiple-views mechanism, which combines all available correspondence irrespective of their origins. The idea of the existence of such ‘amodal’ 3D mechanisms makes the issue of interaction of different depth modalities simple and tractable. Finally, I show that shape perception from mesh plots, shading and texture are likely also fall in the proposed framework of the processing based on partial correspondences.

Acknowledgements

The author thanks Drs Christopher W. Tyler, Richard T. Miller and Andrew Glennerster for valuable comments given on a previous draft of the manuscript. This study was supported by NIH grant #7890 and NIMH grant #49044.

REFERENCES

Barrow, H. G. and Tenenbaum, J. M. (1981). Interpreting line-drawings as three-dimensional surfaces. *Artif. Intel.* **17**, 75–116.

Bennett, B. M., Hoffman, D. D., Nicola, J. E. and Pracash, C. (1989). Structure from two orthographic views of rigid motion. *J. Opt. Soc. Am.* **A10**, 1052–1069.

Binford, T. O. (1981). Inferring surfaces from images. *Artif. Intel.* **17**, 205–244.

Caudek, C. and Proffitt, D. R. (1993). Depth perception in motion parallax and stereokinesis. *J. Exp. Psychol.: Human Percept. Perform.* **19**, 32–47.

Gregory, R. L. (1969). Seeing in depth. In: *Perception. Selected Readings in Science and Phenomenology.* P. Tibbetts (Ed.). Quadrangle Books, Chicago, pp. 137–147.

Hilbert, D. and Cohn-Vossen, S. (1952). *Geometry and the Imagination.* Chelsea Publishing Company, New York.

Ittelson, W. H. (1952). *The Ames Demonstrations in Perception.* Hafner, New York.

Johnston, E. B., Cumming, B. G. and Parker, A. J. (1993). Integration of depth modules: Stereopsis and texture. *Vision Res.* **33**, 813–826.

Johnston, E. B., Cumming, B. G. and Landy, M. S. (1994). Integration of stereopsis and motion shape cues. *Vision Res.* **34**, 2259–2275.

Kanade, T. (1981). Recovery of the three-dimensional shape of an object from a single view. *Artif. Intel.* **17**, 409–460.

Koenderink, J. J., van Doorn, A. J. and Kappers, A. M. L. (1995). Depth relief. *Perception* **24**, 115–126.

Kontsevich, L. L., Kontsevich, M. L. and Shen, A. Kh. (1987). Two algorithms for reconstructing shapes. *Optoelectronics, Instrumentation and Data Processing* **5**, 76–81.

Kontsevich, L. L. (1992). The inference of three-dimensional structure from weak perspective projections based on pairwise comparisons of images. *Proc. SPIE* **1669**, 51–60.

Kontsevich, L. L. (1993). Pairwise comparison technique: a simple solution for depth reconstruction. *J. Opt. Soc. Am.* **A10**, 1129–1135.

Landy, M. S., Maloney, L. T., Johnston, E. B. and Young, M. (1995). Measurement and modeling of depth cue combination: In defense of weak fusion. *Vision Res.* **35**, 389–412.

Lee, C. H. and Huang, T. (1990). Finding point correspondences and determining motion of a rigid object from two weak perspective views. *Comput. Vision, Graphics and Image Processing: Image Understanding* **52**, 302–327.

Lowe, D. G. (1987). Three-dimensional object recognition from two-dimensional shape. *Artif. Intel.* **31**, 355–395.

Luneburg, R. K. (1947). *Mathematical Analysis of Binocular Vision.* The Dartmouth Eye Institute, Hanover.

Malik, J. and Rozenholtz, R. (1993). A differential method for computing local shape-from-texture for planar and curved surfaces. In: *Proc. of 1993 IEEE Computer Society Conference on Computer Vision and Pattern Recognition*, pp. 267–273.

Malik, J. (1987). Interpreting line drawings of curved objects. *Int. J. Comput. Vision* **1**, 73–103.

Necker, L. A. (1832). Observations of some remarkable phaenomena seen in Switzerland and an optical phaenomenon which occurs on viewing of a cristal or geometrical solid. *Phil. Mag. Ser.* **I3**, 329–337.

Parker, A. J., Cumming, B. G., Johnston, E. B. and Hurlbert, A. C. (1995). Multiple cues for three-dimensional shape. In: *The Cognitive Neurosciences.* M. S. Gazzaniga (Ed.). MIT Press, Cambridge, MA, pp. 351–364.

Penrose, L. S. and Penrose, R. (1958). Impossible objects: a special type of illusion. *Brit. J. Psychol.* **49**, 31–33.

Pentland, A. and Kuo, J. (1990). Three-dimensional line interpretation via local processing. *Proc. SPIE* **1249**, 334–344.

Poelman, C. J. and Kanade, T. (1992). A paraperspective factorization method for shape and motion recovery. Carnegie Mellon University Tech. Report, 92–208.

Poggio, T. and Vetter, T. (1992). Recognition and structure from one 2D model view: observations on prototypes, object classes, and symmetries. A. I. Memo No. 1347, Artificial Intelligence Laboratory, MIT.

Singha, P. and Adelson, E. (1993). Recovering 3D shapes from 2D line-drawings. In: *Intelligent Robotics: Proceedings of the International Symposium on Intelligent Robotics*. M. Vidyasagar (Ed.). Bangalore, India (Tata McGraw-Hill, New Delhi), pp. 51–60.

Ullman, S. (1979). *The Interpretation of Visual Motion*. MIT Press, Cambridge, MA.

Vetter, T. and Poggio, T. (1994). Symmetric 3D objects are an easy case for 2D object recognition. *Spatial Vision* **8**, 443–453.

Wagemans, J., Van Gool, L., Swinnen, V. and Van Horebeek, J. (1993). Higher-order structure in regularity detection. *Vision Res.* **33**, 1067–1088.

Young, M. J., Landy, M. S. and Maloney, L. T. (1993). A perturbation analysis of depth perception from combinations of texture and motion cues. *Vision Res.* **33**, 2685–2696.

Human Symmetry Perception, pp. 349–359
C. W. Tyler (Ed.)

Symmetric 3D objects are an easy case for 2D object recognition

THOMAS VETTER[1,*] and TOMASO POGGIO[2]

[1]*Max-Planck-Institut für Biologische Kybernetik, Spemannstr. 38, 72076 Tübingen, Germany*
[2]*Center for Computational and Biological Learning, and Department of Brain Sciences, Massachusetts Institute of Technology, Cambridge, MA 02139, USA*

Abstract—According to the *1.5-views theorem* (Poggio, Technical Report #9005-03, IRST, Povo, 1990; Ullman and Basri, *IEEE Trans. PAMI* **13**, 992–1006, 1991) recognition of a specific 3D object (defined in terms of pointwise features) from a novel 2D view can be achieved from at least two 2D model views (for each object, for orthographic projection). This note considers how recognition can be achieved from a single 2D model view by exploiting prior knowledge of an object's symmetry. It is proved that, for any bilaterally symmetric 3D object, one non-accidental 2D model view is sufficient for recognition since it can be used to generate additional 'virtual' views. It is also proved that, for bilaterally symmetric objects, the correspondence of four points between two views determines the correspondence of *all* other points. Symmetries of higher order allow the recovery of Euclidean structure from a single 2D view.[1]

1. INTRODUCTION

Image-based techniques for object recognition have recently been developed to recognize a specific three-dimensional object after a 'learning' stage, in which a few two-dimensional views of the object are used as training examples (Poggio and Edelman, 1990; Edelman and Poggio, 1992). A theoretical lower bound on the number of views is provided by the *1.5-views theorem* (Poggio, 1990; Ullman and Basri, 1991; for more details see Section 2.1 in this paper). In the orthographic case, this theorem implies that two views — defined in terms of pointwise features — are sufficient for recognition or equivalent to define the affine structure of an object (see also Koenderink and van Doorn, 1991). It is known that, in the case of perspective projection, two views are sufficient to compute projective invariants specific to the object (Faugeras, 1992; Hartley *et al.*, 1992; Shashua, 1993). Under more general conditions (more general definition of 'view', non-uniform transformations, etc.) and, depending on the implementation, many more views may be required (Poggio and

*To whom correspondence should be addressed.

Edelman's estimate is of the order of 100 for the whole viewing sphere using their approximation network).

Though this is an easily-satisfied requirement in many cases, there are situations in which only one 2D view is available as a model. As an example, consider the problem of recognizing a face from just one view: humans can do it, even for different facial expressions (of course an almost-frontal view may not be sufficient for recognizing a profile view and in fact the praxis of person-identification requires usually a frontal *and* a side view).

Clearly one single view of a generic 3D object (if shading is neglected) does not contain sufficient 3D information. If, however, the object belongs to a class of similar objects, it seems possible to infer appropriate transformations for the class and use them to generate other views of the specific object from just one 2D view of it. We certainly are able to recognize faces that are slightly rotated from just one quasi-frontal view, presumably because we exploit our extensive knowledge of the typical 3D structure of faces.

One can pose the following problem: *is it possible from one 2D view of a 3D object to generate other views of that object, exploiting knowledge of the legal transformations associated with objects of the same class?* (We call a 2D transformation of a 2D view *legal* if its result is identical to the projection onto the image-plane of a rigid rotation of the unknown 3D object.) A positive answer would imply (for orthographic projection, uniform affine transformations and in the absence of self-occlusions) that a novel 2D view may be recognized from a single 2D model view, because of the 1.5-views theorem.

In this paper we consider the case in which legal transformations for a specific object (i.e. transformations that generate new correct views from a given one) immediately are available as a property of the class. In particular, we will discuss certain symmetry properties.

The main results of the paper are two.

1. We prove that, for any bilaterally symmetric 3D object (such as a face), one 2D model view is sufficient for recognition of a novel 2D view (for orthographic projection and uniform affine transformations). This result is equivalent to the following statement: for bilaterally symmetric objects, a model-based recognition invariant (as defined by Weinshall, 1993) can be learned from just one model 2D view. It is also closely related to the projective invariant computed on symmetric objects by Rothwell *et al.* (1993).

2. We also prove that for symmetries of higher order (such as two-fold symmetries, i.e. bilateral symmetry with respect to two symmetry planes) it is possible to recover Euclidean structure from one 2D view (see also Kontsevich, 1993).

In the final section, we briefly mention some of the implications of our results for the practical recognition of bilaterally symmetric objects, for human perception of 3D structure from single views of geometric objects and, more generally, for the role of symmetry detection in human vision.

2. RESULTS

2.1. Recognition from one 2D model view

2.1.1. Generating 'virtual' views. Suppose that we have a single 2D model view of a 3D object, which is defined in terms of pointwise features. A 2D view can be represented by a vector $X = (x_1, y_1, x_2, y_2, \ldots, x_n, y_n)$, that is by the x, y-coordinates of its n feature points. Assume further that (a) we know a priori that the object is bilaterally symmetric (for instance, because we identify the class to which it belongs and we know that this class has the property of bilateral symmetry), and (b) we find in the 2D view the correspondence of the symmetric pairs of points. It can be shown that for views of *bilaterally symmetric* objects there exist 2D transformations D on a pair p of symmetric points of the object that yield a *legal view* p^* of the pair. This new view is the projection of a rigid rotation of the unknown 3D object onto the image-plane

$$Dp_{\text{pair}} = p^*_{\text{pair}}. \tag{1}$$

Under the transformations D_1, D_2, and D_3

$$D_1 = \begin{pmatrix} 0 & 0 & -1 & 0 \\ 0 & 0 & 0 & 1 \\ -1 & 0 & 0 & 0 \\ 0 & 1 & 0 & 0 \end{pmatrix}, \quad D_2 = \begin{pmatrix} 0 & 0 & 1 & 0 \\ 0 & 0 & 0 & 1 \\ 1 & 0 & 0 & 0 \\ 0 & 1 & 0 & 0 \end{pmatrix},$$

$$D_3 = \begin{pmatrix} -1 & 0 & 0 & 0 \\ 0 & 1 & 0 & 0 \\ 0 & 0 & -1 & 0 \\ 0 & 0 & 0 & 1 \end{pmatrix},$$

the pair of symmetric points p_0 transforms into p_1, p_2 and p_3

$$p_0 = \begin{pmatrix} x_R \\ y_R \\ x_L \\ y_L \end{pmatrix}, \quad p_1 = \begin{pmatrix} -x_L \\ y_L \\ -x_R \\ y_R \end{pmatrix}, \quad p_2 = \begin{pmatrix} x_L \\ y_L \\ x_R \\ y_R \end{pmatrix}, \quad p_3 = \begin{pmatrix} -x_R \\ y_R \\ -x_L \\ y_L \end{pmatrix}.$$

Each of these transformations applied to all symmetric pairs of points of an image, leads to a new *'virtual'* view of the object under orthographic projection. In the case of perspective projection, only D_1 is a *legal* transformation. Notice that symmetric pairs are the elementary features in this situation and points lying on the symmetry plane are degenerate cases of symmetric pairs.

Geometrically, this analysis means that, for bilaterally symmetric objects, simple transformations of a given 2D view yield other 2D views that are *legal*. It is remarkable that, in order to apply these 2D transformations to a view, it is not necessary to know the spatial orientation of the object or even its 3D structure but only that it is bilaterally symmetric *and* which pairs of features are symmetric. Each transformation

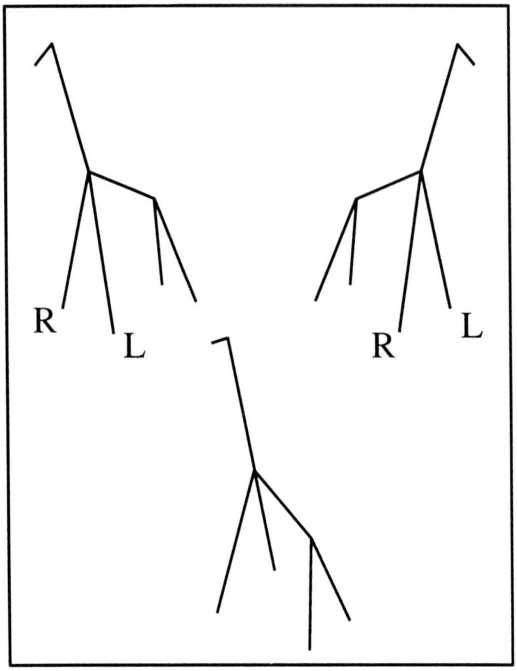

Figure 1. Given a single 2D view (upper left), a new view (upper right) is generated under the assumption of bilateral symmetry. The two views are sufficient to verify that a novel view (second row) corresponds to the same object as the first. (The labels are only used to visualize the applied transformation D_1 between upper left and upper right; for definition see text.)

corresponds to a proper rotation of a rigid 3D object followed by its orthographic projection on the image plane as shown for the transformation D_1 in Fig. 1 top. The transformations are different from a 2D reflection at an axis in the image plane.

In the following, we demonstrate how these 'virtual' views contain additional information, which can be used in object recognition. Let us first point out the difference between recognizing an object and computing its Euclidean 3D structure, which is much harder. We say that an object is *recognized* if its 2D view is an element of the linear vector space $V_{ob_i}^{2N}$, the space of all possible views of a known model object, obtained by 3D linear affine transformations of the 3D model followed by orthographic projection. This definition is equivalent to the affine structure used in Koenderink and van Doorn (1991).

In the case of orthographic projection, we will show that for a bilateral symmetric object a single 2D view (and its 'virtual' view) are sufficient for recognition. Ullman and Basri (1991) showed that the linear vector space $V_{ob_i}^{2N}$ of all possible 2D views of an object with N feature points has 6 dimensions. Their proof is equivalent to saying that $V_{ob_i}^{2N}$ consists of vectors with x and y coordinates and can be written as the direct sum $V_{ob_i}^{2N} = V_x^N \oplus V_y^N$, where V_x^N and V_y^N are non-intersecting linear subspaces, each isomorphic to $\Re^3$. This implies that the x coordinates of a 2D view are the linear combination of the x coordinates of 3 2D views and the y coordinates are the linear combination of the y coordinates of 3 2D views, the two combinations, however, being independent of each other. The 1.5-views theorem (Poggio, 1990) proves that $V_x^N = V_y^N$ and therefore any 4 vectors from V_x^N and V_y^N are linearly dependent. So in general two views are sufficient to span V_{obj}^{2N} by taking two vectors of the x and y coordinates of a first view and a third vector of the x or y coordinates of a second view. Consider now the 2D view of a bilateral symmetric object, which we assume consists of at least four non-collinear feature points. It is easily seen that in general the vectors formed by $p_0 = (x_R, y_R, x_L, y_L)$ and $p_1 = (-x_L, y_L, -x_R, y_R)$ are linearly independent. Only for 'accidental' views, like the perfectly 'frontal' view or the perfect 'side' view, the vectors of the x and y components are linearly dependent.

2.1.2. A recognition algorithm. A single 2D model view together with the knowledge that the object is bilaterally symmetric can be used for recognition in the following way.

1. Take $\mathbf{x}_1$ and $\mathbf{y}_1$ (the vectors of the x and y coordinates of the n feature points) from the available view and generate a third vector $\mathbf{x}_2$ (or $\mathbf{y}_2$) by applying the symmetry transformation D_1 to all pairs of symmetric points.
2. Make a $2n \times 6$ matrix B with its 6 columns representing a basis for $V_{obj}^{2N} = V_x^N \oplus V_y^N$. An explicit form of B is

$$B = \begin{pmatrix} \mathbf{x}_1 & \mathbf{x}_2 & \mathbf{y}_1 & 0 & 0 & 0 \\ 0 & 0 & 0 & \mathbf{x}_1 & \mathbf{x}_2 & \mathbf{y}_1 \end{pmatrix}.$$

3. Check that B is full rank (for instance $(B^T B)^{-1}$ exists). This is equivalent to testing for 'accidental' views.
4. A novel view $\mathbf{t}$ (we assume here that the first n components are the x coordinates followed by n y coordinates) of the same object must be in the space spanned by the columns of B, and therefore must satisfy

$$\mathbf{t} = B\alpha,$$

which implies (since $(B^T B)^{-1}$ exists)

$$\mathbf{t} = B(B^T B)^{-1} B^T \mathbf{t}. \tag{2}$$

B can then be used to check whether $\mathbf{t}$ is a view of the correct object or not, by checking if $\|\mathbf{t} - B(B^T B)^{-1} B^T \mathbf{t}\| = 0$ (an additional test for rigidity may also be applied, if desired, to the three available views). Figure 1 shows the results of using this technique to recognize simple pipe-cleaner animals.

Notice that bilateral symmetry provides from one 2D view a total of four 2D views (image plane rotations not included), each corresponding to a different rotation of the original 3D object. Two of the four views are linearly independent (two linearly-independent vectors of the x coordinates and two for the y coordinates). The results of Shashua (1993) in combination with the virtual views prove even in the perspective case the existence of a projective invariant for bilateral symmetric objects. For recognition functions and projective invariants of symmetric objects see also Moses and Ullman (1991) and Rothwell *et al.* (1993).

2.2. Correspondence and bilateral symmetry

Let us suppose that the correspondence of four non-coplanar points (or more) between a model view (with the pairs of symmetric feature-points already identified) and a novel view is given. Then the following epipolar line argument can be applied separately to each of the two views generated by the model view under the assumption of bilateral symmetry (see Eqn (1)). The x, y coordinates of corresponding points in two images of an object undergoing an affine transformation are linearly dependent, that is

$$\alpha_1\mathbf{x}_1 + \beta_1\mathbf{y}_1 + \alpha_2\mathbf{x}_2 + \beta_2\mathbf{y}_2 = 0.$$

For each point (x_1, y_1) in the model view the corresponding point (x, y) in the novel view then satisfies the two equations:

$$y = mx + A \quad \text{and} \quad y = m'x + A',$$

with $m = -\alpha_2/\beta_2$ and $A = -(\alpha_1 x_1 + \beta_1 y_1)/\beta_2$ and m', A' determined by the virtual view. Therefore (x, y) is uniquely determined (apart from special cases) as

$$y = \frac{m'A - mA'}{m' - m}, \qquad x = \frac{A' - A}{m - m'}.$$

Thus, *the correspondence of four non-coplanar points between two 2D views of a bilateral symmetric object* (*undergoing a uniform affine transformation*) *uniquely determines the correspondence of all other points.*

In the case of occlusions, correspondence of singular points in the novel view can be established as long as the pair of symmetric points is identified in the model view. When full correspondence between the model and novel view is established, any structure-from-motion algorithm can be used to compute the Euclidean structure related to the two views and the assumption of symmetry.

2.3. Euclidean structure from one 2D model view

Suppose, as before, that we have a single 2D view of an object. Assume further that we hypothesize (correctly) that the object is two-fold bilaterally symmetric and that symmetric quadruples can be identified. These are the 'elementary' features in this situation, since any point that is not on a symmetry plane corresponds to three other points. We define an object to be two-fold bilaterally symmetric if the following transformations of any 2D view of a feature quadruple yield *legal views* of the quadruple, that is orthographic projections of rigid rotations of the object:

$$D_{21}\mathbf{q}_{quadr} = \mathbf{q}^1_{quadr}, \tag{3}$$

$$D_{22}\mathbf{q}_{quadr} = \mathbf{q}^2_{quadr}, \tag{4}$$

with

$$\mathbf{q}_{quadr} = \begin{pmatrix} x_1 \\ x_2 \\ x_3 \\ x_4 \\ y_1 \\ y_2 \\ y_3 \\ y_4 \end{pmatrix}, \quad \mathbf{q}^1_{quadr} = \begin{pmatrix} -x_2 \\ -x_1 \\ -x_4 \\ -x_3 \\ y_2 \\ y_1 \\ y_4 \\ y_3 \end{pmatrix} \quad \text{and} \quad \mathbf{q}^2_{quadr} = \begin{pmatrix} x_4 \\ x_3 \\ x_2 \\ x_1 \\ -y_4 \\ -y_3 \\ -y_2 \\ -y_1 \end{pmatrix}.$$

These 3 views are independent apart from special views, such as accidental views (see previous section). Thus the above definition of symmetry provides a way to generate two additional views from the given single view, unless $\mathbf{q}_{quadr}$ is a view which is left invariant by at least one of the symmetry transformations. This is the case, for instance, for exactly frontal views. The same argument can be repeated for all symmetric quadruples. These transformations are the same transformations from the previous section applied to both symmetries.

Thus, these transformations yield in the generic case three independent views of the object (the symmetry yields a total of 16 views, representing 16 different orientations

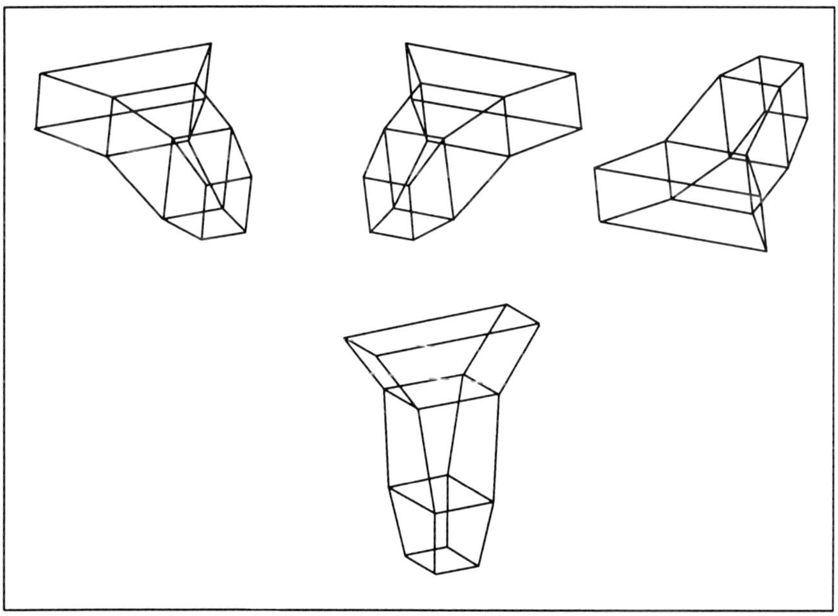

Figure 2. A single 2D view (upper left) of a two-fold bilateral symmetric object can generate additional views (upper center and right) using the symmetric properties of the object. Those three views are sufficient to compute the 3D structure, as indicated in the second row, where we show a new projection of the 3D structure computed from the three views above.

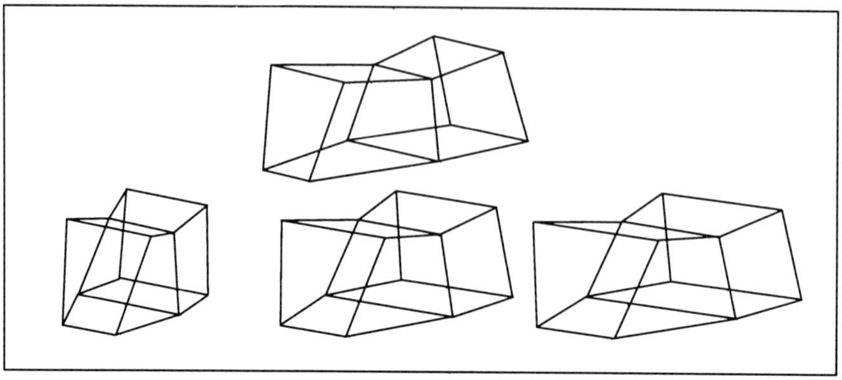

Figure 3. A single 2D view (upper row) of a bilateral symmetric object can be generated by different bilateral symmetric 3D objects. The three objects projected in the second row all generate the 2D view of the first row after a rotation of 20 deg around the vertical axis.

of the object, which span the 6-dimensional viewing space of the object). One can verify that standard structure-from-motion techniques (Huang and Lee, 1989; see also Ullman, 1979) can be applied to conclude that structure is uniquely determined, except for a reflection about the image plane. The matrix defined by Weinshall (1993) to compute an object invariant is full rank in this case; it is, however, rank deficient for simple bilateral symmetry. Using a different approach, the pairwise comparison technique of Kontsevich (1993) comes to a similar result. The following holds:

Given a single 2D orthographic view of a two-fold bilateral symmetric object (with at least two symmetric, nondegenerate quadruple features containing a total of at least four non-coplanar points) the corresponding structure is uniquely determined up to a reflection about the image plane.

In addition, the following results can be easily derived:

1. 3D structure can be obtained from two 2D views of a bilateral symmetric object (see Fig. 2).
2. Structure cannot be uniquely obtained from a single 2D view of a bilateral symmetric object. So a single 2D view of a bilateral symmetric object can be generated by different bilateral symmetric objects (see Fig. 3).

3. DISCUSSION

Exploiting knowledge about the symmetry of an object, recognition is possible from a single view. The geometric constraints of symmetry allow the generation of additional legal views from the given one, using 2D image transformations. This can be done for all non-occluded pairs of symmetric points, without knowing the orientation of the symmetry plane or the camera position. For bilaterally symmetric objects we proved a single view is sufficient for recognition from a different view. For two-fold bilateral symmetric objects the 3D structure can be computed from a single view. Here are some implications of our results:

- *Exact frontal model views should be avoided*

  The results about bilateral symmetry imply that one should avoid using a model view which is a fixed point of the symmetry transformations (since the transformation of it generates an identical new view). In the case of faces, this implies that the model view in the data base should not be an exactly frontal view. Psychophysical evidence supporting this point is given by Schyns and Bülthoff (1993).

- *A symmetry of order higher than bilateral allows recovery of structure from one 2D view*

  Our results imply that even in the absence of 3D cues (such as shading, perspective, texture etc.), an object symmetry of sufficiently high order may provide structure from a single view. An interesting conjecture is that human perception may be biased to impose a symmetry assumption (in the absence of other evidence to the contrary), in order to compute structure.

- *A new algorithm for computing structure from single views of polyedric objects*

  For line drawings, Marrill (1991) and Sinha and Adelson (1993) proposed an iterative algorithm that is capable of recovering structure from single views. Our result on structure-from-1-view may explain some of these results in terms of the underlying algebraic structure induced by symmetry properties. It also yields a new non-iterative algorithm for the recovery of structure since it provides (once symmetric n-tuples are identified) a simple algorithm generating a total of three linearly-independent views to which any of the classical Structure-from-Motion algorithms can be applied, including the recent linear algorithms (Huang and Lee, 1989). It remains an open question to characterize the connection between the minimization principle of Marrill-Sinha and our internal structure constraints. Especially in the case of bilateral symmetric objects, their principle might help to understand which constraints are used from human observers to disambiguate views as shown in Fig. 3.

- *'Virtual' views and image-based object recognition*

  For image-based recognition systems (Poggio and Edelman, 1990), the possibility of generating additional views for objects minimizes the number of necessary example views. In the case of symmetric objects, the image transformations related to rotations in 3D space can be derived directly from the 3D structure of the class of symmetric objects. In the case of nonlinear transformations, the related image transformations have to be approximated from prototypical views of a class of objects. The approach by Beymer *et al.* (1993) demonstrates how this can be done for pose and expressions of human faces. Novel grey-level images, related to changes of facial expression, can be generated from a single image when applying the appropriate image transformation.

- *Psychophysical results on object recognition*

  It is intriguing to speculate about relations between the known human abilities of detecting symmetries and the human tendencies of hypothesizing symmetry in visual perception. There is evidence on spontaneous generalization to left-right reversal in humans and even simpler visual systems (see Sutherland, 1960; Young,

1964; Rock *et al.*, 1989). Our theory offers a simple explanation of these effects as a by-product of a mechanism optimized for the recognition of 3D objects. Thus, visual recognition of 3D objects may be the main reason for the well-known sensitivity of visual systems to bilateral symmetry of 3D objects and 2D patterns. The results found in psychophysical experiments on object recognition (Vetter *et al.*, 1994) are consistent with our theoretical predictions for symmetric objects. Based on a single training view, the generalization performance for novel views is significantly better for symmetric objects than for non-symmetric objects. In contrast to the non-symmetric objects, the generalization field of symmetric objects showed additional peaks of good recognition performance. These additional peaks were in all cases at the location of the virtual views. It is not yet clear in what way the visual system uses symmetry: instead of creating 'explicit' virtual views the system may discover and use symmetry-based features that are view invariant.

Several open questions remain. How does a visual system, natural or artificial, detect symmetric pairs of features of a 3D object, a task which is in general quite different from symmetry detection in a 2D pattern? What are the optimal cues leading to the assumption of symmetry, since it is not possible to prove the symmetry when only a single view is given? In some cases (e.g. line drawings of geometric objects), algorithms capable of identifying feature points likely to be symmetric should be feasible, since all pairs of symmetric points in one view obey to the same epipolar line constraint. In other cases additional information may be available (e.g. once the two eyes are identified as eyes, it is known that they represent a symmetric pair). There the knowledge about the symmetry of the object class can help to establish the correspondence between symmetric feature pairs. Another question which is open is how to extend our approach of using 2D image transformations to geometric constraints other than bilateral symmetry.

Acknowledgements

We are grateful to H. H. Bülthoff, F. Girosi, P. Sinha and S. Ullman for useful discussions and suggestions. This research is sponsored by grants from the Office of Naval Research under contracts N00014-91-J-1270 and N00014-92-J-1879; by a grant from the National Science Foundation under contract ASC-9217041 (including funds from DARPA provided under the HPCC program).

Tomaso Poggio is supported by the Uncas and Helen Whitaker Chair at the Whitaker College, Massachusetts Institute of Technology. Thomas Vetter was supported by a postdoctoral fellowship from the Deutsche Forschungsgemeinschaft (Ve 135/1-1).

NOTE

1. Part of this paper has appeared as MIT AI Lab. Memo No. 1347 (Poggio and Vetter, 1992).

REFERENCES

Beymer, D., Shashua, A. and Poggio, T. (1993). Example-based image anaysis and synthesis. A.I. Memo No. 1431, Artificial Intelligence Laboratory, Massachusetts Institute of Technology.

Edelman, S. and Poggio, T. (1992). Bringing the Grandmother back into the picture: a memory-based view of object recognition. *Int. J. Pattern Recog. Artif. Intell.* **6**, 37–62.

Faugeras, O. (1992). What can be seen in three dimensions with an uncalibrated stereo rig? *Proceedings ECCV*, pp. 563–578.

Hartley, R. I., Gupta, R. and Chang, T. (1992). Stereo from uncalibrated cameras. *Proceedings CVPR92*, pp. 761–764.

Huang, T. and Lee, C. (1989). Motion and structure from orthographic projections. *IEEE Trans. PAMI* **2**, 536–540.

Koenderink, J. and van Doorn, A. (1991). Affine stucture from motion. *J. Opt. Soc. Am.* **8**, 377–385.

Kontsevich, L. L. (1993). Pairwise comparison technique: a simple solution for depth reconstruction. *J. Opt. Soc. Am. A* **10**, 1129–1135.

Marill, T. (1991). Emulating the human interpretation of line-drawings as three-dimensional objects. *Int. J. Comp. Vis.* **6**, 147–161.

Moses, Y. and Ullman, S. (1991). Limitations of non model-based recognition schemes. A.I. Memo No. 1301, Artificial Intelligence Laboratory, Massachusetts Institute of Technology.

Poggio, T. (1990). 3D object recognition: on a result by Basri and Ullman. Technical Report # 9005–03, IRST, Povo, Italy.

Poggio, T. and Edelman, S. (1990). A network that learns to recognize three-dimensional objects. *Nature* **343**, 263–266.

Poggio, T. and Vetter, T. (1992). Recognition and structure from one 2D model view: observations on prototypes, object classes, and symmetries. A.I. Memo No. 1347, Artificial Intelligence Laboratory, Massachusetts Institute of Technology.

Rock, I., Wheeler, D. and Tudor, L. (1989). Can we imagine how objects look from other viewpoints? *Cognitive Psychol.* **21**, 185–210.

Rothwell, C., Forsyth, D., Zisserman, A. and Mundy, J. (1993). Extracting projective structure from single perspective views of 3D point sets. In: *Proceedings of the International Conference on Computer Vision (ICCV)*. Berlin, Germany, pp. 573–582.

Schyns, P. and Bülthoff, H. (1993). Conditions on viewpoint dependent face recognition. A.I. Memo No. 1432, Artificial Intelligence Laboratory, Massachusetts Institute of Technology.

Shashua, A. (1993). Projective depth: A geometric invariant for 3D reconstruction. In: *Proceedings of the International Conference on Computer Vision (ICCV)*. Berlin, Germany, pp. 583–590.

Sinha, P. and Adelson, E. H. (1993). Recovering 3D shapes from 2D line-drawings. In: *Intelligent Robotics: Procs. of the IEEE Int. Symp. on Intell. Robotics*. Bangalore, India.

Sutherland, N. (1960). Visual discrimination of orientation by *octopus*: mirror images. *Brit. J. Psychol.* **51**, 9–18.

Ullman, S. (1979). *The Interpretation of Visual Motion*. MIT Press, Cambridge, MA.

Ullman, S. and Basri, R. (1991). Recognition by linear combinations of models. *IEEE Trans. PAMI* **13**, 992–1006.

Vetter, T., Poggio, T. and Bülthoff, H. (1994). The importance of symmetry and virtual views in three-dimensional object recognition. *Current Biology* **4**, 18–23.

Weinshall, D. (1993). Model-based invariants for 3D vision. *Int. J. Comp. Vis.* **10**, 27–42.

Young, J. Z. (1964). *A Model of the Brain*. Oxford University Press, Oxford.

Human Symmetry Perception, pp. 361–381
C. W. Tyler (Ed.)

Mirror symmetry and parallelism: two opposite rules for the identity transform in space perception and their unified treatment by the Great Circle Model

LEONARD MATIN and WENXUN LI
Department of Psychology, Columbia University, New York, NY 10027, USA

Abstract—Two opposite rules control the contributions of individual lines to the perceptual processing of two different spatial dimensions of egocentric localization and orientation. For lines restricted to the frontal plane, a tilted line on one side of the median plane induces a rotation of the orientation visually perceived as vertical (VPV) identical to that induced by the same tilt on the other side of the median plane, but the influences exerted on the elevation of visually perceived eye level (VPEL) are mirror symmetric. The rule for VPV fits our intuitions; the rule for VPEL does not. However, the reverse peculiarity holds when the inducing lines are rotated within sagittal planes (pitched): Two parallel, pitched-from-vertical lines on opposite sides of the median plane generate identical effects on VPEL but mirror symmetric effects on VPV. These counterintuitive symmetry reversals are reconciled by the Great Circle Model of spatial orientation (GCM), in which line orientations are represented by the great circle coordinates of their images on a sphere centered at the nodal point of the eye via central projection.

1. INTRODUCTION

It would be relatively easy to gain agreement among a group of observers that the two solid lines in Fig. 1a are identically oriented but that the two lines in Fig. 1b are not, that the two solid lines in Fig. 1b are mirror symmetrical with respect to the dashed line as an axis of symmetry. We would also expect to gain quick agreement among workers in spatial vision and space perception to the statement that the two lines in Fig. 1a, presented within a frontal plane, are treated identically by perception while those in Fig. 1b are treated differently. For, certainly, the two lines in Fig. 1a appear more similar than do those in Fig. 1b. The parallelness between the lines in Fig. 1a adds an aspect of similarity to whatever aspects of similarity are perceived between the two lines in Fig. 1b so that at most the two lines in Fig. 1a are only differently located whereas the two lines in Fig. 1b are both differently located and differently oriented.

If some segments of the above statements are interchanged to read that the two lines in Fig. 1b are treated more similarly by perception than are those in Fig. 1a we

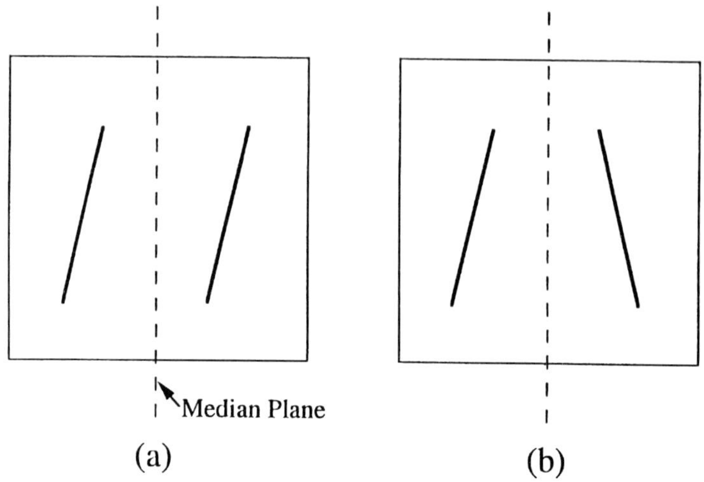

Figure 1. Tilted lines with centers equidistant from the median plane. (a) Parallel lines. (b) Mirror symmetrical lines.

would expect to be greeted with puzzlement if not outright denial. It would be even more difficult to gain agreement if we stated that perception treats the two lines in Fig. 1b identically under conditions for which the two lines in Fig. 1a are treated as opposites. Nevertheless the latter, counterintuitive statement holds for the perception of elevation. For a second dimension of egocentric localization and orientation — perceived orientation within a frontal plane — the reverse, more intuitive relation, holds as it does for the more usual perception of patterns. Thus, for the perception of elevation, mirror symmetry between lines is the identity transform and parallelness the negative transform; for perceived orientation within a frontal plane, parallelness is the identity transform and mirror symmetry the negative transform. The present article will describe the evidence for this and demonstrate that although both perceptual dimensions are controlled by the same fundamental property of individual lines, each is controlled by a separate and different mechanism.

The experiments involve measurements of a norm within each of the two dimensions. For perceived elevation an observer sets the physical elevation of a visual target to appear at eye level (VPEL). For perceived orientation, an observer rotates a test line within a frontal plane to the orientation at which it appears vertical (VPV). The present article will describe measurements of the two norms, VPEL and VPV, exclusively. But, it should be clear that by measuring the physical location corresponding to a single point within a perceptual dimension we also learn about the correspondence between large portions of the physical and perceptual dimensions. For example, a shift in the elevation of VPEL relative to true eye level suggests a linear translation between the physical elevation and perceived elevation of points above and below VPEL, a suggestion supported by manual matches to visual targets at different elevations (Robison, Li, and Matin, unpublished experiments). Similar statements apply to relations between frontal plane orientation and VPV.

VPEL and VPV are set with good accuracy and reliability in total darkness. Thus, standard deviations for VPEL are typically less than 1 deg and constant errors less than

6 deg; Standard deviations and constant errors for VPV are both typically under 2 deg (cf. Neal, 1926; Matin and Li, 1994a). For both discriminations, then, mechanisms must be in place which make use of combinations of information regarding the direction of gravity, the orientation of the eyes in the head, and the location (for VPEL) or orientation (for VPV) of the target on the retina. We refer to the employment of such combinations as 'body-referenced mechanisms'.

However, both VPEL and VPV are strongly influenced by the orientation of individual lines in the visual field, and it is these influences that are of most concern here. The systematic influence of visual pitch on VPEL is only slightly less for a visual field consisting of a single pitched-from-vertical line in darkness than it is for a complexly-structured visual field, and in both cases mislocalizations typically equal about half the angle of visual pitch (Matin and Li, 1989a, 1994a), indicating that the discrimination is based on influences from both the visual field and the body-referenced mechanism.[1] The influence of visual pitch on VPEL does not depend on binocular viewing or on any of several retinal image gradients including variation in line width and blur of the retinal image caused by normal line spread and/or variation in accommodation (Matin and Li, 1989, 1992a, 1994a). The fundamental aspect of the line determining its influence on VPEL is its location and orientation on the retina, as discussed below.

The influence on VPV of the angle of tilt of a large square frame within a frontal plane is also typically somewhat less than that of a complexly-structured visual field (Witkin and Asch, 1948). VPV is generally set at an intermediate orientation between the one that would result from the visual field alone and the one that would result from the body-referenced mechanism alone and indicates that both contribute to the discrimination.

2. THE FOUR STIMULUS LINES AND TWO DISCRIMINATIONS

All experiments in this article were carried out with the erect subject monocularly viewing either one or two of the four lines in Fig. 2 in total darkness at a distance of 1 meter (measured within the median plane of the subject along the normal from the eye to the plane containing the line). A' and B' are parallel lines on the pitched plane π', and would be vertical if π' were erect ($\theta = 0$); A and B are the intersections of the erect plane π with the triangles determined by the nodal point of the eye and the extremities of A' and of B', respectively. Thus A and A' are imaged on the same retinal locus; B and B' are imaged on the same retinal locus. The observer's frontal plane is parallel to π, and the normals from the eye to π and to π' both lie in a midsagittal plane midway between each pair of lines.

In one set of conditions either A' or B' was viewed alone at each of seven angles of pitch extending from topbackward, as in Fig. 2a, to topforward, where the pitch axis is a horizontal line at true eye level within the stimulus plane. In a second set of conditions, either A or B was viewed at one of seven angles of obliquity on the erect plane; each of these settings provided stimulation to the same retinal locus that was stimulated by either A' or B' at one of the seven angles of pitch. When erect,

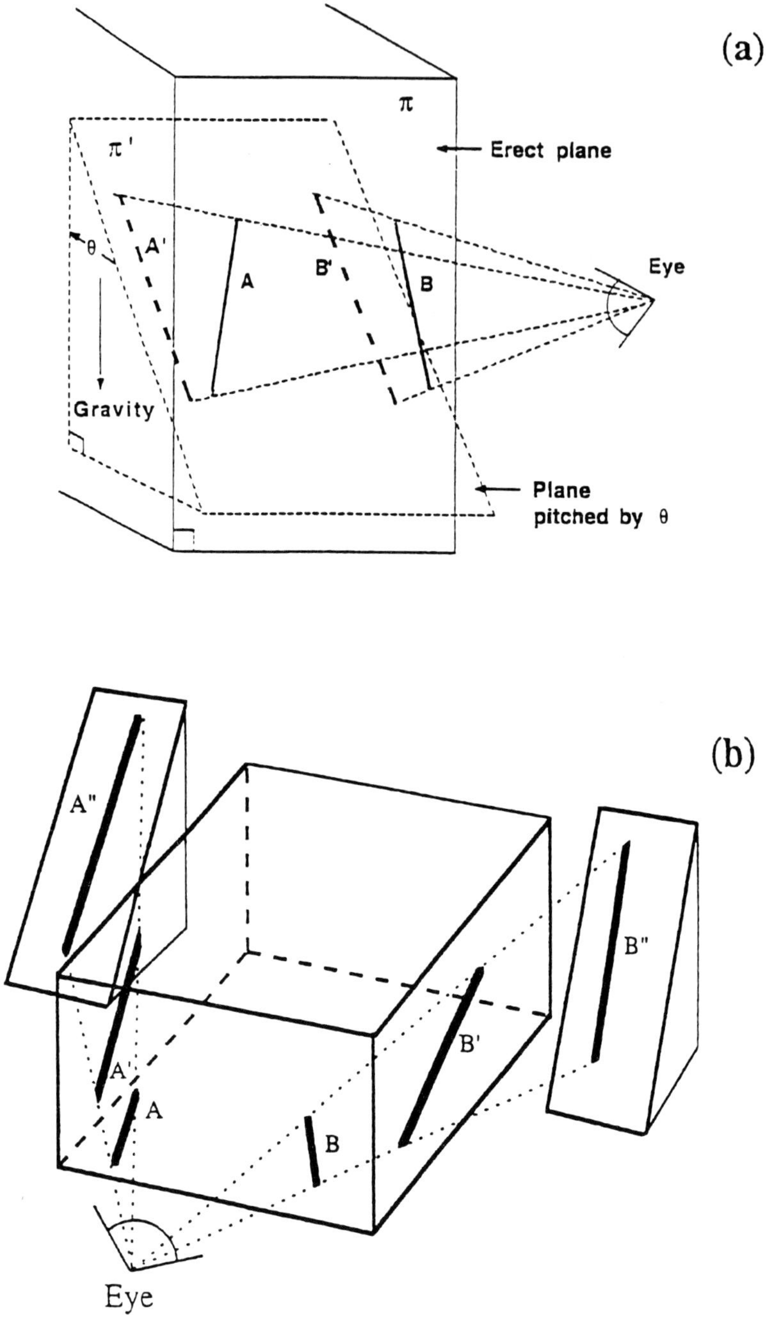

Figure 2. (a) Two parallel pitched-from-vertical lines, A' and B' (dark dashed lines), lie on pitched-only plane π'; two oblique lines, A and B (dark solid lines), lie on erect plane π frontoparallel to the observer. The normal line of visual direction to the erect plane falls halfway between lines A and B on π and between A' and B' on π'. A and B are central projections of A' and B' on the eye, respectively. (b) Lines from three different planes in space that produce identical retinal stimulation; lines A and B are of equal and opposite orientations but lines A' and B' are parallel as are A'' and B''.

A and A' were 25 deg to the left of the median plane of the subject, B and B' were 25 deg to the right of the median plane. Each line was 66.5 deg long by 5.6 arcmin wide with a luminance of 0.01 ml. The monocularly-viewing subject, whose head was stabilized by a chinrest, made VPEL settings in one series of sessions and VPV settings in a separate series of sessions with gaze directed at the test target during each trial. VPEL was measured by the subject's setting of the elevation of a small, dim, red circular target in his/her median plane to appear at eye level. VPV was measured by the subject's setting of the orientation within a frontal plane of a 5.7 deg long line to appear vertical; this test line was centered on his/her median plane at true eye level.

3. THE VPEL DISCRIMINATION

Figure 3 displays average VPELs for five subjects viewing each of the two pitched-from-vertical lines as a function of the pitch of the plane containing the line (Fig. 3a), and for each of the two oblique lines in the erect plane as a function of 'equivalent pitch' (Fig. 3b; the 'equivalent pitch' equals the real pitch of the plane that contains a pitched-from-vertical line that stimulates the same retinal location as the oblique line). The variation of VPEL with either real or equivalent pitch is essentially linear as previously described (Matin and Fox, 1989; Matin and Li, 1990, 1992a,b, 1994a,b) with results for the two symmetrically located lines indistinguishable in each panel. The average standard deviation of the settings for an individual subject at one pitch was 0.87 deg (4 trials/subject with each line). The results for the oblique lines in Fig. 3b are nearly identical to those with the corresponding real pitches in Fig. 3a, with indistinguishable slopes. This near-identity is even clearer in Fig. 4a (a replot of the results in Fig. 3), where the results for the two lines imaged at the same retinal locus from the two planes are superimposed. The results in Fig. 4a are shown separately for the lines on the left of the median plane (open and filled squares) similar to A and A' in Fig. 2 and for the lines on the right of the median plane (open and filled circles) similar to B and B' in Fig. 2.

The closeness of the VPEL values for identical values of real and equivalent pitch is understandable since the loci of retinal stimulation are the same in the two cases and since the various differences in retinal gradients accompanying the difference between real and equivalent pitch play no role (Matin and Li, 1990, 1992a, 1994a). But, an explanation of the identity of result with mirror symmetrical lines requires a deeper analysis.

4. VPEL AND THE GREAT CIRCLE MODEL (GCM)

Figure 5 opens the door to some insight. Each panel contains a sketch of two individual lines in central projection on the inner surface of a sphere centered at the nodal point of the viewing eye in primary position. We refer to the projection sphere as a 'spherical approximation to the eye' and the inner surface as its 'retina', but with

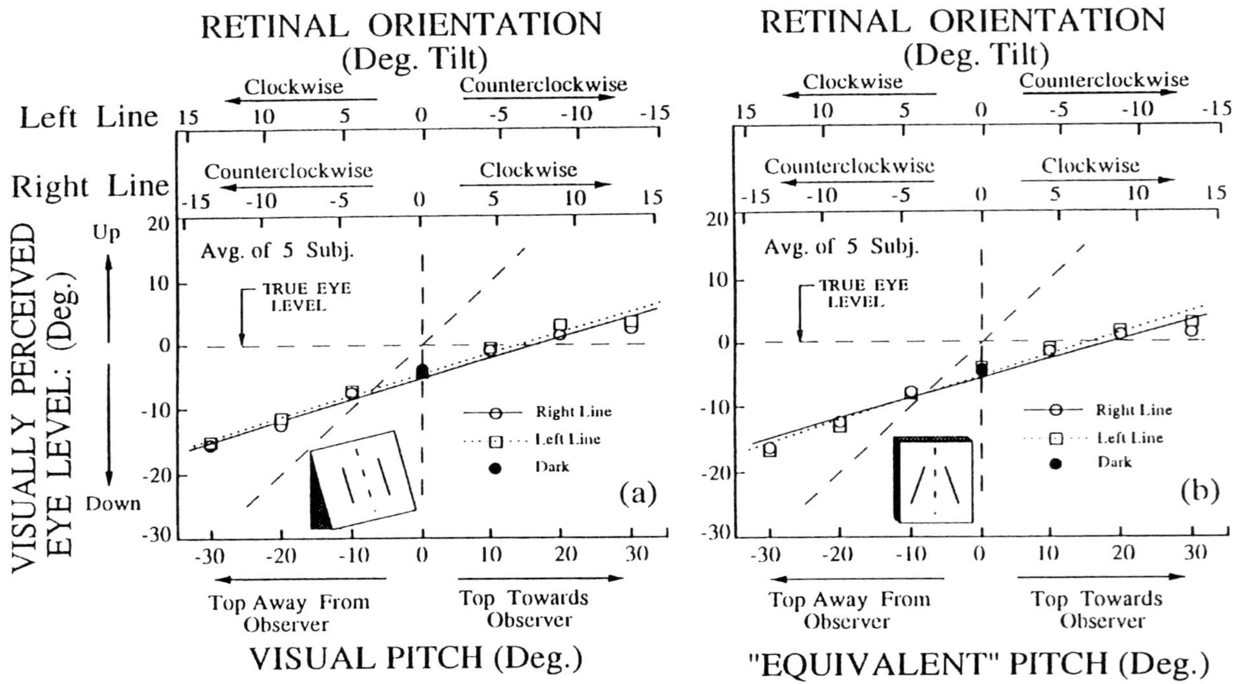

Figure 3. For each of the four sets of VPEL measurements displayed the subject viewed a visual field consisting of a single one of the four lines in Fig. 2 in darkness: (a) results for each of the two pitched-from-vertical lines at each of seven angles of pitch; (b) results for each of the two oblique lines in a frontal plane at each of the seven obliquities. Each orientation of each line in (b) stimulated the same retinal locus as did one of the lines in (a); thus, the lower abscissa in (b) is represented as equivalent pitch. The two abscissa scales of retinal orientation at the top of each panel are in degrees along the circumference of the midfrontal plane of the spherical approximation to the eye (CMFP) at the point that is intersected by the great circle containing the image of the stimulus line (see Figs 5 and 7); the zero on the CMFP is at the intersection with the CVRM at the upper pole of the eye; for the purpose of defining 'clockwise' the view for the reader is behind the spherical eye looking at the stimulus plane. The relation between pitch (or equivalent pitch) and retinal orientation becomes more nonlinear outside the range employed here. VPEL in each case was set by adjustment of the elevation of a single point (projected laser target) within the median plane. Each point is the average of four VPEL measurements on each of the same five subjects.

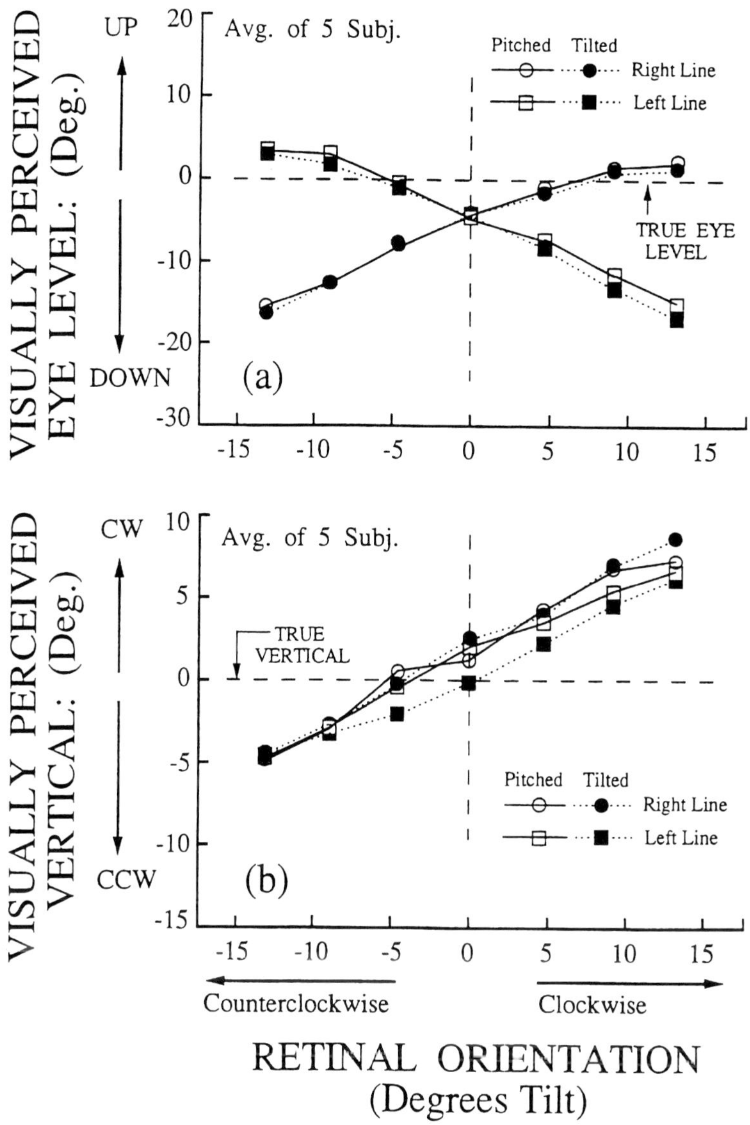

Figure 4. (a) Replot of results of VPEL discrimination from Figs 3a and 3b on a common abscissa for retinal orientation measured in degrees along the circumference of the midfrontal plane of the spherical approximation to the eye (CMFP) at the point that is intersected by the great circle containing the image of the stimulus line (see Figs 5 and 7); the values refer to deviations from a zero at the upper pole (see legend, Fig. 3). The crossed results for the two pitched-from-vertical lines on opposite sides of the median plane reflect the mirror symmetrical influences of lines on opposite sides of the median plane of the subjects. The results for pitched-from-vertical lines in pitched-only planes are indistinguishable from those for the oblique lines in an erect plane for the same value of retinal orientation. (b) The VPV discrimination measured with each of the same four lines at the same pitches and obliquities employed to measure VPEL in (a). VPV values for lines on the left and right sides of the median plane are indistinguishable, reflecting the common influence on VPV of parallel lines within a frontal plane; retinally identical lines from pitched and erect planes yield indistinguishable VPV values. CW refers to clockwise rotation of the stimulus from the viewpoint of the subject.

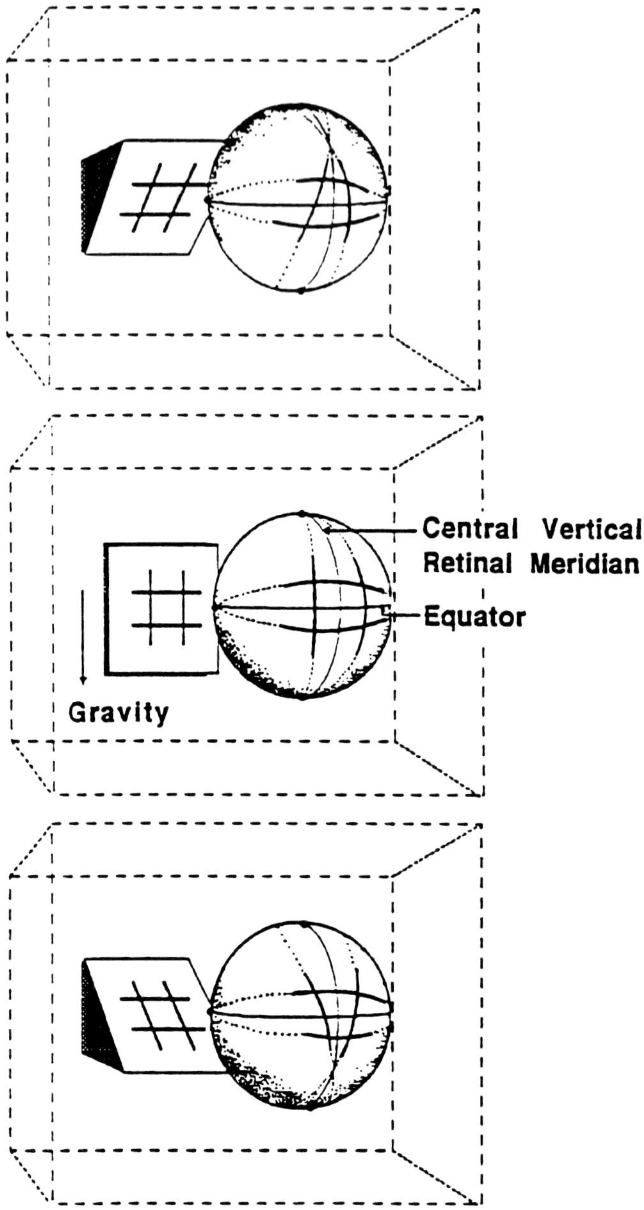

Figure 5. Three sketches of a projection sphere ('spherical approximation to the eye') centered at the nodal point of the eye of a subject viewing (in primary position) a grid consisting of crossed pitched-from-vertical and horizontal 2-line stimuli. The 'retinal' image results from central projection through the pinhole pupil at the center of the sphere. The figures are not drawn to scale; the perspective is intended to display clearly only certain features of the relation between stimulus and image.

The 'eye' is in the same orientation in the 'room' in the three panels. It is erect and faces the center of the stimulus. The equator and central vertical meridian on the surface of the back of the sphere are shown by light solid curves in the three panels. The grid stimulus in the central panel is in a frontoparallel plane. In the upper and lower panels the grid is pitched topforward and topbackward respectively around the horizontal axis at eye level in the stimulus plane.

no intent at approximating the actual shape of the eyeball. In fact, one of two main reasons for employing this approach is an attempt at bypassing the details of both the stimulus-retina and retina-visual-cortex transforms.

The use of spherical projection here derives from the hypothesis that the identity of influence on VPEL by mirror symmetrical lines is a consequence of action by them at a common neural node.[2] This connection builds on the fact that the image of every straight line in object space is imaged on a segment of a great circle on such a projection sphere, and that the images of all members of a set of parallel lines fall on great circles that intersect each other at two nodes unique to the set; the nodes are different for different sets of parallel lines differentiated by their different directions in space. The two nodes for each set represent the point at infinity that lies on the extensions of the line images in the projective plane. As shown in Fig. 5 all members of the set of parallel pitched-from-vertical lines on a pitched-only plane are imaged on segments of great circles that intersect at the same two points on the 'central vertical retinal meridian' (CVRM; only one of these points is visible in the top and bottom panels of Fig. 5). The locations of these intersection points change linearly with pitch and map directly onto the variation in VPEL. Our hypothesis is that they also map topologically on to the neural nodes responsible for the variation of the visual influence on VPEL. This treatment provides the core of the Great Circle Model (GCM; Matin and Li, 1989, 1992a,b, 1994a,b).

The major predictions from GCM regarding perceptual outcomes from stimulation by individual lines and their combinations have been confirmed: Since the focus of GCM is on image location on the projection sphere, it readily accommodates identical results for real and equivalent pitch. Most significant also, the exponential increase with line length of the influence of the pitched line on the slope of the VPEL-vs-pitch function is nearly the same whether the total length of the simultaneously-exposed segment(s) are distributed equally between two symmetrically-placed lines or are coextensive (Fig. 6; Li and Matin, 1991; Matin and Li, 1992c, 1994b).

We employ 'retinal orientation' as a general term to refer to placement of an image on the projection sphere. In the present context specification of retinal orientation for straight lines is most simply done by designating the great circle on which the image falls. Although unique placement of a great circle on the sphere only requires two coordinates, it is useful to refer here to coordinates of intersection on three base great circles: the CVRM, the circumference of the midfrontal plane (CMFP), and the equator (Fig. 5). For VPEL the coordinate of most interest in the present context is the point of intersection with the CVRM; for pitched-from-vertical lines (and for lines of equivalent pitch) this coordinate changes linearly with pitch. The point of intersection with the equator is a measure of the horizontal eccentricity of the line at true eye level. The intersection of a line's great circle with the CMFP is a measure of the orientation within a frontal plane of an oblique line and its retinally-identical pitched-from-vertical line. The intersections on the three base great circles are related by $\tan \mu_i = \tan \mu_j / \tan \mu_k$ where μ_i, μ_j, and μ_k are the intersections on the CVRM, CMFP, and equator, respectively.

The main point of the previous section — that parallel lines from planes of equal pitch or lines of opposite obliquity on opposite sides of the observer's median plane

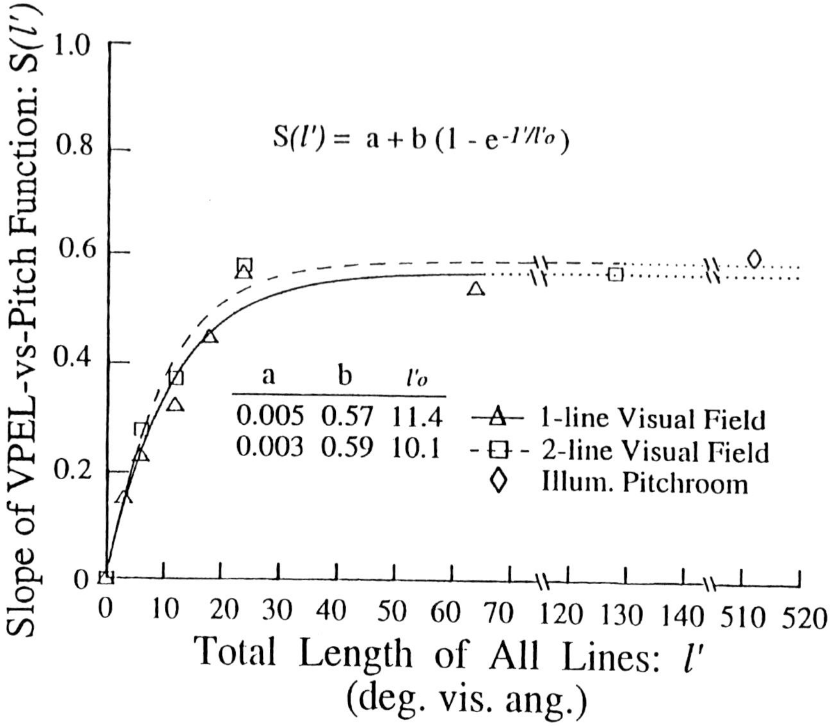

Figure 6. Results from experiments in which line length was systematically varied with one-line and two-line pitched-from-vertical stimuli. The figure plots the slopes of the VPEL-vs-pitch functions against l', the total length of all pitched-from-vertical lines in the visual field. Each data point is the average of the results for the same three subjects. Separate exponentials were fitted to the seven points for the 1-line condition and to the five points for the 2-line condition; one of the points for each visual field was the zero-length (dark) condition shown at $l' = 0$. The difference between exponentials for 1-line and 2-line stimuli is treated by GCM in Matin and Li (1994b).

exert identical influences on VPEL — can be simply expressed in terms of relations on the projection sphere: Indistinguishable VPELs are generated by lines with the same intersection point on the CVRM which have intersection points of equal magnitude but opposite sign on the CMFP (see upper abscissas, Fig. 3). These opposite retinal orientations are stimulated from mirror symmetric locations across the median plane either by oppositely-oriented oblique lines on a frontal plane or by parallel lines on a pitched-only plane.

This treatment in terms of the projection sphere shifts the focus from the external stimulus to stimulation at the eye and permits even broader generalization. For example, as is clear also from Fig. 2, each of the triangles determined by the extremities of a pitched-from-vertical line and the eye's nodal point contains an infinite set of straight and curved lines all of which would be imaged on the same retinal locus as are the pitched-from-vertical line in the pitched-only plane and the oblique line in the erect plane, and it is not a far leap to expect that all of these would yield results identical to those shown. Further generalization will be indicated below.

5. THE VPV DISCRIMINATION AND GCM

Just as two parallel pitched-from-vertical lines on a pitched-only plane projectively transform into mirror symmetrical oblique lines on an erect frontoparallel plane (Fig. 7a, b), so pitched-from-vertical lines from planes of equal and opposite pitch projectively transform into parallel lines on an erect frontoparallel plane which yield images on the projection sphere that are identical to those generated by the pitched-from-vertical lines (Fig. 7c, d). Figure 4b is constructed in a fashion analogous to Fig. 4a, and shows that each of the four lines in Fig. 7 undergoes the same variation of influence on VPV when the retinal orientation of the line is varied systematically. Increasing clockwise rotation of the peripherally located line generates increasing clockwise rotation of VPV. Here, however, parallel lines within a frontal plane that lie on opposite sides of the median plane exert identical influences on VPV, and mirror symmetric lines exert opposite influences — exactly the reverse relation for the influence of individual lines on VPEL in Fig. 4a. A similar relation holds for the retinally identical pitched-from-vertical line pair.

The two lines on opposite sides of the median plane which have identical influences on VPV have images on the projection sphere whose great circles intersect the CVRM at points that are at equal distances on opposite sides of the equator, exactly the relation that yields opposite influences on VPEL. However, these great circles intersect the CMFP at the same two points as do all members of their parallel line set. Thus, the CMFP plays the role for VPV that the CVRM plays for VPEL.

6. SEPARATE GREAT CIRCLE MECHANISMS FOR VPEL AND VPV

Figure 8 summarizes some of the important aspects of the main formal operations of an opponent process model embedded in a three dimensional vector space that treats the results described above regarding the influence of individual lines and predicts the results for pairs (and all other combinations) of lines of different orientations on both VPEL and VPV. Each of the four rectangular boxes at the top of the figure contains a single stimulus line at one of two orientation directions and is located on one of the two sides of the median plane. Although the four stimulus lines shown are oblique lines in an erect plane, the representation applies equally well to pitched-from-vertical lines which strike locations on the projection sphere identical to those struck by the oblique lines. The main qualitative aspects of the results in Fig. 4 are summarized concisely in the arrows directly under the upper set of boxes and show that each of the four combinations of line orientation and laterality produces a different one of the four possible combinations of directional influence on VPEL and VPV. (Figure 10 will quantify this approach to the results in Fig. 4.)

The remainder of Fig. 8 — the connection pattern between the top and bottom rows of boxes (neural flow in downward direction only on the figure) — is the expression of the combining rules of the model for the special case of two simultaneously-presented, equal-length lines that deviate from vertical by equal angular magnitudes and are located at symmetrical horizontal eccentricities on opposite sides of the median

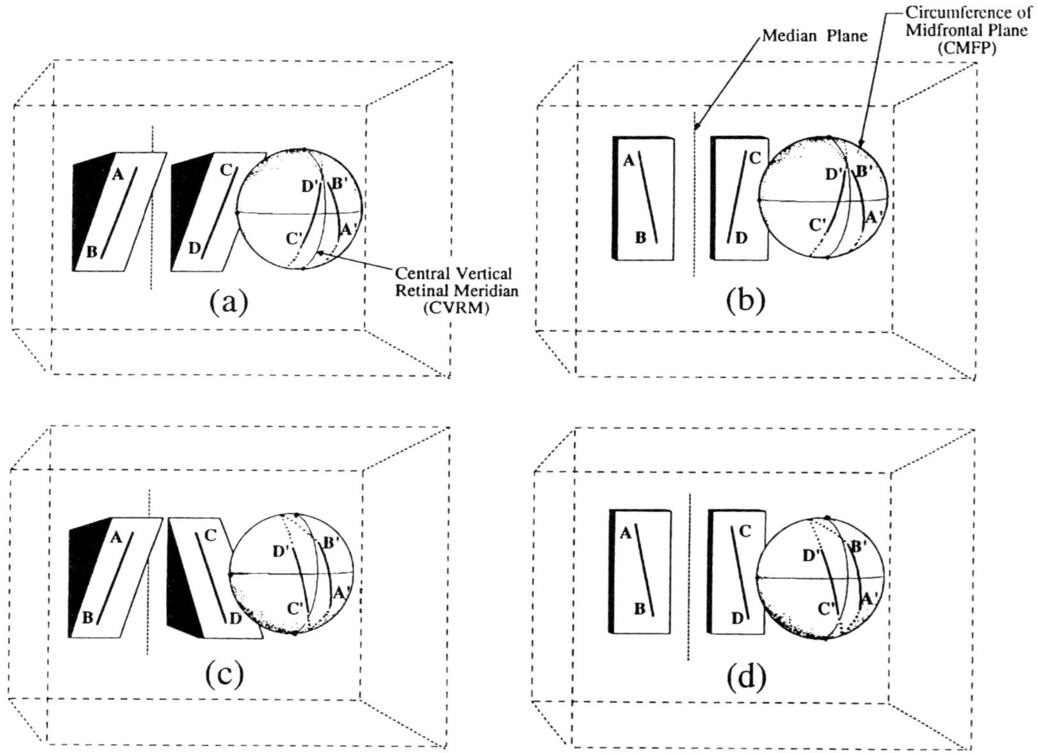

Figure 7. The two upper panels (a, b) display the identity of central projections on the spherical approximation to the eye from two pitched-from-vertical lines lying on the same pitched-only plane and two mirror symmetrical lines on an erect frontal plane. The two lower panels (c, d) display the identity of central projections from two pitched-from-vertical lines lying on planes of equal and opposite pitch and two parallel lines on a frontoparallel plane. In each panel the two lines are at equal horizontal eccentricities on opposite sides of the median plane. The images of the two lines in (a) and in (b) lie on great circles that intersect the same point on the CVRM but intersect mirror symmetrical points on the CMFP. In (c) and (d) the images of the two lines lie on great circles that intersect the same point on the CMFP but intersect points on the CVRM that are equally distant and on opposite sides of the equator.

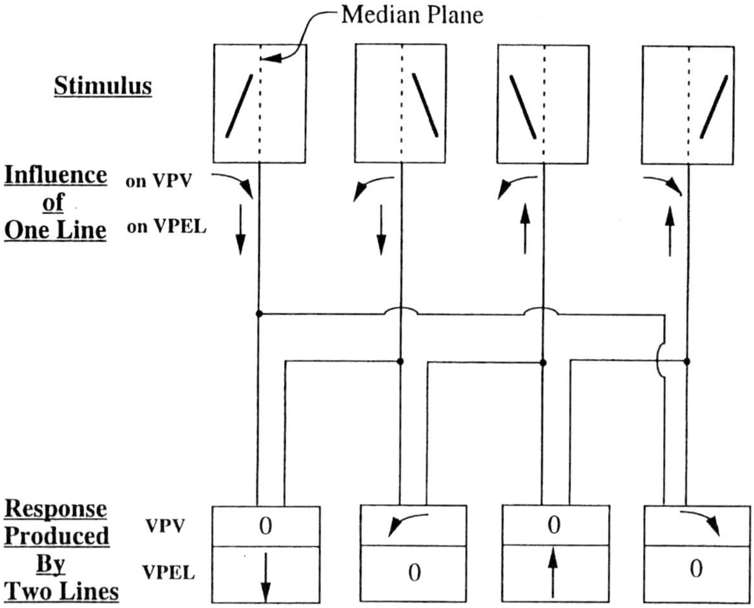

Figure 8. The influence of combinations of lines on VPEL and on VPV as predicted from the Great Circle Model. Each of the upper set of rectangles displays a single line at one of the four combinations of laterality relative to the median plane and orientation relative to vertical. The directions of influences of each of these lines on both VPEL and VPV are displayed directly underneath the upper rectangles by the directions of the pairs of arrows. Although each stimulus line is displayed as lying within a frontal plane, the representation for each of the lines holds also for all other lines that stimulate the same retinal locus as the line displayed here (e.g. a pitched-from-vertical line as in Fig. 2).

The lower set of rectangles displays the predicted consequences from the model of viewing a visual field consisting of the two stimulus lines for both VPEL and VPV as indicated by the connections to the upper rectangles. Although the displayed influences for individual lines and responses produced for two-line combinations are in terms of direction only for VPEL and VPV (with 0 implying no net influence), this is a limitation of the diagram; the model makes quantitative predictions that have been confirmed (see Fig. 10 below) regarding the actual values of VPEL and VPV in the cases shown here.

plane. For this case the rule is that the influence of two nonparallel lines is the average of the influences of the two lines presented separately; for parallel lines the rules is summation along the exponential of Fig. 6. For example, since the influences from the two stimulus lines into the lower leftmost box are of equal but opposite sign for VPV, the predicted value of the average of the two influences is zero as shown; since the influences from the same two lines on VPEL are both downward the predicted direction of the net influence on VPEL is downward as shown. Similar predictions are obtained from the averages of the other three 2-line combinations in the figure.

The predictions for the two-line combinations in Fig. 8 hold for both VPV and VPEL as shown in Fig. 9 for pitched-from-vertical lines in pitched-only planes in the four panels in the left column and for oblique lines in the erect plane in the right column. The individual lines were those employed in the previously-described experiments — 66.5 deg long lines presented at 25 deg horizontal eccentricities. For the VPEL discrimination, a substantial slope of the VPEL-vs-pitch function is obtained with two simultaneously-viewed pitched-from-vertical lines of the same pitch (panel a) but the slope approximates zero when the two planes containing the two pitched-from-vertical lines have equal magnitudes at opposite pitch (panel e). For the VPV discrimination, on the other hand, the reverse pair of results obtain: a substantial slope of the VPV-vs-pitch function is measured with two pitched-from-vertical lines of equal and opposite pitch (panel g) but the slope approximates zero with two lines of the same pitch (panel c). The stimuli with equivalent pitch in the right column yield results that are indistinguishable from those generated by their retinally identical counterparts in the left column (b to a, d to c, f to e, h to g), thus extending the generality of the result in Fig. 4 to the two-line case.

The most telling result that indicates separate processing of VPV and VPEL is the fact that no influence on VPEL is measured for two simultaneously-viewed lines of equal and opposite pitch (Fig. 9e) or for its retinally identical counterpart (Fig. 9f), whereas for VPV no influence is measured for two lines with the same pitch (Fig. 9c) or its retinally identical counterpart (Fig. 9d). In each case the influence generated by one of the lines is nulled by the influence generated by the other simultaneously-viewed line. The slopes of the same-pitch or same-equivalent-pitch VPEL-vs-pitch functions in Figs 9a and 9b are slightly larger than those obtained for the comparable 1-line conditions with the identical group of subjects (not shown); that the magnitude of summation here is not large is a consequence of the approach to saturation for each of the 1-line stimuli employed (Fig. 6). A similar result appears to hold for VPV for parallel lines for what is likely to be a similar reason although the experiments to examine this are not as yet as complete as for the VPEL case.

Nulling by equal and opposite influences from pitches or obliquities of the same magnitude (Fig. 9) is a special case of averaging: When the pitch of one line is fixed and the pitch of the second line on the opposite side of the median plane is varied systematically, averaging is obtained for all 49 combinations of the seven pitches (Li and Matin, 1992). Again the VPEL-vs-pitch function is linear and the measured slope is exactly one half the slope for the case in which the two simultaneously-presented pitched-from-vertical lines are of the same pitch.

Figure 10 is a plane in a three dimensional vector representation of GCM that describes the above results and a number of others for the VPEL and VPV discriminations, and also predicts the outcome of a number of experiments that have not yet been done. The figure displays the visual influence on both VPEL and VPV as two

Figure 9. Visually perceived eye level (VPEL) and visually perceived vertical (VPV) with subject viewing a visual field containing a two-line stimulus in darkness. Both sets of measurements were made with each stimulus, with the VPV measurements displayed in the panel directly beneath the panel displaying the VPEL measurements for the same stimulus. Each data point is the average of four trials on each of the same four subjects. Panels (a)–(d), the 'same-pitch' condition, involved two-line stimuli similar to those in Fig. 2, with panels (a) and (c) employing the pitched-from-vertical line pair (lines A' and B' in Fig. 2) and panels (b) and (d) employing the oblique lines in the erect plane (lines A and B in Fig. 2). Panels (e)–(h), the 'opposing-pitches' condition, involved two-line stimuli similar to those in Figs 7c and 7d, with panels (e) and (g) employing the configuration in Fig. 7c, and panels (f) and (h) employing pitched-from-vertical lines in the configuration in Fig. 7d from an erect plane that strike the identical retinal locus.

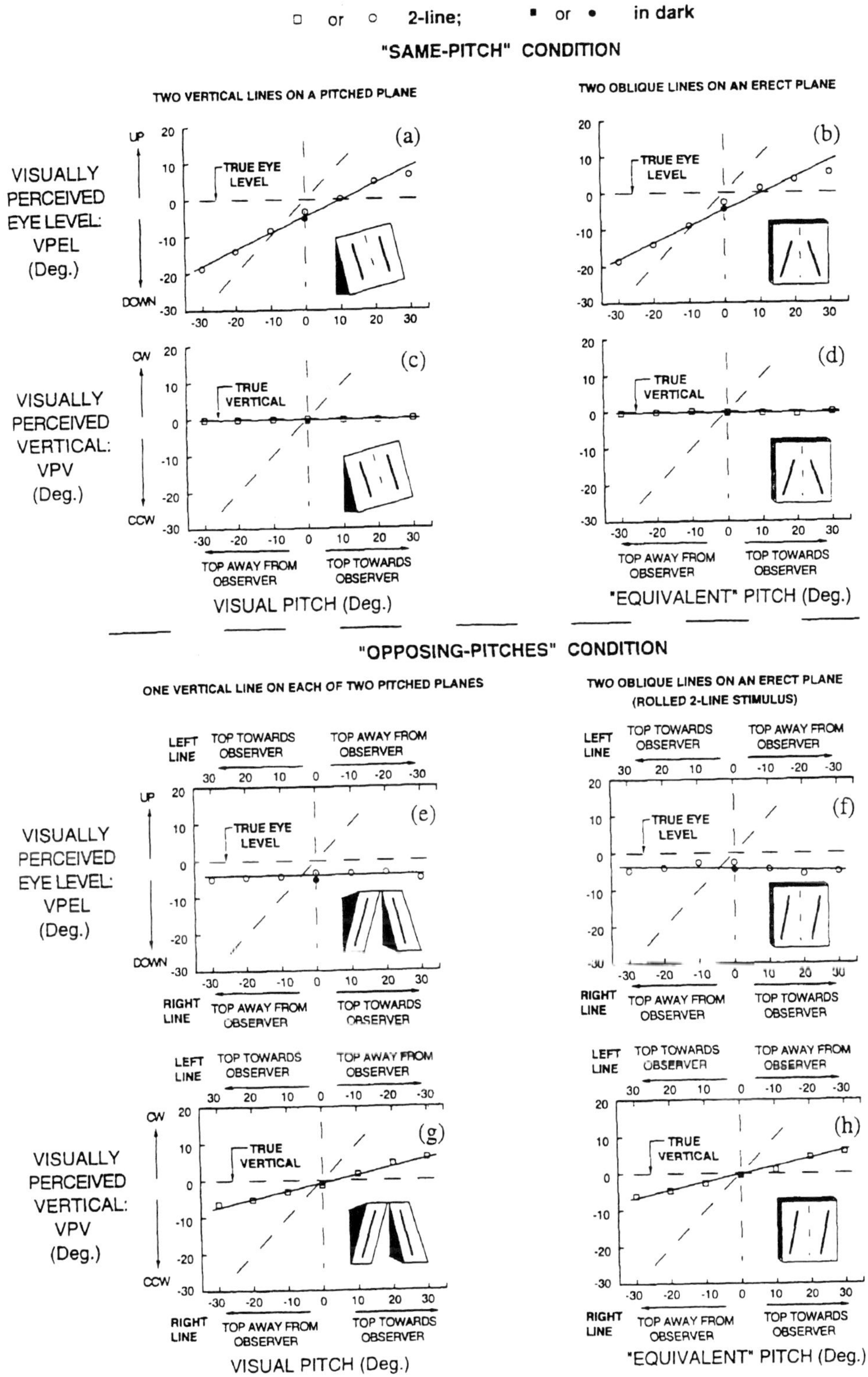
□ or ○ 2-line; ■ or ● in dark
"SAME-PITCH" CONDITION
TWO VERTICAL LINES ON A PITCHED PLANE
TWO OBLIQUE LINES ON AN ERECT PLANE
VISUALLY PERCEIVED EYE LEVEL: VPEL (Deg.)
UP
DOWN
TRUE EYE LEVEL
(a)
(b)
VISUALLY PERCEIVED VERTICAL: VPV (Deg.)
CW
CCW
TRUE VERTICAL
(c)
(d)
TOP AWAY FROM OBSERVER
TOP TOWARDS OBSERVER
VISUAL PITCH (Deg.)
"EQUIVALENT" PITCH (Deg.)
"OPPOSING-PITCHES" CONDITION
ONE VERTICAL LINE ON EACH OF TWO PITCHED PLANES
TWO OBLIQUE LINES ON AN ERECT PLANE (ROLLED 2-LINE STIMULUS)
LEFT LINE
RIGHT LINE
(e)
(f)
(g)
(h)

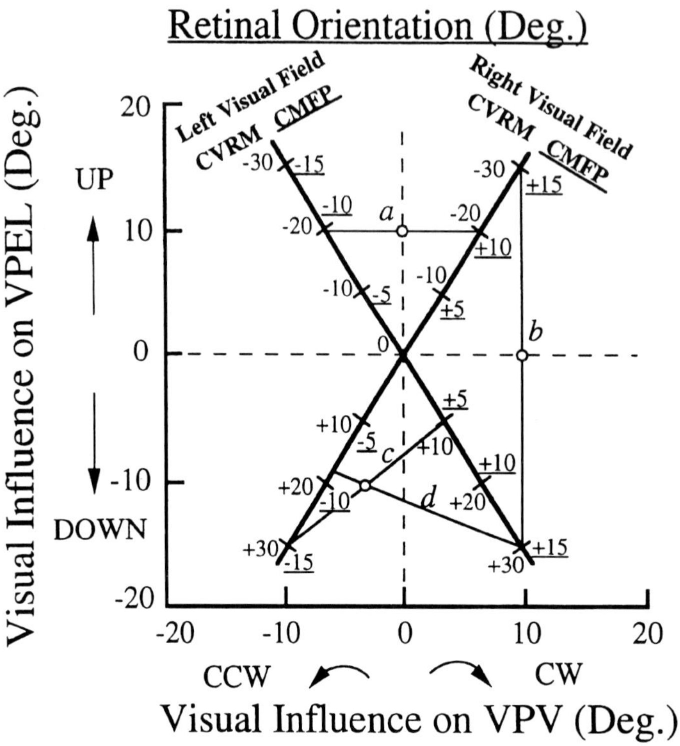

Figure 10. A plane through a three-dimensional vector representation of a generalized opponent process treatment of the Great Circle Model (GCM); the third dimension (not shown) is a function of line length. The linear representation of CVRM and CMFP is an unfolding of the two circles in the projection sphere in Fig. 7; a separate trace is shown for each side of the median plane. The slopes of these traces are determined by the ratios of slopes of the VPEL-vs-retinal-orientation and VPV-vs-retinal-orientation functions in Fig. 4. The scaled relations presented between CVRM and CMFP are shown for 25 deg horizontal eccentricity at true eye level and are set by the tangent relation given in the text. (This differs only slightly from the relation in the experiments in Figs 3 and 4 where eccentricity was measured at the level of the normal from the eye to the plane containing the line stimulus.) VPEL and VPV for the individual stimulus line (Figs 3 and 4) fall closely along the CVRM/CMFP traces following removal of the influence of the body-referenced mechanism. Lines *a*, *b*, and *c* display examples of the independence of the averaging of the individual line influences for VPV and VPEL to obtain the result for the equal-length two-line stimulus. (Lines of saturating length are assumed here; for shorter lines the resultant for parallel lines lies in a different plane in the 3d representation than does the resultant for nonparallel lines of the same length.) The point on line *d* that intersects line *c* (open circle) is obtained by stimulating with two different length lines, one pitches at 18 deg (right visual field) and the other pitched at 30 deg (left visual field); this results in the same combination of VPEL and VPV as does the equal length combination of line *c*.

dimensions within a plane of 'egocentric orientation space'. The bias due to the body-referenced mechanism (cf. dark values for VPEL in Fig. 3 and y-intercepts in Fig. 4) has been eliminated from both dimensions in the figure. The values of both dimensions calculated from the 1-line results fall along two diagonal straight lines whose slopes are equal to the ratios of slopes in Figs 4a and 4b and closely approximate the dual linear unfolding of the CVRM/CMFP traces displayed in Fig. 10.

The representation in Fig. 10 also quantitatively predicts the averaging process that was shown only qualitatively in Fig. 8. For example, each of the two endpoints of lines *a*, *b*, and *c* in Fig. 10 falls on one of the two diagonal traces. The open circle on each line falls halfway between the line's endpoints at the linear average of the endpoint values on each of the two perceptual dimensions. For line *a*, whose endpoints are at equal and opposite intersection points on the CMFP, this average is at a visual influence on VPV of 0 deg as measured (Fig. 9c) and independent of the processing of VPEL; for line *b*, whose endpoints are at equal and opposite values on the CVRM, this average falls at a visual influence on VPEL of 0 deg as measured (Fig. 9e), and is independent of the processing of VPV. For line *c*, for which the endpoints are unequal on both dimensions, this average falls at independent visual influences on VPV and VPEL of −3.3 deg and −10 deg, respectively, halfway between the values at the endpoints on each dimension and conforms to the results mentioned above (Li and Matin, 1992). Any point in the representation can be reached in numerous ways by lines extending from endpoints on the two CVRM/CMFP traces. For example, line *d*, with endpoints at pitches of 18 deg and 30 deg on the left and right visual field traces, respectively, intersects line *c* at the open circle. But, for the match of both VPEL and VPV to occur at the intersections of lines *c* and *d*, the stimulus lines at the endpoints of *d* must be of different lengths, with the line in the right visual field longer than the line in the left visual field. Thus, the averaging process requires weighting by line length.

The third dimension in the representation (not shown) is a function of the total effective length of all lines in an observer's view; two fans of vectors for the different retinal orientations of individual lines extend along this dimension from an origin at 0,0,0 and intersect the CVRM/CMFP traces in different planes of constant effective line length, of which Fig. 10 is only one.

Thus, Fig. 10 provides a description, based on GCM, of the relations at the 25 deg eccentricity employed in the experiments summarized herein, and suggests a number of other relations as well. A separate such representation would be expected to hold at each horizontal retinal eccentricity. Experiments demonstrating a substantial influence of eccentricity on the slope of the VPEL-vs-pitch function require differential weighting of eccentricity in the GCM representation (Li and Matin, 1990), but quantitative experiments on the influence of eccentricity on VPV are not yet in hand, and so we cannot yet tell whether the slopes for the CVRM/CMFP traces in Fig. 10 will be similar or different at different eccentricities. However, there is reason to believe that averaging across different eccentricities will follow the rules we find at 25 deg (Matin and Li, 1993a).

In summary then, the rules controlling the operation of two simultaneously-viewed lines each contain two segments: (a) weighted averaging; (b) summation. The simplest view of matters containing both segments as they operate for VPV and for VPEL is:

(1) VPEL: Lines whose images on the spherical approximation to the eye fall on great circles that intersect the same point on the CVRM generate influences on VPEL that summate along the exponential in Fig. 6; combinations of lines whose

images fall on great circles that intersect the CVRM at different points follow a weighted average rule.

(2) VPV: Lines whose images on the spherical approximation to the eye fall on great circles that intersect the same point on the CMFP generate influences on VPV that summate; combinations of lines whose images fall on great circles that intersect the CMFP at different points follow a weighted average rule.[3]

Some of the features of the representation of GCM described in relation to Fig. 10 will remind the reader of treatments of color space (cf. Schrodinger, 1920/1970; Graham, 1965; Pokorny and Smith, 1986; Wyszecki, 1986) and suggest some generality to the way in which the brain deals with very different perceptual modalities.[4]

7. GAZE DIRECTION, EVOLUTION, AND VPEL

The two relations between lines — parallelness and mirror symmetry relative to the vertical axis in the median plane — are treated in fundamentally different ways by each of two separate mechanisms; one controls the processing of perceived orientation in a frontal plane and the second controls the processing of perceived elevation. Treating parallel lines on opposite sides of the median plane as being related by the identity transform and mirror symmetric lines across the median plane as being related by the negative transform is intuitively reasonable with respect to VPV for lines in the frontal plane and is also concordant with the way perception appears to treat orientation in two-dimensional visual patterns. But treating mirror symmetric orientations within the frontal plane as being related by the identity transform and parallel lines across the median plane as being related by the negative transform is also understandable if we recognize the relation to the processing of elevation. For use in such processing, parallel pitched-from-vertical lines on opposite sides of the viewer would lead to a univocal treatment of a single visual field; it is the projective properties of such lines on to a frontal plane and on to the retina that leads to the mirror symmetry rule regarding frontal plane orientation (Fig. 2b).

The last statement might appear to suggest that mirror symmetry as the identity transform for VPEL processing has some basis in a need for orienting and localizing in the presence of pitched visual fields. However, it is not clear that terrestrial organisms in either natural or carpentered environments encounter planes of substantial pitch or pitched-from-vertical lines as main lines of organization of a visual field sufficiently often for the visual influence on VPEL to have evolved to meet a need for dealing with them. Nor does the visual influence on VPEL that we measure appear to do so usefully. Indeed, the influences of pitched-from-vertical lines that we measure would be maladaptive for most purposes since they would produce mislocalizations from veridical perception instead of reducing them.

A more likely evolutionary basis for the mirror symmetric influence on VPEL is derived from the presence of a mobile eye with foveate vision inhabiting an erect environment. Thus, when gaze is turned upward while viewing an erect visual field, the images from vertical lines on *both* sides of the median plane fall on great circles whose relation to the CVRM is identical to that resulting from topbackward pitch (Fig. 5, bottom panel), whereas when the eye is turned downward the images of vertical lines fall on the same great circles as those resulting from topforward pitch (Fig. 5, top panel). But these changes in the intersection point on the CVRM consequent on change in gaze elevation dictate a visual influence on VPEL from viewing an erect visual field that would tend toward stabilizing VPEL at true eye level for both upward and downward directions of gaze, and in fact, we do find that VPEL is essentially independent of vertical gaze direction for any given orientation of a visual field containing pitched-from-vertical lines, whether accompanied by changes in head orientation or not (Li and Matin, 1993). But it is also necessary that the two opposed rules described herein — mirror symmetry and parallelism — both be accommodated: Elsewhere we have pointed out that the need for the visual system to develop frontal-eyedness in order to deal with the frontal plane more effectively in the face of a prior lateral-eyedness is also likely to have forced the evolution of partial decussation in the mammalian visual system in order to permit the rules of mirror symmetry and parallelness to coexist (Matin and Li, 1993, 1994c, submitted-a).

Acknowledgement

This research was supported by Grants AFOSR 91-0146 and F-49620-94-1-0397 from the Spatial Orientation Program of the US Air Force Office of Scientific Research.

NOTES

1. For treatments of contributions of the body-referenced mechanism see Mittelstaedt (1988), Matin and Fox (1989), Matin and Li (1992a, 1994a,b, submitted-b), Li and Matin (1993), and DiZio *et al.* (submitted).

2. For relations of GCM to neurophysiology see Matin and Li (1992a,b, 1994b).

3. VPEL and VPV are each influenced by intersections on both the CVRM and the CMFP as well as the height above true eye level (Li and Matin, 1990; Matin and Li, 1992b). These contingencies are beyond the scope of the present article.

4. It is of some interest that the representation of color — a modality that is not fundamentally spatial — in spatial terms ('color space') has a considerable history and tradition, but the representation of egocentric spatial orientation — a modality that *is* fundamentally spatial — in spatial terms does not. The similarities of the present treatment to that of color space include: (a) two main independent qualitative dimensions and a single intensive dimension (here line length) which satisfy a requirement in the data structure for three-dimensional representation; (b) opponency on opposite sides of a neutral point; (c) additivity along a single direction within the intensive dimension but failure of intensity additivity among different directions along qualitative dimensions; (d) weighted algebraic averaging of intensities among separately located and qualitatively different stimuli within a constant intensity plane to arrive at a new point in the vector space, averaging that generally conforms to Grassman's laws (equivalent rules in the present case); (e) the neutral point is reached by combining each of numerous complementary pairs of qualitatively different stimuli. It is of further interest that the three perceptual dimensions of the present representation do not correspond 1:1 to the three dimensions of physical space in a Cartesian framework. Although we have made some progress toward incorporating the missing dimension — the horizontal dimension of egocentric orientation and localization — within the present framework (Dallal *et al.*, 1993; Li and Matin, 1995) more work is still needed.

REFERENCES

Dallal, N., Li, W. and Matin, L. (1993). Visually perceived eye level and straight ahead: influences from visual field pitch, visual field slant, observer orientation, and gravity. *Eastern Psychological Association* **64**, 42.

DiZio, P., Li, W., Lackner, J. R. and Matin, L. (submitted for publication). Combined influences of gravitoinertial force level and visual field pitch on visually perceived eye level.

Graham, C. H. (1965). Color mixture and color systems. In: *Vision and Visual Perception*. C. H. Graham (Ed.). Wiley and Sons, New York, pp. 370–394.

Li, W. and Matin, L. (1990). Perceived eye level: sensitivity to pitch of a vertical 2-line stimulus grows with eccentricity but is biased by elevation. *Invest. Ophthalmol. Vis. Sci.* **31** (Suppl.), 84.

Li, W. and Matin, L. (1991). Spatial summation of influences on visually perceived eye level from a single variably-pitched 1-line stimulus. *Invest. Ophthalmol. Vis. Sci.* **32**, 1272.

Li, W. and Matin, L. (1992). Linear averaging of the influences from 2 lines of different pitch or obliquity on visually perceived eye level. *Invest. Ophthalmol. Vis. Sci.* **33**, 1154.

Li, W. and Matin, L. (1993). Eye and head position, visual pitch, and perceived eye level. *Invest. Ophthalmol. Vis. Sci.* **34**, 1311.

Li, W. and Matin, L. (1995). Differences in influence between pitched-from-vertical and slanted-from-frontal horizontal lines on egocentric localization. *Percept. Psychophys.* **57**, 71–83.

Matin, L. and Fox, C. R. (1989). Visually perceived eye level and perceived elevation of objects: linearly additive influences from visual field pitch and from gravity. *Vision Res.* **29**, 315–324.

Matin, L. and Li, W. (1989). A single pitched line in darkness controls elevation of visually perceived eye level. *Invest. Ophthalmol. Vis. Sci.* **30** (Suppl.), 506.

Matin, L. and Li, W. (1990). Identical effects on perceived eye level by oblique lines in erect planes and pitched-from-vertical lines in pitched planes. *Invest. Ophthalmol. Vis. Sci.* **31** (Suppl.), 328.

Matin, L. and Li, W. (1992a). Visually perceived eye level: Changes induced by a pitched-from-vertical 2-line visual field. *J. Exp. Psychol: Human Percept. Perform.* **18**, 257–289.

Matin, L. and Li, W. (1992b). Mislocalizations of visual elevation and visual vertical induced by visual pitch: The Great Circle Model. *Ann. N.Y. Acad. Sci.* **656**, 242–265.

Matin, L. and Li, W. (1992c). Egocentric spatial localization and summation among spatially separated lines. *Bull. Psychonomic Soc.* **30**, 439.

Matin, L. and Li, W. (1993a). Spatial summation among parallel lines for setting visually perceived eye level. *Invest. Ophthalmol. Vis. Sci.* **34**, 778.

Matin, L. and Li, W. (1993b). The perceptual basis for evolution of partial decussation in the visual system. *Bull. Psychonomic Soc.* **31**, 430.

Matin, L. and Li, W. (1994a). The influence of the orientation of a stationary single line in darkness on the visual perception of eye level. *Vision Res.* **34**, 311–330.

Matin, L. and Li, W. (1994b). Spatial summation among parallel lines across wide separation (50°): spatial localization and the great circle model. *Vision Res.* **34**, 2577–2598.

Matin, L. and Li, W. (1994c). Perceptual constraints leading to the evolution of partial decussation. *Invest. Ophthalmol. Vis. Sci.* **35**, 2112.

Matin, L. and Li, W. (submitted for publication-a). Bilateral parity violation in visual processing of egocentric spatial localization: With implications for the evolution of partial decussation.

Matin, L. and Li, W. (submitted for publication-b). Multimodal basis for egocentric spatial location and orientation.

Mittelstaedt, H. (1988). The information processing structure of the subjective vertical. A cybernetic bridge between its psychophysics and its neurobiology. In: *Processing Structures or Perception and Action*. H. Marko, G. Hauske and A. Struppler (Eds). Verlag Chemie, Weinhein, pp. 217–263.

Neal, E. (1926). Visual localization of the vertical. *Am. J. Psychol.* **37**, 287–291.

Pokorny, J. and Smith, V. C. (1986). Colorimetry and color discrimination. In: *Handbook of Perception and Human Performance*, Vol. I. K. Boff, L. Kaufman and J. Thomas (Eds). Wiley and Sons, New York, pp. 8-1–8-51.

Schrödinger, E. (1920). Grundlinien einer Theorie der Farbenmetrik im Tagessehen. *Annual der Physik* **63**, 481. [English translation: Outline of a theory of color measurement for daylight vision. In: *Sources of Color Science.* D. L. MacAdam (Ed. and trans.) (1970). MIT Press, Cambridge, MA, pp. 134–182.]

Witkin, H. A. and Asch, S. E. (1948). Studies in space perception: IV. Further experiments on perception of the upright with displaced visual fields. *J. Exp. Psychol.* **38**, 762–782.

Wyszecki, G. (1986). Color appearance. In: *Handbook of Perception and Human Performance*, Vol. I. K. Boff, L. Kaufman and J. Thomas (Eds). Wiley and Sons, New York, pp. 9-1–9-57.

Human Symmetry Perception, pp. 383–393
C. W. Tyler (Ed.)

The generalized cone in human spatial organization

JOHN R. PANI
Department of Psychology, Emory University, Atlanta, GA 30322, USA

Abstract—The generalized cone is one of the newer concepts useful for describing spatial structures, and it has become popular as a volumetric primitive in models of object recognition. Apart from this use of the concept (or perhaps underlying it), the generalized cone can be considered a species of spatial regularity. In the general definition of symmetry as invariance across transformation, the generalized cone is a combination of translation and dilation symmetry. In such symmetry, there is homogeneity both of the slants of edges and surfaces of an object about an axis and the radial positions of these features about the axis. The results of two research projects are reviewed suggesting that the generalized cone is useful in human spatial organization. In the first instance, each of the three simpler regular polyhedra, the Platonic Solids, are easiest to perceive and imagine when they are organized as generalized cones. In the second instance, people imagine simple rotations best when the symmetric space that would be traced by the motion is aligned with salient spatial reference systems.

1. INTRODUCTION

The generalized cone is a relatively new addition to the set of concepts useful for describing spatial structures (Binford, 1971, 1982). Already, however, it has become popular as the hypothesized structure of volumetric primitives in information systems that represent the shapes of objects (Marr and Nishihara, 1978; Brooks, 1981; Marr, 1982; Biederman, 1987, 1990; Landau and Jackendoff, 1993). In this paper, I discuss the generalized cone in a different, but related, context. The generalized cone is a type of spatial regularity, of symmetry, to which human beings are sensitive in the organization of three-dimensional (3D) structures. In the remainder of this introduction I make more explicit the definition of the generalized cone as a type of symmetry. I then support my claims about the importance of the generalized cone by briefly reviewing two studies that are reported more fully elsewhere. The first concerns the perception and imagination of the three simpler Platonic Solids, the regular polyhedra (Pani *et al.*, 1994). The second study concerns the imagination of simple rotational motion (Pani, 1993; Pani and Dupree, in press). I close by suggesting that the study of symmetry in spatial organization represents continued advance in the study of perception and spatial cognition.

A generalized cone is any shape whose contour could be traced out by moving a cross-section of constant shape along an axis, where the cross-section could change in size (Binford, 1971, 1982; Brooks, 1981). To specify a generalized cone, one specifies a planar shape for the cross-section, a space curve for the axis, an angle between the cross-section and the axis, and a function that describes the size of the cross-section at any point along the axis.

The concept of the generalized cone can be considered a superordinate category for a variety of shapes that have long been described in the mathematical literature. For example, if one restricts the concept to shapes with straight axes, the generalized cone is a generic term for those shapes described by the abstract concepts of cone, pyramid, cylinder, and prism, and their variants (e.g. the frustrum, or truncated cone). These concepts, in turn, include such venerable geometric concepts as the solid of revolution (i.e. any shape that could be made on a lathe). The familiarity of particular classes of generalized cones, however, takes nothing away from the significance of the concept. It is important for cognitive scientists to select from the universe of mathematics what is relevant to cognition and to organize this material in ways useful to cognitive theory. The basis for developing the concept of the generalized cone within psychology is the hypothesis that people organize basic solids in terms of axes and various forms of regularity associated with 'sweeping' cross-sections along those axes (e.g. Marr and Nishihara, 1978). The question is, just what are the regularities that make such shapes a natural class for human perceivers? In the following remarks, I discuss only generalized cones with straight axes. However, I expect that these remarks will apply with few changes to shapes with curved axes.

In the general effort to describe the regularity of physical structures, the term 'symmetry' has come to mean those properties of objects that are invariant across conceivable transformations (Hurlbut and Klein, 1977; Palmer, 1983; Burn, 1985; Smart, 1988; Leyton, 1992; Stewart and Golubitsky, 1992). Different types of symmetry correspond to different transformations, as illustrated in Fig. 1. Some symmetries are associated with multiple transformations. For example, the prototypic flower has both rotation and reflection symmetry. A helix is invariant over a combination

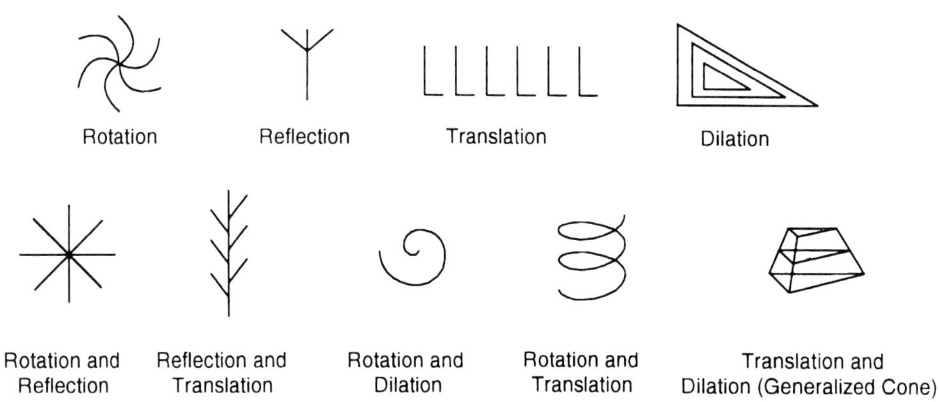

Figure 1. Types of symmetry.

of rotation and translation symmetry (e.g. Weyl, 1952; Hargittai and Pickover, 1992). Generalized cones, by definition, are produced by a combination of translation and dilation (Binford, 1971, 1982). If the function that controls the dilation is linear, then symmetry is produced: the shape is invariant across translation and dilation. If the function that controls dilation is constant, the shape is invariant over translation: the object is a prism or cylinder. But even when the dilation function is nonlinear, there is a type of global symmetry that people appear to find salient. We turn now to considering the minimum regularity of the generalized cone.

One can think of the edges and surfaces of a generalized cone as having spherical coordinates defined relative to the conic axis. To visualize this description, it is helpful to think of the axis as a planetary pole and the edges and surfaces of the object as having latitude and longitude relative to the pole. With this description, there are two critical regularities associated with the generalized cone. The first one pertains to latitudes. At any level up and down the object axis, generalized cones have at least an approximate homogeneity of the *slants* of edges and surfaces. The more equilateral the cross-section of the shape, the more exact the homogeneity of the slants. The second property of the generalized cone pertains to the longitudes of edges and surfaces. At all levels up and down the object axis there is a single set of radial positions of the edges and surfaces of the object about the axis. In other words, one phase structure describes the radial locations of edges and surfaces at all positions along the axis. Thus, the minimum symmetry of a generalized cone includes two properties: more or less homogeneous local slants of edges and surfaces (i.e. at any cross-section through the object) and identical radial positions of object features all along the axis.

My own interest in the generalized cone did not stem from a mathematical or aesthetic appreciation of this structure. Rather, the generalized cone is useful in explaining people's perception and imagination of fundamental 3D structures, motions, and transformations. Consider the imagination of certain simple shapes renowned for their symmetry.

2. SPATIAL ORGANIZATION OF THE PLATONIC SOLIDS

There are five regular polyhedra (e.g. Hilbert and Cohn-Vossen, 1952; Holden, 1971; Kappraff, 1991). They are regular because on each of these objects, every corner and surface is the same as every other. Although they are regular at the level of features of the objects, each of these polyhedra has more than one description at the level of the arrangement of these features. These arrangements are described in standard geometric treatments in terms of rotation and reflection symmetries of the objects relative to particular axes and planes through them (e.g. Hilbert and Cohn-Vossen, 1952; Holden, 1971; Hurlbut and Klein, 1977). The cube and octahedron have axes of fourfold, threefold, and twofold symmetry, and the tetrahedron has axes of threefold and twofold symmetry, as illustrated in Fig. 2. For example, a line running through the middle of opposite faces of the cube is an axis of fourfold rotational symmetry. If the cube is rotated about this axis, it occupies exactly the same space

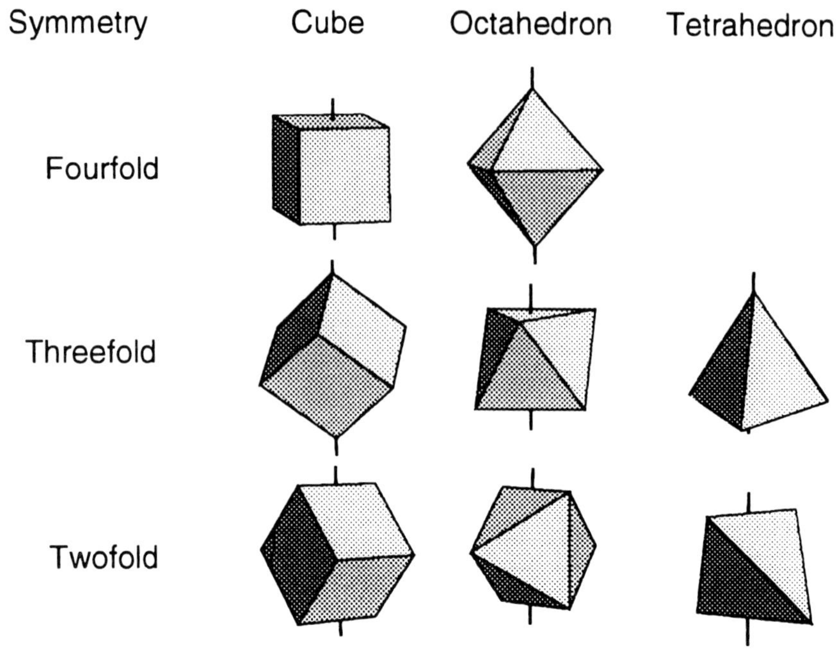

Figure 2. The three simpler Platonic Solids arranged in accordance with their amounts of rotation and reflection symmetry about the vertical.

four times during one revolution. For each of the Platonic Solids, there are sets of planes of reflection symmetry that correspond to the axes of rotational symmetry. For example, the cube has four planes of reflection symmetry coincident with its axis of fourfold rotational symmetry. The orientations that result from aligning the symmetry axes of the three objects to the environmental vertical will be referred to as the canonical orientations of the objects.

The three simpler Platonic Solids are generalized cones (Pani *et al.*, 1994). The cube and octahedron are generalized cones about their fourfold axes of symmetry, and the tetrahedron is a generalized cone about its threefold axis of symmetry, as illustrated in Fig. 3. Note that these objects have the regularity of slant and position discussed above. Consider what happens if one orients one of the generalized conic axes to the vertical and then rotates the object about that axis. As each new surface comes forward, it is at the same orientation to the vertical as the surface that came before. And where there is a surface or edge at the upper part of the object, there is a surface or edge at the lower part.

In contrast to the description of these objects as generalized cones, these objects can be described as being formed of 3D zigzags of oblique edges and surfaces (see Hinton, 1979; Pani *et al.*, 1994). Each object is, or has some of the properties of, an antiprism (or, 'twisted *n*-gram'; Holden, 1971; Kappraff, 1991). The clearest example is the octahedron (see Fig. 3). Relative to its axis of threefold symmetry, the octahedron has a triangular top and bottom in antiphase to each other; it is an antiprism, opposite to the triangular prism. As one goes along the axis, the cross-

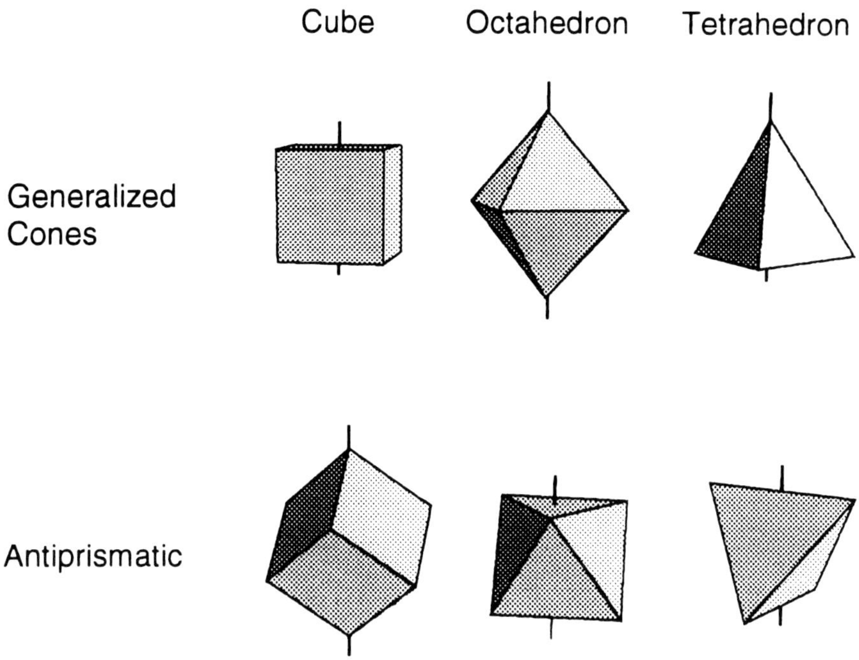

Figure 3. The three simpler Platonic Solids oriented as generalized cones and as antiprismatic shapes about the vertical.

section of the object changes from a triangle in one direction to a hexagon (at the middle of the shape) to a triangle in a different direction. Thus, the octahedron about its threefold symmetry axis is in some ways opposite to the generalized cone. Again, consider the regularities discussed above. If one orients the antiprismatic axis to the vertical and rotates the shape about the axis, one face along the side of the object might be oriented upwards, but then the next face will be oriented downwards (with a 38° difference between them); one set of edges will face right and left and the next set of edges will face left and right. It is clear that the threefold rotation/reflection symmetry does not guarantee that the slants of surfaces about the axis of symmetry will be homogeneous, nor that one radial position about the axis has the same type of feature up and down the axis. The antiprismatic axes of the simpler Platonic Solids are the threefold symmetry axes of the cube and octahedron, and the twofold symmetry axis of the tetrahedron. (The cube is not an antiprism, but it has many of the properties of one, as can be seen in Fig. 3.)

The spatial organization of an object typically is most apparent when that organization is aligned with the environmental vertical (Rock, 1973; Hinton, 1979; Palmer, 1980; Pani, 1993). The simpler Platonic Solids appear simple when they are oriented with their conic axes vertical; they appear complex when they are oriented with their antiprismatic axes vertical. To support and extend this claim to the process of imagination, I discuss research on perception and imagination of the simpler Platonic Solids.

Hinton (1979) has demonstrated that, although it is easy to imagine a cube in the standard orientation, most people are unable to imagine a cube accurately if a main diagonal through the cube is oriented vertically. It is common for people to claim that the task is not difficult, but to report imagining what is actually a regular octahedron. (The cube, of course, is a hexahedron.) Pani *et al.* (1994) reported similar findings for all of the simpler Platonic Solids. A critical aspect of the method was that subjects were permitted to study a shape before an effort to imagine it. The object was displayed as a realistically shaded grayscale computer image of an object rotating slowly about the vertical. Every subject saw each orientation of each object illustrated in Fig. 2. After termination of a display, the subject imagined the object with the same orientation to the vertical, at the location of a styrofoam sphere. One group of subjects inserted pins into the sphere where the corners of the imagined object would be located. A second group of subjects inserted pins where the central surface normals would be located.

Because subjects controlled how long they viewed the displays, viewing time for each orientation of an object was a critical measure. Mean viewing time for each orientation of each object is presented in Fig. 4. The results confirmed expectations. Subjects required relatively short times to study the generalized cones, but relatively long times to study the antiprismatic shapes. Results of the analysis of accuracy provide much the same information as the pattern of times shown in Fig. 4 (Pani *et al.*, 1994).

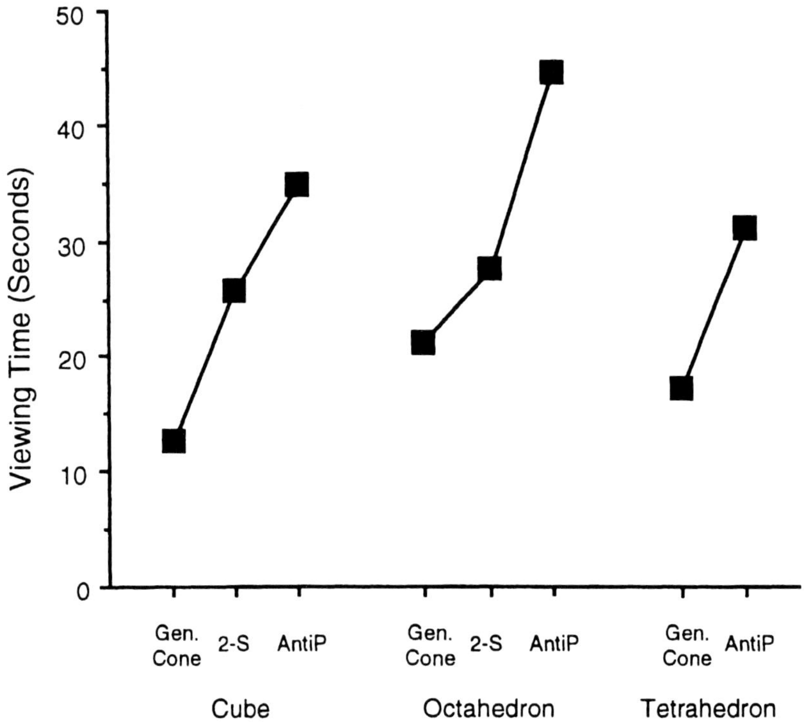

Figure 4. Viewing time as a function of the type and the orientation of the object to the vertical. The abbreviations along the abscissa are for the objects oriented as generalized cones (Gen. Cone), twofold symmetries (2-S), and antiprismatic shapes (AntiP).

The standard description of the regular polyhedra in terms of rotation and reflection symmetry cannot account for the phenomena described here. The antiprismatic cube and octahedron have considerable rotation/reflection symmetry about the vertical, the same threefold symmetry possessed by a triangular prism, but people find these shapes difficult structures to imagine. The threefold symmetry of the tetrahedron, on the other hand, is relatively easy for people to imagine. In addition, people generally find the threefold symmetries of the cube and octahedron more difficult to imagine than the lower, twofold, symmetries. I do not mean to say that rotation/reflection symmetry is irrelevant to perceiving and imagining these objects. However, such symmetry is not sufficient to make the objects well organized for people. Instead, it is the combination of the generalized conic symmetry with rotation/reflection symmetry that makes these objects appear simple.

3. THE IMAGINATION OF ROTATION

As a second example of the importance of symmetry in spatial organization, consider simple rotational motion, the rotation of an object about an axis fixed in space. Basic parameters of rotation include the orientation of the axis of rotation to the environment and the orientation of the rotating object to the axis of rotation (see Shiffrar and Shepard, 1991; Pani, 1993; Pani and Dupree, in press). In Fig. 5, for example, the rods are axes of rotation with fixed directions in space; when the rods spin, the squares rotate about the rods. In the system in Fig. 5A, the rod is aligned with the environmental vertical and the square is perpendicular to the rod. If subjects are asked to indicate what the orientation of the square would be after a rotation of the rod, say 180°, they succeed easily. In Fig. 5B, the rod is vertical, but the square is oblique to the rod. In Fig. 5C, the rod is oblique to the environment, but the square is perpendicular to the rod. In both of these mixed cases, subjects can imagine the rotations rather well. The system in Fig. 5D is double-oblique. It is impossible for the typical subject to accurately imagine the rotation of this system, even when mean response time is over two minutes. When subjects are given a square and are asked to show the result of the hypothetical rotation, typical responses are errors of 45° or more (Pani, 1993; Pani and Dupree, in press; see also Just and Carpenter, 1985; Parsons, 1987; Massironi and Luccio, 1989; Shiffrar and Shepard, 1991). As most readers are unable to predict the outcome of the rotation suggested in Fig. 5D, three orientations from a 180° rotation of this system are illustrated in Fig. 6. These experimental results on the imagination of rotation have been extended to the perception of ongoing motions (Pani *et al.*, in press).

A simple but adequate theory of the constraints on the imagination of rotation has two parts. First, every point on the rotating object moves in a circular path centered at the axis of rotation. The planes of these circles are parallel to each other and normal to the axis. The typical person's cognitive organization of rotational motion emphasizes this structured space. Second, people are able to predict the outcome of a rotation if the planes of rotational motion are aligned with a salient spatial reference system, generally the principal directions of the environment or the intrinsic reference

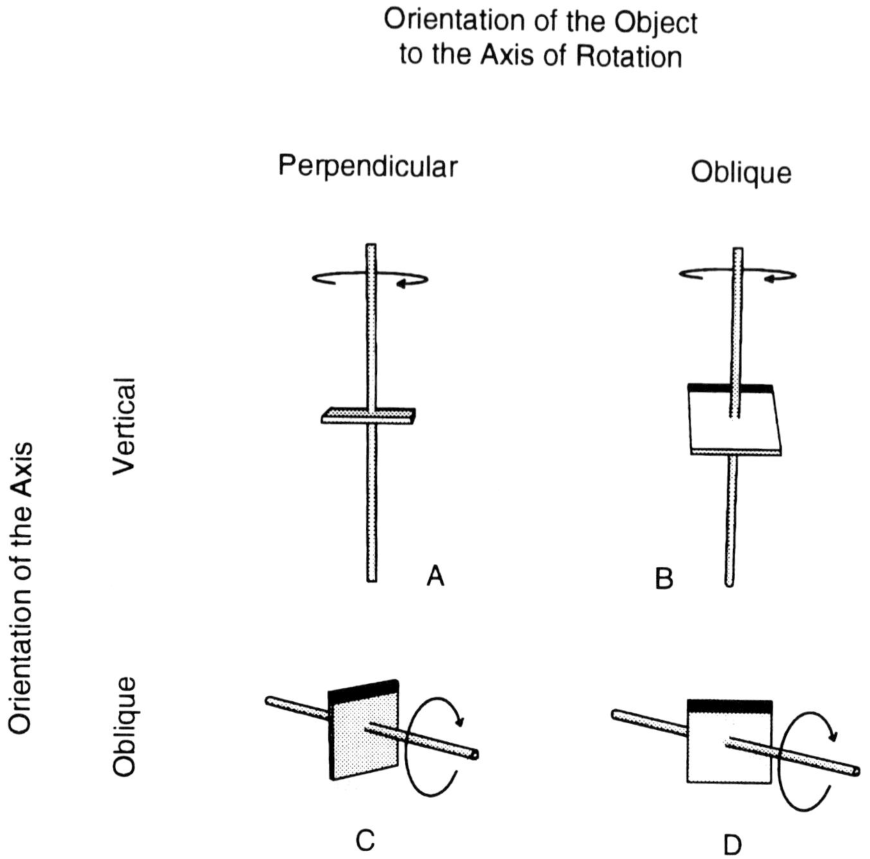

Figure 5. If the rods are axes of rotation with fixed directions, these four assemblies illustrate basic types of rotational motion.

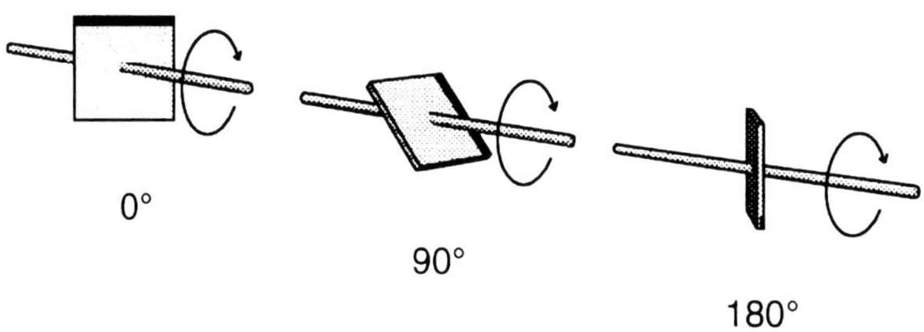

Figure 6. Three orientations of the square from the double-oblique rotation suggested in Fig. 5D.

system of the object (Pani, 1993; Pani and Dupree, in press; see also Massironi and Luccio, 1989; Shiffrar and Shepard, 1991). In the rotation of Fig. 5D, the planes of rotation are aligned with neither basic reference system, and therefore performance is markedly poor.

This view takes account of the fact that the planes of rotation form a circularly symmetric space stretched along an axis. This dynamic space is a solid of revolution, a highly regular variant of the generalized cone (and one with rotation/reflection symmetry). Alignment of this structure to the environment or to the object makes the structure salient in much the same way as static symmetries are more apparent when they are aligned with the environment (Rock, 1973; Palmer and Hemenway, 1978; Palmer, 1980).

4. CONCLUSION

For many years it has been agreed that the Gestalt psychologists made a major contribution to the study of perception by pointing out the importance of perceptual organization (e.g. Wertheimer, 1950). It has been unclear, however, what theoretical perspective best accounts for this organization. A promising approach to this problem was suggested by Attneave (1954) and Leeuwenberg (1971; see also Pomerantz and Kubovy, 1986). Perceptual organization reflects people's efforts to find cognitively simple descriptions of objects and environments. In this view, people look for hierarchical descriptions that include all of the features of an object in a relatively simple description. The next problem, then, has been to specify objectively the terms of a simple description.

One clear starting point for psychology is to consider descriptions of spatial regularity that have come from other analytical disciplines, including mathematics, artificial intelligence, and the physical sciences. Thus, one can advance the study of perception and spatial cognition by noting the various types of symmetry, of invariance across transformation, to which people are sensitive. This is a way of grounding the study of spatial organization in a higher level psychophysics (Garner, 1974; Shaw and Turvey, 1981; Palmer, 1983; Leyton, 1992).

In the present paper, I have offered two examples in which this approach has been helpful. The generalized cone is a type of higher-order symmetry. It is useful in explaining why people perceive and imagine certain orientations of the simpler Platonic Solids well and other orientations poorly. The traditional description of these objects in terms of rotation and reflection symmetry is insufficient for this purpose. The human considers these objects especially simple only when he or she can see the rotation, reflection, and the conic symmetry. The second example concerns the dynamic space formed by simple rotation. Rotations are organized in terms of symmetric spaces. When these spaces are aligned with salient reference systems, the motions are comprehended, just as symmetric structures generally are perceived well when they are aligned with a reference system. Even so basic a motion as simple rotation may be incomprehensible to people when the motion space is misaligned with the rotating object and the environment.

It would not be possible in the long run for a cognitive system to depend on the formation of simple descriptions if these descriptions did not apply well to the physical world. There are very few double-oblique rotations in the world, and our general inability to imagine them causes little difficulty. Quite possibly, then, our dependence in spatial organization on the symmetries discussed here is due to perceptual learning. Perhaps in a different world we would learn to imagine antiprismatic shapes and double-oblique rotations effortlessly. If this is true, the presently observed constraints on spatial organization would be normative descriptions of the typical person's adaptation to the present world. On the other hand, there is a clear logical connection between physical regularity and the efficiency of learning, recognition, and imagination. It would be surprising, indeed if learning could proceed in a world without symmetries as efficiently as it can in a world with symmetries. And, of course, any differences in the efficiency of learning in the two worlds would be magnified if evolution has determined that people are especially sensitive to the regularities that do characterize our world.

Acknowledgements

I would like to thank Carolyn Mervis, Christopher Tyler, and two anonymous reviewers for comments on earlier drafts of this paper.

REFERENCES

Attneave, F. (1954). Some informational aspects of visual perception. *Psychol. Rev.* **61**, 183–193.

Biederman, I. (1987). Recognition by components: A theory of human image understanding. *Psychol. Rev.* **94**, 115–147.

Biederman, I. (1990). Higher-level vision. In: *Visual Cognition and Action*. D. N. Osherson, S. M. Kosslyn and J. M. Hollerbach (Eds). MIT Press, Cambridge, MA, pp. 41–72.

Binford, T. O. (1971, December). *Visual Perception by Computer*. Paper presented at IEEE Systems Science and Cybernetics Conference, Miami, FL.

Binford, T. O. (1982). Survey of model-based image analysis systems. *Int. J. Robotics Res.* **1**, 18–64.

Brooks, R. A. (1981). Symbolic reasoning among 3-D models and 2-D images. *Artificial Intelligence* **17**, 285–348.

Burn, R. P. (1985). *Groups: A Path to Geometry*. Cambridge University Press, New York.

Garner, W. R. (1974). *The Processing of Information and Structure*. Erlbaum, Potomac, MD.

Hargittai, I. and Pickover, C. A. (Eds). (1992). *Spiral Symmetry*. World Scientific, River Edge, NJ.

Hilbert, D. and Cohn-Vossen, S. (1952). *Geometry and the Imagination*. Chelsea, New York.

Hinton, G. (1979). Some demonstrations of the effects of structural descriptions in mental imagery. *Cognitive Sci.* **3**, 231–250.

Holden, A. (1971). *Shapes, Space, and Symmetry*. Columbia University Press, New York.

Hurlbut, C. S. Jr and Klein, C. (1977). *Manual of Mineralogy* (19th edn). Wiley, New York.

Just, M. A. and Carpenter, P. A. (1985). Cognitive coordinate systems: Accounts of mental rotation and individual differences in spatial ability. *Psychol. Rev.* **92**, 137–172.

Kappraff, J. (1991). *Connections: The Geometric Bridge Between Art and Science*. McGraw-Hill, New York.

Landau, B. and Jackendoff, R. (1993). 'What' and 'Where' in spatial language and spatial cognition. *Behav. Brain Sci.* **16**, 217–238.

Leeuwenberg, E. L. J. (1971). A perceptual coding language for visual and auditory patterns. *Am. J. Psychol.* **84**, 307–349.

Leyton, M. (1992). *Symmetry, Causality, Mind*. MIT Press, Cambridge, MA.

Marr, D. (1982). *Vision*. Freeman, San Francisco.

Marr, D. and Nishihara, H. K. (1978). Representation and recognition of the spatial organization of three-dimensional shapes. *Proc. Roy. Soc. Lond. B* **200**, 269–294.

Massironi, M. and Luccio, R. (1989). Organizational versus geometric factors in mental rotation and folding tasks. *Perception* **18**, 321–332.

Palmer, S. E. (1980). What makes triangles point: Local and global effects in configuration of ambiguous triangles. *Cognitive Psychol.* **12**, 285–305.

Palmer, S. E. (1983). The psychology of perceptual organization: A transformational approach. In: *Human and Machine Vision.* J. Beck, B. Hope and A. Rosenfeld (Eds). Academic Press, New York, pp. 269–340.

Palmer, S. E. and Hemenway, K. (1978). Orientation and symmetry: Effects of multiple, rotational, and near symmetries. *J. Exp. Psychol: Human Percept. Perform.* **4**, 691–702.

Pani, J. R. (1993). Limits on the comprehension of rotational motion: Mental imagery of rotations with oblique components. *Perception* **22**, 785–808.

Pani, J. R. and Dupree, D. (in press). Spatial reference systems in the comprehension of rotational motion. *Perception.*

Pani, J. R., William, C. T. and Shippey, G. (in press). Determinants of the perception of rotational motion: Orientation of the motion to the object and to the environment. *J. Exp. Psychol: Human Percept. Perform.*

Pani, J. R., Zhou, H. and Friend, S. M. (1994). Perceiving and imagining Plato's Solids: Spatial organization of fundamental 3D structures. Manuscript submitted for publication.

Parsons, L. M. (1987). Visual discrimination of abstract mirror-reflected three-dimensional objects at many orientations. *Percept. Psychophys.* **42**, 49–59.

Pomerantz, J. R. and Kubovy, M. (1986). Theoretical approaches to perceptual organization: Simplicity and Likelihood Principles. In: *Handbook of Perception and Human Performance. Vol. II: Cognitive Processes and Performance.* K. R. Boff, L. Kaufman and J. P. Thomas (Eds). Wiley, New York, p. 36.

Rock, I. (1973). *Orientation and Form.* Academic Press, New York.

Shaw, R. and Turvey, M. T. (1981). Coalitions as models for ecosystems: A realist perspective on perceptual organization. In: *Perceptual Organization.* M. Kubovy and J. R. Pomerantz (Eds). Erlbaum, Hillsdale, NJ, pp. 343–415.

Shiffrar, M. M. and Shepard, R. N. (1991). Comparison of cube rotations around axes inclined relative to the environment or to the cube. *J. Exp. Psychol: Human Percept. Perform.* **17**, 44–54.

Smart, J. R. (1988). *Modern Geometries* (3rd edn). Brooks/Cole, Pacific Grove, CA.

Stewart, I. and Golubitsky, M. (1992). *Fearful Symmetry: Is God a Geometer?* Blackwell, Cambridge, MA.

Wertheimer, M. (1950). Laws of organization in perceptual forms. In: *A Source Book of Gestalt Psychology.* W. D. Ellis (Ed.). Humanities Press, New York, pp. 71–88 (Abridged translation of a paper originally published in German in 1923).

Weyl, H. (1952). *Symmetry.* Princeton University Press, Princeton, NJ.

1-22-20
Ebay
3.17